T0202803

Lecture Notes in Computer Science　14594

Founding Editors

Gerhard Goos
Juris Hartmanis

The series Lecture Notes in Computer Science (LNCS), including its subseries Lecture Notes in Artificial Intelligence (LNAI) and Lecture Notes in Bioinformatics (LNBI), has established itself as a medium for the publication of new developments in computer science and information technology research, teaching, and education.

LNCS enjoys close cooperation with the computer science R & D community, the series counts many renowned academics among its volume editors and paper authors, and collaborates with prestigious societies. Its mission is to serve this international community by providing an invaluable service, mainly focused on the publication of conference and workshop proceedings and postproceedings. LNCS commenced publication in 1973.

Amitabh Basu · Ali Ridha Mahjoub ·
Juan José Salazar González
Editors

Combinatorial Optimization

8th International Symposium, ISCO 2024
La Laguna, Tenerife, Spain, May 22–24, 2024
Revised Selected Papers

 Springer

Editors
Amitabh Basu 🆔
Johns Hopkins University
Baltimore, MD, USA

Juan José Salazar González 🆔
Universidad de La Laguna
La Laguna, Spain

Ali Ridha Mahjoub 🆔
Paris Dauphine University
Paris, France

Kuwait University
Kuwait City, Kuwait

ISSN 0302-9743   ISSN 1611-3349 (electronic)
Lecture Notes in Computer Science
ISBN 978-3-031-60923-7   ISBN 978-3-031-60924-4 (eBook)
https://doi.org/10.1007/978-3-031-60924-4

Preface

This volume contains 30 regular papers submitted to ISCO 2024, the 8th International Symposium on Combinatorial Optimization, Tenerife on May 22–24, 2024. The conference was held in person at the School of Mathematics, Faculty of Science, University of La Laguna, Tenerife, Spain, attracting more than 110 registered participants from all around the world.

ISCO 2024 was preceded by the doctoral school entitled "Combinatorial Optimization and Machine Learning" given by Amitabh Basu (Johns Hopkins University, Baltimore, USA) and Andrea Lodi (Cornell Tech and Technion, New York, USA). Three eminent invited speakers also gave talks at the symposium: Claudia Archetti (ESSEC, Paris, France), Francisco Barahona (IBM T. J. Watson Research Center, Yorktown Heights, USA), and Jon Lee (University of Michigan, Ann Arbor, USA).

The ISCO series aims to bring together researchers from all communities related to combinatorial optimization, including algorithms and complexity, mathematical programming, operations research, stochastic optimization, graphs, and polyhedral combinatorics, among others. It is intended to be a forum for presenting original research on all aspects of combinatorial optimization, ranging from mathematical foundations and theory of algorithms to computational studies and practical applications, and especially their intersections. In response to the call for papers, ISCO 2024 received 113 submissions in two categories: 46 regular papers (12 pages) and 67 short abstracts (3 pages). Each regular paper was reviewed by three to four Program Committee members with the assistance of external reviewers. The submissions were judged on their originality and technical quality. The review process was extremely selective and many good papers could not be accepted. As a result, 30 regular papers were selected, giving an acceptance rate of 65%. The revised versions of the selected regular papers are included in this LNCS volume. Some of the submitted short abstracts were rejected by a Program Chair and others were withdrawn by the authors. The initial scientific program includes three lectures by the invited speakers, 30 selected regular papers and 54 accepted short abstracts. Overall, 84 talks of 25 minutes each, grouped into 28 sessions, are scheduled at ISCO 2024, with at most three sessions running in parallel. Find more details at http://eventos.ull.es/isco2024.

We thank all the authors who submitted their research work to ISCO 2024, and the Program Committee members and external reviewers for their exceptional work. We also thank the three invited speakers as well as the two invited lecturers of the doctoral school. They all contributed to the scientific quality of the symposium.

Finally, we thank the institutions sponsoring ISCO 2024: Universidad de La Laguna (ULL), Turismo de Tenerife (Cabildo de Tenerife), Ayuntamiento de San Cristóbal de La Laguna, Ministerio de Ciencia (Gobierno de España), Sociedad de Estadística e

Investigación Operativa (SEIO), and the Association of European Operational Research Societies (EURO).

April 2024

Amitabh Basu
Ali Ridha Mahjoub
Juan José Salazar González

Organization

Organizing Committee

José Miguel Gutiérrez Expósito	Universidad de La Laguna, Tenerife, Spain
Hipólito Hernández Pérez	Universidad de La Laguna, Tenerife, Spain
Jorge Riera Ledesma	Universidad de La Laguna, Tenerife, Spain
Juan José Salazar González	Universidad de La Laguna, Tenerife, Spain
Antonio Alberto Sedeño Noda	Universidad de La Laguna, Tenerife, Spain

Steering Committee

Mourad Baïou	CNRS, Clermont-Auvergne University, France
Pierre Fouilhoux	Université Sorbonne Paris Nord, France
Luis Gouveia	University of Lisbon, Portugal
Jon Lee	University of Michigan, USA
Ivana Ljubic	ESSEC, France
Nelson Maculan	Federal University of Rio de Janeiro, Brazil
Ali Ridha Mahjoub	Kuwait University, Kuwait and Paris Dauphine University, France
Vangelis Paschos	Paris Dauphine University, France
Giovanni Rinaldi	IASI, Roma, Italy

Program Committee Chairs

Amitabh Basu	Johns Hopkins University, USA
Ali Ridha Mahjoub	Kuwait University, Kuwait; Paris Dauphine University, France
Juan José Salazar González	Universidad de La Laguna, Tenerife, Spain

Program Committee

Edoardo Amaldi	Polytechnical University of Milan, Italy
Claudia Archetti	ESSEC Business School of Paris, France
Mourad Baiou	Université Blaise Pascal, France
Francisco Barahona	IBM Thomas J. Watson Research Center, USA

Amitabh Basu	Johns Hopkins University, USA
Fatiha Bendali	University of Clermont-Auvergne, France
Tolga Bektas	Liverpool University, UK
Víctor Blanco	University of Granada, Spain
Andreas Bley	Universität Kassel, Germany
Immanuel Bomze	University of Vienna, Austria
Flavia Bonomo	University of Buenos Aires, Argentina
Valentina Cacchiani	University of Bologna, Italy
Hector Cancela	Universidad de La República, Uruguay
Francesco Carrabs	Università di Salerno, Italy
Paula Carroll	University College Dublin, Ireland
Daniele Catanzaro	Catholic University of Louvain, Belgium
Raffaele Cerulli	Università di Salerno, Italy
Karthekeyan Chandrasekaran	University of Illinois at Urbana-Champaign, USA
Bo Chen	Warwick University, UK
Jean-François Cordeau	HEC Montreal, Canada
Marcia Fampa	Universidade Federal do Rio de Janeiro, Brazil
Carlo Filippi	Università di Brescia, Italy
Bernard Fortz	Université libre de Bruxelles, Belgium
Pierre Fouilhoux	Université Sorbonne Paris Nord, France
Ricardo Fukasawa	University of Waterloo, Canada
Elisabeth Gaar	Johannes Kepler University Linz, Austria
Laura Galli	Università di Pisa, Italy
Eric Gourdin	Orange, France
Luis Gouveia	University of Lisbon, Portugal
Mohamed Haouari	Qatar University, Qatar
Robert Hildebrand	Virginia Tech, USA
Ola Jabali	Polytechnical University of Milan, Italy
Naoyuki Kamiyama	Kyushu University, Japan
Aleksandr Kazachkov	University of Florida, USA
Aida Khajavirad	Lehigh University, USA
Burak Kocuk	Sabanci University, Turkey
Jan Kronqvist	KTH Royal Institute of Technology, Sweden
Mercedes Landete	Universidad Miguel Hernández, Spain
Jesper Larsen	Technical University of Denmark, Denmark
Jon Lee	University of Michigan, USA
Markus Leitner	Vrije Universiteit Amsterdam, Netherlands
Adam Letchford	Lancaster University, UK
Ivana Ljubic	ESSEC, France
Andrea Lodi	Cornell University, USA
Irene Loiseau	Universidad de Buenos Aires, Argentina
Marco Lübbecke	RWTH Aachen University, Germany

Nelson Maculan	Federal University of Rio de Janeiro, Brazil
Ridha Mahjoub	Kuwait University, Kuwait and Université Paris Dauphine, France
Jean Mailfert	University Clermont-Auvergne, France
Enrico Malaguti	Università degli Studi di Bologna, Italy
Alfredo Marín	Universidad de Murcia, Spain
Nikolaos Matsatsinis	Technical University of Crete, Greece
Eduardo Moreno	Universidad Adolfo Ibáñez, Chile
Joseph Paat	University of British Columbia, Canada
Sophie Parragh	Johannes Kepler University Linz, Austria
Ulrich Pferschy	University of Graz, Austria
Marc Pfetsch	Technische Universität Darmstadt, Germany
Alain Quilliot	Université Clermont Auvergne, France
Giovanni Rinaldi	Consiglio Nazionale delle Ricerche, Italy
Fabrizio Rossi	University of L'Aquila, Italy
Juan José Salazar González	Universidad de La Laguna, Spain
Domenico Salvagnin	Università degli Studi di Padova Italy
Alberto Santini	ESSEC Business School France
Martin Schmidt	Universität Trier Germany
Ruediger Schultz	University of Duisburg-Essen Germany
Frédéric Semet	Centrale Lille France
Daphne Skipper	U.S. Naval Academy USA
José Soto	University of Chile, Chile
Emily Speakman	University of Colorado, Denver, USA
Frits Spieksma	Eindhoven University of Technology, Netherlands
Paolo Toth	Università degli Studi di Bologna, Italy
Eduardo Uchoa	Universidade Federal Fluminense, Brazil
Stefan Voss	Universität Hamburg, Germany
Stein W. Wallace	Norwegian School of Economics Norway
Stefan Weltge	Technische Universität München, Germany
Angelika Wiegele	Alpen-Adria-Universität Klagenfurt, Austria
Weijun Xie	Georgia Tech, USA
Luze Xu	University of California, Davis, USA

Reviewers

Edoardo Amaldi

Yasmine Beck

Flavia Bonomo

Francesco Carrabs

Raffaele Cerulli

Jean-François Cordeau

Pierre Fouilhoux

Bernard Fortz

Laura Galli

Mohamed Haouari

Naoyuki Kamiyama
Burak Kocuk
Mercedes Landete
Markus Leitner
Irene Loiseau
Alfredo Marín
Pedro Moura
Sophie Parragh
Giovanni Rinaldi
Juan José Salazar González
Martin Schmidt
Domenico Serra
Emily Speakman
Johannes Thürauf
Stefan Voß
Angelika Wiegele
Hannane Yaghoubizade
Manuel Aprile
Tolga Bektas
Valentina Cacchiani
Paula Carroll
Karthekeyan Chandrasekaran
Ahmet Cuerebal
Marcia Fampa
Ricardo Fukasawa
Eric Gourdin
Robert Hildebrand
Aleksandr Kazachkov
Jan Kronqvist
Jesper Larsen
Adam Letchford
Marco Lübbecke
Federico Michelotto
Abtin Nourmohammadzadeh
Ulrich Pferschy

Fabrizio Rossi
Domenico Salvagnin
Ruediger Schultz
Daphne Skipper
Frits Spieksma
Paolo Toth
Stein W. Wallace
Weijun Xie
Shudian Zhao
Claudia Archetti
Andreas Bley
Hector Cancela
Daniele Catanzaro
Bo Chen
Mitre Dourado
Carlo Filippi
Elisabeth Gaar
Luis Gouveia
Ola Jabali
Aida Khajavirad
Sven Krumke
Jon Lee
Andrea Lodi
Enrico Malaguti
Eduardo Moreno
Joseph Paat
Marc Pfetsch
Frédéric Semet
Alberto Santini
Jan Schwiddessen
José Soto
Antonio Sudoso
Eduardo Uchoa
Stefan Weltge
Luze Xu

Contents

Integer Programming for Machine Learning

Applications

Integer Programming

On Disjunction Convex Hulls by Lifting

Yushan Qu and Jon Lee[(⊠)]

University of Michigan, Ann Arbor, USA
{yushanqu,jonxlee}@umich.edu

Abstract. We study the natural extended-variable formulation for the disjunction of $n+1$ polytopes in $\mathbb{R}^d$. We demonstrate that the convex hull $\mathcal{D}$ in the natural extended-variable space $\mathbb{R}^{d+n}$ is given by full optimal big-M lifting (i) when $d \leq 2$ (and that it is not generally true for $d \geq 3$), and also (ii) when the polytopes are all axis-aligned hyper-rectangles. We give further results on the polyhedral structure of $\mathcal{D}$, emphasizing the role of full optimal big-M lifting.

Keywords: mixed-integer optimization · disjunction · big M · lifting · convex hull · facet

1 Introduction

In the context of mathematical-optimization models, the so-called "big-M" method is a classical way to treat disjunctions by using binary variables (see [Dan63, Section 26-3.I, parts (b–d,g)], [Lee22, Sections 8.1.2–3], for example). Perhaps it is the most powerful, feared, and abused modeling technique in mixed-integer optimization. For a simple example, we may have the choice (for $x \in \mathbb{R}$) of $x = 0$ or $\ell \leq x \leq u$ (where the constants satisfy $0 < \ell < u$). In applications, for x in the "operating range" $[\ell, u]$, we might incur a cost $c > 0$, which is commonly modeled with a binary variable z, and cost term cz (in a minimization objective function). It remains to link x and z with some constraints. Carelessly, we could use the continuous relaxation:

$$0 \leq x \leq u,$$
$$(z - m_2)\frac{\ell}{1 - m_2} \leq x \leq \frac{u}{1 - m_1}z, \qquad (1)$$

for any $0 \leq m_i < 1$, for $i = 1, 2$. Note that for $z = 0$, (1) reduces to $-\frac{m_2}{1-m_2}\ell \leq x \leq 0$, the left-hand inequality being redundant with respect to $x \geq 0$, and altogether implying that $x = 0$. And for $z = 1$, (1) reduces to $\ell \leq x \leq \frac{u}{1-m_1}$, the right-hand inequality of which is redundant with respect to $x \leq u$, altogether implying that $\ell \leq x \leq u$. So the model is valid — and not generally recommended!

Now, in this simple example, we can easily check that we get the tightest relaxation possible (the so-called "convex-hull relaxation") if and only if we choose

A. Basu et al. (Eds.): ISCO 2024, LNCS 14594, pp. 3–15, 2024.
https://doi.org/10.1007/978-3-031-60924-4_1

$m_1 = m_2 = 0$, whereupon (1) becomes $\ell z \leq x \leq uz$. Notice that if we take m_1 (resp., m_2) to be near 1, then the coefficient on z on the right-hand (resp., left-hand) inequality of (1) becomes big — a so-called "big-M". In the context of mixed-integer *nonlinear* optimization, the story of this little example does not even end here (see, for example, [LSS22,LSSX21,LSSX23], and the references therein, which look at tradeoffs in models for approximating a convex univariate function on an interval domain).

Although our example is quite simple and perhaps trivial, the phenomenon is serious for modelers and solvers. From [Gur22], we have: "Big-M constraints are a regular source of instability for optimization problems." But it is not just numerical instability; from [Gur22], we also have: "In other words, x can take a [very small] positive value without incurring an expensive fixed charge on [z], which subverts the intent of only allowing a non-zero value for x when the binary variable [z] has the value of 1." This behavior means weak continuous relaxations and often very poor behavior of algorithms like branch-and-bound, branch-and-cut, etc., on such models.

Our Main Contributions: In higher dimension d for x (above it was $d = 1$) and higher number $n + 1$ (above it was $n + 1 = 2$) of polytope regions $\mathcal{P}_i$ for x, the situation is cloudier than what was exposed in the simple example above, and our high-level goal is to carefully investigate the geometry for the cases in which (i) $d \in \{1, 2\}$, with $n \geq 1$ arbitrary, and also the case (ii) the regions are all axis-aligned hyper-rectangles (for arbitrary $d \geq 1$ and $n \geq 1$). We demonstrate that in these cases, the convex hull $\mathcal{D}$ in the natural extended-variable space, i.e., in $\mathbb{R}^{d+n}$, is given by "full optimal big-M lifting" of facet-describing inequalities of the polytope regions $\mathcal{P}_i$. We note that both of theses cases are very natural for applications and within spatial branch-and-bound, the main algorithmic paradigm for mixed-integer nonlinear optimization. Additionally, we give further results on the polyhedral structure of $\mathcal{D}$.

Organization: In Sect. 2, we lay out all of our definitions. In Sect. 3, we state all of our results. In Sect. 4, We present some proofs and proof sketches, interspersing some remarks.

Standard Terminology Concerning Polytopes. We follow the terminology of [Lee04, Chapter 0.5]. If $\alpha^\mathsf{T} x \leq \beta$ is satisfied by all points of a polytope $\mathcal{P} \subset \mathbb{R}^n$, then it is *valid* for $\mathcal{P}$. For a valid inequality $\alpha^\mathsf{T} x \leq \beta$ of $\mathcal{P}$, the *face described* is $\mathcal{P} \cap \{x \in \mathbb{R}^n : \alpha^\mathsf{T} x = \beta\}$. The *dimension* of a polytope $\mathcal{P}$ is one less than the maximum number of affinely independent points in $\mathcal{P}$. A polytope $\mathcal{P} \subset \mathbb{R}^n$ is *full dimensional* if its dimension is n. A *facet* of $\mathcal{P}$ is a face of dimension one less than that of $\mathcal{P}$. If $\alpha^\mathsf{T} x \leq \beta$ describes a facet of a full-dimensional polytope $\mathcal{P}$, then any other inequality that also describes that face is a positive multiple of $\alpha^\mathsf{T} x \leq \beta$. Additionally, for a full-dimensional polytope $\mathcal{P}$, the solution set of its (essentially unique) facet-describing inequalities is precisely $\mathcal{P}$. Also, see [Grü03] and [Zie95] for much more on polytopes.

2 Definitions

2.1 Model and Notation

In what follows, for $n \geq 1$ we let $N_0 := \{0, 1, ..., n\}$ and $N := \{1, ..., n\}$. For $d \geq 1$ and $n \geq 1$, we consider full-dimensional polytopes $\mathcal{P}_i := \{x \in \mathbb{R}^d : \mathbf{A}_i x \leq \mathbf{b}_i\}$, for $i \in N_0$. The number of columns of each matrix $\mathbf{A}_i$ is d, and the number of rows of each $\mathbf{A}_i$ agrees with the number of elements of the corresponding vector $\mathbf{b}_i$. We note that we do not assume that the $\mathcal{P}_i$ are pairwise disjoint, but that could be the case, especially for $d = 1$. For convenience, we assume that each system $\mathbf{A}_i x \leq \mathbf{b}_i$ has no redundant inequalities, and then because of the full-dimensional assumption, we have that each inequality defining $\mathcal{P}_i$ describes a unique facet of $\mathcal{P}_i$.

We define n binary variables z_i, with $z_i = 1$ indicating that the vector of variables x must be in $\mathcal{P}_i$, for $i \in N$. Further, if $z_i = 0$ for all $i \in N$, then the vector of variables x must be in $\mathcal{P}_0$. Finally, we constrain the z_i via

$$\sum_{i \in N} z_i \leq 1, \tag{2}$$

so, overall, x must be in at least one $\mathcal{P}_i$, for $i \in N_0$. Because the $\mathcal{P}_i$ may overlap, it can be that $z_i = 0$ but $x \in \mathcal{P}_i$ for some $i \in N$. Likewise, we can have $z_i = 1$ for some choice of $i \in N$ but $x \in \mathcal{P}_0$.

2.2 Convex Hull

For $i \in N$, let $\mathbf{e}_i$ denote the i-th standard unit vector in $\mathbb{R}^n$, and additionally for convenience, let $\mathbf{e}_0$ denote the zero vector in $\mathbb{R}^n$. For $i \in N_0$, let $\bar{\mathcal{P}}_i := \{\binom{x}{\mathbf{e}_i} : x \in \mathcal{P}_i\}$, which is the polytope $\mathcal{P}_i \subset \mathbb{R}^d$ lifted to $\mathbb{R}^{d+n}$ by setting $z = \mathbf{e}_i$. For $i \in N_0$, let $\mathcal{X}_i$ be the (finite) set of extreme points of $\mathcal{P}_i$, and let $\bar{\mathcal{X}}_i := \{\binom{x}{\mathbf{e}_i} : x \in \mathcal{X}_i\}$, the (finite) set of extreme points of $\bar{\mathcal{P}}_i$.

Let

$$\mathcal{D} := \operatorname{conv}\left\{\bar{\mathcal{P}}_i : i \in N_0\right\} = \operatorname{conv}\left\{\bar{\mathcal{X}}_i : i \in N_0\right\}.$$

The definition of $\mathcal{D}$ is an "inner description", but convenient algorithmic approaches generally work with an "outer description" (i.e., via linear inequalities). See Fig. 1 for an example.

2.3 Full Optimal Big-M Lifting

Lifting is a general well-known technique for extending linear inequalities in disjunctive settings. See, for example, [NW88, Chapter II.2, Sect. 1]. Next, we describe how to do optimal lifting of valid inequalities for individual $\mathcal{P}_i$ to account for the other polytopes.

<u>Case 1</u>: Suppose that

$$\alpha^\mathsf{T} x \leq \beta \tag{$*$}$$

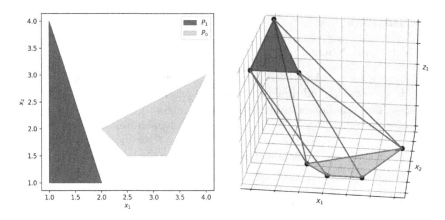

Fig. 1. Example: $d = 2$, $n = 1$

is a valid inequality for $\mathcal{P}_0$. Now, for some $j \in N$, we want to optimally lift $M_j z_j$ into this inequality, by choosing the largest M_j so that the resulting inequality

$$\alpha^\top x + M_j z_j \leq \beta$$

is valid also for $\bar{\mathcal{P}}_j$. We clearly need to choose

$$M_j \leq \beta - \alpha^\top x,$$

for *all* $x \in \mathcal{P}_j$. And so, to get the best such lifting, we set

$$M_j := \min\{\beta - \alpha^\top x \; : \; x \in \mathcal{P}_j\}.$$

It is easy to see that starting with any valid linear inequality for $\mathcal{P}_0$, and lifting one coefficient at a time on the z_j, until we have lifted all of them, the order of lifting does not matter (because $\sum_{j \in N} z_j \leq 1$ is required) — we will arrive at the same "full-lifting inequality" for $\mathcal{D}$, having the form

$$\alpha^\top x + \sum_{i \in N} M_i z_i \leq \beta. \qquad (*_0)$$

Case 2: The above lifting development assumes that we start with the inequality ($*$) being valid for $\mathcal{P}_0$. But instead, let us assume that we start with ($*$) being valid for $\mathcal{P}_k$, for some $k \in N$. To calculate the optimal lifting coefficient for some $j \in N \setminus \{k\}$, we proceed exactly as above. But to lift the inequality so that is valid for $\bar{\mathcal{P}}_0$, we need to proceed a bit differently. We need to choose M_0 so that the resulting inequality

$$\alpha^\top x + M_0 \left(1 - \sum_{i \in N} z_i\right) \leq \beta$$

is valid also for $\bar{\mathcal{P}}_0$. We clearly need to choose

$$M_0 \leq \beta - \alpha^\top x,$$

for *all* $x \in \mathcal{P}_0$. And so, to get the best such lifting, we set

$$M_0 := \min\{\beta - \alpha^\mathsf{T} x \ : \ x \in \mathcal{P}_0\}.$$

Again for this case, it is easy to see that starting with any valid linear inequality for some $\mathcal{P}_k$, and lifting one at a time until we have lifted all, the order of lifting does not matter — we will arrive at the same inequality for $\mathcal{D}$. In this situation, gathering terms together, the resulting "full-lifting inequality" is

$$\alpha^\mathsf{T} x + \sum\nolimits_{i \in N\setminus\{k\}} (M_i - M_0)z_i - M_0 z_k \le \beta - M_0. \qquad (*_k)$$

In both cases, it is clear that for each facet-defining inequality of each $\mathcal{P}_k$, we can compute all of its optimal "big-M" coefficients simultaneously, to lift to an inequality valid for $\mathcal{D}$. In a nod to the fact that in our setting, the sequence of lifting does not matter, and thus the same coefficients are in a modeling sense optimal big-M coefficients, we refer to $(*_0)$ and $(*_k)$ as *full optimal big-M liftings* of $(*)$.

Example 1. Suppose that $d = 1$ and that $\mathcal{P}_i := [a_i, b_i]$, for $i \in N_0$. Consider the valid inequality for $\mathcal{P}_0$: $x \le b_0$. In this case, if we lift all z_i variable, the full optimal big-M lifting inequality is $x + \sum_{i=1}^n M_i z_i \le b_0$, where $M_i = \min\{b_0 - x \ : \ x \in [a_i, b_i]\} = b_0 - b_i$. That is, we obtain: $x + \sum_{i=1}^n (b_0 - b_i)z_i \le b_0$. Similarly, if we take the valid inequality for $\mathcal{P}_0$: $-x_0 \le -a_0$, then $M_i = \min\{-a_0 + x \ : \ x \in [a_i, b_i]\} = a_i - a_0$. That is, we obtain the full optimal big-M lifting inequality $x + \sum_{i=1}^n (a_0 - a_i)z_i \ge a_0$.

Of course disjunctive formulations and the associated liftings that we describe and analyze are not new at all. Disjunctive programming, at a high level, is the subject of concrete convexification of nonconvex sets, in the context of global optimization. This important topic was introduced by Balas in 1971; see [Bal79]. It is a key technique in the well-known integer-programming book [NW88]. Fairly recently, four decades after its introduction, and with lots of progress in the intervening time period, Balas himself, in probably his last major work, finally organized the topic into a major text; see [Bal18]. In closely related work, [Jer88, Bla90, Bal88] studied when the convex hull of polytopes has a simple form, but they do not work with binary variables to manage a disjunction. We prefer to include binary variables because of their modeling power (e.g., carrying fixed costs and logical relationships). [Vie15, Section 6] is closer to our viewpoint, but we have opted to present a self-contained treatment which we hope is broadly accessible to mixed-integer optimization users.

Finally, there is a lot of experience and expectedly many enhancements for methods for handling disjunctions (see, e.g., [BLL+11, LTV23, TG15, KM21]), and with various particularizations (see [KKY15, AHS07], for example). Still, there is a lot to understand about the elementary situation that we described in Sects. 2.1–2.3. In particular, we will investigate conditions for which the convex hull $\mathcal{D}$ is completely described by full optimal big-M lifting of the facet-describing inequalities of the $\mathcal{P}_i$.

3   Results

First, we present elementary valid inequalities for $\mathcal{D}$ and establish (under mild technical conditions) that they describe facets of $\mathcal{D}$.

Theorem 1. Suppose that $d \geq 1$. Then if all $\mathcal{P}_i$ are nonempty, for $i \in N_0$, and at least one is full dimensional, then $\mathcal{D}$ is full dimensional.

Theorem 2. Suppose that $d \geq 1$. Then if all $\mathcal{P}_i$ are nonempty, for $i \in N_0$, then the full optimal big-M lifting of a facet-describing inequality of a full-dimensional $\mathcal{P}_k$ (for any $k \in N_0$) is facet describing for $\mathcal{D}$.

Theorem 3. Suppose that $d \geq 1$. Then if all $\mathcal{P}_i$ are nonempty, for $i \in N_0$, and $\mathcal{P}_0$ is full dimensional, then for all $j \in N$, $z_j \geq 0$ is facet describing for $\mathcal{D}$.

Theorem 4. Suppose that $d \geq 1$. Then if all $\mathcal{P}_i$ are nonempty, for $i \in N_0$, and $\mathcal{P}_k$ is full dimensional for some $k \in N$, then $\sum_{j \in N} z_j \leq 1$ is facet describing for $\mathcal{D}$.

Henceforth, we refer to facets of $\mathcal{D}$ described by $z_j \geq 0$ and $\sum_{j \in N} z_j \leq 1$ as *non-vertical facets*, and other facets of $\mathcal{D}$ as *vertical facets*.

Next, under the same mild technical condition as above, we establish that the facet-describing inequalities identified above (in Theorems 2,3,4) give a complete description of the convex hull $\mathcal{D}$, when $d \in \{1,2\}$.

Theorem 5. Suppose that $n \geq 1$, $d \in \{1,2\}$, all $\mathcal{P}_i$ are nonempty, for $i \in N_0$, and at least one is full dimensional. Then the full optimal big M-lifting of each facet-describing inequality defining each $\mathcal{P}_i$, for $i \in N$, together with the inequalities $\sum_{i \in N} z_i \leq 1$ and $z_i \geq 0$, for $i \in N$, gives the convex hull $\mathcal{D}$.

In fact, Theorem 5 does not extend to $d = 3$, as indicated in the following result.

Proposition 6. For $n = 1$ and $d = 3$, there can be vertical facets of the convex hull $\mathcal{D}$ that are not described by full optimal big-M lifting of facet-describing inequalities defining $\mathcal{P}_0$ and $\mathcal{P}_1$, even for simplices.

Finally, when the polytopes $\mathcal{P}_i$ are all axis-aligned hyper-rectangles, we establish that the facet-describing inequalities identified above (in Theorems 2,3,4) give a complete description of the convex hull $\mathcal{D}$.

Theorem 7. For $n \geq 1$ and $d \geq 1$, we consider $n+1$ hyper-rectangles $\mathcal{P}_j := [a_{j1}, b_{j1}] \times \cdots \times [a_{jd}, b_{jd}]$, for $j \in N_0$. The full optimal big-M lifting inequalities $x_i + \sum_{j \in N}(a_{0i} - a_{ji})z_j \geq a_{0i}$, $x_i + \sum_{j \in N}(b_{0i} - b_{ji})z_j \leq b_{0i}$, for all $i = 1, \cdots, d$, together with $\sum_{j \in N} z_j \leq 1$, and $z_j \geq 0$, for $j \in N$, gives the convex hull $\mathcal{D}$.

Remark 1. Of course we already know that the result is true for $d = 1, 2$ by Theorem 5. So the further contribution of Theorem 7 is for $d \geq 3$.

4   Some Proofs, Proof Sketches, and Remarks

Proof (sketch of Theorem 1). There are two cases to consider. First, we assume that $\mathcal{P}_0$ is full dimensional. Because $\mathcal{P}_0$ is full dimensional, then $\mathcal{P}_0$ contains a set of $d+1$ affinely-independent points, say $\hat{x}^0, \hat{x}^1, \ldots, \hat{x}^d$. Additionally, we assumed that each $\mathcal{P}_i$, for $i \in N$, is nonempty, so it contains a point, say $\tilde{x}^i$. Now, we consider the $n + d + 1$ points (in $\mathbb{R}^{n+d}$):

$$\begin{pmatrix}\hat{x}^0 \\ e_0\end{pmatrix}, \begin{pmatrix}\hat{x}^1 \\ e_0\end{pmatrix}, \ldots, \begin{pmatrix}\hat{x}^d \\ e_0\end{pmatrix}; \begin{pmatrix}\tilde{x}^1 \\ e_1\end{pmatrix}, \begin{pmatrix}\tilde{x}^2 \\ e_2\end{pmatrix}, \ldots, \begin{pmatrix}\tilde{x}^n \\ e_n\end{pmatrix},$$

and it remains only to show that these $n + d + 1$ points are affinely independent (details omitted).

For the other case, we assume instead that $\mathcal{P}_k$ is full dimensional for some $k \in N$. Because $\mathcal{P}_k$ is full dimensional, it contains a set of $d+1$ affinely-independent points, say $\hat{x}^0, \hat{x}^1, \ldots, \hat{x}^d$. Additionally, because each $\mathcal{P}_i$ is nonempty for $i \in N_0 \setminus \{k\}$, it contains a point, say $\tilde{x}^i$. Now, we consider the $n + d + 1$ points (in $\mathbb{R}^{n+d}$):

$$\begin{pmatrix}\hat{x}^0 \\ e_k\end{pmatrix}, \begin{pmatrix}\hat{x}^1 \\ e_k\end{pmatrix}, \ldots, \begin{pmatrix}\hat{x}^d \\ e_k\end{pmatrix}; \begin{pmatrix}\tilde{x}^0 \\ e_0\end{pmatrix}, \begin{pmatrix}\tilde{x}^1 \\ e_1\end{pmatrix}, \ldots, \begin{pmatrix}\tilde{x}^{k-1} \\ e_{k-1}\end{pmatrix}, \begin{pmatrix}\tilde{x}^{k+1} \\ e_{k+1}\end{pmatrix}, \ldots, \begin{pmatrix}\tilde{x}^n \\ e_n\end{pmatrix},$$

and it remains only to show that these $n + d + 1$ points are affinely independent (details omitted). □

Proof (of Theorem 2). There are two cases to consider.

Case 1: First, we assume that $\mathcal{P}_0$ is full dimensional. Let $(*)$ be a facet-describing inequality of $\mathcal{P}_0$. Let $(*_0)$ be its full lifting, where $M_i := \min\{\beta - \alpha^\mathsf{T} x : x \in \mathcal{P}_i\}$. We need to choose $n+d$ affinely-independent points from $\mathcal{D}$ satisfying $(*_0)$ as an equation to demonstrate that $(*_0)$ describes a facet of $\mathcal{D}$. Let $\hat{x}^1, \hat{x}^2, \cdots, \hat{x}^d$ be d affinely-independent points from $\mathcal{P}_0$ satisfying $\alpha^\mathsf{T} x = \beta$. So clearly $\begin{pmatrix}\hat{x}^i \\ e_0\end{pmatrix}$ satisfies $(*_0)$ as an equation, for $1 \le i \le d$. For $i \in N$, let $\tilde{x}^i$ be an optimal solution of the lifting problem: $M_i := \min\{\beta - \alpha^\mathsf{T} x : x \in \mathcal{P}_i\}$. It is easy to see that $\begin{pmatrix}\tilde{x}^i \\ e_i\end{pmatrix}$ satisfies $(*_0)$ as an equation. So, overall, we have $n + d$ points satisfying $(*_0)$ as an equation. In fact, these $n + d$ points satisfy the same properties as the ones using the same notation in the first case of the proof of Theorem 1 (just omitting here the point $\begin{pmatrix}\hat{x}^0 \\ e_0\end{pmatrix}$), so these points are also affinely independent.

Case 2: Next, we assume instead that $\mathcal{P}_k$ is full dimensional for some $k \in N$. Let $(*)$ be a facet-describing inequality of $\mathcal{P}_k$. Let $(*_k)$ be its full lifting, where $M_0 := \min\{\beta - \alpha^\mathsf{T} x : x \in \mathcal{P}_0\}$. Let $\hat{x}^1, \hat{x}^2, \cdots, \hat{x}^d$ be d affinely-independent points from $\mathcal{P}_k$ satisfying $\alpha^\mathsf{T} x = \beta$. So clearly $\begin{pmatrix}\hat{x}^i \\ e_k\end{pmatrix}$ satisfies $(*_k)$ as an equation, for $1 \le i \le d$. For $i \in N_0 \setminus \{k\}$, let $\tilde{x}^i$ be an optimal solution of the lifting problem: $M_i := \min\{\beta - \alpha^\mathsf{T} x : x \in \mathcal{P}_i\}$. It is easy to see that $\begin{pmatrix}\tilde{x}^i \\ e_i\end{pmatrix}$ satisfies $(*_k)$ as an equation. So, overall, we have $n + d$ points satisfying $(*_k)$ as an equation. In fact, these $n+d$ points satisfy the same properties as the ones using the same notation in the second case of the proof of Theorem 1 (just omitting here the point $\begin{pmatrix}\hat{x}^0 \\ e_k\end{pmatrix}$), so these points are also affinely independent. □

Proof (of Theorem 3). We need to choose $n+d$ affinely-independent points from $\mathcal{D}$ satisfying $z_j \geq 0$ as an equation to demonstrate that the $z_j \geq 0$ describes a facet of $\mathcal{D}$. Because $\mathcal{P}_0$ is full dimensional, then $\mathcal{P}_0$ contains a set of $d+1$ affinely-independent points, say $\hat{x}^0, \hat{x}^1, ..., \hat{x}^d$. Clearly, $\left(\begin{smallmatrix}\hat{x}^0\\\mathbf{e}_0\end{smallmatrix}\right), ..., \left(\begin{smallmatrix}\hat{x}^d\\\mathbf{e}_0\end{smallmatrix}\right)$ satisfy $z_j \geq 0$ as an equation. Additionally, because each $\mathcal{P}_i$ is nonempty for $i \in \{1, ..., n\}$, it contains a point, say $\tilde{x}^i$. Now consider $n-1$ points (in $\mathbb{R}^{n+d}$) from $\bar{\mathcal{P}}_i$ for $i \in N \setminus \{j\}$, say $\left(\begin{smallmatrix}\tilde{x}^1\\\mathbf{e}_1\end{smallmatrix}\right), \left(\begin{smallmatrix}\tilde{x}^2\\\mathbf{e}_2\end{smallmatrix}\right), ..., \left(\begin{smallmatrix}\tilde{x}^{j-1}\\\mathbf{e}_{j-1}\end{smallmatrix}\right), \left(\begin{smallmatrix}\tilde{x}^{j+1}\\\mathbf{e}_{j+1}\end{smallmatrix}\right), ..., \left(\begin{smallmatrix}\tilde{x}^n\\\mathbf{e}_n\end{smallmatrix}\right)$. It is obvious that those $n-1$ points satisfy $z_j \geq 0$ as an equation. So overall we have $n+d$ points satisfy $z_j \geq 0$ as an equation. We want to prove that those $n+d$ points are affinely independent. Because these $n+d$ points satisfy the same properties as the ones using the same notation in the first case of the proof of Theorem 1 (just omitting here the point $\left(\begin{smallmatrix}\tilde{x}^j\\\mathbf{e}_j\end{smallmatrix}\right)$), these points are also affinely independent. □

Proof (of Theorem 4). We need to choose $n+d$ affinely-independent points from $\mathcal{D}$ satisfying $\sum_{j \in N} z_j \leq 1$ as an equation to demonstrate that the $\sum_{j \in N} z_j \leq 1$ describes a facet of $\mathcal{D}$. Because $\mathcal{P}_k$ is full dimensional, then $\mathcal{P}_k$ contains a set of $d+1$ affinely-independent points, say $\hat{x}^0, \hat{x}^1, ..., \hat{x}^d$. Clearly, $\left(\begin{smallmatrix}\hat{x}^0\\\mathbf{e}_k\end{smallmatrix}\right), ..., \left(\begin{smallmatrix}\hat{x}^d\\\mathbf{e}_k\end{smallmatrix}\right)$ satisfy $\sum_{j \in N} z_j \leq 1$ as an equation. Additionally, because each $\mathcal{P}_i$ is nonempty for $i \in N_0 \setminus \{k\}$, it contains a point, say $\tilde{x}^i$. We consider $n-1$ points from $\bar{\mathcal{P}}_i$ for $i \in N \setminus \{k\}$, say $\left(\begin{smallmatrix}\tilde{x}^1\\\mathbf{e}_1\end{smallmatrix}\right), \left(\begin{smallmatrix}\tilde{x}^2\\\mathbf{e}_2\end{smallmatrix}\right), ..., \left(\begin{smallmatrix}\tilde{x}^{k-1}\\\mathbf{e}_{k-1}\end{smallmatrix}\right), \left(\begin{smallmatrix}\tilde{x}^{k+1}\\\mathbf{e}_{k+1}\end{smallmatrix}\right), ..., \left(\begin{smallmatrix}\tilde{x}^n\\\mathbf{e}_n\end{smallmatrix}\right)$. It is obvious that $\left(\begin{smallmatrix}\tilde{x}^i\\\mathbf{e}_i\end{smallmatrix}\right), i \in N \setminus \{k\}$ satisfy $\sum_{j \in N} z_j \leq 1$ as an equation. So we have $n+d$ points satisfy $\sum_{j \in N} z_j \leq 1$ as an equation. In fact, these $n+d$ points satisfy the same properties as the ones using the same notation in the second case of the proof of Theorem 1 (just omitting here the point $\left(\begin{smallmatrix}\tilde{x}^0\\\mathbf{e}_0\end{smallmatrix}\right)$), so these points are also affinely independent. □

Proof (sketch of Theorem 5). By Theorem 1, $\mathcal{D}$ is a full-dimensional polytope in $\mathbb{R}^{n+d}$, so every one of its facets contains $n+d$ affinely independent points of the form $\left(\begin{smallmatrix}\hat{x}^i\\\mathbf{e}_i\end{smallmatrix}\right)$, where each such point has $\hat{x}^i$ being an extreme point of a $\mathcal{P}_i$. In fact, at most $d+1$ of these can be associated with the same i, for each $i \in N_0$ as the different $\hat{x}^i$ for the same i must be affinely independent.

For $d = 1$, we have $n+d = n+1$, and one possibility is that the $n+1$ points partition, one for each $i \in N_0$. In this case, we can view the facet as an optimal sequential lifting, starting from a facet of $\mathcal{P}_0$ (details omitted). If, instead, the points do not partition this way, then some $\mathcal{P}_i$ is missed. If $i \in N$, then the facet must be described by $z_i \geq 0$. If, instead, $i = 0$, then the facet must be described by $\sum_{i \in N} z_i \leq 1$.

For $d = 2$, we have $n+d = n+2$, and one possibility is that the $n+2$ points partition as two for some $\ell \in N_0$, and one for each of the remaining $i \in N_0 \setminus \{\ell\}$. Without loss of generality, we can take the two points to be the extreme points of a facet of $\mathcal{P}_\ell$. In this case, we can see that the facet is an optimal sequential lifting, starting from a facet of $\mathcal{P}_\ell$ (details omitted). If some i is missed, then we reason exactly as we did above for $d = 1$ (details omitted). □

Remark 2. Considering the case of $d = 2$ in the sketch of the proof of Theorem 5, we can refer to Fig. 1, where we have $n = 1$. We can observe that generally for $d = 2$ and $n = 1$, facets of $\mathcal{D}$ are either triangles or trapezoids, with a trapezoid (there is one in the example) arising by lifting from a facet of one polytope $\mathcal{P}_i$ that is parallel to a facet of the other polytope.

Remark 3. The argument in the proof of Theorem 5 falls apart already for $d = 3$ and $n = 1$. In such a situation, a vertical facet must contain 4 affinely-independent points ($\mathcal{D}$ is in $\mathbb{R}^4$), but 2 could come from each of $\mathcal{P}_0$ and $\mathcal{P}_1$. In such a case, they do not affinely span a facet of either (we need 3 affinely independent points on a facet of $\mathcal{P}_0$ and of $\mathcal{P}_1$). So such a vertical facet will not be a lift of a facet-describing inequality of $\mathcal{P}_0$ or $\mathcal{P}_1$.

Proof (of Proposition 6). This only requires an example. Let

$$\mathcal{P}_0 := \{x \in \mathbb{R}^3 : x_1, x_2, x_3 \leq 5, \ x_1 + x_2 + x_3 \geq 14\},$$

and let

$$\mathcal{P}_1 := \{x \in \mathbb{R}^3 : x_1, x_2, x_3 \geq 0, \ x_1 + x_2 + x_3 \leq 1\}.$$

It can be checked (using software like [Fuk22]) that $\mathcal{D}$ has 14 vertical facets, but there cannot be more than 8 full optimal big-M liftings of facet-describing inequalities, because the total number of facet-describing inequalities for $\mathcal{P}_0$ and $\mathcal{P}_1$ is 8.   □

Remark 4. Analyzing the example in the proof of Proposition 6 in a bit more detail, there are 6 facets that do not come from lifting facet-describing inequalities of the $\mathcal{P}_i$. Namely, those described by $-9z_1 + 9 \leq x_i + x_j \leq 10 - 9z_1$, for each of the three choices of distinct pairs $i, j \in \{1, 2, 3\}$. In fact, they can be seen as coming from *face*-describing inequalities, namely: $x_i + x_j \leq 1$ and $x_i + x_j \geq -9$. This is not a good thing! Solving appropriate linear-optimization problems, we can check full-dimensionality (see [FRT85]) and remove redundant inequalities (obvious). But we cannot realistically handle (for large d) all face-describing inequalities for each $\mathcal{P}_i$.

It turns out that for this example, the facet-describing inequalities of $\mathcal{D}$ that are not optimal liftings of facet-describing inequalities of the $\mathcal{P}_i$ are MIR (mixed-integer rounding) inequalities relative to weighted combinations of optimal liftings of facet-describing inequalities of the $\mathcal{P}_i$.

Proof (of Theorem 7). Consider the linear-optimization problem with *arbitrary* $c \in \mathbb{R}^d$, $g \in \mathbb{R}^n$:

$$\max \ c^\mathsf{T} x + g^\mathsf{T} z \qquad\qquad \text{dual} \qquad (P_H)$$

subject to   var

$$x_i + \sum_{j \in N}(a_{0i} - a_{ji})z_j \geq a_{0i}, \ \text{for all } i = 1, \cdots, d; \qquad \omega_i$$

$$x_i + \sum_{j \in N}(b_{0i} - b_{ji})z_j \leq b_{0i}, \ \text{for all } i = 1, \cdots, d; \qquad \nu_i$$

$$\sum_{j \in N} z_j \leq 1; \qquad\qquad\qquad\qquad\qquad\qquad\qquad \pi$$

$$z_j \geq 0, \ \text{for } j \in N.$$

Our goal is to give a recipe that solves the linear-optimization problem P_H, using always an extreme point of $\mathcal{D}$. If so, then we show that the lifted inequalities are facet-describing inequalities for the convex hull.

For $k \in N_0$, let

$$\hat{x}_i^k := \begin{cases} a_{ki}, & \text{if } c_i \leq 0; \\ b_{ki}, & \text{if } c_i > 0. \end{cases}$$

So, $\hat{x}^k$ maximizes $c^\mathsf{T} x$ on $\mathcal{P}_k$. Now, let $\hat{k} := \operatorname{argmax}_{k \in N_0}\{c^\mathsf{T}\hat{x}^k + g^\mathsf{T}\mathbf{e}_k\}$. Therefore, $\binom{\hat{x}^{\hat{k}}}{\mathbf{e}_{\hat{k}}}$ maximizes $c^\mathsf{T} x + g^\mathsf{T} z$ on $\mathcal{D}$.

We can immediately observe that $\binom{\hat{x}^{\hat{k}}}{\mathbf{e}_{\hat{k}}}$ is feasible for P_H, which is a relaxation of $\mathcal{D}$. To prove that $\binom{\hat{x}^{\hat{k}}}{\mathbf{e}_{\hat{k}}}$ is optimal for P_H, we consider its dual:

$$\min \ \textstyle\sum_{i=1}^d a_{0i}\omega_i + \sum_{i=1}^d b_{0i}\nu_i + \pi \tag{D_H}$$

subject to

$\omega_i + \nu_i = c_i$, for $i = 1, \cdots, d$;

$\left(\sum_{i=1}^d (a_{0i} - a_{ji})\right)\omega_j + \left(\sum_{i=1}^d (b_{0i} - b_{ji})\right)\nu_j + \pi \geq g_j$, for $j \in N$;

$\omega_i \leq 0$, for all $1 \leq i \leq d$;

$\nu_i \geq 0$, for all $1 \leq i \leq d$;

$\pi \geq 0$.

Next, we construct a dual solution $(\hat{\omega}, \hat{u}, \hat{\pi})$ of D_H, defined by

$$\hat{\omega}_i := \begin{cases} c_i, & \text{if } c_i \leq 0, \\ 0, & \text{if } c_i > 0, \end{cases} \quad \text{for } i = 1, \cdots, d;$$

$$\hat{\nu}_i := \begin{cases} 0, & \text{if } c_i \leq 0, \\ c_i, & \text{if } c_i > 0, \end{cases} \quad \text{for } i = 1, \cdots, d;$$

$$\hat{\pi} := \begin{cases} 0, & \text{if } \hat{k} = 0; \\ g_{\hat{k}} + \displaystyle\sum_{i\,:\,c_i<0}(a_{\hat{k}i} - a_{0i})c_i + \sum_{i\,:\,c_i>0}(b_{\hat{k}i} - b_{0i})c_i, & \text{if } \hat{k} \in N. \end{cases}$$

It remains only to check that this solution is feasible for D_H and has the same objective value in D_H as $\binom{\hat{x}^{\hat{k}}}{\mathbf{e}_{\hat{k}}}$ has in P_H. Then, by weak duality, it follows that $\binom{\hat{x}^{\hat{k}}}{\mathbf{e}_{\hat{k}}}$ is optimal for P_H. Because we find always an extreme point of $\mathcal{D}$ to solve P_H, we can conclude that the $\mathcal{D}$ is precisely the feasible region of P_H.

We will check that these solutions are feasible for D_H. It is easy to see that $\hat{\omega} \leq 0$ and $\hat{\nu} \geq 0$, and that $\hat{\omega} + \hat{\nu} = c$.

Next, we check that $\hat{\pi} \geq 0$. If $\hat{k} = 0$, then $\hat{\pi} \geq 0$, so we can assume that $\hat{k} \in N$, and then we have

$$\hat{\pi} = g_{\hat{k}} + \sum_{i\,:\,c_i<0} (a_{\hat{k}i} - a_{0i})c_i + \sum_{i\,:\,c_i>0} (b_{\hat{k}i} - b_{0i})c_i$$

$$= c^{\mathsf{T}}\hat{x}^{\hat{k}} + g^{\mathsf{T}}e^{\hat{k}} - c^{\mathsf{T}}\hat{x}^0,$$

which is nonnegative due to the choice of $\hat{k}$. So we have $\hat{\pi} \geq 0$.

Next, if $\hat{k} = 0$, we have

$$\sum_{i\,:\,c_i<0} (a_{0i} - a_{ki})c_i + \sum_{i\,:\,c_i>0} (b_{0i} - b_{ki})c_i + 0 \geq g_k$$

$$\Longleftrightarrow \sum_{i\,:\,c_i<0} a_{0i}c_i + \sum_{i\,:\,c_i>0} b_{0i}c_i \geq \sum_{i\,:\,c_i<0} a_{ki}c_i + \sum_{i\,:\,c_i>0} b_{ki}c_i + g_k$$

$$\Longleftrightarrow c^{\mathsf{T}}\hat{x}^0 \geq c^{\mathsf{T}}\hat{x}^k + g_k .$$

Otherwise, we have $\hat{k} \in N$, and

$$\sum_{i=1}^d (a_{0i} - a_{ki})\hat{\omega}_i + \sum_{i=1}^d (b_{0i} - b_{ki})\hat{\nu}_i + \hat{\pi} \geq g_k$$

$$\Longleftrightarrow \sum_{i\,:\,c_i<0} (a_{0i} - a_{ki})c_i + \sum_{i\,:\,c_i>0} (b_{0i} - b_{ki})c_i$$

$$+ \left(g_{\hat{k}} + \sum_{i\,:\,c_i<0} (a_{\hat{k}i} - a_{0i})c_i + \sum_{i\,:\,c_i>0} (b_{\hat{k}i} - b_{0i})c_i \right) \geq g_k$$

$$\Longleftrightarrow g_{\hat{k}} + \sum_{i\,:\,c_i<0} (a_{\hat{k}i} - a_{0i})c_i + \sum_{i\,:\,c_i>0} (b_{\hat{k}i} - b_{0i})c_i$$

$$\geq g_k + \sum_{i\,:\,c_i<0} (a_{ki} - a_{0i})c_i + \sum_{i\,:\,c_i>0} (b_{ki} - b_{0i})c_i$$

$$\Longleftrightarrow c^{\mathsf{T}}\hat{x}^{\hat{k}} + g_{\hat{k}} \geq c^{\mathsf{T}}\hat{x}^k + g_k .$$

Therefore, $(\hat{\omega}, \hat{\nu}, \hat{\pi})$ is feasible for D_H for $k \in N_0$.

Finally, we check that the objective value of $\binom{\hat{x}^{\hat{k}}}{e_{\hat{k}}}$ in P_H and $(\hat{\omega}, \hat{\nu}, \hat{\pi})$ in D_H are the same. If $\hat{k} = 0$, we have

$$\sum_{i=1}^d a_{0i}\hat{\omega}_i + \sum_{i=1}^d b_{0i}\hat{\nu}_i + \hat{\pi} = \sum_{i\,:\,c_i<0} a_{0i}c_i + \sum_{i\,:\,c_i>0} b_{0i}c_i = c^{\mathsf{T}}\hat{x}^0.$$

Otherwise, we have $\hat{k} \in N$, and

$$\sum_{i=1}^d a_{0i}\hat{\omega}_i + \sum_{i=1}^d b_{0i}\hat{\nu}_i + \hat{\pi}$$

$$= \sum_{i\,:\,c_i<0} a_{0i}c_i + \sum_{i\,:\,c_i>0} b_{0i}c_i + g_{\hat{k}} + \sum_{i\,:\,c_i<0} (a_{\hat{k}i} - a_{0i})c_i + \sum_{i\,:\,c_i>0} (b_{\hat{k}i} - b_{0i})c_i$$

$$= \sum_{i\,:\,c_i<0} a_{\hat{k}i}c_i + \sum_{i\,:\,c_i>0} b_{\hat{k}i}c_i + g_{\hat{k}}$$

$$= c^{\mathsf{T}}\hat{x}^{\hat{k}} + g^{\mathsf{T}}e_{\hat{k}} .$$

In both cases, we have equality of primal and dual objective values.   □

Acknowledgments. The work of J. Lee was supported in part by Office of Naval Research grant N00014-21-1-2135, the Gaspard Monge Visiting Professor Program at École Polytechnique (Palaiseau), and the National Science Foundation under grant DMS-1929284, while he was in residence at the Institute for Computational and Experimental Research in Mathematics at Providence, RI, during the Discrete Optimization program, 2023.

Disclosure of Interests. The authors have no competing interests to declare that are relevant to the content of this article.

References

[AHS07] Audet, C., Haddad, J., Savard, G.: Disjunctive cuts for continuous linear bilevel programming. Optim. Lett. **1**(3), 259–267 (2007). https://doi.org/10.1007/s11590-006-0024-3

[Bal79] Balas, E.: Disjunctive programming. In: Hammer, P., Johnson, E., Korte, B. (eds.) Annals of Discrete Mathematics 5: Discrete Optimization, pp. 3–51. North Holland (1979). https://doi.org/10.1016/S0167-5060(08)70342-X

[Bal88] Balas, E.: On the convex hull of the union of certain polyhedra. Oper. Res. Lett. **7**(6), 279–283 (1988). https://doi.org/10.1016/0167-6377(88)90058-2

[Bal18] Balas, E.: Disjunctive Programming. Springer, Heidelberg (2018). https://doi.org/10.1007/978-3-030-00148-3

[Bla90] Blair, C.: Representation for multiple right-hand sides. Math. Program. **49**, 1–5 (1990). https://doi.org/10.1007/BF01588775

[BLL+11] Belotti, P., Liberti, L., Lodi, A., Nannicini, G., Tramontani, A.: Disjunctive inequalities: applications and extensions. Wiley Encycl. Oper. Res. Manag. Sci. **2**, 1441–1450 (2011). https://doi.org/10.1002/9780470400531.eorms0537

[Dan63] Dantzig, G.B.: Linear Programming and Extensions. RAND Corporation, Santa Monica, CA (1963). https://www.rand.org/pubs/reports/R366.html

[FRT85] Freund, R.M., Roundy, R., Todd, M.J.: Identifying the set of always-active constraints in a system of linear inequalities by a single linear program. Sloan W.P. No. 1674–85 (Rev) (1985). https://dspace.mit.edu/handle/1721.1/2111

[Fuk22] Fukuda, K.: CDD (2022). https://github.com/cddlib/cddlib

[Grü03] Grünbaum, B.: Convex Polytopes (2nd edition prepared by Volker Kaibel, Victor Klee, and Günter M. Ziegler). Springer, New York (2003). https://doi.org/10.1007/978-1-4613-0019-9

[Gur22] Gurobi Optimization. Dealing with big-M constraints (2022). https://www.gurobi.com/documentation/9.5/refman/dealing_with_big_m_constra.html. Accessed 12 Sept 2022

[Jer88] Jeroslow, R.G.: A simplification for some disjunctive formulations. Eur. J. Oper. Res. **36**(1), 116–121 (1988). https://doi.org/10.1016/0377-2217(88)90013-6

[KKY15] Kılınç-Karzan, F., Yıldız, S.: Two-term disjunctions on the second-order cone. Math. Program. **154**(1), 463–491 (2015). https://doi.org/10.1007/s10107-015-0903-4

[KM21] Kronqvist, J., Misener, R.: A disjunctive cut strengthening technique for convex MINLP. Optim. Eng. **22**, 1315–1345 (2021). https://doi.org/10.1007/s11081-020-09551-6

[Lee22] Lee, J.: A First Course in Linear Optimization (Fourth Edition, version 4.07). Reex Press, (2013–22). https://github.com/jon77lee/JLee_LinearOptimizationBook

[Lee04] Lee, J.: A First Course in Combinatorial Optimization. Cambridge Texts in Applied Mathematics. Cambridge University Press, Cambridge (2004). https://doi.org/10.1017/CBO9780511616655

[LSS22] Lee, J., Skipper, D., Speakman, E.: Gaining or losing perspective. J. Global Optim. **82**, 835–862 (2022). https://doi.org/10.1007/s10898-021-01055-6

[LSSX21] Lee, J., Skipper, D., Speakman, E., Xu, L.: Gaining or losing perspective for piecewise-linear under-estimators of convex univariate functions. In: Gentile, C., Stecca, G., Ventura, P. (eds.) Graphs and Combinatorial Optimization: from Theory to Applications, CTW 2020. AIRO Springer Series, vol. 5, pp. 349–360 (2021). https://doi.org/10.1007/978-3-030-63072-0_27

[LSSX23] Lee, J., Skipper, D., Speakman, E., Xu, L.: Gaining or losing perspective for piecewise-linear under-estimators of convex univariate functions. J. Optim. Theory Appl. **196**, 1–35 (2023). https://doi.org/10.1007/s10957-022-02144-6

[LTV23] Lodi, A., Tanneau, M., Vielma, J.-P.: Disjunctive cuts in mixed-integer conic optimization. Math. Program. **199**, 671–719 (2023). https://doi.org/10.1007/s10107-022-01844-1

[NW88] Nemhauser, G.L., Wolsey, L.A.: Integer and combinatorial optimization. In: Wiley Interscience Series in Discrete Mathematics and Optimization. Wiley (1988). ISBN: 978-0-471-35943-2, https://doi.org/10.1002/9781118627372

[TG15] Trespalacios, F., Grossmann, I.E.: Improved big-m reformulation for generalized disjunctive programs. Comput. Chem. Eng. **76**, 98–103 (2015). https://doi.org/10.1016/j.compchemeng.2015.02.013

[Vie15] Vielma, J.P.: Mixed integer linear programming formulation techniques. SIAM Rev. **57**, 3–57 (2015). https://doi.org/10.1137/130915303

[Zie95] Ziegler, G.M.: Lectures on Polytopes: Graduate Texts in Mathematics, vol. 152. Springer, Heidelberg (1995). https://doi.org/10.1007/978-1-4613-8431-1

On a Geometric Graph-Covering Problem Related to Optimal Safety-Landing-Site Location

Claudia D'Ambrosio[1], Marcia Fampa[2], Jon Lee[3($\boxtimes$)], and Felipe Sinnecker[2]

[1] LIX CNRS, École Polytechnique, Institut Polytechnique de Paris, Palaiseau, France
[2] Federal University of Rio de Janeiro, Rio de Janeiro, Brazil
[3] University of Michigan, Ann Arbor, USA
`jonxlee@umich.edu`

Abstract. We develop a set-cover based integer-programming approach to an optimal safety-landing-site location arising in the design of urban air-transportation networks. We link our minimum-weight set-cover problems to efficiently-solvable cases of minimum-weight set covering that have been studied. We were able to solve large random instances to optimality using our modeling approach. We carried out *strong fixing*, a technique that generalizes reduced-cost fixing, and which we found to be very effective in reducing the size of our integer-programming instances.

Keywords: set covering · 0/1 linear programming · reduced-cost fixing · strong fixing · safety landing site · urban air mobility

1 Introduction

In the last few years, different actors all around the world have been pushing for the development of Urban Air Mobility (UAM). The idea is to integrate into the current transportation system, new ways to move people and merchandise. In particular, drones are already a reality, and they have the potential to be highly exploited for last-mile deliveries (for example, Amazon, UPS, DHL, and FedEx, just to mention a few involved operators). Concerning passenger transportation, several companies are competing to produce the first commercial electric Vertical Take-Off and Landing (eVTOLs) vehicles, which will be used to move passengers between vertiports of sprawling cities. Several aspects have to be taken into account for this kind of service, the most important one being safety. Air-traffic management (ATM) provides and adapts flight planning to guarantee a proper separation of the trajectories of the flights; see e.g., [4,13,14]. In the case of UAM, some infrastructure has to be built to provide safe landing locations in case of failure or damage of drones/eVTOLs. These locations are called "Safety Landing Sites" (SLSs) and should be organized to cover the trajectory of eVTOLs for emergency landings at any position along flight paths.

In what follows, we study the optimal placement of SLSs in the air-transportation network. We aim at installing the minimum-cost set of SLSs, such that

A. Basu et al. (Eds.): ISCO 2024, LNCS 14594, pp. 16–29, 2024.
https://doi.org/10.1007/978-3-031-60924-4_2

all the drones/eVTOLs trajectories are covered. We show an example in Fig. 1. It represents an aerial 2D view of a part of a city, where the rectangles are the roofs of existing buildings. The black crosses represent potential sites for SLSs. The two red segments are the trajectory of the flights in this portion of the space. The trajectory is fully covered thanks to the installation of 3 SLS over 5 potential sites, namely the ones corresponding to the center of the green circles. The latter represents the points in space that are at a distance that is smaller than the safety distance for an emergency landing. Note that every point along the trajectory is inside at least one circle, thus the trajectory is fully covered.

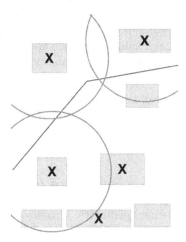

Fig. 1. Example of full covering provided by installing 3 SLSs

The problem of finding an optimal placement of SLSs has not received a lot of attention. In fact, there appears to be no published work on the variant that we are considering. In the literature, we can find the masters thesis of Xu [19], where he studied the problem of SLS placement coupled with the routing problem of drones/eVTOLs for each origin-destination pair. The potential SLSs are assumed to fully cover a subset of the arcs of the considered network, i.e., a partial covering is not allowed. In [15], Pelegín and Xu consider a variant of the problem, which can be formulated as a continuous covering problem. In particular, they interpret the problem as a set-covering/location problem where both candidate locations and demand points are continuous on a network. In this work, we consider the demand points continuous on the network, but the candidate-location set is finite and composed of points not restricted to be in the network; see e.g., Fig. 1.

More attention has been accorded to the optimal placement of vertiports; see e.g., [17,18]. However, their location depends on the estimated service demand and, based on the decisions made on the vertiport location, the UAM network is identified. In contrast, we suppose that these decisions were already made.

In fact, despite the scarcity of literature on the topic, the main actors in the UAM field assert that pre-identified emergency landing sites are necessary to guarantee safety in UAM; see e.g., [8,12].

Organization and Contributions. In Sect. 2, we describe our new mathematical model for the optimal safety-landing-site location problem, a generally NP-hard minimum-weight set covering problem. In Sect. 3, we see what kinds of set-covering matrices can arise from our setting, linking to the literature on ideal matrices. In Sect. 4, we identify three classes of efficiently-solvable cases for our setting. In Sect. 5, toward practical optimal solution of instances, we carry out "strong fixing", which enhances the classical technique of reduced-cost fixing. In Sect. 6, we present results of computational experiments, demonstrating the value of strong fixing for reducing model size and as a useful tool for solving difficult instances to optimality. In Sect. 7, we identify some potential next steps.

Notation. We denote the set of real numbers by $\mathbb{R}$ and the set of positive real numbers by $\mathbb{R}_{++}$. We denote an all-ones vector by $\mathbf{e}$. For a vector $x \in \mathbb{R}^n$, we denote its 2-norm by $\|x\| := \sqrt{\sum_{i=1}^{n} x_i^2}$. For a matrix A, we denote the transpose of A by A^{T} and column j of A by $A_{\cdot j}$.

2   Covering Edges with a Subset of a Finite Set of Balls

We begin with a formally defined geometric optimization problem. Let G be a straight-line embedding of a graph in $\mathbb{R}^d$, $d \geq 1$ (although our main interest is $d = 2$, with $d = 3$ possibly also having some applied interest), where we denote the vertex set of G by $\mathcal{V}(G)$, and the edge set of G by $\mathcal{I}(G)$, which is a finite set of intervals, which we regard as *closed*, thus containing its end vertices. Note that an interval can be a single point (even though this might not be useful for our motivating application). We are further given a finite set N of n points in $\mathbb{R}^d$, a weight function $w : N \to \mathbb{R}_{++}$, and covering radii $r : N \to \mathbb{R}_{++}$ (we emphasize that points in N may have differing covering radii). A point $x \in N$ $r(x)$-covers all points in the $r(x)$-ball $B(x, r(x)) := \{y \in \mathbb{R}^d : \|x - y\|_2 \leq r(x)\}$. A subset $S \subset N$ r-covers G if every point y in every edge $I \in \mathcal{I}(G)$ is $r(x)$-covered by some point $x \in S$. We may as well assume, for feasibility, that N r-covers G. Our goal is to find a minimum w-weight r-covering of G.

Connecting this geometric problem with our motivating application, we observe that any realistic road network can be approximated to arbitrary precision by a straight-line embedded graph, using extra vertices, in addition to road junctions; this is just the standard technique of piecewise-linear approximation of curves. The point set N corresponds to the set of potential SLSs. In our application, a constant radius for each SLS is rather natural, but our methodology does not require this. We also allow for cost to depend on SLSs, which can be natural if sites are rented, for example.

Associated with each ball $B(x, r(x))$ is its boundary, the sphere $\bar{B}(x, r(x))$. Each such sphere $\bar{B}(x, r(x))$ intersects each interval $I \in \mathcal{I}(G)$ at most twice.

Collecting all of these at most $2n$ subdivision points, as we let x vary over N, the interval I is subdivided into at most $1+2n$ closed subintervals, the collection of which we denote as $\mathcal{C}(I)$. It is clear that all points in a given one of these subintervals are covered by the same set of balls.

With all of this notation, we can re-cast the problem of finding a minimum w-weight r-covering of G as the 0/1-linear optimization problem

$$\min \ \sum_{x \in N} w(x)z(x)$$

$$\sum_{\substack{x \in N : \\ J \subset B(x,r(x))}} z(x) \geq 1, \ \forall \ J \in \mathcal{C}(I), \ I \in \mathcal{I}(G); \qquad \text{(CP)}$$

$$z(x) \in \{0,1\}, \ \forall \ x \in N,$$

where each $z(x)$ is a binary indicator variable associated with selecting the point $x \in N$ (equivalently, the ball $B(x,r(x))$). Because $|\mathcal{C}(I)| \leq 1+2n$, for each edge $I \in \mathcal{I}(G)$, the number of covering constraints, which we will denote by m, is at most $(1+2n)|\mathcal{I}(G)|$. Of course we can view this formulation in matrix format as

$$\min\{w^\top z \ : \ Az \geq \mathbf{e}, \ z \in \{0,1\}^n\}, \qquad \text{(SCP)}$$

for an appropriate 0/1-valued $m \times n$ matrix A, and it is this view that we mainly work with in what follows.

The problem is already NP-Hard for $d > 1$, when all balls have identical radius, the weights are all unity, $N := \mathcal{V}(G)$, and the edges of G are simply the points N; see [9, Theorem 4]. Of course, this type of graph (with only degenerate edges) is not directly relevant to our motivating application, and anyway we are aiming at exact algorithms for practical instances of moderate size.

3   What Kind of Constraint Matrices Can We Get?

There is a big theory on when set covering LPs have integer optima (for all objectives). It is the theory of *ideal* matrices; see [5]. A 0/1 matrix is *balanced* if it does not contain a square submatrix of odd order with two ones per row and per column. In fact, the 0/1 TU (totally unimodular; see [16], for example) matrices are a proper subclass of the balanced matrices.

Berge [3] showed that, if A is balanced, then both the packing and covering systems associated with A have integer vertices. Fulkerson, Hoffman, and Oppenheim [10] showed that, if A is balanced, then the covering system is TDI (totally dual integral; see [16], for example). So balanced 0/1 matrices are a subclass of ideal 0/1 matrices.

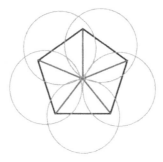

Fig. 2. Does not have an ideal covering matrix

We can observe that the matrices that can arise for us are not generally balanced, already for a simple example, depicted in Fig. 2; the drawing is for $n = 5$, but it could have been for any odd $n \geq 3$. The edges of the graph are indicated with red. The points of N are at the midpoints of the black lines (which are not themselves edges). The (green) circle for each point of N goes through the center. It is easy to see that edges are not subdivided by circles, and each circle covers a pair of edges in a cyclic fashion. The constraint matrix of the covering problem is an odd-order (5 in this case) $0/1$ matrix violating the definition of balanced. In fact, the matrix is not even ideal as the covering LP has an all-$\frac{1}{2}$ extreme point.

We can also get a counterexample to idealness for the covering matrix with respect to unit-grid graphs. In particular, it is well known that the "circulant matrix" of Fig. 3 is non-ideal, see [6]. Now, we can realize this matrix from an 8-edge "unit-grid graph" (or the reader may prefer to see it as a 4-edge "unit-grid graph"), see Fig. 4, and eight well-designed covering disks, each covering an "L", namely $\{a, b, c\}$, $\{b, c, d\}$, $\{c, d, e\}$, $\{d, e, f\}$, $\{e, f, g\}$, $\{f, g, h\}$, $\{g, h, a\}$, $\{h, a, b\}$. As a sanity check, referring to Fig. 5, we see that $\mathcal{C}_8^3$ is not balanced.

$$\mathcal{C}_8^3 := \begin{pmatrix} 1\,1\,1\,0\,0\,0\,0\,0 \\ 0\,1\,1\,1\,0\,0\,0\,0 \\ 0\,0\,1\,1\,1\,0\,0\,0 \\ 0\,0\,0\,1\,1\,1\,0\,0 \\ 0\,0\,0\,0\,1\,1\,1\,0 \\ 0\,0\,0\,0\,0\,1\,1\,1 \\ 1\,0\,0\,0\,0\,0\,1\,1 \\ 1\,1\,0\,0\,0\,0\,0\,1 \end{pmatrix}$$

Fig. 3. Non-ideal circulant matrix

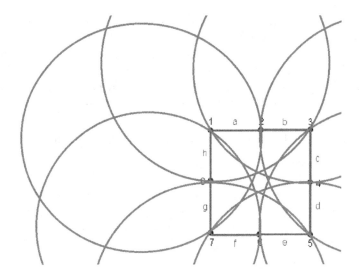

Fig. 4. Yields the non-ideal covering matrix $\mathcal{C}_8^3$.

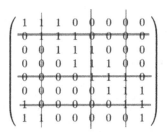

Fig. 5. $\mathcal{C}_8^3$ is not balanced.

4 Efficiently Solvable Cases

In this section, we present three situations for which CP/SCP is efficiently solvable.

4.1 When G Is a Unit-Grid Graph in $\mathbb{R}^d$, $d \geq 2$, $N \subset \mathcal{V}(G)$, and $1 \leq r(x) < \sqrt[d]{5/4}$, for All $x \in N$

Briefly, a *unit-grid graph* is a finite subgraph of the standard integer lattice graph. For this case, we will observe that our minimum-weight r-covering problem on G is equivalently an ordinary minimum-weight vertex covering problem on G, with vertices in $\mathcal{V}(G) \setminus N$ disallowed. Choosing a vertex x as a covering vertex fully r-covers all of its incident edges (because $r(x) \geq 1$), but it will not r-cover the midpoint of any other edge (because $r(x) < \sqrt[d]{5/4}$, which is the minimum distance between x and the midpoint of an edge that is not incident to x). The only way to r-cover a midpoint of an edge is to choose one of its endpoints, in which case the entire edge is r-covered (because $r(x) \geq 1$).

The efficient solvability easily follows, because unit-grid graphs are bipartite, and the ordinary formulation of minimum-weight vertex covering (with variables corresponding to vertices in $\mathcal{V}(G) \setminus N$ set to 0) has a totally-unimodular constraint matrix; so the SCP is efficiently solved by linear optimization.

4.2 When G Is a Path Intersecting Each Ball on a Subpath

When G is a *straight* path (in any dimension $d \geq 1$), ordering the subintervals naturally, as we traverse the path, we can see that in this case the constraint matrix for SCP is a (column-wise) consecutive-ones matrix, and hence is totally unimodular (and so SCP is polynomially solvable in such cases). In fact, this is true as long as the path (not necessarily straight) intersects $B(x, r(x))$ on a subpath, for each $x \in N$. For example, if G is monotone in each coordinate, then G enters and leaves each ball at most once each.

4.3 A Fork-Free Set of Subtrees

Given a tree T, a pair of subtrees T_1 and T_2 has a *fork* if there is a path P_1 with end-vertices in T_1 but not T_2, and a path P_2 with end-vertices in T_2 but not T_1, such that P_1 and P_2 have a vertex in common. We consider the problem of finding a minimum-weight covering of a tree by a given set of subtrees; see [2, called problem "C_0"]. This problem admits a polynomial-time algorithm when the family of subtrees is fork free, using some problem transformations and then a recursive algorithm. This problem and algorithm is relevant to our situation when: G is a tree, each ball intersects G on a subtree (easily checked), and the set of these subtrees is fork free (easily checked). A simple special case is the situation considered in Sect. 4.1. Much more broadly, if the degrees of a tree are ≤ 3, then any family of subtrees is fork free.

Now, it is easy to make a simple example arising from our situation (which is even a unit-grid graph), where a fork arises; see Fig. 6 (The tree is the red graph, and the pair of green covering disks define the two subtrees). So the algorithm from [2] does not apply to this example. On the other hand, the constraint matrix of this instance of SCP is balanced, so this is an easy instance for linear programming.

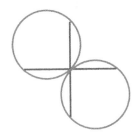

Fig. 6. A forking configuration

Referring back to Fig. 2, we also find forks. Also see [2, Sec. 5] which raises the interesting question on the relationship between *totally*-balanced matrices[1] and covering matrices of fork-free families. These notions are incomparable; [2] has a very simple example that is fork free but not *totally* balanced. But we can even get fork free coming from our context and not balanced, returning again to our example of Fig. 4 (it is not a tree, but we can break an edge).

5   Strong Fixing

We consider the linear relaxation of SCP, that is $\min\{w^\mathsf{T}z \;:\; Az \geq \mathbf{e},\ z \geq 0\}$, and the associated dual problem

$$\max\{u^\mathsf{T}\mathbf{e} \;:\; u^\mathsf{T}A \leq w^\mathsf{T},\ u \geq 0\}. \tag{D}$$

An optimal solution of D is commonly used in the application of *reduced-cost fixing*, see, e.g. [7], a classical technique in integer programming that uses upper bounds on the optimal solution values of minimization problems for inferring variables whose values can be fixed while preserving the optimal solutions. The well-known technique is based on Theorem 1 (a general theorem, which we state specifically for our situation).

Theorem 1. *Let* UB *be the objective-function value of a feasible solution for SCP, and let* $\hat{u}$ *be a feasible solution for D. Then, for every optimal solution* $z^\star$ *for SCP, we have:*

$$z_j^\star \leq \left\lfloor \frac{\mathrm{UB} - \hat{u}^\mathsf{T}\mathbf{e}}{w_j - \hat{u}^\mathsf{T}A_{.j}} \right\rfloor, \quad \forall\, j \in \{1, \ldots, n\} \text{ such that } w_j - \hat{u}^\mathsf{T}A_{.j} > 0. \tag{1}$$

For a given $j \in \{1, \ldots, n\}$, we should have the right-hand side in (1) equal to zero to be able to fix the variable z_j at zero in SCP. Equivalently, we should have $\hat{u}^\mathsf{T}(\mathbf{e} - A_{.j}) > \mathrm{UB} - w_j$.

We observe that any feasible solution $\hat{u}$ can be used in (1). Then, for all $j \in \{1, \ldots, n\}$, we propose the solution of

$$\mathfrak{z}_j := \max\{u^\mathsf{T}(\mathbf{e} - A_{.j}) \;:\; u^\mathsf{T}A \leq w^\mathsf{T},\ u \geq 0\}. \tag{F_j}$$

Note that, for each $j \in \{1, \ldots, n\}$, if there is a feasible solution $\hat{u}$ to D that can be used in (1) to fix z_j at zero, then the optimal solution of F_j has objective value greater than $\mathrm{UB} - w_j$ and can be used as well.

We call *strong fixing*, the procedure that, for a given upper bound UB on the optimal value of SCP, fixes the maximum number of variables in SCP at zero, by solving problems F_j and applying Theorem 1, for all $j \in \{1, \ldots, n\}$. In fact, [1] (and probably many others) considered this approach (they call a model "relaxed consistent" after no variable fixing of this type is possible) but discarded the idea as probably being computationally prohibitive.

---

[1] a 0/1 matrix is *totally balanced* if it has no square submatrix *(of any order)* with two ones per row and per column, thus a subclass of balanced 0/1 matrices; see [11].

6   Computational Experiments

We have implemented a framework to generate 25 random instances of CP and formulate them as the set covering problem SCP. We solve the instances applying the following procedures in the given order.

(a) Reduce the number of constraints in SCP by eliminating dominated rows of the associated constraint matrix A.
(b) Fix variables in the reduced SCP by applying *reduced-cost fixing*, i.e., using Theorem 1, taking $\hat{u}$ as an optimal solution of D. If it was possible to fix variables, reapply (a) to reduce the number of constraints further.
(c) Apply *strong fixing* (see Sect. 5). We note that the only difference between problems F_j, for $j = 1 \ldots, n$, is the objective function, so we warm start the solution of each problem (except the first one solved) with the optimal solution of the previous one solved. After solving F_j for a given j we fix all possible variables using its optimal solution. Then, we select the next problem to solve, $F_{\hat{j}}$, among all the possible options associated to the unfixed variables. $F_{\hat{j}}$ is such that $A._{\hat{j}}$ is the closest column to $A._j$, according to the 'Jaccard similarity' (i.e. for a pair of columns, the cardinality of the intersection of the supports divided by the cardinality of the union of the supports). In case it was possible to fix variables, reapply procedure (a) to reduce the number of constraints in the remaining problem.
(d) Solve the last problem obtained with `Gurobi`.

Our implementation is in `Python`, using `Gurobi` v. 10.0.2. We ran `Gurobi` with one thread per core, default parameter settings (with the presolve option on). We ran on an 8-core machine (under Ubuntu): Intel i9-9900K CPU running at 3.60GHz, with 32 GB of memory.

In Algorithm 1, we show how we construct random instances for CP, for a given number of points n, and a given interval $[R_{\min}, R_{\max}]$ in which the covering radii for the points must be. We construct a graph $G = (V, E)$ with node-set V given by $\nu := 0.03n$ points randomly generated in the unit square in the plane, and the edges in E initially given by the edges of the minimum spanning

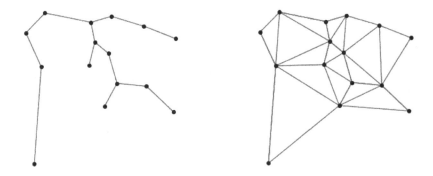

Fig. 7. Constructing an instance of CP - MST of V (left) and graph G (right)

tree (MST) of V, which guarantees connectivity of G. Finally, we compute the Delaunay triangulation of the ν points in V, and add to E the edges from the triangulation that do not belong to its convex hull. All other details of the instance generation can be seen in Algorithm 1. In Fig. 7, we show an example of the MST of V and of the graph G.

Algorithm 1: Instance generator

Input: n, $R_{\min}$, $R_{\max}$ ($0.1 < R_{\min} < R_{\max} < 0.2$).
Output: instance of CP/SCP.

1 randomly generate a set V of $\nu := 0.03n$ points in $Q := \{[0,1] \times [0,1]\}$;
2 use the Python package `networkx` to compute the edge set E_{ST} of an MST of V, letting the weight for edge (i,j) as the Euclidean distance between i and j;
3 compute the Delaunay triangulation of V, and denote the subset of its edges that are not in the convex hull by E_{DT} ;
4 let $G = (V, E)$, where $E := E_{\mathrm{ST}} \cup E_{\mathrm{DT}}$;
5 randomly generate a set N of n points x^k in Q, $k = 1, \ldots, n$;
6 randomly generate r_k in $[R_{\min}, R_{\max}]$, $k = 1, \ldots, n$;
7 randomly generate w_k in $[0.5r_k^2, 1.5r_k^2]$, $k = 1, \ldots, n$;
8 let c_k be the circle centered at x^k with radius r_k, for $k = 1, \ldots, n$;
9 for each $e \in E$, compute the intersections (0, 1 or 2) of c_k and e, for $k = 1, \ldots, n$;
10 compute all the subintervals defined on each edge by the intersections, and let m be the total number of subintervals for all edges;

We note that the instance constructed by Algorithm 1 may be infeasible, if any part of an edge of G is not covered by any point. In this case, we iteratively increase all radii r_k, for $k = 1, \ldots, n$, by 10%, until the instance is feasible. For feasible instances, we check if there are circles that do not intercept any edges of the graph. If so, we iteratively increase the radius of each of those circles by 10%, until they intercept an edge. By this last procedure, we avoid zero columns in the constraint matrices A associated to our instances of SCP. In Fig. 8, we represent the data for an instance of CP and its optimal solution. In the optimal solution we see nine points/circles selected.

In Table 1 we present numerical results aiming at observing the impact of strong fixing in reducing the size of SCP, after having already applied standard reduced-cost fixing and eliminating redundant constraints. Because one of our main goals is to see the full power of strong fixing, in applying Theorem 1, we always set UB to the optimal objective value of SCP. We display the number of rows (m) and columns (n) of the constraint matrix A after applying each procedure described in (a–d) and their elapsed time (seconds). We have under 'Gurobi' the size of the original A and the time to solve with Gurobi, and under 'Gurobi presolve', the size of the matrix A after Gurobi's presolve is applied and the time to apply it. Finally, we have in the next three columns the size of the matrix A after applying the procedures described in (a–c) and the times to apply them. Finally, in the last column, we have the time to solve

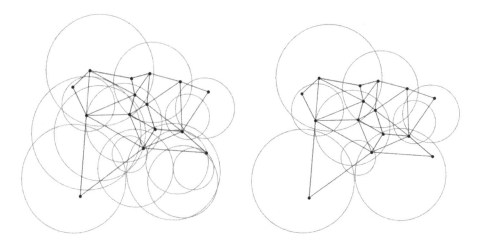

Fig. 8. Instance data (left) and its optimal solution (right)

the last problem obtained with Gurobi, as described in (d). We see that our own presolve, corresponding to procedures (a–c), is effective in reducing the size of the problem, and does not lead in general to problems bigger than the ones obtained with Gurobi's presolve (even though, from the increase in the number of variables when compared to the original problem, we see that Gurobi's presolve implements different procedures, such as sparsification on the equation system after adding slack variables). We also see that *strong fixing* is very effective in reducing the size of the problem. Compared to the problems to which it is applied, we have an average decrease of 49% in m and 44% in n. Of course, our presolve is very time consuming compared to Gurobi's. Nevertheless, for 22 out of the 25 instances tested, there is an improvement in the final Gurobi time to solve the problem, and we note that all the steps of reduction and fixing can still be further improved. This shows that *strong fixing* is a promising tool to be adopted in the solution of difficult problems, as is the case of the well-known *strong-branching* procedure.

We can also see that with our set-covering modeling methodology, even without strong fixing, a modern integer-programming solver is capable of solving to optimality large instances of CP in the plane. Specifically, our largest five instances all have 75 vertices and ≈ 273 edges, and we choose from a set of $n = 75/.03 = 2500$ points/circles.

7   Outlook

Extensions. Our integer-programming approach can be extended to other geometric settings. We can use other metrics, or replace balls $B(x, r(x))$ with arbitrary convex sets $B(x)$ for which we can compute the intersection of with each edge $I \in \mathcal{I}(G)$. In this way, N is just an index set, and we only need a pair of

Table 1. Numerical results

	Gurobi			Gurobi presolve			matrix reduction (a)			reduced-cost fixing (b)			strong fixing (c)			Gurobi reduced
I	m	n	time	m	n	time	m	n	time	m	n	time	m	n	time	time (d)
1	7164	500	0.28	372	257	0.17	531	500	24.49	214	167	0.50	12	19	5.39	0.003
2	4078	500	0.37	427	327	0.10	574	500	12.08	344	267	0.94	49	63	16.20	0.01
3	3865	500	0.34	399	331	0.10	537	500	10.22	342	291	0.83	46	71	17.90	0.007
4	4399	500	0.27	412	300	0.10	559	500	13.15	310	236	0.79	6	14	11.35	0.007
5	4098	500	0.22	446	359	0.13	546	500	11.92	146	132	0.35	16	16	2.33	0.008
6	13254	1000	2.90	1279	1119	0.73	1728	1000	136.00	1331	705	15.92	244	202	96.34	0.035
7	9709	1000	3.91	1263	1125	0.77	1581	1000	81.86	1443	837	13.33	909	562	140.72	2.29
8	10847	1000	2.53	1307	1140	0.80	1672	1000	94.13	1351	712	15.22	263	196	94.20	0.06
9	9720	1000	2.81	1335	1149	0.69	1700	1000	84.27	1247	667	12.14	209	193	89.13	0.03
10	9377	1000	3.20	1300	1193	0.67	1621	1000	75.76	1329	764	12.17	322	246	103.94	0.08
11	17970	1500	29.46	2720	2222	2.75	3175	1500	278.33	2862	1304	62.73	1061	609	486.60	2.31
12	22022	1500	65.65	2838	2208	3.01	3316	1500	425.03	3115	1336	68.69	1725	844	595.06	9.24
13	18716	1500	42.05	2944	2219	2.86	3466	1500	315.96	3311	1379	82.08	2184	983	619.66	24.44
14	18464	1500	14.30	2545	1988	2.44	3114	1500	306.49	2856	1270	59.32	1548	762	441.53	6.17
15	19053	1500	33.95	2908	2207	2.82	3376	1500	327.13	3162	1361	72.56	1745	851	581.55	7.57
16	29227	2000	170.78	4194	3037	6.59	4988	2000	832.81	4691	1793	177.51	2959	1221	1493.63	83.46
17	29206	2000	212.07	4365	3186	7.04	4993	2000	823.83	4793	1828	188.71	3182	1318	1547.75	102.57
18	28927	2000	762.26	4120	2971	7.22	5016	2000	799.22	4889	1862	192.64	4060	1557	1717.11	497.09
19	30746	2000	6970.90	4943	3645	7.22	5243	2000	842.61	5243	1998	229.23	5109	1944	2879.75	7421.27
20	31606	2000	1914.49	5158	3711	8.20	5378	2000	885.00	5374	1997	241.15	5088	1897	2924.22	1034.66
21	42137	2500	30111.51	6928	4656	14.99	7606	2500	1785.99	7580	2470	517.05	7308	2356	7588.82	28318.53
22	41229	2500	594.63	6511	4454	15.07	7454	2500	1689.45	7127	2344	432.95	3979	1450	4391.32	142.18
23	43100	2500	614.62	6236	4209	15.16	7381	2500	1856.51	7139	2313	462.52	5477	1828	4212.32	494.37
24	42053	2500	51978.94	6654	4776	16.84	7031	2500	1685.95	7031	2500	433.37	6958	2467	7493.67	56012.21
25	42641	2500	3594.00	6639	4723	16.43	7038	2500	1615.20	7032	2497	412.83	6604	2360	6135.45	6540.81

line searches, to determine the endpoints of the intersection of $B(x)$ with each $I \in \mathcal{I}(G)$. We still have $|\mathcal{C}(I)| \leq 1+2n$, for each $I \in \mathcal{I}(G)$, and so the number of covering constraints in CP is at most $(1+2n)|\mathcal{I}(G)|$. Finally, we could take G to be geodesically embedded on a sphere and the balls replaced by geodesic balls.

Solvable Cases. We may seek to generalize the algorithm mentioned in Sect. 4.3 to arbitrary families of subtrees of unit-grid-graph trees (i.e., get rid of the fork-free condition but restrict to unit-grid graphs — which can have degree-4 vertices).

Extending Our Computational Work. An important direction is to reduce the time for strong fixing, so as to get a large number of variables fixed without solving all of the F_j. Additionally, the strong-fixing methodology is very general and could work well for other specific classes of mixed-integer optimization problems.

Acknowledgments. This work was supported by NSF grant DMS-1929284 while C. D'Ambrosio, M. Fampa and J. Lee were in residence at ICERM (the Institute for Computational and Experimental Research in Mathematics; Providence, RI) during the Discrete Optimization program, 2023. C. D'Ambrosio was supported by the Chair "Integrated Urban Mobility", backed by L'X - École Polytechnique and La Fondation de lâĂŹÉcole Polytechnique (The Partners of the Chair accept no liability related to this publication, for which the chair holder is solely liable). M. Fampa was supported by CNPq grant 307167/2022-4. J. Lee was supported by the Gaspard Monge Visiting Professor Program, École Polytechnique, and from ONR grant N00014-21-1-2135. F. Sinnecker was supported on a masters scholarship from CNPq. J. Lee and M. Fampa acknowledge (i) interesting conversations at ICERM with Zhongzhu Chen on variable fixing, and (ii) some helpful information from Tobias Achterberg on `Gurobi` presolve.

Disclosure of Interests. The authors have no competing interests to declare.

References

1. Bajgiran, O.S., Cire, A.A., Rousseau, L.M.: A first look at picking dual variables for maximizing reduced cost fixing. In: Salvagnin, D., Lombardi, M. (eds.) Integration of AI and OR Techniques in Constraint Programming, pp. 221–228. Springer, Heidelberg (2017). https://doi.org/10.1007/978-3-319-59776-8_18
2. Bárány, I., Edmonds, J., Wolsey, L.A.: Packing and covering a tree by subtrees. Combinatorica **6**(3), 221–233 (1986). https://doi.org/10.1007/BF02579383
3. Berge, C.: Balanced matrices. Math. Program. **2**(1), 19–31 (1972). https://doi.org/10.1007/BF01584535
4. Cerulli, M., D'Ambrosio, C., Liberti, L., Pelegrín, M.: Detecting and solving aircraft conflicts using bilevel programming. J. Glob. Optim. **81**(2), 529–557 (2021). https://doi.org/10.1007/s10898-021-00997-1
5. Cornuéjols, G.: Combinatorial Optimization: Packing and Covering. SIAM (2001). https://doi.org/10.1137/1.9780898717105
6. Cornuéjols, G., Novick, B.: Ideal 0,1 matrices. J. Comb. Theory Series B **60**(1), 145–157 (1994). https://doi.org/10.1006/jctb.1994.1009
7. Crowder, H., Johnson, E.L., Padberg, M.: Solving large-scale zero-one linear programming problems. Oper. Res. **31**(5), 803–834 (1983). https://doi.org/10.1287/opre.31.5.803
8. FAA: Urban Air Mobility (UAM) Concept of Operations, version 2.0 (2023). https://www.faa.gov/air-taxis/uam_blueprint
9. Fowler, R.J., Paterson, M.S., Tanimoto, S.L.: Optimal packing and covering in the plane are NP-complete. Inf. Process. Lett. **12**(3), 133–137 (1981). https://doi.org/10.1016/0020-0190(81)90111-3
10. Fulkerson, D.R., Hoffman, A.J., Oppenheim, R.: On balanced matrices. In: Balinski, M.L. (ed.) Pivoting and Extensions. Mathematical Programming Studies, vol. 1, pp. 120–132 (1974). https://doi.org/10.1007/BFb0121244
11. Hoffman, A.J., Kolen, A.W., Sakarovitch, M.: Totally-balanced and greedy matrices. SIAM J. Algebr. Disc. Methods **6**(4), 721–730 (1985). https://doi.org/10.1137/0606070
12. NASA: UAM Vision Concept of Operations (ConOps) UAM Maturity Level (UML) 4 (2021). https://ntrs.nasa.gov/citations/20205011091
13. Pelegrín, M., D'Ambrosio, C.: Aircraft deconfliction via mathematical programming: review and insights. Transport. Sci. **56**(1), 118–140 (2022). https://doi.org/10.1287/trsc.2021.1056
14. Pelegrín, M., D'Ambrosio, C., Delmas, R., Hamadi, Y.: Urban air mobility: from complex tactical conflict resolution to network design and fairness insights. Optim. Methods Softw. **38**(6), 1311–1343 (2023). https://doi.org/10.1080/10556788.2023.2241148

15. Pelegrín, M., Xu, L.: Continuous covering on networks: improved mixed integer programming formulations. Omega **117**, 102835 (2023). https://doi.org/10.1016/j.omega.2023.102835
16. Schrijver, A.: Theory of linear and integer programming. In: Wiley-Interscience Series in Discrete Mathematics. John Wiley & Sons, Ltd., Chichester (1986). ISBN: 0-471-90854-1
17. Serhal, K., Macias, J.J.E., Angeloudis, P.: Demand modelling and optimal vertiport placement for airport-purposed eVTOL services (2021). https://optimization-online.org/?p=18217
18. Villa, A.: Hub Location Problem For Optimal Vertiport Selection In Urban Air Mobility — Chicago Case Study. Masters Thesis, Politecnico di Milano (2020). https://doi.org/10.25417/uic.22226494.v1
19. Xu, L.: Optimal location of safety landing sites. Research Report, LIX, École Polytechnique (2021). https://hal.science/hal-03286640

Quadratically Constrained Reformulation, Strong Semidefinite Programming Bounds, and Algorithms for the Chordless Cycle Problem

Dilson Lucas Pereira[1($\boxtimes$)], Dilson Almeida Guimarães[2], Alexandre Salles da Cunha[2], and Abílio Lucena[3]

[1] Universidade Federal de Lavras, Lavras, Brazil
dilson.pereira@ufla.br
[2] Universidade Federal de Minas Gerais, Belo Horizonte, Brazil
acunha@dcc.ufmg.br
[3] Universidade Federal do Rio de Janeiro, Rio de Janeiro, Brazil
abiliolucena@cos.ufrj.br

Abstract. Given a connected undirected graph $G = (V, E)$, let $G[S]$ be the subgraph of G induced by the set of vertices $S \subseteq V$. The Chordless Cycle Problem (CCP) consists in finding a subset $S \subseteq V$ of maximum cardinality such that $G[S]$ is a chordless cycle. We present a Quadratically Constrained reformulation for the CCP, derive a Semidefinite Programming (SDP) relaxation for it and solve that relaxation by Lagrangian Relaxation (LR). Compared to previously available dual bounds, our SDP bounds resulted to be quite strong. We then introduce a hybrid algorithm involving two combined phases: the LR scheme, which acts as a warm starter for a Branch-and-cut (BC) algorithm that follows it. In short, the LR algorithm allows us to formulate a finite set of SDP cuts that can be used to retrieve the SDP bounds in a Linear Programming relaxation for the CCP. Such cuts are not ready to be used by the BC as they are formulated in an extended variable space. Thus, the BC projects them back onto the original space of variables and separates them by solving a Linear Program. On dense input graphs, our proposed BC algorithm, in its current preliminary state of development, already outperforms its competitors in the literature.

Keywords: Semidefinite programming · Lagrangian relaxation · Induced cycle · Branch-and-cut

1 Introduction

Given a simple undirected graph $G = (V, E)$ with a set of vertices $V = \{1, \ldots, n\}$ and set of edges E ($m = |E|$), denote by $E(S) \subseteq E$ the edges of G with both

A. S. da Cunha—Supported by CNPq grant 305357/2021-2 and FAPEMIG grants CEX - PPM-00164/17 and RED-00119-21.
A. Lucena—Supported by CNPq grant 310185/2021-1.

A. Basu et al. (Eds.): ISCO 2024, LNCS 14594, pp. 30–42, 2024.
https://doi.org/10.1007/978-3-031-60924-4_3

endpoints in $S \subseteq V$. Additionally, denote by $G[S] = (S, E(S))$ the subgraph of G *induced* by S. An *induced subgraph* $G[S]$ is defined by all edges of G having both endpoints in S and not by a strict subset of $E(S)$. Accordingly, $S \subseteq V$ induces a chordless cycle of G if $G[S]$ is a cycle. Finally, the Chordless Cycle Problem (CCP) asks for a maximum cardinality chordless cycle of G.

Numerous induced structures have been investigated in the literature. Among others, trees, forests, cliques, independent sets, as well as cycles [1,3], the focus of our paper. Previous chordless cycle investigations concentrated mostly on computational complexity issues. However, CCP formulations and algorithms were also investigated. For a broad up to date review on chordless cycles, we refer the reader to [5].

An Integer Programming (IP) formulation, valid inequalities, heuristics, and Branch-and-cut (BC) algorithms for the CCP are investigated in [5]. Building upon the formulation proposed in that reference, we define a Quadratically Constrained (QC) reformulation for the CCP. As such, the CCP becomes amenable to be tackled by Semidefinite Programming (SDP) techniques [6]. SDP based dual bounds for the CCP are computed here by means of a Lagrangian Relaxation (LR) algorithm, pretty much in the spirit of a recent study for the Quadratic Minimum Spanning Tree Problem [2]. In short, starting with an SDP relaxation for the CCP, we relax and dualize the positive semidefiniteness condition it imposes, together with some additional constraints. In turn, the associated Lagrangian Dual is solved by a specialization of the Subgradient Method. The SDP bounds thus obtained resulted to be much stronger than Linear Programming Relaxation (LPR) bounds available in the literature [5].

Another contribution of this paper is the use of an SDP LR scheme as a warm starter for the best performing BC algorithm proposed in [5]. The key idea behind this proposal is to use SDP cuts to reinforce LPR bounds at the root node of the BC tree. Such an approach is not as straightforward as one might think since additional decision variables are required to attain the SDP relaxation. Therefore, our SDP cuts must be projected back onto the space of variables in which the BC algorithm operates on. In summary, the CCP formulation reinforced by our SDP cuts paves the way for a BC algorithm that, in its current preliminary stage of development, already outperforms its competitors in [5] on dense graphs. The ideas we explore here are quite straightforward to be extended to other induced subgraph problems.

This paper is organized as follows. In Sect. 2, we present the QC reformulation and an SDP relaxation for the CCP. The LR scheme and the hybrid algorithm are described in Sects. 3 and 4, respectively. Some preliminary numerical results are reported in Sect. 5. We conclude the paper in Sect. 6, where directions for future research are offered.

2   QC Reformulation and SDP Relaxation

Let $\mathbb{B} := \{0, 1\}$. The formulation in [5] uses two sets of decision variables: $\mathbf{y} = \{y_v \in \mathbb{B} : v \in V\}$, to identify the vertices of our chordless cycle, and $\mathbf{x} =$

$\{x_{uv} \in \mathbb{B} : \{u, v\} \in E\}$, to indicate the cycle edges. Prior to presenting the formulation, some additional notation is required. Accordingly, we denote by $\delta(S)$ the set of edges with one endpoint in S and the other in $V \setminus S$. If S contains a single vertex, say v, we denote $\delta(S)$ by $\delta(v)$. Finally, the shorthands $y(S)$, for $S \subseteq V$, and $x(M)$, for $M \subseteq E$, are used to represent the sums $\sum_{v \in S} y_v$ and $\sum_{\{u,v\} \in M} x_{uv}$, respectively. The CCP formulation in [5] is given by

$$\max \{y(V) : (\mathbf{x}, \mathbf{y}) \in F \cap \mathbb{B}^{m+n}\}, \tag{1}$$

where polytope $F \subset \mathbb{R}^{m+n}$ corresponds to the intersection of

$$x(E) = y(V), \tag{2}$$
$$x(\delta(v)) = 2y_v, \qquad v \in V, \tag{3}$$
$$x(\delta(S)) \geq 2(y_u + y_v - 1), \qquad S \subset V, u \in S, v \in V \setminus S, \tag{4}$$
$$x_{vu} \leq y_v, \qquad v \in V, \{v, u\} \in E, \tag{5}$$
$$x_{uv} \geq y_u + y_v - 1, \qquad e = \{u, v\} \in E, \tag{6}$$
$$\mathbf{x} \geq \mathbf{0}, \tag{7}$$
$$\mathbf{y} \geq \mathbf{0}, \tag{8}$$
$$\mathbf{y} \leq \mathbf{1}. \tag{9}$$

Degree constraints (3) and undirected cutset inequalities (4) impose that feasible solutions to (1) must correspond to elementary cycles. Inequalities (6) enforce that whenever u and v are cycle vertices, an eventual edge of G connecting them, $e = \{u, v\}$, must necessarily be part of the solution. Accordingly, only chordless cycles of G would satisfy the constraints imposed by the formulation.

Pereira et al. [5] also proposed two sets of valid inequalities that proved useful to reinforce the LPR of the CCP, namely clique-y (10) and clique-bqp (11) inequalities:

$$\sum_{v \in Q} y_v \leq 2, \tag{10}$$

$$\sum_{u,v \in Q, u < v} x_{uv} \geq \sum_{v \in Q} y_v - 1. \tag{11}$$

These are defined for cliques of G, i.e., subsets Q of V that contain at least 3 vertices and induce complete subgraphs of G.

As implied by (5) and (6), our upcoming QC reformulation for the CCP exploits the fact that every variable x_{uv} represents the product $y_u y_v$. In fact, a Quadratically Constrained formulation for the CCP can be obtained by replacing (5) and (6) by $x_{uv} = y_u y_v$, $\{u, v\} \in E$.

For convenience, let F_1 be the CCP formulation obtained from F by replacing variables y_v by x_{vv} for every $v \in V$. Now, let $X = (x_{uv})_{u,v \in V} \in \{0, 1\}^{n \times n}$ be a binary matrix where: i) for every vertex v, x_{vv} is a decision variable indicating whether or not v is in the solution, ii) for each edge $\{u, v\} \in E$, x_{uv} and x_{vu} are a pair of variables indicating whether or not $\{u, v\}$ is included in the solution cycle, and iii) the other entries of X correspond to variables associated to pairs $\{u, v\}$, $\{v, u\}$ that are not assigned to the edges of E.

We stress that F_1 does not use all binary variables of X. Thus, the next step of our reformulation procedure consists in introducing these missing variables. To that aim, define $\widetilde{E} = \{\{u, v\} : u, v \in V, u \neq v\}$ as the edge set of a complete simple graph with n vertices. Now, extending linearization constraints (5) and (6) to the missing edges and variables and imposing the symmetry of $X \in \mathbb{B}^{n \times n}$, we have

$$x_{vu} \leq x_{vv}, \qquad\qquad v \in V, \{v, u\} \in \widetilde{E}, \qquad (12)$$

$$x_{uv} \geq x_{vv} + x_{uu} - 1, \qquad \{u, v\} \in \widetilde{E}, \qquad (13)$$

$$x_{uv} = x_{vu}, \qquad\qquad \{u, v\} \in \widetilde{E}, \qquad (14)$$

$$x_{vv} \in \{0, 1\}, \qquad\qquad v \in V, \qquad (15)$$

$$x_{uv} \in \{0, 1\}, \qquad\qquad \{u, v\} \in \widetilde{E}. \qquad (16)$$

Now notice that CCP solutions must also satisfy the rank-one constraint

$$X = \mathrm{diag}(X)\mathrm{diag}(X)^T, \qquad (17)$$

where $\mathrm{diag}(X)$ is the n-dimensional vector $(x_{11}, x_{22}, \ldots, x_{nn})^T$. Since X is symmetric and positive semidefinite, one could relax (17) to $X \succeq 0$. Instead, by using the Schur complement, we impose a tighter SDP relaxation [6]

$$X' = \begin{pmatrix} 1 & \mathrm{diag}(X)^T \\ \mathrm{diag}(X) & X \end{pmatrix} \succeq 0, \qquad (18)$$

which is equivalent to enforcing $X - \mathrm{diag}(X)\mathrm{diag}(X)^T \succeq 0$. Denote by F_{SDP} the (non-polyhedral) convex set defined by constraints (18), (12)-(14), the continuous relaxation of (15)-(16), and

$$x(\delta(v)) = 2x_{vv}, \qquad\qquad v \in V, \qquad (19)$$
$$x(\delta(S)) \geq 2(x_{uu} + x_{vv} - 1), \qquad S \subset V, \delta(S) \subset E, u \in S, v \in V \setminus S. \qquad (20)$$

Our goal now is to derive SDP bounds for the CCP by solving the program

$$\max\left\{ \sum_{v \in V} x_{vv} : X \in F_{SDP} \right\}.$$

3   SDP Lagrangian Relaxation Bounds

Our LR scheme relaxes (18), all constraints (20) and (13), as well as a subset of constraints (12) and (14). Among constraints (12), only those associated with artificial edges (i.e., those in $\widetilde{E} \setminus E$) are relaxed; the others, $x_{vu} \leq x_{vv}$, $v \in V, \{v, u\} \in E$, are kept in the Lagrangian subproblem. The opposite applies to constraints (14). Those associated to edges in E are relaxed while the remaining ones are kept in the subproblem. This specific choice of constraints to relax is made as an attempt to obtain a relaxed problem that can be solved efficiently, while at the same time dualizing as few constraints as possible.

Our procedure relaxes constraints (20) but never dualizes them in the objective function. The reasons are twofold. Since exponentially many of them are available, their dualization requires the implementation of a more expensive algorithm of the Relax-and-cut type [4] for updating Lagrangian multipliers. Additionally, computational experience reported in [5] indicates that the separation of cutset constraints has little impact on the quality of CCP LPR bounds. Similarly, we relax symmetry constraints (14) for edges in E but do not dualize them directly. As will be shown later, these are indirectly taken care of in our approach.

All other relaxed constraints are attached to Lagrangian multipliers and dualized, in the usual Lagrangian way. Lagrangian multipliers are organized in matrices Π, Θ, and Λ, defined below. For SM application purposes, any unspecified entry of such matrices is always set to 0. In our exposition, we assume that the first row and column of X' are indexed by 0, i.e., $X' = (x'_{uv})_{u,v \in V \cup \{0\}}$.

Our Lagrangian multiplier matrices are defined as follows:

- $\Pi = (\pi_{uv})_{u,v \in V}$ is an $n \times n$ matrix, such that for $v \in V$ and an artificial edge $\{v, u\} \in \tilde{E} \setminus E$, π_{vu} is a non-negative multiplier attached to (12).
- $\Theta = (\theta_{uv})_{u,v \in V}$ is a $n \times n$ strictly upper triangular matrix such that θ_{uv}, $u < v$, is a non-negative multiplier attached to (13).
- Finally, $\Lambda = (\lambda_{uv})_{u,v \in V \cup \{0\}} \succeq 0$ is an $(n+1) \times (n+1)$ positive semidefinite, not necessarily symmetric, matrix attached to constraint (18). Here, we use the symbol $\succeq$ to indicate positive semidefiniteness (symmetry is not necessarily implied).

Since $X' \succeq 0$ (and symmetric) and $\Lambda \succeq 0$ applies, one must have $\langle \Lambda, X' \rangle \geq 0$. Therefore, the dualization of constraint (18) is carried out by adding the inner product $\langle \Lambda, X' \rangle = \sum_{v \in V} (\lambda_{vv} + \lambda_{0v} + \lambda_{v0}) x_{vv} + \sum_{\{u,v\} \in \tilde{E}} (\lambda_{uv} x_{uv} + \lambda_{vu} x_{vu}) + \lambda_{00}$ to the objective function (see [2] for further details).

Assume now that Lagrangian modified costs are denoted by $\{p_{uv} : u, v \in V\}$. These costs are defined as follows: For every $v \in V$, $p_{vv} = 1 + \sum_{\{v,u\} \in E} (\pi_{vu} - \theta_{vu} - \theta_{uv}) + \lambda_{vv} + \lambda_{0v} + \lambda_{v0}$. For every $u, v \in V$, $u \neq v$, $p_{uv} = -\pi_{uv} + \theta_{uv} + \lambda_{uv}$. Subject to these definitions, the LR Subproblem is defined by

$$L(\Pi, \Theta, \Lambda) = \max \sum_{u,v \in V} (p_{uv} x_{uv} + \theta_{uv}) + \lambda_{00}, \tag{21}$$

$$\text{s.t.} \quad (15), (16), (19),$$

$$x_{vu} \leq x_{vv}, \qquad\qquad v \in V, u \in \delta(v),$$

$$x_{uv} = x_{vu}, \qquad\qquad \{u, v\} \in \tilde{E} \setminus E.$$

And its Lagrangian Dual is defined by

$$\min \quad L(\Pi, \Theta, \Lambda), \tag{22}$$

$$\text{s.t.} \quad \Pi \geq \mathbf{0}, \Theta \geq \mathbf{0}, \Lambda \succeq 0.$$

The LR subproblem (21) can be efficiently solved as follows. For every artificial edge $e = \{u, v\} \in \widetilde{E} \setminus E$, set $x_{uv} = x_{vu} = 1$ if $p_{uv} + p_{vu} > 0$ holds. Otherwise, set x_{uv} and x_{vu} to 0. For every vertex $v \in V$, let u_1 and u_2 be the two neighbors of v (considering only the edges in E), with the two largest costs. If $p_{vv} + p_{vu_1} + p_{vu_2} > 0$ applies, set $x_{vv} = x_{vu_1} = x_{vu_2} = 1$, and all other entries x_{vu} to 0. Otherwise, i.e., if $p_{vv} + p_{vu_1} + p_{vu_2} \leq 0$ applies, set $x_{vv} = 0$ and $x_{vu} = 0$, for all $\{v, u\} \in E$. The subproblem can be solved in $O(n^2)$ time independently of the density of G.

Let F_{SDP}^- denote the convex set F_{SDP} without the inclusion of cutset inequalities (20). It is not difficult to see that the LR Subproblem always has an integer optimal solution. Consequently, the optimal value of the Lagrangian Dual (22) matches the bound provided by F_{SDP}^-.

3.1   Subgradient Optimization

SM initially sets all multipliers to zero and then updates the entries of Π, Θ, Λ by taking a positive step in the opposite direction of a subgradient of (22). In particular, assuming that X^k denotes an optimal solution to the LR Subproblem (21), formulated and solved at SM iteration k, Λ^k is updated by taking a positive step in the direction of $-X^k$. If the step along $-X^k$ is too large, the resulting matrix Λ^{k+1} may no longer be positive semidefinite. Thus, we need to project Λ^{k+1} back onto the set of positive semidefinite matrices. To be more specific, let Λ^{k+1} be the matrix of multipliers obtained after moving a step along $-X^k$ at SM iteration k. The projection can be carried out in two steps (see [2] for details). First, we decompose Λ^{k+1} as $\Lambda^{k+1} = \Lambda_A^{k+1} + \Lambda_S^{k+1}$, the sum of its symmetric, Λ_S^{k+1}, and anti-symmetric part, $\Lambda_A^{k+1} = \frac{\Lambda^{k+1} - (\Lambda^{k+1})^T}{2}$. Then, we compute the spectral decomposition of $\Lambda_S^{k+1} = QDQ^T$, where Q is an orthogonal matrix whose i-th column contains the i-th eigenvector of Λ_S^{k+1}, and D is the corresponding diagonal matrix containing the eigenvalues. Assuming that D_+ is the matrix obtained by replacing any negative diagonal entry of D by zero, we set $\Lambda^{k+1} = QD_+Q^T + \Lambda_A^{k+1}$ as the new positive semidefinite matrix of Lagrangian multipliers to be used in the next SM iteration. After a step along the negative subgradient, any negative entry of Π^{k+1} and Θ^{k+1} is simply set to zero.

Our algorithm for solving (22) performs SMITERMAX$= 30000$ SM iterations. A step of size $\frac{10}{\sqrt{k}}$ in the opposite direction of the normalized subgradient is taken at the $k - th$ iteration.

4   Lagrangian Relaxation as a Warm Starter to BC

In this section, we discuss how the LR procedure presented earlier can be used as a warm starter to an improved BC algorithm. Our aim is to replicate, in the Linear Programming context, the strong lower bounds provided by F_{SDP}^- and benefit from the features available in modern Mixed Integer Programming (MIP) packages to solve the CCP.

Note that $X' \succeq 0$ if and only if $q^T X'q \geq 0$ for all real valued $(n+1)$-dimensional vectors $q = (q_v)_{v \in V \cup \{0\}}$. The latter is a linear inequality that can be written as:

$$q^T X'q = \sum_{u,v \in V \cup \{0\}} q_u q_v x'_{uv} \geq 0, \qquad \forall q \in \mathbb{R}^{n+1}, \tag{23}$$

which is the same as

$$\sum_{v \in V} (q_0 q_v + q_v q_0 + q_v q_v) x_{vv} + \sum_{\{u,v\} \in \tilde{E}} q_u q_v (x_{uv} + x_{vu}) \geq -q_0^2, \quad \forall q \in \mathbb{R}^{n+1}. \tag{24}$$

Thus, the same continuous relaxation bounds of F_{SDP}^- are attained by the Linear Semi-infinite Program obtained by replacing (18) by (24). One approach to evaluate these bounds is through dynamic separation of (24), a technique that suffers from slow convergence to the point of being impractical.

We will now show that only a tiny subset of (24) are actually necessary to attain the F_{SDP}^- bounds in a Linear Program. First, we show that dualization of (14) is implied by the asymmetry of Λ. To see that, note that $\langle \Lambda, X' \rangle = \langle \Lambda_S, X' \rangle + \langle \Lambda_A, X' \rangle$, where Λ_S and Λ_A are the symmetric and skew symmetric parts of Λ. The term $\langle \Lambda_A, X' \rangle$ is precisely the term one gets when (14) is dualized.

Thus, if (14) is dualized, we may assume that Λ is symmetric. Consider that case and let Λ^* be the optimal positive semidefinite matrix of multipliers for the corresponding Lagrangian Dual. Assume as well that its spectral decomposition is $\Lambda^* = QDQ^T$, and let $(q^w)_{w \in V \cup \{0\}}$ denote the columns of Q (the eigenvectors of Λ^*) while $(d_w)_{w \in V \cup \{0\}}$ denotes the diagonal entries of D (the eigenvalues of Λ^*). Consider also a restricted version of (24), defined only for the eigenvectors $\{q^w : w \in V \cup \{0\}\}$ of Λ^*:

$$\sum_{v \in V} (q_0^w q_v^w + q_v^w q_0^w + q_v^w q_v^w) x_{vv} + \sum_{\{u,v\} \in \tilde{E}} q_u^w q_v^w (x_{uv} + x_{vu}) \geq -(q_0^w)^2,$$

$$w \in V \cup \{0\}. \tag{25}$$

Now, note that

$$\langle \Lambda^*, X' \rangle = \langle QDQ^T, X' \rangle = \sum_{w \in V \cup \{0\}} \langle d_w q^w (q^w)^T, X' \rangle = \sum_{w \in V \cup \{0\}} d_w (q^w)^T X' q^w. \tag{26}$$

The rightmost term of equation (26) is the penalty term for dualizing (25) for each q^w with Lagrangian multiplier d_w. This discussion shows that the continuous relaxation bounds of F_{SDP}^- are attained by the Linear Program obtained by replacing (18) by (25) in its definition.

A strategy for obtaining the F_{SDP}^- bounds in a Linear Programming setting is thus the following. Apply the Lagrangian relaxation method of Sect. 3. Suppose that Λ_S^* denotes the symmetric part of Λ^*, the positive semidefinite matrix of multipliers associated with the best bound $L(\Pi^*, \Theta^*, \Lambda^*)$ found during the application of the SM to solve (22). Perform the spectral decomposition of Λ_S^* to identify the necessary (q^w). Finally, the Linear Program given by F_{SDP}^- with (25) in place of (18), i.e., (12)-(14), the continuous relaxation of (15)-(16), (19), and (25).

Note that (25) are defined on the extended space that considers the artificial variables. Next, we project the artificial variables out of (25), to obtain inequalities that can be used to reinforce F_1.

Using (14), (25) can be written as:

$$\sum_{v\in V}(q_0^w q_v^w + q_v^w q_0^w + q_v^w q_v^w)x_{vv} + \sum_{\{u,v\}\in E} 2q_u^w q_v^w x_{uv}$$

$$+ \sum_{\{u,v\}\in \widetilde{E}\setminus E} 2q_u^w q_v^w x_{uv} \geq -(q_0^w)^2, \quad w\in V\cup\{0\}. \quad (27)$$

In order to formulate the projection cone of (27), plus the constraints $0\leq x_{uv}\leq 1, \{u,v\}\in \widetilde{E}\setminus E$, with respect to the artificial variables, consider dual multipliers $\gamma = \{\gamma_w \geq 0 : w\in V\cup\{0\}\}$, corresponding to (27), and $\beta = \{\beta_{uv} \geq 0 : \{u,v\}\in \widetilde{E}\setminus E\}$, $\beta_{uv} := \beta_{vu}$, corresponding to $x_{uv}\leq 1$. The projection cone is the set of vectors (γ, β) satisfying:

$$\sum_{w\in V\cup\{0\}} 2q_u^w q_v^w \gamma_w \leq \beta_{uv}, \qquad \{u,v\}\in \widetilde{E}\setminus E, \qquad (28)$$

$$\gamma, \beta \geq \mathbf{0}. \qquad (29)$$

Inequalities (27) and $\{0\leq x_{uv}\leq 1, \{u,v\}\in \widetilde{E}\setminus E\}$ may then be replaced by the equivalent set of exponentially many inequalities (projection cuts)

$$\sum_{w\in V\cup\{0\}} \overline{\gamma}_w \left(\sum_{v\in V}(q_0^w q_v^w + q_v^w q_0^w + q_v^w q_v^w)x_{vv} + \sum_{\{u,v\}\in E} 2q_u^w q_v^w x_{uv}\right)$$

$$\geq - \sum_{w\in V\cup\{0\}} \overline{\gamma}_w (q_0^w)^2 - \sum_{\{u,v\}\in \widetilde{E}\setminus E} \overline{\beta}_{uv}, \quad (30)$$

where $(\overline{\gamma}, \overline{\beta})$ are the extreme rays of the cone (28)–(29). Denote by F_1^+ the polyhedral set F_1 strengthened with projection cuts (30).

The F_1^+ Linear Relaxation bounds approximate the F_{SDP}^- continuous relaxation bounds. In fact, the bounds we get are not the best possible LR bounds themselves because some of the inequalities involving the artificial variables (namely (12) and (13)) were not considered in the projection step to simplify the process. However, the approximation attained is very close, as indicated in Sect. 5.

In practice, the inclusion of projection cuts (30) to reinforce F_1 is carried out dynamically, only when needed. Given a solution $\overline{X}\in \mathbb{R}^{n\times n}$ to a relaxation of F_1^+, the separation problem associated to (30) can be formulated as the Linear Program

$$\max \quad - \sum_{w \in V \cup \{0\}} \left(\sum_{v \in V} (q_0^w q_v^w + q_v^w q_0^w + q_v^w q_v^w) \overline{x}_{vv} \right.$$

$$\left. + \sum_{\{u,v\} \in E} 2q_u^w q_v^w \overline{x}_{uv} + (q_0^w)^2 \right) \gamma_w - \sum_{\{u,v\} \in \widetilde{E} \setminus E} \beta_{uv}, \quad (31)$$

$$\text{s.t.} \quad (28) - (29),$$

$$\gamma \leq 1, \beta \leq 1, \quad (32)$$

where (32) are normalization constraints, to bound the projection cone. In case the objective function in (31) is positive, an optimal solution $(\overline{\gamma}, \overline{\beta})$ to it defines a violated projection cut (30), to be appended to the current relaxation of F_1^+.

4.1   An Improved BC Algorithm

A new CCP BC algorithm, BC_PROJ_SDP, is introduced in this section. BC_PROJ_SDP benefits from the strong SDP bounds provided by the LR scheme outlined in Sect. 3. BC_PROJ_SDP involves two phases: an LR phase, which computes Λ^* and works as a warm starter to the second phase, the BC algorithm itself. The BC phase is implemented on top of the BC algorithm BC1, introduced in [5].

In [5], formulation F and several valid inequalities are presented for the MICP. Among these, the clique based inequalities (10) and (11) were the ones proven most useful. Three BB algorithms stand out in that reference: BC1, BC2, and BC3, with no clear dominance relation between them. BC1, with some minor modifications explained below, will be used as the basis of comparison for the BB algorithms introduced here. In preliminary experiments, computational times for BC1, BC2, and BC3 did not differ by much on the sparser instances, while the latter two were clearly outperformed in the denser instances.

BC1 is a simple Branch-and-cut algorithm obtained by directly solving F in a commercial MIP solver. Cutset inequalities (20) are dynamically separated, but only for integer solutions, as lazy constraints. The implementation in [5] and the one used here differ in that here, when violated cutsets (20) are found, only a single inequality, the first one identified, is added to the model, while Pereira et al. [5] add all violated ones.

All the implementation details that apply to BC1 are replicated by BC_PROJ_SDP. The key difference between BC1 and the BC phase of BC_PROJ_SDP is that the latter separates projection cuts at the root node of the search tree. With the optimal matrix Λ^* at hand, BC_PROJ_SDP separates SDP projection cuts (30) by solving the Linear Program (31) for every LPR solution computed at the root node. Whenever a violated projection cut is found, it is appended to a new reinforced relaxation of F_1^+ at the root node.

We also implemented and tested a second hybrid BC algorithm, BC_LP_SDP. It is based on formulation F_1 extended by all artificial variables defining matrix

X. BC_LP_SDP also calls the SM in a warm start phase. Since BC_LP_SDP operates on the extended variable space, SDP cuts in the extended variable space are ready to be used. Therefore, all inequalities (27) are added to the model outright at the beginning of the BC phase. Compared to BC_PROJ_SDP, BC_LP_SDP is not competitive with respect to CPU times. For this reason, no computational results are presented for it in Sect. 5.

5   Preliminary Computational Experiments

In this section, we discuss the strength of our SDP Lagrangian Relaxation bounds as well as how BC_PROJ_SDP benefits from them in practice. Our experiments were carried out on an Intel XEON machine with 8GB of RAM and eight 3.5GHz cores using a Linux operating system. All algorithms tested and compared here, including those coming from the literature [5], were implemented in **Python** 3. All BC algorithms were implemented on top of the CPLEX Mixed Integer Programming (MIP) solver, release 12.71.

Our numerical experiments were conducted with the same two sets of instances tested in [5]. One of them comprises a set of randomly generated graphs while the other set comes from the Maximum Clique literature. The latter instances are available at https://iridia.ulb.ac.be/~fmascia/maximum_clique. Since the conclusions we have drawn from these experiments are the same for both sets of instances, we only present results for the first one. Each instance in that set is defined by two parameters: the number of vertices n and the density d (in %) of the input graph G. In [5], 10 different instances were generated for every combination of $n \in \{50, 60, \ldots, 100\}$ and $d \in \{10\%, 30\%\}$. In this paper, we enlarge that set and consider new instances with additional values of density $d \in \{50\%, 70\%\}$.

Table 1 presents the average primal bounds in its first row, followed by a comparison of dual bounding strategies in its subsequent rows. For each graph density, the table provides an average *Bound* and $t(s)$, the average CPU time needed to evaluate the bound. Results in that table are averaged over the 10 instances with $n = 100$, for each value of d. Results provided in the row labeled *Best known lb* are the average best lower (primal) bounds for each instance. For that row, no average CPU times are provided. Subsequent rows provide average dual bounds and CPU times, obtained by different strategies. Rows with label F_1 report the bounds (and time) associated to formulation F_1 (identical to F), separating cutsets (20) only for integer points. Row $F_1 + (10)$-(11) does the same for F_1 strengthened with clique-y and clique-bqp inequalities. Row F_{SDP} gives results for the Lagrangian Relaxation scheme outlined in Sect. 3. Row F_1^+ gives upper bounds associated with formulation F_1 strengthened with projected SDP cuts (30). Finally, row $F_1^+ + (10)$-(11) reports results when clique inequalities are appended to F_1^+.

As noted in [5], very weak lower bounds are provided by F_1. It yields an optimality gap of 48% for the $d = 10\%$ instances and, as d increases, this gap progressively worsens: 192%, 292%, and 463%, respectively, for $d = 30\%$, 50%,

Table 1. The first row provides average CCP primal bounds. Subsequent rows give dual bounds, computed with different CCP formulations. Values are averaged for 10 randomly generated instances with $n = 100$ vertices.

Dual bounding Strategy	$d = 10\%$		$d = 30\%$		$d = 50\%$		$d = 70\%$	
	Bound	$t(s)$	Bound	$t(s)$	Bound	$t(s)$	Bound	$t(s)$
Best known lb	38.9	–	17.7	–	13.0	–	9.0	–
F_1	57.6	0.0	51.8	1.4	51.0	1.0	50.7	0.3
$F_1 + (10)\text{--}(11)$	51.4	0.1	29.3	10.2	26.6	155.5	26.1	2036.2
F_{SDP}	49.3	42.7	25.8	42.2	16.5	41.8	11.13	41.0
F_1^+	52.3	97.9	25.9	68.0	16.5	55.7	11.1	51.1
$F_1^+ + (10)\text{-}(11)$	51.3	117.1	25.9	116	16.5	73.2	11.1	81.8

and 70%. Bounds provided by F_1 are substantially improved when clique-y (10) and clique-bqp (11) come into play, the respective gaps as d increases are: 32%, 66%, 105%, 190%. Even so, the quality of the bounds still worsens as the graph densities increase.

F_{SDP} bounds, evaluated by the Lagrangian relaxation strategy are much stronger overall. Even for the $d = 10\%$ instances, where a large number of artificial edges are needed in order to formulate F_{SDP}, its bounds are stronger than those of the reinforced F_1. This scheme also does not exhibit the pattern of progressively getting worse as d increases, its respective gaps are: 26%, 46%, 27%, and 23%. For this scheme, subgradient optimization takes about 40s to run, independently of d.

Table 1 also shows that the bounds provided by formulation F_1^+, i.e., LPR bounds given by F_1 strengthened with projection cuts (30), are very strong. For the $d = 10\%$ case, they are slightly weaker then their F_{SDP} counterparts. That applies because not all inequalities involving the artificial variables were used to formulate the projection cone. However, as d increases, the difference between the dual bounds provided by F_1^+ and F_{SDP} becomes neglectable. The impact of clique inequalities in strengthening the bounds F_1^+ is small for the $d = 10\%$ instances and, as Table 1 also suggests, decreases as d increases. That is the reason why using them in BC_PROJ_SDP did not result in a better algorithm.

Bounds provided by F_1^+ are significantly stronger than those implied by F_1, the formulation on which BC1 relies. However, their numerical evaluation is substantially more CPU time demanding. That applies because the SM method needs to be executed before the BC phase itself. Additionally, projection cuts (30) are also dense, yielding harder to solve Linear Programs.

In what follows, we discuss computational results attained by BC_PROJ_SDP to address whether or not the warm start phase of our new BC algorithm, in its current preliminary state of development, pays off. That matter is addressed by Table 2, which presents numerical results for BC_PROJ_SDP and BC1. For each (n, d) pair tested here, the table presents the average total computational time (in seconds, $t(s)$) and the number of Branch-and-bound nodes investigated during the search. Again, values reported by the table are averaged for 10 instances for each (n, d) pair. A time limit of one hour is imposed for each algorithm. A symbol "*" indicates that the algorithm did not solve all 10 instances of a given pair (n, d).

BC1 is about 3 times faster than BC_PROJ_SDP for $d = 10\%$, and such an advantage is reduced as d increases. For example, it is about 30% faster for $d = 30\%$. For the $d = 50\%$ instances, they provide results of similar quality, with a small advantage in favor of BC1. Now, for the $d = 70\%$ instances, results lean in favor of BC_PROJ_SDP, which is twice as fast as BC1.

Although BC_PROJ_SDP relies on a much stronger formulation, it usually investigates more Branch-and-bound nodes than BC1 to conclude the search. The reason for that remains unclear for us, as both algorithms use the same MIP package, CPLEX, to solve LPs and to manage the search tree. We believe that, once such a matter is better understood and more specific branching and search polices are designed for BC_PROJ_SDP, numerical results provided by that algorithm should improve significantly.

Table 2. Comparison of Branch-and-cut algorithms.

Algorithm	n	$d = 10\%$		$d = 30\%$		$d = 50\%$		$d = 70\%$	
		t(s)	nodes	t(s)	nodes	t(s)	nodes	t(s)	nodes
BC1	60	1.5	266.6	40.5	6896.2	39.2	2614.2	62.7	835.8
BC_PROJ_SDP		25.7	266.3	71.9	7132.4	66.7	3259.3	46.5	1301.2
BC1	70	6.4	1566.4	211.5	23553.7	143.2	6895.8	161.7	1560.1
BC_PROJ_SDP		36.0	1584.9	278.7	26453.5	165.1	9211.6	89.9	2806.2
BC1	80	31.0	65169.9	752.2	65169.9	353.7	11630.5	344.0	2149.6
BC_PROJ_SDP		94.9	10617.8	1008.9	62337.5	373.6	14971.7	151.8	4211.5
BC1	90	124.3	20054.5	2226.1	162983	981.1	30069.6	645.0	2851
BC_PROJ_SDP		428.5	49919.9	*3165.0	129404.2	1160.1	36454.7	298.8	6308.3
BC1	100	974.1	116320.5	*3600.0	121797.9	1819.9	37948.9	1245.0	3885.8
BC_PROJ_SDP		*3070.4	299455.4	*3600.0	97387.1	2553.5	65866.4	533.7	10797.6

6 Conclusions

Building upon an Integer Programming formulation coming from the literature, we reformulated the Chordless Cycle Problem as a Quadratically Constrained Problem and derived an SDP relaxation for it, whose bounds were approximated by a Lagrangian Relaxation scheme. Compared to previously available Linear Programming bounds, our SDP bounds are quite strong. Additionally, these SDP bounds do not worsen as the density of G increases. Motivated by the quality of these bounds, we introduced a hybrid Branch-and-cut algorithm involving two combined phases. The first is the Lagrangian Relaxation algorithm, which is used as a warm starter for the second, the best BC algorithm coming from [5]. More precisely, the Lagrangian Relaxation algorithm approximates the optimal positive semidefinite matrix of Lagrangian multipliers associated with the dualization of the semidefiniteness constraint. That (near) optimal matrix allows us to formulate a finite set of linear inequalities, SDP cuts, that can be used to attain the Lagrangian SDP bounds in a Linear Programming Relaxation for the CCP. These SDP cuts are not ready to use though, since they are written in an extended variable space. Thus, we project these cuts back onto the original space of variables and separate them, within the root node of our BC algorithm, by solving a Linear Program. In summary, the new hybrid BC algorithm turns out to be better than the previous methods in the literature for dense instances. An aspect that raised our attention is that our BC algorithm enumerates more nodes than its contender, despite the fact that it relies on stronger dual bounds. We believe that, by deriving specific branching rules, numerical results provided by our BC algorithm should improve. From an algorithmic standpoint, we also plan to implement a full blow Branch-and-bound algorithm based on the Lagrangian Relaxation algorithm.

The reformulation step we conducted here for the CCP can be applied for solving other induced subgraph problems. Similar observations apply for the algorithms we introduced here, including the SDP Lagrangian Relaxation scheme and its use as a warm starter for specific Branch-and-cut algorithms, dedicated to solving other induced subgraph problems.

References

1. Chen, W.Y.C.: Induced cycle structures of the hyperoctahedral group. SIAM J. Disc. Math. **6**(3), 353–362 (1993)
2. Guimarães, D., da Cunha, A.S., Pereira, D.L.: Semidefinite programming lower bounds and branch-and-bound algorithms for the quadratic minimum spanning tree problem. Eur. J. Oper. Res. **280**(1), 46–58 (2020)
3. Henning, M.A., Joos, F., Löwenstein, C., Sasse, T.: Induced cycles in graphs. Graphs Comb. **32**, 2425–2441 (2016)
4. Lucena, A.: Non delayed relax-and-cut algorithms. Ann. Oper. Res. **140**, 375–410 (2005)
5. Pereira, D.L., Lucena, A., Cunha, A.S., Simonetti, L.: Exact solution approaches for the chordless cycle problem. INFORMS J. Comput. **34**(4), 1970–1986 (2022)
6. Tunçel, L.: Polyhedral and Semidefinite Programming Methods in Combinatorial Optimization. Fields Institute Monographs, Ontario, Canada (2010)

A Family of Spanning-Tree Formulations for the Maximum Cut Problem

Sven Mallach[(✉)] [iD]

High Performance Computing and Analytics Lab, University of Bonn,
Friedrich-Hirzebruch-Allee 8, 53115 Bonn, Germany
`sven.mallach@cs.uni-bonn.de`

Abstract. We present a family of integer programming formulations for the maximum cut problem. These formulations encode the incidence vectors of the cuts of a connected graph by employing a subset of the odd-cycle inequalities that relate to a spanning tree, and they require only the corresponding edge variables to be integral explicitly. They so describe sufficient restrictions of the classic integer linear program by Barahona and Mahjoub. In addition, we characterize according formulations comprising facet-defining inequalities only. Trade-offs and comparisons to prevalent formulations concerning size and relaxation strength are subject to an experimental study.

Keywords: Maximum Cut Problem · Integer Programming · Binary Quadratic Programming · Polyhedral Combinatorics

1 Introduction

Given an undirected graph $G = (V, E)$, a *cut* in G is an edge set $\delta(U) := \{ \{u, v\} \in E : u \in U, v \in V \setminus U \}$ induced by some *bi-partitioning* of V into $U \subseteq V$ and $V \setminus U$. Considering edge weights $c : E \to \mathbb{R}$, let $c(\delta(U)) := \sum_{e \in \delta(U)} c_e$ for any $U \subseteq V$. Then the *Maximum Cut Problem* (MaxCut) is to determine $U^* \subseteq V$ such that $c(\delta(U^*)) \geq c(\delta(U))$ for all $U \subseteq V$. Since a collection of maximum cuts in the connected components of an undirected graph G gives a maximum cut in G, we will assume throughout this paper that G is connected.

MaxCut is a classical combinatorial optimization problem that is $\mathcal{NP}$-hard in the general case [14], and that receives increasing interest in recent years. Besides direct applications in e.g. Image Segmentation [10], Frequency Assignment [6] and VLSI Design [1], a major reason for this interest is its direct correspondence to the Unconstrained Binary Quadratic Programming problem (UBQP) [2,9,11] that by itself has numerous applications in research and economy.

A common approach to solve MaxCut instances to proven optimality is based on integer linear programming and polyhedral relaxations. Especially for sparse graph instances, substantial progress regarding the effectiveness of such exact methods has been achieved only recently [7,13,18].

A. Basu et al. (Eds.): ISCO 2024, LNCS 14594, pp. 43–55, 2024.
https://doi.org/10.1007/978-3-031-60924-4_4

In this paper, we deduce a family of integer programming formulations for MaxCut that is based on spanning trees. These formulations constitute proper restrictions of the classical edge-variable formulation by Barahona and Mahjoub [3]. In particular, in order to still encode precisely the cuts in $G = (V, E)$, our derivations imply that only $|V| - 1$ of the edge variables in the classic formulation need an explicit enforcement of integrality, and that one may confine oneself to an appropriate subset of its only class of non-trivial constraints, called *odd-cycle inequalities*, at the same time. We characterize natural approaches to determine such appropriate subsets based on cycles related to a spanning tree of G, and we discuss the corresponding trade-offs that arise regarding formulation size and quality. In addition, we address how to obtain violated odd-cycle inequalities from these cycles by means of an efficient separation. Besides that, the presented family of spanning-tree formulations generalizes on the so-called root-triangulated model for MaxCut recently introduced in [8] as well as on the further classic formulation with node- and edge-variables that has been obtained in the light of UBQP transformations [2,9,17]. In our computational experiments, we compare the different formulations regarding their size and their strength in terms of their linear programming (LP) relaxations.

This paper is organized as follows: In Sect. 2, we briefly summarize fundaments and related work. The family of spanning-tree formulations for MaxCut is presented in Sect. 3 along with a correctness proof and improvements concerning the size and the strength of the formulations. In Sect. 4, we report on our computational experiments, and finally, a conclusion is given in Sect. 5.

2   Preliminaries and Related Work

As is common in (integer) linear programming approaches to MaxCut, we will identify edge subsets $S \subseteq E$ in $G = (V, E)$ with their *incidence vectors* $\chi^S \in \mathbb{R}^E$ where $\chi_e^S \in \{0, 1\}$ and $\chi_e^S = 1$ if and only if $e \in S$. Based on this notion, the *cut polytope* $P_{\mathrm{CUT}}(G)$ with respect to G, as defined by Barahona and Mahjoub in [3], is the convex hull of all the incidence vectors of cuts in G, i.e.,

$$P_{\mathrm{CUT}}(G) := \mathrm{conv}\{\chi^S \mid S \subseteq E \text{ is a cut in } G\}.$$

It is well-known that an edge subset $S \subseteq E$ is a cut in G if and only if S intersects with every cycle in G in an *even* number (possibly zero) of edges. For variables x_e, $e \in E$, where $x_e = 1$ if e is a cut-edge and $x_e = 0$ otherwise, this condition is established by the constraints (1), called *odd-cycle inequalities*, in the seminal binary linear programming formulation by Barahona and Mahjoub [3] that we refer to as the edge model (E).

(E) $\max \sum_{e \in E} c_e x_e$

s.t. $\sum_{e \in S} x_e - \sum_{e \in C \setminus S} x_e \leq |S| - 1$ for all cycles $C \subseteq E$

$$\text{and all } S \subseteq C, |S| \text{ odd} \quad (1)$$

$$0 \leq x_e \leq 1 \qquad \text{for all } e \in E \quad (2)$$

$$x_e \in \{0, 1\} \qquad \text{for all } e \in E$$

Given a cycle $C \subseteq E$, an edge $e \in E \setminus C$ that is incident to two nodes of C is called a *chord of* C. An odd-cycle inequality defines a facet of $P_{\mathrm{CUT}}(G)$ if and only if the corresponding cycle C is chordless [3]. We also recall from the same reference that, although we do not explicitly state this in (E), the lower and upper bounds on the variables (2) are implied by (1) for those edges which are part of a triangle in G. When removing the integrality restrictions, we obtain

$$P_{\mathrm{OC}} := \{x \in \mathbb{R}^E : x \text{ satisfies } (1), (2)\}$$

as the feasible set, giving rise to the LP relaxation $\max\{c^\mathsf{T} x : x \in P_{\mathrm{OC}}\}$ of (E) which provides an upper bound on the value of a maximum cut in G.

An alternative and also well-known integer linear programming formulation for MaxCut [2,9,17] is given by the following node-edge model (NE):

(NE) $\max \sum_{\{i,j\} \in E} c_{ij} x_{ij}$

s.t. $x_{ij} + z_i + z_j \leq 2$ \qquad for all $\{i,j\} \in E$ \quad (3)

$\qquad x_{ij} - z_i - z_j \leq 0$ \qquad for all $\{i,j\} \in E$ \quad (4)

$\qquad -x_{ij} + z_i - z_j \leq 0$ \qquad for all $\{i,j\} \in E$ \quad (5)

$\qquad -x_{ij} - z_i + z_j \leq 0$ \qquad for all $\{i,j\} \in E$ \quad (6)

$\qquad\qquad z_i \in \{0,1\}$ \qquad for all $i \in V$

In (NE), the integrality restrictions are imposed only on the additional node-variables $z \in \mathbb{R}^V$ which express the partition-assignments while the consistency and integrality of the variables $x \in \mathbb{R}^E$ is then established via the constraints (3)–(6). These inequalities also imply $0 \leq z \leq 1$ and $0 \leq x \leq 1$.

The node variables z_i, $i \in V$, in (NE) may also be regarded as edge variables in terms of i and an additional auxiliary node that is connected to all original ones. In [8], Charfreitag et al. take on this perspective and propose to refine (NE) by choosing an original node $r \in V$ to adopt the role of the auxiliary one. Consequently, in the resulting root-triangulated model (RT), the original edge set is (only) extended to $E' := E \cup \{\{r,v\} : v \in V, \{r,v\} \notin E\}$.

(RT) $\max \sum\limits_{\{i,j\}\in E} c_{ij}x_{ij}$

$\quad$ s.t. $\quad x_{ij} + x_{ri} + x_{rj} \leq 2$ $\qquad$ for all $\{i,j\} \in E, r \notin \{i,j\}$ $\qquad$ (7)

$\qquad\qquad x_{ij} - x_{ri} - x_{rj} \leq 0$ $\qquad$ for all $\{i,j\} \in E, r \notin \{i,j\}$ $\qquad$ (8)

$\qquad\qquad -x_{ij} + x_{ri} - x_{rj} \leq 0$ $\qquad$ for all $\{i,j\} \in E, r \notin \{i,j\}$ $\qquad$ (9)

$\qquad\qquad -x_{ij} - x_{ri} + x_{rj} \leq 0$ $\qquad$ for all $\{i,j\} \in E, r \notin \{i,j\}$ $\qquad$ (10)

$\qquad\qquad\qquad x_{ri} \in \{0,1\}$ $\qquad$ for all $\{r,i\} \in E'$

In (RT), the inequalities (7)–(10), substituting for (3)–(6), now appear as triangle inequalities (i.e., odd-cycle inequalities for $|C| = 3$) for each $\{i,j\} \in E$, $r \notin \{i,j\}$, and the corresponding unique "root-triangle" in E'. These triangle inequalities again imply $0 \leq x_{ij} \leq 1$ for all $\{i,j\} \in E'$, and their presence combined with the integrality of the variables x_{ri}, for all $i \in V \setminus \{r\}$, suffices to establish the integrality of all variables. Indeed if $x_{ri}, x_{rj} \in \{0,1\}$ one has $x_{ij} \in \{0,1\}$ because the corresponding instances of (7)–(10) then reduce to $x_{ij} = 0$ (if $x_{ri} = x_{rj}$) or $x_{ij} = 1$ (if $x_{ri} = 1 - x_{rj}$) [8,16].

Compared to (E), (NE) and (RT) are extended by at most $|V|$ variables, and they contain only a small subset of the odd-cycle or, more precisely, triangle inequalities. As such, their LP relaxations provide only a weak bound on the value of a maximum cut, but they may serve as a starting point to be successively enriched by further odd-cycle inequalities, and thus promoted to be as strong as the one of (E) [8]. From a converse perspective, the strength of P_{OC} can principally be retained even solely by triangle inequalities at the expense of augmenting G with (zero-weight) edges, respectively variables, to a chordal graph, like e.g. in [16]. This is because then the resulting triangle inequalities become the only facet-defining, i.e., irredundant, odd-cycle inequalities.

3   A Family of Spanning-Tree Formulations for MaxCut

In this section, we characterize a family of integer programming formulations for MaxCut that is based on the selection of a spanning edge subset of the original graph G, and whose sets of constraints and variables with an explicit integrality requirement are (usually strict) subsets of those in (E). In contrast to (NE) and (RT) from Sect. 2, the spanning-tree formulations do not involve any further variables and they do not restrict to triangle inequalities. Instead, other appropriate subsets of the odd-cycle inequalities (1) are identified that prove sufficient to restrict the feasible set to the incidence vectors of cuts in G, provided that (only) the variables corresponding to the spanning edge set are (explicitly enforced to be) integral. More precisely, the inequalities transfer the integrality of these variables to the others with the effect that all odd-cycle inequalities for G are satisfied, exactly as the triangle inequalities do in (RT).

We will use the following basic terminology with respect to spanning trees and extend it appropriately during the course of the discussion.

Definition 1. *Let $T = (V, E_T)$ be a spanning tree of a connected undirected graph $G = (V, E)$, i.e., $E_T \subseteq E$, $|E_T| = |V| - 1$, and T is connected. Let $e = \{i, j\} \in E \setminus E_T$, and let T_e be the unique i-j-path in T. Then, we call $C_e = T_e \cup \{e\} \subseteq E$ an* elementary cycle *(with respect to T).*

3.1 Basic Spanning-Tree Formulations

Given a graph $G = (V, E)$ and some spanning tree $T = (V, E_T)$ with $E_T \subseteq E$, we obtain a first family of new edge-variable formulations (ST) by restricting to the odd-cycle inequalities associated with the elementary cycles C_e, $e \in E \setminus E_T$.

$$(\text{ST}) \quad \max \sum_{e \in E} c_e x_e$$

$$\text{s.t.} \sum_{e \in S} x_e - \sum_{e \in C_f \setminus S} x_e \le |S| - 1 \quad \text{for all elem. cycles } C_f \subseteq E, f \in E \setminus E_T,$$

$$\text{and all } S \subseteq C_f, |S| \text{ odd} \qquad (11)$$

$$0 \le x_e \le 1 \quad \text{for all } e \in E$$

$$x_e \in \{0, 1\} \quad \text{for all } e \in E_T$$

Before we show that (ST) indeed encodes the incidence vectors of cuts in G, we remark that, in analogy to (E), the lower and upper bounds on the variables in (ST) are implied by (11) for those edges which take part in a triangle that serves as one of the corresponding elementary cycles $C_f \subseteq E$, $f \in E \setminus E_T$.

Theorem 1. *Let $\bar{x} \in \mathbb{R}^E$ be a solution to (an instance of) (ST) associated with $G = (V, E)$. Then $\bar{x}$ is integral and the incidence vector of a cut in G.*

Proof. Concerning integrality, since $\bar{x}_e \in \{0, 1\}$ is enforced explicitly in (ST) for all $e \in E_T$, it suffices to show that $\bar{x}_e \in \{0, 1\}$ for all $e \in E \setminus E_T$. For any such edge $e \in E \setminus E_T$, consider the corresponding elementary cycle C_e which consists of $|C_e| - 1$ edges from E_T and e. The odd-cycle inequalities for C_e are satisfied by $\bar{x}$. Each of them can be written as $\sum_{f \in S}(1 - x_f) + \sum_{f \in C_e \setminus S} x_f \ge 1$ for all $S \subseteq C_e$, $|S|$ odd. Let $O := \{f \in C_e \setminus \{e\} : \bar{x}_f = 1\}$ and let $k := |O|$. If k is odd, consider that $\sum_{f \in O}(1 - \bar{x}_f) + \sum_{f \in C_e \setminus O} \bar{x}_f \ge 1$ reduces to $\bar{x}_e \ge 1$. If k is even, consider that $\sum_{f \in O \cup \{e\}}(1 - \bar{x}_f) + \sum_{f \in C_e \setminus (O \cup \{e\})} \bar{x}_f \ge 1$ reduces to $(1 - \bar{x}_e) \ge 1 \Leftrightarrow -\bar{x}_e \ge 0 \Leftrightarrow \bar{x}_e \le 0$. Since $0 \le \bar{x}_e \le 1$ is enforced explicitly in (ST), we have $\bar{x}_e = 1$ in the first, and $\bar{x}_e = 0$ in the second case. Thus, $\bar{x}_e$ is integral for all $e \in E \setminus E_T$. Moreover, since the edge set $C_e \setminus \{e\} \subseteq E_T$ corresponds to the unique simple path between the two endpoints of each $e = \{i, j\}$ in T, the previous arguments show that $\bar{x}_e = 1$ ($\bar{x}_e = 0$) if, when traversing this simple path, the partition-assignments of the nodes change an odd (even) number of times. i.e., if i and j belong to different (the same) partition(s). Thus, the contradiction-free bi-partitioning of $V(T)$ given by the restriction of $\bar{x}$ to the components for $E_T{}^1$ is consistently imposed on the components for $E \setminus E_T$ by (11). In other words, $\bar{x}$ is the incidence vector of a cut.   □

Each choice of the spanning edge set $E_T \subseteq E$ for (ST) gives rise to a concrete integer program which leads to different variables with an explicit integrality requirement, and to different elementary cycles whose lengths influence the size of the formulation as well as the upper bound on the value of a maximum cut provided by its LP relaxation[2]. Thereby, the number of odd-cycle inequalities associated with each of the $|E|-|V|+1$ elementary cycles C_f in (11) is $2^{|C_f|-1}$ and thus strongly increases with their length. Moreover, the upper bound provided by (ST) is expected to be comparatively weak (irrespective of the choice of E_T), as is also visible from the experiments in Sect. 4. In particular, the selection (11) of odd-cycle inequalities is not necessarily facet-defining since the associated cycles need not be chordless. This can in turn be used to reduce the size of the formulation and to strengthen its LP relaxation at the same time, as is exposed in the following.

3.2   Improved Spanning-Tree Formulations

We discuss two strategies to improve over (ST) by focusing on facet-defining odd-cycle inequalities deduced from a given spanning tree $T = (V, E_T)$ of $G = (V, E)$. A first idea could be to replace (11) by all the (not necessarily elementary) chordless cycles that are composed from the edges of C_e and its (possibly empty set of) chords. Indeed, the corresponding set $\mathcal{C}_e$ of chordless cycles is of relevance and thus formalized in the first part of Definition 2 below. However, the mentioned characteristic property may apply to a single chordless cycle for more than one $e \in E \setminus E_T$. In particular, each chord $f \in E$ of an elementary cycle C_e, $e \in E \setminus E_T$, gives rise to an elementary cycle C_f (with $C_f \cap E_T \subsetneq C_e \cap E_T$) itself, and some chords of C_e may also be chords of C_f (this is illustrated in Fig. 1). As opposed to that, the particular chordless cycles in $\mathcal{C}_e$ that *involve* e, formalized as the set $\mathcal{C}_e^e$ in the subsequent Definition 2 uniquely relate to the elementary cycles C_e, $e \in E \setminus E_T$.

Definition 2. *Let $C_e \subseteq E$ be the unique elementary cycle of $G = (V, E)$ w.r.t. the spanning tree $T = (V, E_T)$ and $e \in E \setminus E_T$. Moreover, let $D_e \subseteq E \setminus C_e$ be the set of chords of C_e. Then the set of chordless cycles associated with C_e is the set $\mathcal{C}_e := \{F \subseteq C_e \cup D_e : F \text{ chordless cycle}, F \cap C_e \neq \emptyset\}$. Further, the subset of these cycles containing e is referred to as $\mathcal{C}_e^e := \{F \subseteq C_e \cup D_e : F \text{ chordless cycle}, e \in F \cap C_e\}$.*

---

[1] Indeed, one may choose an arbitrary root $r \in V$, assign it to any of the two partitions, and then determine the partition of every $v \in V \setminus \{r\}$ based on the $\bar{x}_e$ for the edges e on the unique r-v-path in T.

[2] Analogously, (RT) defines a family of integer programs as well, with varying edges and associated variables to be added and different triangle inequality sets based on the choice of the root node $r \in V$. However, the corresponding number of triangle inequalities is always $4 \cdot (|E| - d(r))$ where $d(r)$ is the degree of r in G.

The sets $\mathcal{C}_e^e$ facilitate to generate chordless cycles and associated odd-cycle inequalities only once during a step-wise consideration of elementary cycles. Even more, it turns out that they are sufficient to replace the inequalities (11) of (ST).

A central ingredient to see this is that the odd-cycle inequalities for C_e are satisfied by $x \in \mathbb{R}^E$ as soon as x satisfies the odd-cycle inequalities corresponding to a set of (chordless) cycles whose union contains C_e [3,13,16]. We formalize such cycle sets in the present context as follows.

Definition 3. *Let $\mathcal{C}_e$ be the set of chordless cycles associated with the unique elementary cycle $C_e \subseteq E$ of $G = (V, E)$ w.r.t. the spanning tree $T = (V, E_T)$ and $e \in E \setminus E_T$. Then a subset $\mathcal{C}_e^* \subseteq \mathcal{C}_e$ such that $C_e \subseteq \bigcup_{C \in \mathcal{C}_e^*} C$ is called a* chordless composition *of C_e. Moreover, if $\mathcal{C}_e^*$ is a chordless composition of C_e but $\mathcal{C}_e^* \setminus C$ is not for any $C \in \mathcal{C}_e^*$, then $\mathcal{C}_e^*$ is called* irreducible.

While it is clear from Definition 3 that $\mathcal{C}_e$ is by itself a chordless composition of C_e, it is instructive to observe that one can obtain a (typically smaller) chordless composition of C_e by unifying (and thus restricting to) the sets $\mathcal{C}_f^f$ where either $f = e$ or $f \in D_e$ according to Definition 2. This is due to the aforementioned fact that each chord of C_e defines an elementary cycle - which also needs to be covered by a chordless decomposition - itself. That is, after generating the chordless cycles in $\mathcal{C}_e^e$, we can rely on the sets $\mathcal{C}_f^f$, $f \in D_e$, to gradually and jointly obtain a chordless composition for C_e, as is exemplified in Fig. 1.

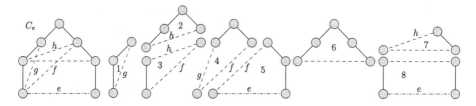

Fig. 1. An elementary cycle C_e (left) with $C_e \cap E_T$ solid, e dash-dotted, and chords D_e dashed, three of which are labeled (f, g, and h). Several chordless cycles, labeled 1-8 and building C_e as of Definition 2, could be extracted from their union. In turn, any composition (union) of these chordless cycles covering all edges of C_e, like e.g. 1-4-5, is referred to as a *chordless composition of C_e* according to Definition 3. To obtain one for C_e, it suffices to consider the union of (even each time a single element of) the *chordless cycle sets* $\mathcal{C}_d^d$ according to Definition 2 where either $d = e$ or $d \in D_e$. For example, for C_e itself one may choose from $\mathcal{C}_e^e$ either the cycle 5 or 8. When choosing cycle 5, one may then rely on a chordless cycle from $\mathcal{C}_f^f$ (either 3 or 4), and then on either cycle 1 (from $\mathcal{C}_g^g$) or 2 (from $\mathcal{C}_h^h$), respectively, to jointly obtain a complete composition for C_e.

The first improved formulation, referred to as (ST_{cl}), is then naturally obtained from (ST) by replacing the odd-cycle inequalities (11) for the elementary cycles C_e, $e \in E \setminus E_T$, with those for the cycles contained in the sets $\mathcal{C}_e^e$.

$$(\text{ST}_{\text{cl}}) \quad \max \sum_{e \in E} c_e x_e$$

$$\text{s.t.} \sum_{e \in S} x_e - \sum_{e \in C \setminus S} x_e \leq |S| - 1 \quad \text{for all } C \in \mathcal{C}_f^f, f \in E \setminus E_T,$$

$$\text{and all } S \subseteq C, |S| \text{ odd} \quad (12)$$

$$0 \leq x_e \leq 1 \qquad \text{for all } e \in E$$

$$x_e \in \{0,1\} \qquad \text{for all } e \in E_T$$

(ST$_{\text{cl}}$) replaces each C_e by possibly multiple chordless cycles $\mathcal{C}_e^e$ which are then however shorter (at most two chordless cycles of length $|C_e| - 1$ involving e are possible, each of which could be substituted for by at most two such cycles of length $|C_e| - 2$, and so on). Due to their binomial nature, (ST$_{\text{cl}}$) has thus (typically strictly) less inequalities than (ST) while its LP relaxation provides a (strictly) better upper bound on the value of a maximum cut in G, as is also demonstrated in Sect. 4.

In practice, given an elementary cycle C_e, $e \in E \setminus E_T$, the set $\mathcal{C}_e^e$ can be derived e.g. by recursive split operations based on the chords of C_e as described in [13] combined with backtracking. Since only cycles containing e are of interest, one may thereby neglect any chord of C_e that is also a chord of a shorter elementary cycle, and truncate the search using the node numbering described below.

At the expense of a weakening of the LP relaxation compared to (ST$_{\text{cl}}$) but still improving over (ST), a further reduction in size is possible, since a chordless composition $\mathcal{C}_e^*$ of each C_e, $e \in E \setminus E_T$, is already obtained by taking only one element (chordless cycle) from each set $\mathcal{C}_f^f$ where either $f = e$ or $f \in D_e$. This way, one is guaranteed to obtain a minimum *total number* of chordless cycles (by the necessity to have one per C_e, $e \in E \setminus E_T$) while these cycles may still, and possibly inevitably, induce a chordless composition for (larger) elementary cycles that is not irreducible in the sense of Definition 3.

Consequently, we obtain another sufficient formulation that we refer to as (ST$_{\text{cl}}^{\text{co}}$) by replacing the odd-cycle inequalities (11) for the elementary cycles C_e, $e \in E \setminus E_T$ in (ST), by the odd-cycle inequalities for exactly one cycle $C \in \mathcal{C}_e^e$.

$$(\text{ST}_{\text{cl}}^{\text{co}}) \quad \max \sum_{e \in E} c_e x_e$$

$$\text{s.t.} \sum_{e \in S} x_e - \sum_{e \in C \setminus S} x_e \leq |S| - 1 \quad \text{for } \textbf{one } C \in \mathcal{C}_f^f, f \in E \setminus E_T,$$

$$\text{and all } S \subseteq C, |S| \text{ odd} \quad (13)$$

$$0 \leq x_e \leq 1 \qquad \text{for all } e \in E$$

$$x_e \in \{0,1\} \qquad \text{for all } e \in E_T$$

In order to determine the desired chordless cycle for each elementary cycle efficiently, no backtracking is required anymore and one may proceed as follows:

For each $e = \{i, j\} \in E \setminus E_T$, consider the unique i-j-path T_e in T that builds C_e with e. Enumerate the nodes of T_e increasingly, starting at i. Then construct a path $R_e \subseteq T_e \cup D_e$, where D_e are the chords of C_e, as follows. Initialize $v \in V(T_e)$ with i. Determine $w \in V(T_e)$ among the nodes which are $(T_e \cup D_e)$-adjacent to v such that w has the maximum index w.r.t. to the created numbering. Add the respective edge to R_e, and continue with replacing v by w unless $w = j$. Finally, create the chordless cycle $C := R_e \cup \{e\} \in \mathcal{C}_e^e$.

3.3 Separation of Odd-Cycle Inequalities for a Given Cycle

Given $x \in \mathbb{R}^E$, $0 \leq x \leq 1$, the general separation problem for the odd-cycle inequalities asks for either an odd-cycle inequality that is violated by x, or for a proof that no such inequality exists. Barahona and Mahjoub [3] showed that it can be solved in polynomial time. Decades later, the algorithm has been refined to produce only facet-defining inequalities in terms of an a posteriori extraction described by Jünger and Mallach [12,13]. Further ideas on such an extraction have been described by Rehfeldt, Koch, and Shinano [18].

Besides that, the special separation problem associated with a *given* cycle $C \subseteq E$ is not well addressed by the general separation approach but of interest in the present context and beyond, e.g., for spin-glass problems [15]. Here, one may exploit that at most one of the odd-cycle inequalities associated with C can be violated at a time. Based on this, given $x \in \mathbb{R}^E$, $0 \leq x \leq 1$, the separation problem w.r.t. C can be solved in linear time as follows [8,15,19]:

1. If $\sum_{e \in C : x_e > \frac{1}{2}} (1 - x_e) + \sum_{e \in C : x_e \leq \frac{1}{2}} x_e \geq 1$, there is no odd-cycle inequality for C violated by x. STOP
2. If $\{e \in C : x_e > \frac{1}{2}\}$ is odd, return the violated inequality. STOP
3. Switch the role of $e^* = \mathrm{argmin}_{e \in C} \left| \frac{1}{2} - x_e \right|$. If violated, return the inequality obtained, otherwise no odd-cycle inequality for C is violated by x.

Thereby, the condition in the first statement is a necessary one for the cycle C to admit a violated odd-cycle inequality. If $\{e \in C : x_e > \frac{1}{2}\}$ is odd, it is also sufficient. Otherwise, the increase of the left-hand side when switching the role of any edge $e \in C$ is precisely $2 \left| \frac{1}{2} - x_e \right|$. Thus, selecting for this purpose any edge that minimizes this increase gives another necessary and sufficient condition [19].

In particular, one may employ this procedure to obtain a cutting plane algorithm for (ST), (ST$_{\mathrm{cl}}$), and (ST$_{\mathrm{cl}}^{\mathrm{co}}$) after pre-computing the corresponding cycles defining (11), (12), and (13), respectively. Moreover, in case of (ST) and (ST$_{\mathrm{cl}}^{\mathrm{co}}$), at most $|E| - |V| + 1$ calls to the separation procedure are necessary per iteration.

4   Computational Experiments

The family of spanning-tree-based MaxCut formulations presented in Sect. 3 naturally relates to the - in terms of odd-cycle inequalities - complete edge-based integer program (E), and the model (RT) that potentially introduces additional variables but restricts to triangle inequalities. None of these formulations suits for the solution of a broader class of (sparse) MaxCut instances "per se", but requires a sophisticated branch-and-cut approach involving at least the dynamic separation of odd-cycle inequalities and further ingredients like e.g. in [7,8,18].

A particularly interesting question in this respect as well as in the focus and context of this paper is how the qualities of the linear programming relaxations of (ST), (ST$_{cl}$), and (ST$_{cl}^{co}$) compare with each other and with those of (E) and (RT). That is, we strive to quantify the impact of the respective odd-cycle inequality selection in terms of the obtained upper bounds on the value of a maximum cut, and to relate this quality to the actual size of the respective linear program.

To carry out such an experiment, a reasonable choice needs to be made for the several degrees of freedom provided by the different formulations. To this end, we adopt the rule from [8] to choose a node of maximum degree as the root node for (RT), and we sort the edges non-increasingly w.r.t. their absolute weight in order to iteratively select them if suitable to build a spanning tree for (ST), (ST$_{cl}$) or (ST$_{cl}^{co}$). Finally, for (ST$_{cl}^{co}$), we compute the chordless compositions exactly by means of the procedure described in Sect. 3.2.

For these representatives of each formulation family, the experiments are carried out with the following established instances from the Biq Mac Library [5]:

- g05_60, 80, 100: Graphs with 60, 80, or 100 nodes, unit edge weights.
- pm1d_80, 100, pm1s_100: Graphs with 80 or 100 nodes, ± 1 edge weights.
- pw01_100, pw05_100, pw09_100: Graphs with 100 nodes, and integer edge weights from $[1, 10]$.
- be200.8: Graphs corresponding to the transformation of UBQP instances with $n = 200$ generated by Billionnet and Elloumi [4] where the entries of the original cost matrices are integer from $[-100, 100]$ for the diagonal entries and $[-50, 50]$ for off-diagonal entries, and their non-zero density is about 0.8.

Each set consists of ten instances. Their selection is based on combining different sizes, densities, and edge weight ranges, among graphs with known and varying maximum chordless cycle length, denoted as $\max |C_{cl}|$ in Table 1. We remark that the latter is only an upper bound on the maximal chordless cycle length observed when constructing (ST$_{cl}$) or (ST$_{cl}^{co}$). The first three sets comprise native MaxCut instances generated as Erdös-Renyi-Gilbert random graphs [5]. We only include one set of native UBQP instances since their transformation to MaxCut (involving a node adjacent to all others via edges of relatively high absolute weight) leads to the situation that many of the edges forming the spanning trees of (RT), and (ST), (ST$_{cl}$), or (ST$_{cl}^{co}$), coincide.

The results in Table 1 demonstrate well the trade-offs arising between the (chosen representatives) of the different formulation families. The formulation (RT) and the largest formulation (ST), which however typically comprises a large fraction of odd-cycle inequalities that are not facet-defining, provide the weakest LP relaxations, with an upper bound often only slightly below the sum of positive edge weights. Constructing instead all facet-defining odd-cycle inequalities involving the non-tree edge closing an elementary cycle lets (ST$_{cl}$) admit the strongest LP relaxation while its size is considerably reduced over (ST) but

Table 1. Model Size and LP Relaxation Comparison: For each instance set, the rows show minimum, mean, and maximum values, respectively in this order, rounded to two decimal digits. The density (column "Dsty") shown is defined as $|E|/\binom{|V|}{2}$. The upper bounds provided by the LP relaxations of (ST), (ST$_{cl}$), (ST$_{cl}^{co}$), and (RT), as well as the sum of positive edge weights (+W), are shown as a factor in terms of the bound provided by P_{OC}. The number of constraints of each of the formulations is displayed as a factor of the number of edges of the instances (Constraint-to-Edge Ratios).

Group	$|V|$	Dsty	max $|C_{cl}|$	(ST)	(ST$_{cl}$)	(ST$_{cl}^{co}$)	(RT)	+W	(ST)	(ST$_{cl}$)	(ST$_{cl}^{co}$)	(RT)	$\frac{|E'|}{|E|}$ (RT)
g05	60	0.50	11	1.45	1.27	1.31	1.46	1.50	22.63	7.44	4.85	3.81	1.02
g05	60	0.50	11	1.46	1.31	1.34	1.47	1.50	78.18	11.33	5.12	3.83	1.02
g05	60	0.50	11	1.46	1.37	1.39	1.47	1.50	234.22	13.76	5.52	3.84	1.03
g05	80	0.50	12	1.47	1.32	1.34	1.47	1.50	44.44	10.14	4.79	3.87	1.02
g05	80	0.50	12	1.47	1.35	1.37	1.48	1.50	86.51	12.07	5.11	3.87	1.02
g05	80	0.50	12	1.47	1.37	1.39	1.48	1.50	221.88	14.94	5.32	3.88	1.02
g05	100	0.50	13	1.47	1.36	1.38	1.48	1.50	32.48	9.28	4.96	3.90	1.01
g05	100	0.50	13	1.47	1.37	1.39	1.48	1.50	105.67	11.85	5.33	3.90	1.01
g05	100	0.50	13	1.47	1.39	1.40	1.48	1.50	506.03	17.46	5.62	3.90	1.02
pm1d	80	0.99	4	2.80	1.85	2.52	2.80	2.87	13.93	8.82	3.90	3.90	1.00
pm1d	80	0.99	4	2.94	1.94	2.66	2.93	3.01	23.36	10.13	3.90	3.90	1.00
pm1d	80	0.99	4	3.14	2.10	2.87	3.14	3.23	65.73	12.37	3.90	3.90	1.00
pm1d	100	0.99	4	2.79	1.89	2.51	2.79	2.85	26.00	10.98	3.92	3.92	1.00
pm1d	100	0.99	4	2.91	1.98	2.62	2.91	2.97	46.12	12.79	3.92	3.92	1.00
pm1d	100	0.99	4	3.02	2.09	2.71	3.01	3.08	107.03	14.53	3.92	3.92	1.00
pm1s	100	0.10	37	1.76	1.43	1.50	1.83	1.90	541.63	54.87	30.88	3.81	1.15
pm1s	100	0.10	38.8	1.81	1.51	1.57	1.88	1.95	1167.65	69.04	46.06	3.85	1.16
pm1s	100	0.10	40	1.88	1.58	1.64	1.98	2.05	2179.39	103.19	65.23	3.88	1.17
pw01	100	0.10	37	1.25	1.11	1.14	1.28	1.32	3063.08	57.96	22.12	3.81	1.15
pw01	100	0.10	38.8	1.25	1.12	1.15	1.29	1.32	14764.12	113.81	41.99	3.85	1.16
pw01	100	0.10	40	1.26	1.13	1.16	1.30	1.33	289171.06	670.26	168.09	3.88	1.17
pw05	100	0.50	12	1.46	1.21	1.27	1.48	1.50	496.27	19.81	4.91	3.90	1.01
pw05	100	0.50	12.9	1.46	1.25	1.28	1.48	1.50	3288.65	26.12	5.15	3.90	1.02
pw05	100	0.50	13	1.46	1.28	1.30	1.48	1.50	21141.55	36.76	5.61	3.90	1.02
pw09	100	0.90	6	1.47	1.27	1.32	1.48	1.50	395.50	18.79	3.92	3.91	1.00
pw09	100	0.90	6	1.47	1.29	1.34	1.48	1.50	1499.83	22.14	3.94	3.91	1.00
pw09	100	0.90	6	1.47	1.32	1.35	1.49	1.50	6832.52	25.62	3.96	3.92	1.00
be200.8	201	0.79	9	2.56	2.28	2.42	2.56	2.93	4.67	4.49	3.95	3.95	1.00
be200.8	201	0.79	9	2.61	2.34	2.47	2.61	3.00	5.04	4.69	3.95	3.95	1.00
be200.8	201	0.79	9	2.64	2.37	2.51	2.64	3.06	5.53	4.93	3.96	3.95	1.00

often still rather large, especially for the sparse instance sets. The quality gap of the relaxation of (ST$_{cl}$) compared to the one of (E), i.e., P_{OC}, can still be significant. Finally, (ST$_{cl}^{co}$) provides an effective reduction to only one chordless cycle per non-tree edge. Its number of odd-cycle inequalities typically only slightly exceeds the one of (RT) for the dense instances while providing a significantly better upper bound. At the same time, $(\mathrm{ST}^{\mathrm{co}}_{\mathrm{cl}})$ can still become quite large (pm1s, pw01) and lose strength (pm1d) compared to $(\mathrm{ST}_{\mathrm{cl}})$. Fortunately, for all the other instances, the size of $(\mathrm{ST}^{\mathrm{co}}_{\mathrm{cl}})$ as well as the loss in terms of the upper bound compared to $(\mathrm{ST}_{\mathrm{cl}})$ is moderate.

5   Conclusion

We have presented families of spanning-tree formulations for the Maximum Cut Problem that encode incidence vectors of cuts via a subset of the odd-cycle inequalities associated with a connected graph $G = (V, E)$, and that require only $|V| - 1$ edge variables to be integral explicitly. Further, two variants have been described which are reduced in size and consist of facet-defining inequalities only. In an experimental study, it has been shown that for rather dense problem instances, one may obtain a formulation that moderately exceeds the size of a common integer program comprising only triangle inequalities while providing a stronger linear programming relaxation. For sparse instances, respectively for graphs with long chordless cycles, even the reduced spanning-tree formulations can become quite large. Generally, the experimental results show an interesting trade-off in terms of size and relaxation quality which motivates further research regarding the relevance of certain subsets of the odd-cycle inequalities. Moreover, they underline the importance of selecting facet-defining inequalities to obtain a significantly better and more compact relaxation at the same time.

References

1. Barahona, F., Grötschel, M., Jünger, M., Reinelt, G.: An application of combinatorial optimization to statistical physics and circuit layout design. Oper. Res. **36**(3), 493–513 (1988)
2. Barahona, F., Jünger, M., Reinelt, G.: Experiments in quadratic 0–1 programming. Math. Program. **44**(1), 127–137 (1989)
3. Barahona, F., Mahjoub, A.R.: On the cut polytope. Math. Program. **36**(2), 157–173 (1986)
4. Billionnet, A., Elloumi, S.: Using a mixed integer quadratic programming solver for the unconstrained quadratic 0–1 problem. Math. Program. **109**(1), 55–68 (2007)
5. Biq Mac Library – Binary quadratic and Max cut Library (2009). https://biqmac.aau.at/biqmaclib.html
6. Bonato, T., Jünger, M., Reinelt, G., Rinaldi, G.: Lifting and separation procedures for the cut polytope. Math. Program. A **146**, 351–378 (2013)
7. Charfreitag, J., Jünger, M., Mallach, S., Mutzel, P.: McSparse: exact solutions of sparse maximum cut and sparse unconstrained binary quadratic optimization problems. In: Phillips, C.A., Speckmann, B. (eds.) 2022 Proceedings of the Symposium on Algorithm Engineering and Experiments (ALENEX), pp. 54–66. SIAM (2022)
8. Charfreitag, J., Mallach, S., Mutzel, P.: Integer programming for the maximum cut problem: a refined model and implications for branching. In: Berry, J., Shmoys, D.B. (eds.) Proceedings of the 2023 SIAM Conference on Applied and Computational Discrete Algorithms (ACDA23), pp. 63–74 (2023)

9. De Simone, C.: The cut polytope and the Boolean quadric polytope. Disc. Math. **79**(1), 71–75 (1990)
10. de Sousa, S., Haxhimusa, Y., Kropatsch, W.G.: Estimation of distribution algorithm for the max-cut problem. In: Kropatsch, W.G., Artner, N.M., Haxhimusa, Y., Jiang, X. (eds.) GbRPR 2013. LNCS, vol. 7877, pp. 244–253. Springer, Heidelberg (2013)
11. Hammer, P.L.: Some network flow problems solved with pseudo-Boolean programming. Oper. Res. **13**(3), 388–399 (1965)
12. Jünger, M., Mallach, S.: Odd-cycle separation for maximum cut and binary quadratic optimization. In: Bender, M.A., Svensson, O., Herman, G. (eds.) 27th Annual European Symposium on Algorithms (ESA 2019), vol. 144 of Leibniz International Proceedings in Informatics (LIPIcs), Dagstuhl, Germany, pp. 63:1–63:13 (2019)
13. Jünger, M., Mallach, S.: Exact facial odd-cycle separation for maximum cut and binary quadratic optimization. INFORMS J. Comput. **33**(4), 1419–1430 (2021)
14. Karp, R.M.: Reducibility among combinatorial problems. In: Miller, R.E., Thatcher, J.W. (eds.) Proceedings of a Symposium on the Complexity of Computer Computations, New York, The IBM Research Symposia Series, pp. 85–103. Plenum Press, New York (1972)
15. Liers, F.: Contributions to determining exact ground-states of Ising spin-glasses and to their physics. PhD thesis, Universität zu Köln (2004)
16. Nguyen, V.H., Minoux, M.: Linear size MIP formulation of max-cut: new properties, links with cycle inequalities and computational results. Optim. Lett. **15**(4), 1041–1060 (2021)
17. Padberg, M.: The Boolean quadric polytope: some characteristics, facets and relatives. Math. Program. **45**(1), 139–172 (1989)
18. Rehfeldt, D., Koch, T., Shinano, Y.: Faster exact solution of sparse maxCut and QUBO problems. Math. Program. Comput. **15**, 445–470 (2023)
19. Zhang, X., Siegel, P.H.: Adaptive cut generation algorithm for improved linear programming decoding of binary linear codes. IEEE Trans. Inf. Theory **58**(10), 6581–6594 (2012)

Optimal Cycle Selections: An Experimental Assessment of Integer Programming Formulations

Marie Baratto[✉] and Yves Crama

QuantOM, HEC Management School of the University of Liege,
Rue Louvrex 14, 4000 Liège, Belgium
{marie.baratto,yves.crama}@uliege.be

Abstract. In this paper, we conduct numerical experiments to test the effectiveness of several integer programming formulations of the cycle selection problem. Specifically, we carry out experiments to identify a maximum weighted cycle selection in random or in structured digraphs. The results show that random instances are relatively easy and that two formulations outperform the other ones in terms of total running time. We also examine variants of the problem obtained by adding a budget constraint and/or a maximum cycle length constraint. These variants are more challenging, especially when a budget constraint is imposed. To investigate the cycle selection problem with a maximum cycle length equal to 3, we provide an arc-based formulation with an exponential number of constraints that can be separated in polynomial time. All inequalities in the formulation are facet-defining for complete digraphs.

Keywords: Cycle selections · Integer programming · Kidney exchanges

1 Introduction

The notion of *cycle selection* has been introduced in [2]: for a (loopless) digraph $G = (V, A)$, where V is the set of vertices and A is the set of arcs of G, a subset of arcs $B \subseteq A$ is a *cycle selection* if the arcs of B form a union (possibly empty) of directed cycles. As an illustration, Fig. 1 displays a digraph and the collection of its nonempty cycle selections. As explained in [2], cycle selections arise in connection with stochastic versions of *kidney exchange problems*: they model those subsets of arcs which can potentially support a cycle packing associated with a feasible kidney exchange; we refer to [1,2,6] for details. Contrary to cycle packings, however, cycle selections have not been much studied in the literature.

When a weight $w_{i,j} \in \mathbb{R}$ is assigned to each arc $(i, j) \in A$, the *maximum weighted cycle selection problem* (**MWCS**) is to identify a cycle selection B which maximizes $w(B) = \sum_{(i,j) \in B} w_{i,j}$. If all arc weights are non-negative, a maximum weighted cycle selection simply consists of all arcs contained in strongly connected components of the digraph G. Therefore, in this case, **MWCS** is easily solvable in linear time [7]. For general weights, however, **MWCS** was proved in [2] to be NP-hard.

The goal of this paper is to numerically assess the respective effectiveness of different integer linear programming (ILP) formulations of the cycle selection problem. We also investigate, both theoretically and experimentally, some variants of the problem obtained by adding a budget constraint and/or a maximum cycle length constraint to the models.

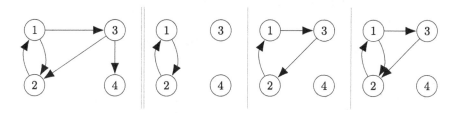

Fig. 1. A directed graph (left) and all its nonempty cycle selections (right)

2 Formulations and Implementations

Six integer linear programming formulations of **MWCS** have been introduced in [2], namely: the arc (**ARC**) formulation, the simple extended arc (**SEA**) formulation, the extended arc (**EA**) formulation, the modified extended arc (**MEA**) formulation, the position-indexed (**PI**) formulation, and the cycle (**CY**) formulation. Let us briefly recall all six formulations.

We introduce the arc variables $\beta_{i,j} \in \{0,1\}$, where $\beta_{i,j} = 1$ if arc (i,j) is selected and $\beta_{i,j} = 0$ otherwise, for all $(i,j) \in A$. Then, the **ARC** formulation is as follows: $\beta \in \{0,1\}^{|A|}$ defines a cycle selection if and only if it satisfies the *return inequalities*:

$$\beta_{i,j} \leq \sum_{(l,k)\in A:l\in V\setminus S, k\in S} \beta_{l,k} \qquad \forall(i,j) \in A, \forall S \subseteq V : i \in S, j \in V \setminus S. \qquad (1)$$

The **SEA** formulation is as follows: vector $\beta \in \{0,1\}^{|A|}$ defines a selection if and only there exists a *circulation* $x \in \mathbb{R}_+^{|A|}$ such that

$$x_{i,j} \leq m\,\beta_{i,j} \qquad\qquad\qquad \forall(i,j) \in A \qquad (2)$$

$$\beta_{i,j} \leq x_{i,j} \qquad\qquad\qquad \forall(i,j) \in A \qquad (3)$$

$$\sum_{h:(h,k)\in A} x_{h,k} = \sum_{h:(k,h)\in A} x_{k,h} \qquad\qquad \forall k \in V. \qquad (4)$$

The **EA** and **MEA** formulations rely on expressing that each arc (u,v) of a cycle selection must be contained in the support $C^{(u,v)}$ of a representative binary circulation $x^{(u,v)}$. We define $x_{i,j}^{(u,v)} \in \{0,1\}$ where $x_{i,j}^{(u,v)} = 1$ if $(i,j) \in C^{(u,v)}$ and

$x_{i,j}^{(u,v)} = 0$ otherwise for all $(i,j) \in A$ and for all $(u,v) \in A$. We identify $x_{i,j}^{(i,j)}$ with $\beta_{i,j}$. The **EA** formulation is as follows:

$$x_{i,j}^{(u,v)} \leq x_{i,j}^{(i,j)} \qquad\qquad \forall(i,j) \in A, \forall(u,v) \in A \qquad (5)$$

$$\sum_{h:(h,k)\in A} x_{h,k}^{(u,v)} = \sum_{h:(k,h)\in A} x_{k,h}^{(u,v)} \qquad\qquad \forall k \in V, \forall(u,v) \in A \qquad (6)$$

$$x_{i,j}^{(u,v)} \in \{0,1\} \qquad\qquad \forall(i,j) \in A, \forall(u,v) \in A. \qquad (7)$$

We can also add to **EA** the following constraints which impose that the support of each binary circulation $x^{(u,v)}$ must consist of vertex-disjoint cycles:

$$\sum_{h:(k,h)\in A} x_{k,h}^{(u,v)} \leq 1 \qquad\qquad \forall k \in V, \forall(u,v) \in A. \qquad (8)$$

Constraints (5)–(8) define the **MEA** formulation for cycle selections.

Additionally to the **SEA**, **EA** and **MEA**, another extended formulation has been proposed in [2], inspired by a formulation in [6]. Assuming that the vertex-set of the digraph $G = (V, A)$ is $V = \{1, \ldots, n\}$, let V^l be the subset of vertices $\{l, \ldots, n\}$, for each l in V. Given binary values for the arc variables $\beta_{i,j}$, define $B^l = \{(i,j) \in A : i \in V^l, j \in V^l, \beta_{i,j} = 1\}$. We introduce the position-indexed binary variables:

$$\phi_{i,j,k}^l \text{ for all } (i,j) \in A, l \in V, k \in \kappa(i,j,l) \text{ where } \kappa(i,j,l) = \begin{cases} \{1\} \text{ if } i = l \\ \{2, \ldots, n\} \text{ if } j = l \\ \{2, \ldots, n-1\} \text{ if } i, j > l \end{cases}$$

where $\phi_{i,j,k}^l = 1$ if arc (i,j) is in position k in a cycle of the digraph (V^l, B^l) containing vertex l, and 0 otherwise. The **PI** formulation is as follows :

$$\beta_{i,j} \leq \sum_{l\in V} \sum_{k\in\kappa(i,j,l)} \phi_{i,j,k}^l \qquad\qquad \forall(i,j) \in A \qquad (9)$$

$$\phi_{i,j,k}^l \leq \beta_{i,j} \qquad\qquad \forall l \in V, (i,j) \in A^l, k \in \kappa(i,j,l) \qquad (10)$$

$$\phi_{i,j,k}^l \leq \sum_{h:(h,i)\in A^l \wedge k-1\in\kappa(h,i,l)} \phi_{h,i,k-1}^l \qquad \forall l \in V, (i,j) \in A^l, k \in \kappa(i,j,l), k > 1 \qquad (11)$$

$$\phi_{i,j,k}^l \leq \sum_{h:(j,h)\in A^l \wedge k+1\in\kappa(j,h,l)} \phi_{j,h,k+1}^l \qquad \forall l \in V, (i,j) \in A^l, j \neq l, k \in \kappa(i,j,l) \qquad (12)$$

$$\phi_{i,j,k}^l \in \{0,1\} \qquad\qquad \forall l \in V, (i,j) \in A^l, k \in \kappa(i,j,l) \qquad (13)$$

$$\beta_{i,j} \in \{0,1\} \qquad\qquad \forall(i,j) \in A \qquad (14)$$

Finally, let Γ_G (or simply, Γ) denote the set of all directed cycles of the digraph G. We introduce the cycle variables z_C, $C \in \Gamma$, where $z_C = 1$ if cycle C is contained in the selection, and 0 otherwise. Then, the **CY** formulation is:

$$z_C \leq \beta_{i,j} \qquad\qquad \forall C \in \Gamma, \forall(i,j) \in C \qquad (15)$$

$$\beta_{i,j} \leq \sum_{C\in\Gamma:(i,j)\in C} z_C \qquad\qquad \forall(i,j) \in A \qquad (16)$$

$$z_C \in \{0,1\} \qquad\qquad \forall C \in \Gamma \qquad (17)$$

$$\beta_{i,j} \in \{0,1\} \qquad\qquad \forall(i,j) \in A. \qquad (18)$$

The relative strength of the above formulations has been compared from a theoretical perspective in [2]: **ARC**, **EA** and **MEA** are the tightest formulations in the sense that their linear relaxations always have the same optimal value, which is never worse than the optimal value of the relaxations of **SEA**, **PI** or **CY**. Our first goal in this paper is to experimentally compare the effectiveness of the formulations.

Note that the formulations **SEA**, **EA**, **MEA**, and **PI** are compact (i.e., of polynomial size), whereas the **ARC** formulation has an exponential number of constraints (1) and the **CY** formulation has an exponential number of variables (one for each cycle of G). To use the **CY** formulation, either a branch-and-price process must be implemented to generate the variables gradually, or all the cycles of the digraph G must be initially generated. We implemented the formulation by generating all cycles. Given that the maximum cycle length is not limited in **MWCS**, generating the models took over an hour in our preliminary tests, even for the smallest instances. Therefore, in this work, we decided not to pursue the study of the **CY** formulation. Note, however, that a more advanced branch-and-price implementation may deliver better results. The results for the **PI** formulation are not displayed either, as the running times are also too large compared to the other formulations in spite of the compactness of **PI**. On the other hand, when the maximum length of cycles is limited, the **PI** and **CY** formulations become more efficient as illustrated in Sect. 5 when the maximum cycle length is bounded by 2 or 3.

In order to test the **ARC** formulation, a branch-and-cut procedure has been implemented. Because of the exponential number of return inequalities (1), it is not advisable to include them all in the model given to the solver. The idea is instead to include part of these inequalities in the initial model and to gradually add return inequalities to the model through a separation procedure. We chose to only include the *predecessor inequalities* and the *successor inequalities* in the initial formulation. These inequalities are the special cases of the return inequalities (1) obtained by setting $S = \{i\}$ or $S = V \setminus \{j\}$, respectively, for each arc (i, j). Thus, the initial ILP model P_0 passed to the solver is:

$$\max \sum_{(i,j) \in A} w_{i,j} \beta_{i,j} \tag{19}$$

$$\text{subject to} \quad \beta_{i,j} \leq \sum_{(k,i) \in A} \beta_{k,i} \qquad \forall (i,j) \in A \tag{20}$$

$$\beta_{i,j} \leq \sum_{(j,k) \in A} \beta_{j,k} \qquad \forall (i,j) \in A \tag{21}$$

$$\beta_{i,j} \in \{0, 1\} \qquad \forall (i,j) \in A. \tag{22}$$

When the ILP solver performing a branch-and-bound procedure finds a feasible solution of P_0, it is necessary to check whether this solution indeed defines a cycle selection. A separation procedure must be run in order to identify some return inequalities violated by the current solution (if any) and to add them to the model. In our implementation, the separation procedure is implemented in the following way: each time an integer solution β' is found during the classical branch-and-bound process, Tarjan's algorithm [7] is called on the digraph $G_{B'} = (V, B')$, where $B' = \{(i, j) \in A : \beta'_{i,j} = 1\}$, in order to identify its strongly connected components. Let us say that an arc (i, j) is a *link* of $G_{B'}$ if i and j are in different strongly connected components of $G_{B'}$. Clearly, if B' is a selection, then $G_{B'}$ cannot have any link. So, for each link (i, j), we add the following return inequality to the model:

$$\beta_{i,j} \leq \sum_{(l,k)\in A: l\in V\setminus S, k\in S} \beta_{l,k}, \qquad (23)$$

where S is the set of vertices $l \in V$ such that there does not exist a path from j to l in $G_{B'} = (V, B')$. Note that $i \in S$, $j \in V \setminus S$, and the inequality (23) is violated by the solution β' since (i, j) is a link of $G_{B'}$.

3   Experimental Results

3.1   Random Instances

We first compare the ILP formulations mentioned above by solving **MWCS** on random weighted digraphs. Given three parameters $n \in \mathbb{N}$, $p \in [0, 100]$, $d \in [0, 100]$, we generate random digraphs $G = (V, A)$ with weights $w_{i,j}$, $(i, j) \in A$, such that: $|V| = n$; for each ordered pair of distinct vertices (i, j), $(i, j) \in A$ with probability $\frac{p}{100}$ independently of the other arcs; for each arc $(i, j) \in A$, $w_{i,j}$ is uniformly distributed in $[0, 1]$ with probability $\frac{d}{100}$, and $w_{i,j}$ is uniformly distributed in $[-1, 0]$ with probability $1 - \frac{d}{100}$, independently of the other arcs.

We have conducted computational experiments for various values of n in $\{50, 100, 150, 200, 250, 300\}$, p in $\{10, 20, 40, 90\}$, and d in $\{10, 80, 90\}$ as illustrated in several figures and tables throughout the paper. For each configuration of n, p and d, we have generated 30 random instances. For a given (n, p, d) configuration, the same instances are used across all the tests performed.

All numerical experiments have been implemented using C++ 14 as programming language and CPLEX 12.10.0 as generic MILP solver. The tests have been performed on a Dell Latitude 7490 running Windows 10 64Bit with an Intel Core i5-7300U CPU and two cores at 2.60GHz and 16 GB of RAM.

In terms of total running time, that is, the time needed to construct the model and to solve it, the **ARC** formulation and the **SEA** formulation clearly outperform the other formulations, as illustrated by the performance profiles in Fig. 2 for the parameter values $n = 50$, $p = 20$, and $d = 80$. Indeed, all of the 30 instances generated with these parameters are solved in less than two seconds when using the formulations **ARC** and **SEA**. On the other hand, with either the **EA** or the **MEA** formulation, the fastest running time is around 10 s, and some instances require up to 22 s (in spite of the fact that **ARC**, **EA** and **MEA** yield the same linear relaxation bound, which can be better than the bound provided by **SEA**). The results for the **PI** formulation are not displayed in the

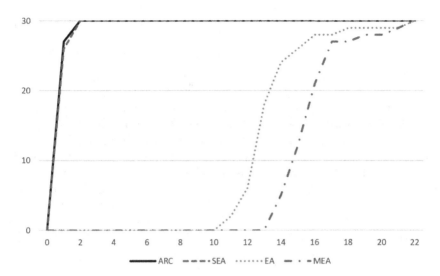

Fig. 2. Comparison of total running time of the **MWCS** formulations when $n = 50$, $p = 20$ $d = 80$. The horizontal axis displays running times (in seconds), and the vertical axis displays the number of instances solved within a given time, for each formulation.

figure; with this formulation, the fastest running time for the same instances is 156 s, and the longest one is 667 s.

Beyond the differences in running times, we observed that for all formulations and for all the instances tested, the solver always immediately finds a maximum weighted selection. By "immediately", we mean that the linear relaxation of each model under consideration has a binary optimal solution which defines a maximum cycle selection. This observation even holds for the linear relaxation of the incomplete **ARC** formulation P_0, which does not include all return inequalities. More precisely, for all the instances handled with the **ARC** formulation, the solution process goes as follows: At the root node, the solver solves the linear relaxation of P_0. It finds an integer solution and runs the callback procedure to verify that this solution defines a selection. This is always the case, so that no cut is ever added and the optimization process stops without any branching. For all these instances, we further observed that the average optimal value is very close to the expected total weight of the positive arcs, that is, $\frac{n\,(n-1)}{2}\,\frac{p}{100}\,\frac{d}{100}$.

These seemingly intriguing observations can actually be explained by theoretical properties of random graphs. To this effect, consider a random digraph $G = (V, A)$ where $|V| = n$ and each ordered pair of distinct vertices (i, j) is in A with probability $u(n)$ independently of the other pairs. A result of [5], which extends a result of [3] regarding undirected graphs, implies that when $u(n)$ is "asymptotically larger" than $\frac{\ln n}{n}$, the probability that G is strongly connected tends to 1 as n tends to infinity. (Here, "asymptotically larger" means that $\lim_{n\to\infty} \frac{\ln n}{n\,u(n)} = 0$. We refer to [5] for details.)

For our random instances, we can apply this result to the random digraph $G' = (V, A')$, where A' is the subset of arcs with positive weight; so here,

$$u(n) = \frac{p}{100} \frac{d}{100}.$$

We obtain that for fixed p, d, and for n large enough, G' is almost surely strongly connected. This means, in particular, that the set A' of positive arcs is (almost surely) a cycle selection, in which case it is necessarily optimal for **MWCS**.

These observations intuitively suggest that in order to generate non trivial instances, we should take values of p and d small enough with respect to the Erdős-Rényi threshold $\frac{\ln n}{n}$, which leads to very sparse digraphs. With this in mind, we have further generated random digraphs with parameters $p = 10$ and $d = 10$, so that $u(n) = 0.01$ is smaller than $\frac{\ln n}{n}$ when $n \leq 300$. **MWCS** has been solved on these digraphs using the **ARC** and the **SEA** formulations. For these sparse instances, the optimization process is not as straightforward as for the earlier ones. Some violated return inequalities are added to the model for the **ARC** formulation, and some branching occurs for the **SEA** formulation, especially for the smaller instances with $n \in \{50, 100\}$. Still, the process remains really fast: all 180 instances are solved in less than 3 s with the **SEA** formulation; as for the **ARC** formulation, 173 instances are solved in less than 3 s, and the seven remaining ones in less than 22 s.

3.2   Instances Associated with Steiner Triples

In [2], **MWCS** is proved to be strongly NP-hard by a reduction from the hitting set problem. Based on this proof, we have generated instances of **MWCS** by applying the same reduction to well-known hard instances of the hitting set problem, namely, the so-called Steiner triple instances [4].

For the sake of completeness, let us describe the reduction. A Steiner triple instance of the hitting set problem is defined by a finite set X and a collection $T = \{T_1, \ldots, T_r\}$ of subsets of X, with $r = \frac{1}{6}|X|(|X| - 1)$ and $|T_j| = 3$ for $j = 1, \ldots, r$. The set T has a special (Steiner) structure which is not of direct interest here. Given such an instance we construct an associated weighted digraph $G = (V, A)$ as follows: $V = X \cup T \cup \{0\}$, and

- for all $j = 1, \ldots, r$, $(T_j, 0) \in A$ with weight $w(T_j, 0) = r$,
- for all $i \in X$, $(0, i) \in A$ with weight $w(0, i) = -1$,
- for all $j = 1, \ldots, r$ and for all $i \in T_j$, $(i, T_j) \in A$ with weight $w(i, T_j) = 0$.

As follows from the complexity proof of **MWCS** in [2], the optimal size of a hitting set of T is h^{opt} if and only if the maximum weight of an optimal selection of G is $w^{opt} = r^2 - h^{opt}$. Note that the resulting instances of **MWCS** are quite sparse: they contain $n = |X| + r + 1$ vertices and $|X| + 4r \leq 4n$ arcs.

We tried to solve these instances of **MWCS** using the formulations **ARC** and **SEA** for different sizes of X. Instances with $|X|$ equal to 9, 15, 27, 45, 81, 135, 243 are from the online OR Library of J.E. Beasley (http://people.brunel.

ac.uk/~mastjjb/jeb/info.html). Additional instances with $|X|$ equal to 49 and 55 have been additionally generated based on https://www.dmgordon.org/cover.

For values of $|X|$ up to 55, CPLEX was able to terminate in less than three hours of computing time. Table 1 provides information about the size of these Steiner instances and the associated digraphs: size of X, number r of triples, number of vertices $|V|$ and number of arcs $|A|$. The middle section of the table displays the size h^{opt} of a minimum hitting set, as well as the optimal value w^{opt} of the maximum weighted cycle selection returned by CPLEX. The right part of the table displays the total running time t (in seconds) and the number of nodes explored by the solver for each of the **ARC** and **SEA** formulations.

Table 1. Steiner triple instances (part 1)

$|X|$	r	$|V|$	$|A|$	h^{opt}	w^{opt}	t_{ARC}	Nodes$_{ARC}$	t_{SEA}	Nodes$_{SEA}$
9	12	22	57	5	139	1	0	0	0
15	35	51	155	9	1216	0	83	1	256
27	117	145	495	18	13671	3	4634	2	3287
45	330	376	1365	30	108870	109	68490	627	278419
49	392	442	1617	32	153632	160	74043	1022	220053
51	425	477	1751	34	180591	401	147400	5351	533609
55	495	551	2035	37	244988	823	309300	9513	998079

The **ARC** formulation generally has a shorter total running time than the **SEA** formulation, except for a few instances where the opposite is observed. As expected, the instances derived from the Steiner triple instances are more challenging than the random ones, and many nodes are explored in the branch-and-bound tree. Note that no separation cuts were added during the resolution process for the **ARC** formulation, for any of the instances.

Three larger instances with $|X|$ in $\{81, 135, 243\}$ have also been tested. The time limit of three hours was reached with both formulations before the solver could find an optimal solution. However, as shown in Table 2, the solver was able to produce feasible solutions that are very close to optimality using both formulations. In fact, the **ARC** formulation quickly provided the optimal solution for the instance with $|X| = 81$, although the solver could not prove optimality within the given time limit.

Table 2. Steiner triple instances (part 2)

$|X|$	r	$|V|$	$|A|$	h^{opt}	w^{opt}	w^{best}_{ARC}	w^{best}_{SEA}
81	1080	1162	4401	61	1166339	1166339	1166338
135	3015	3151	12195	103	9090122	9090119	9090118
243	9801	10045	39447	198	96059403	96059392	96059392

4   MWCS with budget constraint

In some applications of **MWCS** (see, e.g., [6]), it makes sense to impose a *budget constraint* on the cardinality of a cycle selection, that is, a constraint of the form:

$$\sum_{(i,j)\in A} \beta_{i,j} \le b,$$

where $b \in \mathbb{N}$. To assess the difficulty of the resulting problem, we have used the same random instances as in Sect. 3.1 with various values for the budget b. As in Sect. 3.1, it can be shown that if b is too large, then the set of b arcs with the largest positive weights is almost surely strongly connected and hence defines a cycle selection (that is, the optimal solution is obtained by setting $\beta_{i,j} = 1$ for those b arcs with the highest positive weight and $\beta_{i,j} = 0$ otherwise). Such instances are excessively easy to solve. Therefore, as a rule of thumb, we chose budget values such that $\frac{b}{n(n-1)}$ is small with respect to the Erdős-Rényi threshold $\frac{\ln n}{n}$.

Table 3. MWCS with budget constraint: results of computational experiments

ARC		time			nodes			cuts		I
$n{=}200$	av	min	max	av	min	max	av	min	max	
$p = 20, d = 80, b = 100$	73.06	2	TL	559.00	0	7115	3398.90	0	44358	3
$p = 40, d = 90, b = 100$	138.1	6	TL	285.53	0	2304	1524.77	0	15105	3
$p = 20, d = 80, b = 50$	39.93	2	592	371.07	0	7322	417.23	0	6397	4
$p = 40, d = 90, b = 50$	73.03	5	TL	739.53	0	18457	744.47	0	16392	8
SEA		time			nodes					I
$n{=}200$	av	min	max	av	min	max				
$p = 20, d = 80, b = 100$	27.87	6	62	1327.53	0	4752				3
$p = 40, d = 90, b = 100$	84.43	25	303	1620.37	0	6387				7
$p = 20, d = 80, b = 50$	39.63	13	85	2342.83	0	11521				0
$p = 40, d = 90, b = 50$	59.63	30	131	1321.63	0	4994				9

Table 3 illustrates some of the results of our computational experiments on **MWCS** with a budget constraint. The top part of the table refers to the **ARC** formulation and the bottom part to the **SEA** formulation. Each line refers to a fixed configuration of parameter values n, p, d, b, and to the solution of 30 random instances generated with these parameters. Each line displays:

⋄ The average, minimum and maximum total running time (in seconds) over 30 instances. A time limit of 10 min was given to the solver. If at least one instance reached the time limit, then the maximum total running time is indicated as TL.

⋄ The average, minimum and maximum number of nodes explored during the solution process.

⋄ For the **ARC** formulation only, the average, minimum and maximum number of separation cuts added during the solution process.

⋄ The number I of instances for which the optimal solution of the linear relaxation is binary, and hence is optimal for **MWCS**.

One can observe that, whatever formulation is used, **MWCS** becomes harder to solve when a budget constraint is imposed on the number of selected arcs. With the **ARC** formulation, cuts are added during the solution process and branching occurs for most of the instances. With the **SEA** formulation, branching also occurs for most instances. The average total running time is higher for the **ARC** formulation. However, Fig. 3 shows that the **ARC** formulation is faster for at least two-thirds of the instances but takes more time for the last third of the instances. Some of the instances are not solved in 10 min. By way of comparison, for **MWCS** *without* a budget constraint, the instances with parameter values $n = 200$, $p = 20$, $d = 80$ are all solved in less than 3 s with the **ARC** formulation and in less than 2 s with the **SEA** formulation; the unconstrained instances with parameter values $n = 200$, $p = 40$, $d = 90$ are all solved in less than 7 s with the **ARC** formulation and in less than 4 s with the **SEA** formulation.

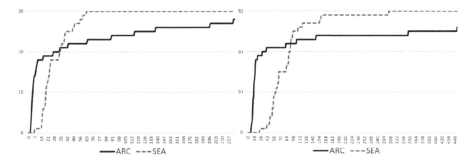

Fig. 3. Comparison of total running time of the **ARC** and **SEA** formulations when $n = 200$, $p = 20$, $d = 80$, $b = 100$ (left) and $n = 200$, $p = 40$, $d = 90$, $b = 100$ (right). The horizontal axis displays running times (in seconds), and the vertical axis displays the number of instances solved within a given time.

5   MWCS with maximum cycle length constraint

In this section, we consider another constrained version of cycle selections which naturally arises in their application to kidney exchange problems [6]. For any $K \in \mathbb{N}$, we say that a subset $B \subseteq A$ is a K-*cycle selection* if the arcs of B form a union of directed cycles, each of length at most K. The problem K-**MWCS** is the version of **MWCS** asking for a maximum weighted K-cycle selection.

Whereas it was easy to incorporate a budget constraint into all the **MWCS** formulations, imposing a constraint on the maximum cycle length requires more work for some formulations. The **PI** formulations of **MWCS** can be rather easily adapted to account for such a constraint, as shown in [6]. For the **CY** formulation, it suffices to disregard all cycles with length larger than K. On the other hand, arc-based formulations are more difficult to deal with when K is arbitrary. In the following section, we propose an adaptation of the **ARC** formulation that exactly describes the 3-cycle selections.

5.1   Formulation of 3-Cycle Selections

By a small abuse of language, we say that (S, T) is a partition of a set W if $S \cup T = W$ and $S \cap T = \emptyset$, where either S or T may be empty. We denote by $\mathcal{P}(W)$ the set of all partitions of W.

Consider now the following constraints, for all arcs $(i, j) \in A$ and for all partitions $(S, T) \in \mathcal{P}(V \setminus \{i, j\})$:

$$\beta_{i,j} \leq \beta_{j,i} + \sum_{k \in S:(j,k) \in A} \beta_{j,k} + \sum_{k \in T:(k,i) \in A} \beta_{k,i} \qquad \text{if}(j, i) \in A \qquad (24)$$

$$\beta_{i,j} \leq \sum_{k \in S:(j,k) \in A} \beta_{j,k} + \sum_{k \in T:(k,i) \in A} \beta_{k,i} \qquad \text{if}(j, i) \notin A \qquad (25)$$

$$\beta_{i,j} \in \{0, 1\}. \qquad (26)$$

We denote by P_3 the set of solutions of constraints (24)–(26). Note that when either S or T is empty, (24) and (25) boil down to the *predecessor* and *successor* *inequalities* (20)–(21), respectively.

Theorem 1. *The constraints (24)–(26) provide a correct formulation of the set of 3-cycle selections.*

Proof. Consider a cycle selection β with cycles of length at most 3. For each arc (i, j) such that $\beta_{i,j} = 1$, either $(j, i) \in A$ and $\beta_{j,i} = 1$ (2-cycle), or there is a vertex $k \neq i, j$ such that $\beta_{j,k} = \beta_{k,i} = 1$ (3-cycle). For each partition (S, T) of $V \setminus \{i, j\}$, k is either in S or in T. Hence, β is in P_3.

Conversely, assume that $\beta \in P_3$, let B be the corresponding set of arcs, and consider an arc $(i, j) \in B$ ($\beta_{i,j} = 1$). We must show that (i, j) is in a 2-cycle or in a 3-cycle of G_B. Define $S = \{k \in V : (k, i) \in A \text{ and } \beta_{k,i} = 1\}$.

– If $j \in S$, then $\beta_{j,i} = 1$ and (i, j) is in a 2-cycle.
– If $j \notin S$, then $S \subseteq V \setminus \{i, j\}$. Let $T = (V \setminus \{i, j\}) \setminus S$. Consider Constraints (24) and (25) associated with (S, T). Note that they are identical in this case, because $j \notin S$ means that either $(j, i) \notin A$ or $\beta_{j,i} = 0$. Furthermore, by definition of S, $\beta_{k,i} = 0$ for all $(k, i) \in A$, $k \in T$. So, by (24)–(25), there must exist $k \in S$ such that $(j, k) \in A$ and $\beta_{j,k} = 1$. Then, (i, j, k) forms a 3-cycle in B.   □

Theorem 2. *The separation problem for the linear relaxation of (24)–(26) can be solved in polynomial time.*

Proof. The separation problem is the following: given a vector $\beta \in [0,1]^{|A|}$, is there $(i,j) \in A$ and a partition (S,T) of $V \setminus \{i,j\}$ such that $\beta_{i,j} > \beta_{j,i} + \sum_{k \in S} \beta_{j,k} + \sum_{k \in T} \beta_{k,i}$ if $(j,i) \in A$ or $\beta_{i,j} > \sum_{k \in S} \beta_{j,k} + \sum_{k \in T} \beta_{k,i}$ if $(j,i) \notin A$?

We can ask the question for each arc (i,j) successively. When (i,j) is fixed, we know whether $(j,i) \in A$ or not. We just need to identify the partition (S,T) which minimizes $\sum_{k \in S} \beta_{j,k} + \sum_{k \in T} \beta_{k,i}$ and check whether the relevant strict inequality mentioned above is satisfied. In order to identify this partition, we compare $\beta_{j,k}$ and $\beta_{k,i}$ for each $k \in V \setminus \{i,j\}$. If $\beta_{j,k} > \beta_{k,i}$, then we assign k to T, and otherwise we assign k to S. □

5.2 Polyhedral Study

In this subsection, we restrict our attention to the case of the *complete digraph* K_n on n vertices, where A contains all possible $n(n-1)$ arcs. In this case, Constraints (25) are vacuous. We denote by P_3^* the convex hull of the set P_3. The next results provide information about the structure of P_3^* for K_n.

Theorem 3. *When $n = 2$, the dimension of P_3^* is 1. When $n \geq 3$, P_3^* is full dimensional, that is, $dim(P_3^*) = n(n-1)$.*

Theorem 4. *For all $(i,j) \in A$, the inequalities $\beta_{i,j} \geq 0$ and $\beta_{i,j} \leq 1$ define facets of P_3^*.*

The validity of Theorems 3–4 has been established in [2].

Theorem 5. *For $|V| \geq 3$, for all $(i,j) \in A$, for all partitions (S,T) of $V \setminus \{i,j\}$, the inequality (24) defines a facet of P_3^*.*

Proof. Fix $(i,j) \in A$ and (S,T) a partition of $V \setminus \{i,j\}$. Let F be the face of P_3^* defined as

$$F = \left\{ \beta \in P_3^* : \beta_{i,j} = \beta_{j,i} + \sum_{k \in S} \beta_{j,k} + \sum_{k \in T} \beta_{k,i} \right\}.$$

Suppose that F is included in a hyperplane defined by the equation

$$\sum_{(u,v) \in A} b_{u,v} \beta_{u,v} = b_0 \tag{27}$$

and consider the following points $\beta^1, \ldots, \beta^{10}$ which are all in F (we only specify the nonzero components):

1. $\beta^1 = 0$.
2. $\beta_{i,j}^2 = \beta_{j,i}^2 = 1$.
3. If $|S| \geq 1$, fix $k \in S$, let $\beta_{i,j}^3 = \beta_{j,k}^3 = \beta_{k,i}^3 = 1$, let $\beta_{i,j}^{3'} = \beta_{j,k}^{3'} = \beta_{k,i}^{3'} = \beta_{i,k}^{3'} = 1$, and let $\beta_{i,j}^{3''} = \beta_{j,k}^{3''} = \beta_{k,i}^{3''} = \beta_{k,j}^{3''} = 1$.
4. If $|S| \geq 1$, fix $k \in S$, and let $\beta_{i,k}^4 = \beta_{k,i}^4 = 1$.

5. If $|T| \geq 1$, let β^5, $\beta^{5'}$ and $\beta^{5''}$ be defined like β^3, $\beta^{3'}$ and $\beta^{3''}$, but with $k \in T$.
6. If $|T| \geq 1$, fix $k \in T$, and let $\beta^6_{j,k} = \beta^6_{k,j} = 1$.
7. If $|S| \geq 2$, fix $h, k \in S$, and let $\beta^7_{i,k} = \beta^7_{k,h} = \beta^7_{h,i} = 1$.
8. If $|T| \geq 2$, fix $h, k \in T$, and let $\beta^8_{i,k} = \beta^8_{k,h} = \beta^8_{h,i} = 1$.
9. If $|S| \geq 1$ and $|T| \geq 1$, fix $k \in S$, $h \in T$, and let $\beta^9_{k,j} = \beta^9_{j,h} = \beta^9_{h,k}$.
10. If $|S| \geq 1$ and $|T| \geq 1$, fix $k \in S$, $h \in T$, and let $\beta^{10}_{k,h} = \beta^{10}_{h,k} = 1$.

By successively substituting the points $\beta^1, \ldots, \beta^{10}$ in (27), one can easily conclude that, up to a multiplicative constant, (27) is equivalent to the equation defining F. This proves that F is a facet of P_3^*.   □

5.3   Numerical Experiments

In this section, we numerically compare the formulation defined by the inequalities (24)–(26), that we denote as **ARC3**, with the **PI** and the **CY** formulations with maximum cycle length equal to 3, denoted as **PI3** and **CY3**, respectively by solving 3-**MWCS** on random digraphs.

Figure 4 displays, on the left, the performance profiles of the three formulations for the instances with parameter values $n = 100$, $p = 20$, $d = 80$. It is clear that **ARC3** and **CY3** are more computationally efficient than **PI3**: the **CY3** formulation solves all 30 instances in less than a second, whereas the **ARC3** formulation takes 8 s for some instances, and the **PI3** formulation requires 95 to 130 s, depending on the instance. Regarding the optimization process, the optimal solution of the linear relaxation of the **PI3** and **CY3** formulations is binary for all instances, so that no branching is performed. For the **ARC3** formulation, no separation cuts are added, but an average of 116.53 nodes are explored before obtaining an optimal solution.

A budget constraint can be incorporated into the formulations in addition to the maximum cycle length constraint. For the same parameter values as above, that is, $n = 100$, $p = 20$, $d = 80$, $K = 3$, and $b = 50$, Fig. 4 displays, on the right, the performance profiles of the three formulations with the budget constraint. We can compare the results with and without the budget constraint. Surprisingly, a budget constraint does not make the problem significantly harder in terms of running time, especially when using the **ARC3** and **CY3** formulations. The optimal binary solution is found by solving the linear relaxation of the **PI3** formulation for all the instances, and the linear relaxation of the **CY3** formulation for 27 instances. For the three remaining instances, a few nodes are explored to find the optimal binary solution. For the **ARC3** formulation, an average of 142.56 separation cuts are added to the model, and an average of 335.80 nodes are explored before obtaining an optimal solution. Similar observations apply when $b = 25$.

Finally, since the digraphs associated with the Steiner triple instances have cycles of length 3 only (cf. Sect. 3.2), we also evaluated the **ARC3** and **CY3** formulations on these digraphs and compared them with the **ARC** and **SEA**

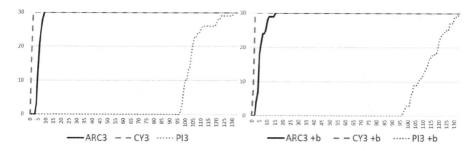

Fig. 4. Comparison of total running time of the **MWCS** formulations when $n = 50$, $p = 20$ $d = 80$, $K = 3$ without a budget constraint (left) or with $b = 50$ (right). The horizontal axis displays running times (in seconds), and the vertical axis displays the number of instances solved within a given time.

formulations. Although the **ARC** and **ARC3** formulations have similar total running times, the most efficient formulation in terms of total running time is **CY3** as illustrated in Table 4 for $|X|$ in $\{45, 49, 51, 55\}$.

Table 4. Comparison of total running time on the digraphs associated with the Steiner triple instances, in seconds

$|X|$	**ARC**	**SEA**	**ARC3**	**CY3**
45	109	627	256	156
49	160	1022	175	113
51	401	5351	322	185
55	823	9513	812	486

6   Conclusions

In this paper, we have examined the computational performance of various formulations that describe cycle selections in directed graphs, with a focus on the maximum weighted cycle selection problem. In [2], six ILP formulations were presented, and the **ARC** formulation was studied in more detail. Our experiments indicate that the **ARC** formulation and the **SEA** formulation are most efficient among the six formulations. While **MWCS** is computationally easy to solve on random graphs, we have also investigated structured instances associated with Steiner triples, as well as variants that arise by imposing a budget constraint and/or a maximum cycle length constraint. These variants proved more challenging, especially in the presence of a budget constraint.

References

1. Baratto, M.: Optimization models and methods for kidney transplantation programs, Ph.D. thesis, University of Liège (2024). https://hdl.handle.net/2268/314151
2. Baratto, M., Crama, Y.: Cycle selections. Discret. Appl. Math. **335**, 4–24 (2023)
3. Erdős, P., Rényi, A.: On random graphs I. Publicationes Mathematicae Debrecen **6**, 290–297 (1959)
4. Fulkerson, D.R., Nemhauser, G.L., Trotter, L.E.: Two computationally difficult set covering problems that arise in computing the 1-width of incidence matrices of Steiner triple systems. In: Balinski, M.L. (ed.) Approaches to Integer Programming. Mathematical Programming Studies, vol. 2, pp. 72–81. Springer, Berlin (1974). https://doi.org/10.1007/BFb0120689
5. Graham, A.J., Pike, D.A.: A note on thresholds and connectivity in random directed graphs. Atlantic Electron. J. Math. **3**, 1–5 (2008)
6. Smeulders, B., Bartier, V., Crama, Y., Spieksma, F.: Recourse in kidney exchange programs. INFORMS J. Comput. **335**, 4–24 (2021)
7. Tarjan, R.E.: Depth-first search and linear graph algorithms. SIAM J. Comput. **1**, 146–160 (1972)

1-Persistency of the Clique Relaxation of the Stable Set Polytope

Diego Delle Donne[1], Mariana Escalante[2,3], Pablo Fekete[3(✉)], and Lucía Moroni[2,3]

[1] ESSEC Business School, Cergy, France
delledonne@essec.edu
[2] CONICET, Rosario, Argentina
[3] FCEIA-Universidad Nacional de Rosario, Rosario, Argentina
{mariana,fekete,lmoroni}@fceia.unr.edu.ar

Abstract. A polytope $P \subset [0,1]^n$ is said to have the *persistency* property if for every vector $c \in \mathbb{R}^n$ and every c-optimal point $x \in P$, there exists a c-optimal integer point $y \in P \cap \{0,1\}^n$ such that $x_i = y_i$ for each $i \in \{1,\ldots,n\}$ with $x_i \in \{0,1\}$. In this paper, we consider a relaxation of the persistency property, called 1-persistency, over the clique relaxation of the stable set polytope in graphs. In particular, we study the family $\mathcal{Q}$ of graphs whose clique relaxation of the stable set polytope has 1-persistency. The main objective of this contribution is to analyze forbidden structures for a given graph to belong to $\mathcal{Q}$. The graphs given by these structures are denoted here as mn$\mathcal{Q}$. On one hand, we provide sufficient conditions for a graph to belong to $\mathcal{Q}$, and identify several graph classes of this family. On the other hand, we give two different infinite families of forbidden minimal structures for this class of graphs. We conclude the paper by suggesting an interesting future line of work about the persistency-preservation property of valid inequalities and its potential practical applications.

Keywords: Stable set polytope · Persistency · Integer programming

1 Introduction

Given a polyhedron $P \subset \mathbb{R}^n$ and a vector $c \in \mathbb{R}^n$, the point $x \in P$ is *c-optimal* if $cx \geq cx'$ for every $x' \in P$. A polytope $P \subset [0,1]^n$ is said to have the *persistency* property if for every vector $c \in \mathbb{R}^n$ and every c-optimal point $x \in P$, there exists a c-optimal integer point $y \in P \cap \{0,1\}^n$ such that $x_i = y_i$ for each $i \in \{1,\ldots,n\}$ with $x_i \in \{0,1\}$. Nemhauser and Trotter [4] proved that the *edge relaxation* of the stable set polytope (see Sect. 2 for further definitions) has this property for any graph. This result may be useful in practice, as it allows us to reduce the size of the problem by fixing some variables to provable optimal integer values. In addition, these variable fixings may be incorporated into classical cutting-planes

Partially supported by grant PIP0227 - CONICET and grant PICT 3032-ANPCyT.

or branch-and-cut algorithms, thus speeding up the solution process (we give more details about this aspect in Sect. 4). Unfortunately, the edge relaxation is known to be very weak and it is not likely to find c-optimal solutions with many integer values. Additionally, this is the only proper relaxation of the stable set polytope (under some mild conditions) satisfying the persistency property [5].

In this contribution, we study a relaxation of the persistency property, which we define as *1-persistency*, where we focus on c-optimal vertices preserving the components at value 1 (instead of considering both 1 and 0). Although this gives a weaker property, we found families of 1-persistent graphs when considering the well-known *clique relaxation* of the stable set polytope (stronger than the edge relaxation).

2   Definitions and Preliminary Results

Throughout this work, $\mathbb{0}$ stands for the vector of all 0's and $\mathbb{1}$ the vector of all ones, both of appropriate dimension. For simplicity, we use $[\![n]\!]$ as a shortcut for the set $\{1, \ldots, n\}$. Given $x \in \mathbb{R}^n$ and $U \subset [\![n]\!]$, $x(U) = \sum_{i \in U} x_i$.

Let $G = (V, E)$ be a graph with node set V and edge set E. Two vertices u, v of G are *adjacent*, or *neighbours*, if $uv \in E$. If G has n vertices pairwise adjacent, then G is the *complete* graph K_n and, in particular, a K_3 is called a *triangle*. The *complementary graph* of G, denoted as $\overline{G}$, has the same node set as G and two nodes are adjacent in $\overline{G}$ if and only if they are not adjacent in G. The *open neighbourhood* of a node u in G is the set $N(u) = \{v \in V : uv \in E\}$ and the *closed neighbourhood* is $N[u] = N(u) \cup \{u\}$. More generally, for $U \subset V$, $N(U) = (\cup_{u \in U} N(u)) - U$ and $N[U] = \cup_{v \in U} N[v]$. Given $U \subset V$, the *subgraph induced* by U is the graph with node set U and edge set $\{uv \in E : u, v \in U\}$. We denote it by $G[U]$. If $G' = G[U]$ for some $U \subset V$ then G' is a *node-induced subgraph* of G and we denote it $G' \subset G$. Given a node $u \in V$, the graph obtained by *deleting* the node u is $G[V - \{u\}]$, and we denote it by $G - u$. If $U \subset V$ then $G - U = G[V - U]$. The graph obtained by *destruction of a node* u is $G \ominus u = G - N[u]$.

A *clique* in a graph G is a subset of nodes of G inducing a complete graph. A *stable set* is a subset of pairwise nonadjacent nodes in G. The *stability number* of G is the cardinality of a stable set of maximum cardinality and is denoted by $\alpha(G)$. The set $K(G)$ denotes the family of maximal cliques in G and $K^t(G)$ the maximal cliques of size equal to t. When the graph is clear from the context, we can simply denote them by K and K^t.

In this contribution we assume that the chordless cycle of $2k+1$ nodes, C_{2k+1} has node set $\{v_i, i \in [\![2k+1]\!]\}$ and edges in $\{v_i v_{i+1} : i \in [\![2k+1]\!]\}$ (sum of indices mod. $2k+1$). An *odd hole* in a graph G is an induced chordless cycle of odd length at least 5. The complement of an odd hole is called *odd antihole*. A *perfect* graph is a graph having neither an odd hole nor an odd antihole as a node-induced subgraph. Although this is not the original definition, it holds from the Perfect Graph Theorem [1]. A graph G is *near-bipartite* if for all $v \in V(G)$, $G - N(v)$ can be partitioned into two stable sets.

Let $G_1 = (V_1, E_1)$ and $G_2 = (V_2, E_2)$ be two graphs such that $V_1 \cap V_2 = \{v\}$. The *1-sum* of G_1 and G_2 at v, is the graph $G_1 \oplus G_2 = (V_1 \cup V_2, E_1 \cup E_2)$.

A *paw* P_u is a graph with 4 nodes (including u) where $P_u - u$ is a triangle and u has degree one. A *bad paw* for a graph G is an induced paw such that $G \ominus u$ is imperfect. When a graph has no bad paw, it is *bad-paw-free*.

Given a graph $G = (V, E)$ and $c \in \mathbb{R}^V$, the (weighted) stable-set problem asks for finding a stable set S that maximizes $c(S) = \sum_{v \in S} c_v$. Given a set $S \subset V$, the *characteristic vector* of S is the vector $\chi^S \in \mathbb{R}^V$ such that, $\chi_i^S = 1$ if $i \in S$, and $\chi_i^S = 0$ otherwise. If $x \in \mathbb{R}^V$ and $U \subset V$, $x_U \in \mathbb{R}^U$ is the restriction of x to U, i.e., $(x_U)_i = x_i$ for $i \in U$. The *stable set polytope* STAB(G) of a graph G is defined as the convex hull of the characteristic vectors of all stable sets of G. Two well-known relaxations of the polytope of stable sets are the *edge relaxation* FRAC(G) and the *clique relaxation* QSTAB(G) respectively given by

$$\text{FRAC}(G) = \{x \in [0,1]^V : x_v + x_w \leq 1, \; vw \in E\}, \text{ and}$$

$$\text{QSTAB}(G) = \{x \in [0,1]^V : \sum_{i \in Q} x_i \leq 1, \; Q \text{ clique in } G\}.$$

While it is clear that STAB$(G) \subset$ QSTAB$(G) \subset$ FRAC(G) for every graph G, the equality STAB$(G) =$ QSTAB(G) holds if and only if G is perfect ([2]). Equivalently, QSTAB(G) has a non-integer vertex if and only if G is imperfect.

A graph G is *rank-perfect* [6] if

$$\text{STAB}(G) = \{x \in [0,1]^V : \sum_{v \in U} x_v \leq \alpha(G[U]), \; U \subset V\}.$$

A polyhedron $P \subset \mathbb{R}_+^n$ is *lower-comprehensive* if $0 \leq y \leq x$ with $x \in P$ implies $y \in P$. Note that the above-considered relaxations of the stable set polytope are lower-comprehensive.

To present the main results of this paper we introduce some technical results which proof is omitted due to space limitations.

Lemma 1. *Given a graph* $G = (V, E)$ *and* $u \in V$,

i) $x = (x_u, x_{V-\{u\}})$ *with* $x_u = 0$ *is a vertex of* QSTAB(G) *if and only if* $x_{V-\{u\}}$ *is a vertex of* QSTAB$(G - u)$.

ii) $x = (x_u, x_{N(u)}, x_{V-N[u]})$ *with* $x_u = 1$ *and* $x_{N(u)} = \mathbb{0}$ *is a vertex of* QSTAB(G) *if and only if* $x_{V-N[u]}$ *is a vertex of* QSTAB$(G \ominus u)$.

Lemma 2. *Let* G *be the 1-sum of two graphs* $G_1 = (V_1, E_1)$ *and* $G_2 = (V_2, E_2)$ *at* v, *where* $V_1 \cap V_2 = \{v\}$. *If* $\bar{x}$ *is an extreme point of* QSTAB(G) *then* $\bar{x}_{V_i}$ *is an extreme point of* QSTAB$(G[V_i])$, *for* $i = 1$ *or* $i = 2$.

3   1-Persistency on Relaxations of the Stable Set Polytope

It is proven in [5] that the edge relaxation is the only proper relaxation of the stable set polytope (under some mild conditions) satisfying the persistency property, as it is stated implicitly in [4]. In addition, this relaxation is known to be very weak and it is not likely to find c-optimal solutions with many integer values. Driven by these facts, we study a relaxation of the persistency property, which we define as 1-persistency, in which we focus on c-optimal points preserving only the components at value 1.

Definition 1. *A polyhedron $P \subset [0,1]^n$ has the 1-persistency property if for every $c \in \mathbb{R}^n$ and $x \in P$ c-optimal there exists an integer point y, which is c-optimal in $P \cap \{0,1\}^n$, such that $y_i = x_i$ whenever $x_i = 1$.*

If a polyhedron $P \subset [0,1]^n$ *does not have* 1-persistency we say that a pair (c,x) for which the property is not valid, *breaks* the 1-persistency of P. For simplicity, we say that x breaks 1-persistency when there exists c such that (c,x) breaks it. To analyze the 1-persistency of a polyhedron we only need to look at those vertices having both, integer and non-integer components. More precisely,

Definition 2. *A point $x \in [0,1]^n$ is a* mixed-integer *point if its components can be partitioned into three non-empty sets $I_0(x) = \{i : x_i = 0\}$, $I_1(x) = \{i : x_i = 1\}$ and $I_f(x) = \{i : 0 < x_i < 1\}$.*

The next two results show that, in order to study the 1-persistency property on a lower-comprehensive polytope, it is sufficient to analyze non-negative costs c and *maximal* mixed-integer vertices.

Lemma 3. *Given a lower-comprehensive polytope $P \subset [0,1]^n$, if $(c,x) \in \mathbb{R}^n \times P$ breaks 1-persistency of P then there exists $\tilde{c} \geq \mathbb{0}$ such that $(\tilde{c},x)$ also breaks it.*

Proof. Let $(c,x) \in \mathbb{R}^n \times P$ such that it breaks 1-persistency of P. Consider $I_c = \{i : c_i < 0\} \subset [\![n]\!]$. Since P is lower-comprehensive, $x_i = 0$ for $i \in I_c$. For fixed c, define the function p_c such that

$$p_c(z)_i = \begin{cases} 0 \text{ if } i \in I_c, \\ z_i \text{ otherwise,} \end{cases}$$

for $i \in [\![n]\!]$ and $z \in \mathbb{R}^n$. Let $\tilde{c} = p_c(c)$. If $z \in P$, $\tilde{c}z = \tilde{c}p_c(z) = cp_c(z) \leq cx = \tilde{c}x$. Then, x is $\tilde{c}$-optimal in P.

If $(\tilde{c},x)$ does not break 1-persistency, there exists a $\tilde{c}$-optimal point $y \in P \cap \{0,1\}^n$, with $I_1(x) \subset I_1(y)$. Then, $cz \leq \tilde{c}z \leq \tilde{c}y = \tilde{c}p_c(y) = cp_c(y)$, for all $z \in P \cap \{0,1\}^n$ and therefore $p_c(y)$ is c-optimal in $P \cap \{0,1\}^n$.

For $i \in I_1(x)$, $i \notin I_c$ and $p_c(y)_i = y_i = x_i = 1$. Then, $I_1(x) \subset I_1(p_c(y))$, and (c,x) does not break 1-persistency, a contradiction. This shows that $(\tilde{c},x)$ breaks 1-persistency of P. □

Lemma 4. *Given x^1 and x^2 mixed-integer vertices of a lower-comprehensive polytope $P \subset [0,1]^n$, if $x^1 \leq x^2$ and x^1 breaks 1-persistency, then x^2 also breaks it.*

Proof. Let $c \in \mathbb{R}^n$ such that (c, x^1) breaks 1-persistency. By Lemma 3 there exists $\tilde{c} \geq 0$ such that $(\tilde{c}, x^1)$ also breaks it. Since x^1 is $\tilde{c}$-optimal and $x^1 \leq x^2$, x^2 is also $\tilde{c}$-optimal. Moreover, $I_1(x^1) \subset I_1(x^2)$ implies that the pair $(\tilde{c}, x^2)$ breaks 1-persistency of P.   □

In this contribution, we focus on the study of 1-persistency on the clique relaxation of the stable set problem. Therefore, we introduce the following definition.

Definition 3. *We say that a graph G is Q-persistent if QSTAB(G) has the 1-persistency property, and denote $\mathcal{Q}$ the family of all Q-persistent graphs.*

There are some trivial members of $\mathcal{Q}$ as triangle-free and perfect graphs. To see this, note that the clique relaxation of a triangle-free graph coincides with the fractional stable set polytope and the one corresponding to a perfect graph, with the stable set polytope. However, not every graph is Q-persistent as it will become clear after the forthcoming results.

3.1   Basic Results on the Family of Q-Persistent Graphs

When the clique relaxation of the stable set polytope of a graph does not have mixed-integer vertices, the graph clearly belongs to $\mathcal{Q}$. This is the case for a near-bipartite graph and, due to results in [3], also for the complementary graph of a rank-perfect graph. We state these facts in the next lemma, whose proof is omitted due to space limitation.

Lemma 5. *If a graph is near-bipartite or its complementary graph is rank-perfect then it is Q-persistent.*

The following result proves that 1-persistency is a hereditary property.

Theorem 1. *If G is Q-persistent then G' is Q-persistent, for every $G' \subset G$.*

Proof. Let $G \in \mathcal{Q}$ and $G' \subset G$ with node sets $V = \{v_1, \ldots, v_n\}$ and w.l.o.g. $V' = \{v_1, \ldots, v_m\}$ with $m < n$. Let $c' \in \mathbb{R}^m$ and x' a c'-optimal mixed-integer vertex of QSTAB(G'). If $c = (c', 0) \in \mathbb{R}^m \times \mathbb{R}^{n-m}$ then $x = (x', 0) \in$ QSTAB(G) and x is c-optimal in QSTAB(G). Since $G \in \mathcal{Q}$ there exists $y \in$ QSTAB$(G) \cap \{0, 1\}^n$ c-optimal such that $y_i = x_i$ whenever $x_i = 1$. Then, $y = \chi^S$ for S stable set in G. If $S' = S \cap V'$ the point $\chi^{S'}$ is c'-optimal in STAB(G') thus proving that $G' \in \mathcal{Q}$.   □

In [4] the authors proved that FRAC(G) has 1-persistency for every graph G. To do so, they establish the following results concerning optimal stable sets.

Theorem 2 (Nemhauser and Trotter, [4]).

i) If $S \subset V$ is a stable set, then S is c-optimal if and only if there is no stable set $I \subset V - S$ such that $c(S \cap N(I)) < c(I)$.

ii) If S is a c-optimal stable set of the induced subgraph G[N[S]], then there exist a c-optimal stable set S′ of G such that S ⊂ S′.

In the following theorem, we make use of the previous results to give a sufficient condition for a graph to be Q-persistent.

Theorem 3. *Every bad-paw-free graph is Q-persistent.*

Proof. Let $G = (V, E)$ be a bad-paw-free graph and x a c-optimal mixed-integer point of QSTAB(G) for a given $c \in \mathbb{R}_+^V$. To prove that the graph is Q-persistent, we will show that $S = I_1(x)$ is a subset of a c-optimal stable set of G. By Theorem 2 ii), it will suffice to prove that S is c-optimal in $G[N[S]]$. Assume it is not. Then, there exists $I \subset N(S)$ such that $c(S \cap N(I)) < c(I)$. Define $x' \in \mathbb{R}^V$ as follows:

$$x'_v = \begin{cases} 1 - \epsilon & \text{if } v \in S \cap N(I), \\ \epsilon & \text{if } v \in I, \\ x_v & \text{otherwise,} \end{cases}$$

where $\epsilon = \min\{1 - x_v : v \notin N[S]\} \in (0, 1)$. It is clear that $x' \geq 0$. In order to show that $x' \in $ QSTAB(G), it remains to prove that $x'(Q) \leq 1$ for all $Q \in K$. We divide our analysis into four different cases:

1. $Q \cap I = \emptyset$ and $Q \cap (S \cap N(I)) = \emptyset$.
   Then, $x'_v = x_v$ for all $v \in Q$ and $x'(Q) = x(Q) \leq 1$.
2. $Q \cap I \neq \emptyset$ and $Q \cap (S \cap N(I)) \neq \emptyset$.
   Let $v, u \in V$ such that $Q \cap I = \{v\}$ and $Q \cap (S \cap N(I)) = \{u\}$. The node u is not adjacent to any node in $R = V - N[S]$, hence $R \cap Q = \emptyset$. Since $x_w = 0$ for all $w \in N(S)$, we have $x'(Q) = x'_v + x'_u + x'(Q \cap (N(S) - I)) = 1$.
3. $Q \cap I = \emptyset$ and $Q \cap (S \cap N(I)) \neq \emptyset$.
   The proof follows the same reasoning as before.
4. $Q \cap I \neq \emptyset$ and $Q \cap (S \cap N(I)) = \emptyset$.
   In this case, $Q \cap S = \emptyset$. Let $Q \cap I = \{v\}$. Suppose $w_1, w_2 \in Q \cap R$, with $w_1 \neq w_2$. Since $v \in I \subset N(S)$, there exists $u \in S \cap N(I)$ adjacent to v. Hence, $\{u, v, w_1, w_2\}$ induce a paw P_u in G. Moreover, since $I_f(x) \neq \emptyset$, $x_{V-N[u]}$ is a fractional vertex of QSTAB$(G \ominus u)$ by Lemma 1 ii), which implies $G \ominus u$ is an imperfect graph and P_u a bad-paw in G. Since G is a bad-paw-free graph, $|Q \cap R| \leq 1$.
   If $|Q \cap R| = 0$, then $Q \subset N(S)$ and $x'(Q) = x'_v + x'(N(S) - I) = \epsilon + 0$.
   If $|Q \cap R| = 1$, then $Q \subset N(S) \cup \{w\}$, where $\{w\} = Q \cap R$. Hence, $x'(Q) = x'_w + x'_v + x'(Q \cap (N(S) - I)) = x_w + \epsilon + 0$. In any case, $x'(Q) \leq 1$.

Thus, $x' \in $ QSTAB(G). Moreover, from the definition of I,

$$cx' - cx = \epsilon c(I) + (1 - \epsilon) c(S \cap N(I)) - c(S \cap N(I))$$
$$= \epsilon (c(I) - c(S \cap N(I))) > 0.$$

Hence, x is not c-optimal, a contradiction. Therefore, S is c-optimal in $G[N[S]]$, and then G is Q-persistent. □

3.2   Forbidden Structures for the Family of Q-Persistent Graphs

Theorem 1 inspires the definition of minimal forbidden structures for $\mathcal{Q}$.

Definition 4. *A graph G is minimally not Q-persistent (mnQ for short) if $G \notin \mathcal{Q}$ but $G' \in \mathcal{Q}$ for every $G' \subset G$.*

We present two infinite families of mn$\mathcal{Q}$ graphs.

Definition 5. *Given $k \geq 2$, $i \in [\![2k+1]\!]$ and two nodes, u and v, $\mathcal{H}(k,u,v,i)$ is the graph with node set $\{u,v,v_1,\ldots,v_{2k+1}\}$ where $\{v_i : i \in [\![2k+1]\!]\}$ induces C_{2k+1}, $N(u) = \{v\}$, and $N(v) = \{u,v_i,v_{i+1}\}$. Similarly, given $k \geq 2$, a maximal clique Q in $\overline{C}_{2k+1}$ and two nodes, u and v, we call $\mathcal{A}(k,u,v,Q)$ the graph with node set $\{u,v,v_1,\ldots,v_{2k+1}\}$, where $\{v_i : i \in [\![2k+1]\!]\}$ induces $\overline{C}_{2k+1}$, $N(u) = \{v\}$ and $N(v) = \{u\} \cup Q$. When it is clear from the context we simply denote these graphs as $\mathcal{H}_k$ and $\mathcal{A}_k$, respectively.*

Theorem 3 gives rise to the following result.

Theorem 4. *The graphs $\mathcal{H}(k,u,v,i)$ and $\mathcal{A}(k,u,v,Q)$ are mn$\mathcal{Q}$.*

Proof. For $G = \mathcal{H}_k$ or $G = \mathcal{A}_k$, denote by $x = (x_u, x_v, x_{V-\{u,v\}})$ the points in $\mathbb{R}^{V(G)}$.

First, consider $c = (0,1,2 \cdot \mathbb{1})$. If $x \in \mathrm{QSTAB}(\mathcal{H}_k)$ then

$$cx = x_v + 2x(V - \{u,v\}) \leq 1 + 2k$$

since there are exactly $2k$ cliques in K^2 and one in K^3.

If $\bar{x} = (1,0,\frac{1}{2} \cdot \mathbb{1})$ then $\bar{x} \in \mathrm{QSTAB}(\mathcal{H}_k)$ and $c\bar{x} = 2k+1$. Therefore, $\bar{x}$ is c−optimal in $\mathrm{QSTAB}(\mathcal{H}_k)$.

We next prove that there is no c-optimal stable set containing u. Let S_1 be a maximal stable set in $\mathcal{H}_k \ominus v$. Then, $|S_1| = k$ and $S_1 \cup \{v\}$ forms a stable set in $\mathcal{H}_k$. Therefore,

$$c(S_1 \cup \{v\}) = 2|S_1| + 1 = 2k+1.$$

Given that $\mathcal{H}_k \ominus u = C_{2k+1}$, if S_2 is a stable set containing u, $|S_2 - \{u\}| \leq k$ and then

$$c(S_2) = 2|S_2 - \{u\}| + 0 \leq 2k.$$

This implies that $c(S_1 \cup \{v\}) > c(S_2)$ for any stable set S_2 containing u.

In this way, we have proved that $\bar{x}$ is c-optimal in $\mathrm{QSTAB}(\mathcal{H}_k)$ and there is no c-optimal stable set containing u. Then, $\mathcal{H}_k \notin \mathcal{Q}$.

The fact that every proper node-induced subgraph belongs to $\mathcal{Q}$ holds since $\mathcal{H}_k - \{w\}$ is bad-paw-free for every $w \in V(\mathcal{H}_k)$ (Theorem 3). Hence, $\mathcal{H}_k$ is mn$\mathcal{Q}$.

Let us now consider the graph $\mathcal{A}_k = \mathcal{A}(k,u,v,Q)$ where $\mathcal{A}_k \ominus u = \overline{C}_{2k+1}$. Following a similar reasoning, it holds that the point $\bar{x} = (1,0,\frac{1}{k} \cdot \mathbb{1})$ is c-optimal in $\mathrm{QSTAB}(\mathcal{A}_k)$ for $c = (1,2,2k \cdot \mathbb{1})$.

Let $v_i \in V(\overline{C}_{2k+1})$ such that $v_i, v_{i+1} \notin Q$. Then $S_1 = \{v, v_i, v_{i+1}\}$ is a stable set with $c(S_1) = 4k+2$.

If S_2 is a stable set containing u, $c(S_2) = c(S_2 - \{u\}) + c_u \leq 4k + 1$.

This implies that the pair $(c, \bar{x})$ breaks the 1-persistency of QSTAB($\mathcal{A}_k$). Again, since $\mathcal{A}_k - \{w\}$ is bad-paw-free for every $w \in V(\mathcal{A}_k)$, it holds that $\mathcal{A}_k$ is mn$\mathcal{Q}$. □

It is natural to ask if there are some other mn$\mathcal{Q}$ graphs. To further study them, based on Theorem 3, we first allow other connections between v and the odd hole C_{2k+1} in the graphs $\mathcal{H}(k, u, v, i)$.

Definition 6. *Given $k \geq 2$, a set of nodes $U \subset \{v_1, \ldots, v_{2k+1}\}$ (possibly empty) and two extra nodes u, v, the graph $G(k, u, v, U)$ is an* umbrella *if $\{v_1, \ldots, v_{2k+1}\}$ induces C_{2k+1}, $N(u) = \{v\}$ and $N(v) = \{u\} \cup U$.*

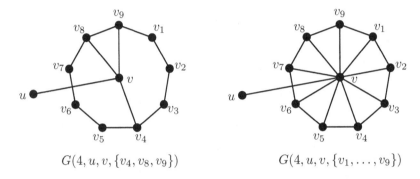

$$G(4, u, v, \{v_4, v_8, v_9\}) \qquad\qquad G(4, u, v, \{v_1, \ldots, v_9\})$$

Remark 1. The graph $\mathcal{H}(k, u, v, i)$ is the umbrella graph $G(k, u, v, \{v_i, v_{i+1}\})$. Notice that for any maximal clique Q in an umbrella graph $G = G(k, u, v, U)$, it is either $|Q| = 2$ or $|Q| = 3$. Moreover, if $s = \alpha(G \ominus v)$ then $|K^3(G)| \leq 2(k-s)+1$.

Remark 2. Given an umbrella graph $G = G(k, u, v, U)$, we denote the points in $\mathbb{R}^{V(G)}$ by $x = (x_u, x_v, x_{V-\{u,v\}})$. From Lemma 1 we have that $\bar{x} = (1, 0, \frac{1}{2} \cdot \mathbb{1})$ is a mixed-integer vertex of QSTAB(G).

When the number of cliques in K^3 attains its maximum value for umbrella graphs, we have the following result.

Theorem 5. *If $G = G(k, u, v, U)$ is an umbrella graph such that $|K^3(G)| = 2(k-s)+1$ then G is not $\mathcal{Q}$-persistent.*

Proof. First, consider $c = (0, 2(k-s)+1, 2 \cdot \mathbb{1})$. If $x \in$ QSTAB(G) then

$$cx = (2(k-s)+1)x_v + 2x(V - \{u, v\}) = \sum_{Q \in K^3} x(Q) + \sum_{Q \in K^2} x(Q)$$

$$\leq |K^3| + 2s = 2k + 1$$

If $\bar{x} = (1, 0, \frac{1}{2} \cdot \mathbb{1})$ then $\bar{x} \in$ QSTAB(G) and $c\bar{x} = 2k + 1$. Therefore, $\bar{x}$ is c−optimal in QSTAB(G).

We next prove that there is no c-optimal stable set containing u. Let S_1 be a maximal stable set in $G \ominus v$. Then, $S_1 \cup \{v\}$ forms a stable set in G and

$$c(S_1 \cup \{v\}) = 2|S_1| + 2(k - s) + 1 = 2k + 1.$$

Given that $G \ominus u = C_{2k+1}$, if S_2 is a stable set containing u, $|S_1 - \{u\}| \leq k$ and then

$$c(S_2) = 2|S_2 - \{u\}| \leq 2k.$$

Thus, $c(S_1 \cup \{v\}) > c(S_2)$ for any stable set S_2 containing u. Therefore, the pair $(c, \bar{x})$ breaks the 1-persistency of QSTAB(G).   □

As opposed to the case above, when the number of cliques in K^3 is not maximum, we have that some umbrella graphs belong to the family $\mathcal{Q}$.

Theorem 6. *Let $G = G(k, u, v, U)$ be an umbrella graph such that $|K^3(G)| < 2(k - s) + 1$. If QSTAB(G) has only one mixed-integer vertex then G is $\mathcal{Q}$-persistent.*

Proof. From hypothesis and Remark 2, $\bar{x}$ is the only mixed-integer vertex of QSTAB(G). Recalling Lemma 3, it suffices to show that for all $c \in \mathbb{R}^{V(G)}$, $c \geq 0$, such that $\bar{x}$ is c-optimal in QSTAB(G) there exists a c-optimal stable set containing $I_1(\bar{x}) = \{u\}$.

Denote $Q_{i,i+1}$ the maximal clique containing both v_i and v_{i+1}, for $i \in [\![2k+1]\!]$ (indices mod. $2k + 1$), i.e., $Q_{i,i+1} = \{v_i, v_{i+1}\}$ or $Q_{i,i+1} = \{v, v_i, v_{i+1}\}$. Let $c \in \mathbb{R}^{V(G)}$ be such that $c \geq 0$ and $c\bar{x} = \max\{cx : x \in \text{QSTAB}(G)\}$. By the complementary slackness theorem, there exists a dual solution $y = (y_Q)_{Q \in K}$ satisfying

$$y_{Q_{i-1,i}} + y_{Q_{i,i+1}} = c_{v_i} \quad \text{for } i \in [\![2k + 1]\!],$$
$$y_{\{u,v\}} = c_u,$$
$$y_{\{v,v_i\}} = 0 \quad \text{for } \{v, v_i\} \in K^2,$$
$$\sum_{Q \in K^3} y_Q + y_{\{u,v\}} \geq c_v,$$
$$y_Q \geq 0 \quad \text{for } Q \in K.$$

Suppose $S \subset V$ is a c-optimal stable set such that $u \notin S$. Assume $v \in S$, since otherwise $S \cup \{u\}$ would also be a c-optimal stable set. Let $S' = S - \{v\}$. Then, $S' \subset V(G \ominus v)$, hence $|S'| \leq s$. Notice that if $v_i \in S'$ then $Q_{i,i+1} = \{v_i, v_{i+1}\}$ and $Q_{i-1,i} = \{v_{i-1}, v_i\}$ because v_i is not adjacent to v in that case.

For $U \subset V$, denote K_U the family of cliques Q such that $Q \cap U \neq \emptyset$. In particular, $K_{\{v_i\}} = \{Q_{i-1,i}, Q_{i,i+1}\}$ and $K_{\{v_i\}} \cap K_{\{v_j\}} = \emptyset$ for all $v_i, v_j \in S'$, $v_i \neq v_j$. Thus,

$$c(S') = \sum_{v_i \in S'} c_{v_i} = \sum_{v_i \in S'} (y_{Q_{i-1,i}} + y_{Q_{i,i+1}}) = \sum_{Q \in K_{S'}} y_Q.$$

Notice that $K_S = K_{S'} \cup K_{\{v\}}$, with $|K_{S'}| = 2|S'|$ and $K_{\{v\}} = K^3 \cup \{\{u, v\}\}$. Moreover, $|K^3| + |K_{S'}| \leq 2(k - s) + 2s = 2k$, which implies that there exist

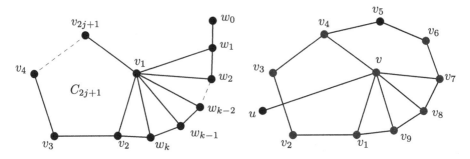

Fig. 1. A whale graph $W(j,k)$ (on the left) and the umbrella graph $G(4,u,v,$ $\{v_1,v_4,v_7,v_8,v_9\})$ (on the right). The blue vertices in the umbrella identify a whale $W(2,3)$ as a node-induced subgraph of it.

$\{v_t, v_{t+1}\} \in K^2 - K_S$. Let $\bar{S}$ be the only stable set such that $\bar{S} \subset \{v_1, \ldots, v_{2k+1}\} - \{v_t, v_{t+1}\}$ and $|\bar{S}| = k$. Notice that $K_{\bar{S}} = K - \{\{u,v\}, \{v_t, v_{t+1}\}\} \supset K_{S'} \cup K^3$. Then,

$$c(S) = c(S') + c_v \leq \sum_{Q \in K_{S'}} y_Q + \sum_{Q \in K^3} y_Q + y_{\{u,v\}} \leq \sum_{Q \in K_{\bar{S}}} y_Q + y_{\{u,v\}} = c(\bar{S}) + c_u.$$

Therefore, the stable set $\bar{S} \cup \{u\}$ is c-optimal too, thus completing the proof.
□

Let us now introduce a family of graphs, called *whale graphs*, which will become important for determining forbidden structures in $\mathcal{Q}$.

Definition 7. *Given $j \geq 2$ and $k \geq 1$, $W(j,k)$ is a* whale graph *with nodes $\{w_0, w_1, \ldots, w_k, v_1, \ldots, v_{2j+1}\}$ if $\{v_i : i \in [\![2j+1]\!]\}$ induces C_{2j+1}, and $N(w_0) = \{w_1\}$, $N(w_i) = \{w_{i-1}, w_{i+1}, v_1\}$ for $i \in [\![k-1]\!]$ and $N(w_k) = \{w_{k-1}, v_1, v_2\}$.*

Figure 1 shows a whale graph and an umbrella graph. The blue vertices in the umbrella identify a whale as a node-induced subgraph of it. Note also that $W(j,1) = \mathcal{H}(j, w_0, w_1, 1)$, for every $j \geq 2$.

In Theorem 7 and Theorem 8 we analyze the 1-persistency property of QSTAB(W) when W is a whale graph. For this, we need the following technical result on the structure of mixed-integer vertices of the clique relaxation of whale graphs.

Lemma 6. *If x is a mixed-integer vertex of QSTAB($W(j,k)$) then $x_{w_0} = 1$, $x_{w_1} = 0$, $x_{w_i} \in \{0, \frac{1}{2}\}$ for all $i \in \{2, \ldots, k-1\}$, $x_{w_k} = 0$ and $x_{v_i} = \frac{1}{2}$ for $i \in [\![2j+1]\!]$.*

Proof. Denote $W = W(j,k)$. Since the only node of W whose destruction results in an imperfect graph is w_0, Lemma 1 ii) implies that if x is a mixed-integer vertex of QSTAB(W) then $x_{w_0} = 1$ and $x_{w_1} = 0$. Also, since $W - v_i$ is a perfect graph, $0 < x_{v_i} < 1$ for all $i \in [\![2j+1]\!]$ by Lemma 1 i) (otherwise x would not be a mixed-integer vertex).

If $p = \max\{j : x_{w_j} = 0\}$ and $V' = V(W) - \{w_p\}$, $x_{V'}$ is a vertex of QSTAB($W[V']$). Note that $W[V']$ is the 1-sum of $W[V_1]$ and $W[V_2]$ at v_1, where $V_1 = \{v_1, w_0, \ldots, w_{p-1}\}$ and $V_2 = (V' - V_1) \cup \{v_1\}$. Lemma 2 shows that x_{V_i} is a vertex of QSTAB($W[V_i]$) for some $i \in \{1, 2\}$. Suppose x_{V_1} is a vertex of QSTAB($W[V_1]$). Since $W[V_1]$ is perfect, x_{V_1} is an integral point contradicting the fact that $0 < x_{v_1} < 1$. Hence, x_{V_2} is a (non-integer) vertex of QSTAB($W[V_2]$). The fact that $x_{V_2} > 0$ implies that it satisfies at equality $|V_2| = 2j + 1 + k - p$ linearly independent clique-inequalities. However, if $p < k$, the number of maximal cliques in $W[V_2]$ is $2j + k - p$. Then, $p = k$ and $W[V_2] = C_{2j+1}$, implying $x_{V_2} = \frac{1}{2} \cdot 1$.

It remains to show $x_{w_i} \in \{0, \frac{1}{2}\}$ for all $i \in \{2, \ldots, k\}$. Assume that i, i' satisfy $1 < i < i' < k$, $x_{w_{i-1}} = x_{w_{i'+1}} = 0$ and $x_{w_h} > 0$ if $i \le h \le i'$. Let us partition the set of these indices into $I_o = \{h : i \le h \le i', h \text{ odd}\}$ and $I_e = \{h : i \le h \le i', h \text{ even}\}$. Consider $0 < \epsilon = \min\{x_{w_h} : i \le h \le i'\}$. From the clique inequalities containing v_1 we have that $\epsilon \le \frac{1}{2}$. Then, $y = x + \epsilon \chi^{I_o} - \epsilon \chi^{I_e} \in$ QSTAB(W) and $z = x - \epsilon \chi^{I_o} + \epsilon \chi^{I_e} \in$ QSTAB(W). Therefore, we have a contradiction of the fact that x is a vertex of QSTAB(W) as $x = \frac{1}{2}y + \frac{1}{2}z$. Therefore, $x_{w_i} > 0$ implies $x_{w_{i-1}} = x_{w_{i+1}} = 0$ for $i \in \{2, \ldots, k - 1\}$. Moreover, if $x_{w_i} > 0$ we must have $x_{w_i} = \frac{1}{2}$ since otherwise, considering $\epsilon' = \min\{x_{w_i}, \frac{1}{2} - x_{w_i}\}$, x would be a convex combination of two distinct points of QSTAB(W), $x + \epsilon' \chi^{\{w_i\}}$ and $x - \epsilon' \chi^{\{w_i\}}$. □

Theorem 7. *For every $j \ge 2$ and $m \ge 1$, the graph $W(j, 2m)$ is Q-persistent.*

Proof. Denote $W = W(j, 2m)$. Let $x = (x_{w_0}, x_{w_1}, \ldots, x_{w_{2m}}, x_{v_1}, \ldots, x_{v_{2j+1}}) \in$ QSTAB(W). According to Lemma 4, it is enough to consider only *maximal* mixed-integer vertices. Let $\bar{x}$ be such a vertex, and let c be such that $\bar{x}$ is c-optimal in QSTAB(W). By Lemma 3 we can assume $c \ge 0$ and from Lemma 6, $\bar{x}_{w_0} = 1$, $\bar{x}_{w_1} = 0$, $\bar{x}_{w_i} \in \{0, \frac{1}{2}\}$ for all $i \in \{2, \ldots, k - 1\}$, $\bar{x}_{w_k} = 0$ and $\bar{x}_{v_i} = \frac{1}{2}$ for $i \in [\![2j + 1]\!]$.

Suppose $\bar{x}_{w_2} = 0$. Let y be such

$$y_i = \begin{cases} \frac{1}{2} & \text{if } i = w_1 \text{ or } i = w_0, \\ \bar{x}_i & \text{if } i \in V(W) - \{w_0, w_1\}. \end{cases}$$

Clearly, $y \in$ QSTAB(W) and since $\bar{x}$ is c-optimal, $c\bar{x} \ge cy$. Then, $c_{w_0} \ge c_{w_1}$. After Theorem 2 ii), the pair $(c, \bar{x})$ does not break 1-persistency.

Consider now $\bar{x}_{w_2} = \frac{1}{2}$. Since $\bar{x}_{w_{2m}} = 0$ and, as shown in the proof of Lemma 6, $\bar{x}_{w_i} = \frac{1}{2}$ implies $\bar{x}_{w_{i+1}} = 0$, there exists $i \in \{3, \ldots, 2m - 1\}$ such that $\bar{x}_{w_i} = \bar{x}_{w_{i+1}} = 0$. Also, the first index i where this holds has to be odd, say $2r + 1$.

Let S be a c-optimal stable set in W, which we can assume to be maximal. If $w_0 \in S$, then $(c, \bar{x})$ does not break 1-persistency. Suppose $w_0 \notin S$ and, consequently, $w_1 \in S$.

Let t be such that $2t + 1 = \min\{h \in [\![2m]\!] : w_h \notin S, h \text{ odd}\}$. If $2t + 1 \le 2r + 1$, consider y defined as follows:

$$y_i = \begin{cases} \frac{1}{2} & \text{if } i = w_0 \text{ or } i = w_{2q-1} \text{ for } q \in [\![t]\!], \\ 0 & \text{if } i = w_{2q} \text{ for } q \in [\![t]\!], \\ \bar{x}_i & \text{if } i \in V(W) - \{w_0, \ldots, w_{2t}\}. \end{cases}$$

Clearly, $y \in \text{QSTAB}(W)$. Since $\bar{x}$ is c-optimal, $cy \leq c\bar{x}$ and then

$$\sum_{q=1}^{t} c_{w_{2q-1}} \leq c_{w_0} + \sum_{q=1}^{t} c_{w_{2q}}. \tag{1}$$

After this, we can define another stable set S' by just interchanging the nodes w_i with odd indices by the ones with even indices in S up to $2t$ and adding w_0 to it. That is,

$$S' = (S - \{w_{2q-1} : q \in [\![t]\!]\}) \cup \{w_{2q} : q \in [\![t]\!]\} \cup \{w_0\}.$$

From (1), $c(S) \leq c(S')$, hence S' is c-optimal, with $w_0 \in S'$.

If $2t + 1 > 2r + 1$ we follow a similar reasoning defining y as follows:

$$y_i = \begin{cases} \frac{1}{2} & \text{if } i = w_0 \text{ or } i = w_{2q-1} \text{ for } q \in [\![r+1]\!], \\ 0 & \text{if } i = w_{2q} \text{ for } q \in [\![r]\!], \\ \bar{x}_i & \text{if } i \in V(W) - \{w_0, \ldots, w_{2r}\}, \end{cases}$$

and clearly $y \in \text{QSTAB}(W)$. Since $\bar{x}$ is c-optimal, we have $cy \leq c\bar{x}$ and then

$$\sum_{q=1}^{r+1} c_{w_{2q-1}} \leq c_{w_0} + \sum_{q=1}^{r} c_{w_{2q}}. \tag{2}$$

Again, we can define a stable set

$$S' = (S - \{w_{2q-1} : q \in [\![r+1]\!]\}) \cup \{w_{2q} : q \in [\![r]\!]\} \cup \{w_0\}$$

which is c-optimal by (2), and $w_0 \in S'$. In any case, $(c, \bar{x})$ does not break 1-persistency and therefore, $W \in \mathcal{Q}$. $\qquad\square$

Despite the above result, it is not hard to prove that not every whale graph belongs to $\mathcal{Q}$. Indeed, we show that there are minimal forbidden structures for 1-persistency in the family of whale graphs.

Theorem 8. *For every $j \geq 2$ and $m \geq 1$, the graph $W(j, 2m+1)$ is mn$\mathcal{Q}$.*

Proof. Let $W = W(j, 2m+1)$ and consider c defined as

$$c_i = \begin{cases} 0 & \text{if } i = w_0, \\ 1 & \text{if } i = w_1, \\ 2m+2 & \text{if } i = v_1, \\ 2 & \text{if } i \in V(W) - \{w_0, w_1, v_1\}. \end{cases}$$

Let $\bar{x}$ be such that

$$\bar{x}_i = \begin{cases} 1 & \text{if } i = w_0, \\ 0 & \text{if } i = w_{2q-1} \text{ for } q \in [\![m+1]\!], \\ \frac{1}{2} & \text{if } i = w_{2q} \text{ for } q \in [\![m]\!] \text{ or } i = v_i \text{ for } i \in [\![2j+1]\!], \end{cases}$$

Now, observe that for any $x \in \text{QSTAB}(W)$ the clique inequalities imply that $cx \leq 2(j + m) + 1$. Also, since $c\bar{x} = 2(j + m) + 1$ we have that $\bar{x}$ is c-optimal.

Let S' be a stable set with $w_0 \in S'$ and let us define $S_1 = S' \cap \{v_i : i \in [\![2j+1]\!]\}$ and $S_2 = S' \cap \{w_2, \ldots, w_{2m+1}\}$. Note that $|S_1| \leq j$ and $|S_2| \leq m$. If $v_1 \notin S'$,

$$c(S') = c(S_1) + c(S_2) + c_{w_0} \leq 2(j + m).$$

On the other hand, if $v_1 \in S'$,

$$c(S') = c(S_1 - \{v_1\}) + c_{v_1} + c_{w_0} \leq 2(j - 1) + (2m + 2) = 2(j + m).$$

Therefore, $c(S') \leq 2(j + m)$ for any stable set S' with $w_0 \in S'$.

If $S = \{w_{2q-1} : q \in [\![m + 1]\!]\} \cup \{v_{2q+1} : q \in [\![j]\!]\}$ then S is a stable set in W such that $w_0 \notin S$ and $c(S) = 2(j + m) + 1$. Consequently, $(c, \bar{x})$ breaks 1-persistency and then $W \notin \mathcal{Q}$.

It remains to check that every proper induced subgraph of W belongs to $\mathcal{Q}$. The graphs $W - v_i$ for $i \in [\![2j + 1]\!]$ and $W - w_q$ for $q \in \{0, 1\}$ are bad-paw-free and then are Q-persistent from Theorem 3.

The graph $W - w_i$ for $i \in \{2, \ldots, 2m + 1\}$ is an induced subgraph of an even whale $W(j, 2m+2)$. Then, by Theorem 7 and Theorem 1, $W - w_i$ is Q-persistent. □

4   Final Remarks

In this work, we present a variant of the persistency property studied in [5] which we call the 1-persistency. We analyze this property on a stronger relaxation than the edge relaxation of the stable set polytope and provide sufficient conditions for a graph to belong to the family of Q-persistent graphs. Based on the fact that this property is hereditary we found two different infinite families of forbidden minimal structures for it. Our next step is to continue in this line with the aim of fully characterizing Q-persistent graphs.

As we briefly mention in Sect. 1, studying persistency properties on polytopes may be useful in practice, as it allows us to reduce the size of the problem by fixing some variables to provable optimal integer values. These variable fixings may be used whenever a fractional optimal solution for a persistent relaxation of the problem contains integer values. It is worth noting that these fixings may be further incorporated into classical branch-and-bound algorithms, provided that the subproblems created by the branching rules preserve the persistency property (e.g., for the continuous relaxation). To this end, after finding an optimal fractional solution on a relaxation, variable fixings may be safely applied even before performing the branching step. Afterwards, the branch-and-bound algorithm may continue as usual, thus repeating the variable fixing step before every branching.

An Interesting Novel Line of Work. Variable fixings due to persistency could also be incorporated into a classical cutting-planes procedure: after finding an optimal fractional solution on a relaxation, variable fixings may be applied before the addition of valid inequalities (to cut-off the fractional solution) and the re-optimizing step. However, in order to iterate this idea safely, we should ensure that every added inequality preserves the persistency property of the obtained relaxation. Since different valid inequalities may lead to different relaxations, this suggests an interesting line of work, namely, *the study of the persistency-preservation property of a valid inequality with respect to a given relaxation.* If a cutting-planes algorithm only uses valid inequalities that preserve the persistency property of the relaxation, then variable fixings may be safely applied at every cutting round. We propose to refer to this scheme as a *cut-and-fix algorithm* (or *branch-and-cut-and-fix* if it is combined with a branch-and-bound technique). In this work, we focus on the study of the persistency property on known relaxations, not on individual valid inequalities. However, we believe that our results may be used as a starting point for this novel line of work in the polyhedral combinatorics field.

References

1. Chudnovski, M., Robertson, N., Seymour, P., Thomas, R.: The strong perfect graph theorem. Ann. Math. **164**(1), 51–229 (2006)
2. Chvátal, V.: On certain polytopes associated with graphs. J. Comb. Theory Ser. B **18**(2), 138–154 (1975)
3. Koster, A., Wagler, A.: The extreme points of QSTAB(G) and its implications, ZIB-Report 06-30 (2006)
4. Nemhauser, G., Trotter, L.: Vertex packings: structural properties and algorithms. Math. Prog. **8**(1), 232–248 (1975)
5. Rodríguez-Heck, E., Stickler, K., Walter, M., Weltge, S.: Persistency of linear programming relaxations for the stable set problem. Math. Prog. **192**, 387–407 (2022)
6. Wagler, A.K.: Rank-perfect and weakly rank-perfect graphs. Math. Meth. Oper. Res. **56**, 127–149 (2002)

Alternating Direction Method and Deep Learning for Discrete Control with Storage

Sophie Demassey[1(✉)] , Valentina Sessa[1] , and Amirhossein Tavakoli[1,2]

[1] Centre for Applied Mathematics, Mines Paris-PSL, Sophia-Antipolis, Paris, France
{sophie.demassey,valentina.sessa,amirhossein.tavakoli}@minesparis.psl.eu
[2] University Côte d'Azur, Sophia-Antipolis, Nice, France

Abstract. This paper deals with scheduling the operations in systems with storage modeled as a mixed integer nonlinear program (MINLP). Due to time interdependency induced by storage, discrete control, and nonlinear operational conditions, computing even a feasible solution may require an unaffordable computational burden. We exploit a property common to a broad class of these problems to devise a decomposition algorithm related to alternating direction methods, which progressively adjusts the operations to the storage state profile. We also design a deep learning model to predict the continuous storage states to start the algorithm instead of the discrete decisions, as commonly done in the literature. This enables search diversification through a multi-start mechanism and prediction using scaling in the absence of a training set. Numerical experiments on the pump scheduling problem in water networks show the effectiveness of this hybrid learning/decomposition algorithm in computing near-optimal strict-feasible solutions in more reasonable times than other approaches.

Keywords: Mixed Integer Nonlinear Programming · Variable splitting · Deep Learning

1 Introduction

Operating any system governed by nonlinear laws and discrete controls leads to complex optimization problems. If memory or storage capacities are present, load shifting allows for a further reduction of operating costs. However, the problem of planning on a timeline the static operations interrelated by the storage balance and capacity limits is not only challenging to optimize but even to satisfy.

Consider the operation of a pressurized drinking water distribution network: the problem is to plan the activation of the pumps on a discrete time horizon to realize, at minimum cost, a hydraulic head/flow equilibrium meeting the demand in water on each time step. Elevated water tanks enable shifting the pumping ahead of time, resulting in substantial savings in energy consumption and cost, given the nonlinear efficiency of the pumps and a dynamic incentive electricity tariff. Nonconvex MINLPs modeling this problem are hard to optimize and even

A. Basu et al. (Eds.): ISCO 2024, LNCS 14594, pp. 85–96, 2024.
https://doi.org/10.1007/978-3-031-60924-4_7

to satisfy, as the feasible set is usually sparse and scarce in the binary search space, especially when the tank capacities are tight. Dedicated exact or heuristic algorithms attempt in various ways to simplify the hydraulic relations [2,14] or the storage constraints [1,8,19], by trading off accuracy for speed. Still, even for small-size problems, they struggle to reach feasible solutions, and little seeks to fully repair the approximate solutions. The latter is mainly addressed by flipping the binary variables within a local [1,15] or global search [2,8]. Also, most existing approaches rely on fast simulation, using Newton methods, to compute the unique possible static hydraulic equilibrium at a time step when both the status of the pumps and the level of the tanks are known [18].

This property – fast computation of the static operations given a storage state and no condition on the resulting state – is not specific to hydraulic systems. In particular, it arises alike in other potential-flow networks in various contexts, ranging from electric circuits or thermodynamic systems to traffic congestion [16, p.350]. It also holds in long-term planning problems split into a sequence of smaller periods, where the planning subproblems become independent and easy once the initial states are fixed, and the final state conditions dropped. This paper focuses on computing feasible solutions for discrete control of systems with storage satisfying this property by combining two independent but complementary models: a data model built from deep learning to predict multiple approximate solutions, followed by a feasibility recovery phase based on the storage/control variable splitting in the MINLP formulation. The whole approach revolves around storage state profiles as partial solutions: it searches through these partial solutions to gradually match a feasible control decision.

It then differs from previous applications of machine learning to optimize combinatorial problems like TSP [13] or to predict partial discrete solutions of MILPs [5] where repairing/completing a feasible full assignment is less of an issue. Working in the continuous space of the storage state variables instead of the discrete control variables has advantages. First, our data model is regression and not classification. In particular, we combined a convolutional neural network (CNN) and bidirectional long short-term memory (Bi-LSTM) to capture local patterns in the input (loads and costs) and output (storage states) time series. Furthermore, we employ Monte Carlo dropout [6] to obtain multiple predictions. Working with continuous variables allows for smoother moves (both in the neighborhood exploration and in the multi-start mechanism) to address the feasibility recovery issue. It also enables us to develop a scaling mechanism as a substitute for missing data: if no training dataset is available for a given system, we propose to build it by solving, for a sufficient number of inputs, a tractable variant of the problem with a coarser temporal resolution. Once trained, the data model predicts coarse-grained storage state profiles, which are easily interpolated into fine-grained profiles, possibly without much loss in the prediction accuracy.

In the feasibility recovery phase, continuity allows for smoother moves in the search space, and the property above offers a mechanism to map storage solutions to control decisions. To leverage this property, we investigate an alternating direction method (ADM), restricting and solving the MINLP in these two spaces alternatively. Precisely, we penalize the time-coupling storage balance constraints, then iterate between the control and storage subproblems: at each iteration, the penalized MINLP is restricted to the partial solution obtained from one subproblem, hence defining the next subproblem.

As we make no assumption on the analytic nature of the coupling between storage and control in the static subproblems, our algorithm differs from other ADM schemes. In particular, it differs from ADMM [20] and PADM [12], as we choose not to dualize this coupling. These algorithms have known theoretical convergence guarantees, even in the nonconvex case, but only if the coupling is linear, which is not the case in pump scheduling, for example. Instead, we can easily enforce this structural constraint in the control subproblem thanks to the property, but then we need to relax it in the storage subproblem. In doing so, we may lose some mild theoretical convergence guarantees [10], but we expect a fast and practical convergence to feasible solutions without much altering the cost. Besides, we enforce search diversification by multi-start, using the multiple data model predictions as trial initial points.

We illustrate and experiment with our approach to the pump scheduling problem and build an extended test set based on the benchmark water network *Van Zyl* [19]. Both deep learning and ADM are original approaches for this well-studied problem. While independent, they appear to be strongly complementary: ADM is able to recover feasible decisions with a reasonable optimality gap from predictions of the data model, which, conversely, provides good approximate solutions to initialize ADM.

2 MINLP for Discrete Control with Storage

The optimal control problem is modeled as a sequence of steady states over a discretized time horizon $\mathcal{T} = \{0, 1, ..., T-1\}$ as follows:

$$(P): \min_{x,y,s} \sum_{t \in \mathcal{T}} f_t(x_t, y_t, s_t, C_t) \tag{1a}$$

$$\text{s.t.: } g_t(x_t, y_t, s_t, L_t) = 0 \qquad\qquad \forall t \in \mathcal{T} \tag{1b}$$

$$s_{t+1} = s_t + y_t^I \qquad\qquad \forall t \in \mathcal{T} \tag{1c}$$

$$s_t \in [\underline{S}_t, \overline{S}_t] \subseteq \mathbb{R}^I \qquad\qquad \forall t \in \mathcal{T} \cup T \tag{1d}$$

$$x_t \in \mathcal{X}_t \subseteq \{0, 1\}^N, y_t \in \mathbb{R}^M \qquad\qquad \forall t \in \mathcal{T}. \tag{1e}$$

Vector $s_t \in \mathbb{R}^I$ denotes the state variables figuring the state/level of the I storage devices at $t \in \mathcal{T}$ or at the end of the horizon $t = T$. Vector x_t represents N discrete control variables at $t \in \mathcal{T}$. W.l.o.g. We assume they are binary variables figuring the on/off status of N controllers, and $\mathcal{X}_t$ represents the allowed combinations. Vector $y_t \in \mathbb{R}^M$ includes all other continuous (control or implied) variables. Together with variables x_t under Constraints (1b), they model the steady operation of the system on period $t \in \mathcal{T}$ to serve a given load $L_t \in \mathbb{R}^J$ starting from the storage state s_t. Note that y_t includes, as a subvector denoted $y_t^I \in \mathbb{R}^I$, the positive or negative contribution of the system operation to every storage at $t \in \mathcal{T}$, as modeled by Constraints (1c). Constraints (1d) define the minimum $\underline{S}$ and maximum $\overline{S}$ storage capacities. These limits are allowed to vary in time, and $\underline{S}_0 = \overline{S}_0$, i.e., the storage state is fixed at $t = 0$. Finally, the objective (1a) is to minimize the global operation cost, where $C_t \in \mathbb{R}^\ell$ denotes exogenous price signals for the different elements operated at $t \in \mathcal{T}$. We make no assumption on the analytic nature of the real-valued functions in objective f_t and in constraints g_t, but the following property must hold.

Property 1. For all time $t \in \mathcal{T}$, solving or optimizing over $\{(x, y) \in \mathcal{X}_t \times \mathbb{R}^M : g_t(x, y, S, L_t) = 0\}$ is computationally easy, given any initial state $S \in \mathbb{R}^I$.

With the above property, we assume the subproblems can be solved quickly enough to iterate on them.

3 A Concrete Model: Pump Scheduling

The standard nonconvex MINLP for the pump scheduling problem in water distribution networks without maintenance constraints [2] is a concrete implementation of (P) through the following identification:

$$f_t(x_t, y_t, s_t, C_t) \equiv C_t^0 x_t + C_t^1 y_t^Q \quad \text{(pumping electric cost)},$$

$$g_t(x_t, y_t, s_t, L_t) \equiv \begin{cases} E_I y_t^Q - y_t^I & \text{(tank inflow)} \\ E_J y_t^Q - L_t & \text{(load satisfaction)} \\ s_t - B y_t^H & \text{(tank head)} \\ (E^\top y_t^H + \phi(y_t^Q)) x_t & \text{(head loss)} \end{cases}$$

where variables x represent the activity of the network arcs (i.e., pumps and pipes with or without valves), y^Q the flow through the arcs, y^H the pressure (hydraulic head) at the network nodes (i.e., tanks and service nodes), and s the tank levels. Given E the network incidence matrix, the operational constraints g_t include, in order: flow conservation at tanks, then at service nodes, tank volume/head linear relation B, nonlinear relations ϕ between head loss and flow on each arc.

For fixed $s_t = S \in \mathbb{R}^I$ and $x_t = X \in \{0,1\}^N$, the system of equations $g_t(X, y, S, L_t) = 0$, known as the *network analysis problem*, admits at most one solution y, easy to compute with Newton methods [18]. Furthermore, it breaks into independent subsystems following the network partition along the tanks [1]. Usually, each subnetwork (resp. subsystem) $p \in P$ contains a number N_p of controllable arcs (resp. binary variables), small enough to envisage to enumerate all possible combinations $X \in \{0,1\}^{N_p}$ and compute the unique corresponding solution denoted $y^p(X, S)$. The solutions of system $g_t(X, y, S, L_t) = 0$ are thus any product of such vectors $(X_p, y^p(X_p, S))$ over $p \in P$ with $X_p \in \{0,1\}^{N_p}$. Hence, Property 1 holds for pump scheduling.

Finally, note that the system $g_t(x_t, y_t, s_t, L_t) = 0$ defined above in the context of hydraulic networks, is an instance of the *nonlinear network equilibrium problem* [16, p.350] arising in many contexts, ranging from thermodynamic systems to traffic congestion. This provides insight into the broad applicability of model (P) with Property 1.

4   ADM for Local Optimization by Decomposition

Model (P) admits a time-decomposition after dualizing or penalizing the storage balance constraints (1c) as in the following model, given ℓ_1 penalty and multiplier $\rho_{ti} \geq 0$ for each $d_{ti}(y, s) = |s_{(t+1)i} - (s_{ti} + y_{ti}^I)|$ at storage i and time t:

$$(L_\rho):\ \min_{x,y,s}\{l(x, y, s, \rho):(1b),(1d),(1e)\},\ \text{with}$$
$$l(x, y, s, \rho) = \sum_{t \in \mathcal{T}} f_t(x_t, y_t, s_t, C_t) + \sum_{i=1}^{I} \rho_{ti} d_{ti}(y, s).$$

Even if separable as T independent subproblems, such relaxations remain difficult to solve because of the complexity of each system (1b). However, they become tractable for fixed values of s according to Property 1.

We thus propose solving approximately (L_ρ) following the storage/control variable split: we first solve (L_ρ) with respect to the control variables (x, y), given a storage state profile s, then solve (L_ρ) with respect to the state variables s for fixed (x, y), and iterate, using the solution of one subproblem to define the restriction in the next subproblem. Contrarily to other alternating direction methods, like ADMM or PADM, we do not dualize the variable-coupling constraints (1b) but keep them as structural constraints in the control subproblems to leverage Property 1, and relax them from the storage subproblems to ensure feasibility. As a consequence, the algorithm loses the mild guarantee of [10] to converge to a stationary point, not even a local minimum, of the relaxed problem. Still, it generates a sequence of solutions closer to being feasible for (P).

Algorithm 1. Partial storage/control splitting for (P)

1: choose storage profile $s^0 \in \mathbb{R}^{TI}$, penalty $\rho \in \mathbb{R}_+^{TI}$ and set $k = 0$
2: given (s^k, ρ), solve $(x^{k+1}, y^{k+1}) \in \arg \min_{(x,y)}\{l(x, y, s^k, \rho):(1b),(1e)\}$
3: given (x^{k+1}, y^{k+1}, ρ), solve $s^{k+1} \in \arg \min_s\{l(x^{k+1}, y^{k+1}, s, \rho):(1d)\}$
4: If $\|d(y^{k+1}, s^{k+1})\|_\infty < \varepsilon$ then $(x^{k+1}, y^{k+1}, s^{k+1})$ is ε-feasible for (P), then STOP.
   Otherwise, if $\|s^{k+1} - s^k\|_\infty < \varepsilon'$, then update ρ. Increment k and go to Step 2.

Conceptually, Algorithm 1 searches through the s-space of the storage state profiles satisfying the capacity limits and attempts to derive a matching control profile (x, y) by gradually reconciling the storage states $s_t + y_t^I$ and s_{t+1} at each time t. Following the framework for PADM [7], the penalty ρ is updated when the algorithm stalls or after a given number of iterations.

The efficiency of this algorithm is highly dependent on the initial storage state profile s^0. We design a deep learning model to predict near-feasible near-optimal profiles. This model is capable of deriving multiple predictions, allowing us to add diversification to the local optimization algorithm: starting from various initial points s^0 increases the chance of reaching a feasible solution.

5   The Deep Learning Model

Our deep learning algorithm aims at building a hypothesis function $\mathcal{H}$ mapping to each input $(L, C) \in \mathbb{R}^{T(j+\ell)}$ (loads and costs), the storage profile

$s(L, C) \in \mathbb{R}^{TI}$ in an optimal solution of problem (P) with input (L, C). The map is built, given a precomputed collection $\{((L_l, C_l), s_l)\}_{l=1}^{N_D}$ of input/target tuples, by approximately minimizing a loss function $\mathcal{L}_{loss}(\mathcal{H}(L, C), s(L, C))$, which is the mean square error in our case, as commonly used in regression problems.

To compute $\mathcal{H}$, we choose to combine a Convolutional Neural Network (CNNs) and a Bidirectional Long Short-Term Memory Network (Bi-LSTM) with dropout layers, as found in several works, e.g., in [3]. The capability of Bi-LSTM lies in its ability to handle information in both forward and backward temporal directions. This attribute is crucial in our application due to the dynamic relations spanning the entire planning horizon. On the other hand, our CNN comprises convolutional (conv1D) layers of different sizes, located in parallel to each other with zero padding, and with their outputs exclusively linked to neighboring areas within the input (i.e., loads and costs). This arrangement is accomplished by moving a filter (i.e., a weight matrix) along the input vectors, and this design enables the model to acquire filters to discover distinct patterns within the dataset [9]. To handle the temporal dimension of the target, the output of the CNN layers, implied as extracted features, are passed through a hidden layer with ReLU activation functions, and their outputs are concatenated, reshaped, and fed into the Bi-LSTM. The final prediction $\mathcal{H}(L, C)$ results from a fully connected layer with a linear activation function placed after the Bi-LSTM unit.

Finally, we implement the Monte Carlo dropout technique [17] to force our model to make multiple predictions. While this technique has been proposed to estimate the prediction uncertainty by randomly masking neurons in the neural network, we employ it for a diversification purpose. It is particularly suitable in our context, given the epistemic uncertainty due to the model complexity and the relatively small size of the training dataset. The dropout layers are placed just before and after the Bi-LSTM unit, and, in alignment with [6], we apply dropout (i.e., random neuron masking) not only during training but also during the testing phase. As a result, for a given input (L, C), the dropouts yield as many distinct solution samples $\mathcal{H}(L, C)$.

6   Experimental Evaluation

We run experiments in the context of pump scheduling on *van Zyl*, a benchmark hydraulic network. We first evaluate the predictive accuracy of our deep learning model (denoted CNN-LSTM below) and compare it to a conventional feedforward neural network (FFW) having a similar number of parameters. We then evaluate the performance of our hybrid algorithm (HA) and compare it to the open-source branch-and-check global optimizer (BC) from [2] dedicated to pump scheduling, and the enhanced variant from [4] (BCpre). BC solves the MILP relaxation of (P) with an LP-Branch-and-Bound combined with hydraulic simulation to check the actual feasibility and cost of the integer solutions. BCpre implements a heavy preprocessing to tighten the MILP formulation before running BC.

All algorithms are implemented in Python and experiments are executed on an Intel(R) Xeon(R) 6148 2.40GHz and 128 GB memory. The deep learning models CNN-LSTM and FFW are built using Tensorflow API v2.12.0 on Jupyter notebook in Google Colab with GPU A100, and the Adam optimizer [11] with the default initial learning rate of 10^{-3}. Algorithms BC and BCpre are implemented on top of the MILP solver Gurobi v10.0.1.

6.1   Benchmark, Dataset Generation, and Scaling

The *Van Zyl* network [19] is characterized by $I = 2$ water tanks of different capacities, $J = 2$ service nodes with individual loads, and $N = 4$ controllable elements (2 symmetrical pumps and a boost pump operating parallel to a check valve). This is a medium-sized but difficult network, often used for empirical evaluations. However, the available test set is small.

We present a new dataset for implementing deep learning algorithms. It consists of a collection of 2113 unique daily observations (L, C, S), each given as $T = 12$ time steps profiles: the input features are the water demand and the electricity tariff profiles $(L, C) \in \mathbb{R}^{T(J+1)}$, and the target is the storage state profile $S \in \mathbb{R}^{TK}$ in the optimal solution of (P) for input (L, C) obtained with algorithm BCpre. Tariffs C are taken from the Belgian spot market data, considering a reference period from 2007 to 2013. Loads L are drawn from a 3-year highly seasonal history of actual consumption sourced from a network based in a touristic zone in Brittany, France. The data is split into 75 percent training, 15 percent validation, and 10 percent evaluation. Given $T = 12$, pumps and valves are scheduled on 2-hour periods, aligned with the resolution of the load and cost forecasts. Because of the nonlinear behavior and the limited capacity of the tanks, it is advantageous to enforce a finer-grained schedule on periods of 1 h $(T = 24)$ or $1/2$ h $(T = 48)$. However, when the time horizon length increases, it quickly becomes difficult to compute even feasible solutions in a reasonable time. This prevents the creation of a dataset of reasonable size to train our data model on daily instances for $T = 24$ or $T = 48$. To apply our algorithm to such instances, we then use the data model trained for $T = 12$ and scale down and up the input and output time series, using linear interpolation, to get the predictions for starting ADM. We expect that such a simple transformation does not deteriorate the quality (feasibility and optimality) of the predicted storage state profiles. We generate the two benchmark sets of 50 instances for $T = 24$ and $T = 48$ also by resampling instances with coarser-time discretization $T = 12$ and high variability. The three benchmark sets (denoted VZ12, VZ24, and VZ48), together with the training set and the code of the algorithms, are available online[1].

6.2   Hyperparameters of the Deep Learning Models

To train CNN-LSTM, we set the batch size to 32, and the maximum number of epochs to 1350. For all conv1D and feedforward layers, we use l_2 regularization

---

[1] https://github.com/sofdem/gopslpnlpbb

with a coefficient equal to 0.02. The conv1D layers have 32 filters, and their kernel sizes range from 4 to 10. The number of hidden layers of Bi-LSTM is set to 16. Due to the complexity of the model and the relatively small training set, and for more diversification, we select a high dropout rate of 0.75, then generate 80 Monte Carlo dropout samples.

For comparison, we design a FFW network with 8 hidden layers, $ReLU$ activation functions, and a pyramid architecture. We set the maximum number of epochs to 250, and the regularization coefficient to 0.01. We also add dropout layers with a rate between 0.2 and 0.5 after each hidden layer to mitigate overfitting and generate multiple starting points for ADM. In the end, CNN-LSTM and FFW have 45k and 63k parameters, respectively.

6.3   Parameters of the Optimization Algorithms

We run Algorithm 1 sequentially, although it could be parallelized, from at most 35 trial points s^0: the first trial is the component-wise mean prediction, other trials are just picked randomly from the 80 predictions samples. Both ADM and BC algorithms are stopped when reaching a first feasible solution, given the feasibility tolerance $\varepsilon = 10^{-6}$, or a global time limit.

All other parameter values were chosen after a brief numerical evaluation process. In Algorithm 1, we set $\varepsilon' = 10^{-3}$. For a given trial s^0, the penalty vector ρ is updated at most four times, and the number of iterations k is limited to 85 between each update. At each update, penalties are increased according to the update number $u \in \{0, \ldots, 4\}$ and to the constraint violations. Precisely, for each time $t \in \mathcal{T}$ and storage $i \in \{1, \ldots, I\}$, we update ρ_{ti} to $1 + 5a$ if $d_{ti}(y^{k+1}, s^{k+1}) > \varepsilon$, and to $1 + 2a$ otherwise, given a random value $\xi \in [0.75, 1]$ and $a = \xi e^{(\frac{-u}{10})}\rho_{ti}$. We initialize the penalty terms ρ_{ti}, uniformly, either to value 2 or to value 50, corresponding to algorithms denoted HA2 and HA50 below.

6.4   Performance of Deep Learning

Prediction error. To assess the effectiveness of the CNN-LSTM, we first compute the prediction error between the outcome $\mathcal{H}(L, C)$ (the component-wise mean of the 80 prediction samples) and the target $s(L, C)$ over the test set, using the classical metrics for time series in regression problems. With CNN-LSTM, the Mean Square Error between predictions and targets is MSE= 0.64, and the Mean Absolute Error is MAE= 0.53 or nMAE= 9.2% after normalization (divided by the range of target values). The Pearson coefficient, which measures the linear correlation between predictions and targets, is $R = 0.772$. In comparison, FFW yields MSE= 1.56, MAE= 0.89, nMAE= 12.7%, and $R = 0.724$. These metrics show that CNN-LSTM performs well and better than FFW. Figure 1 depicts the storage state profiles in each tank for one random instance in the test set, showing the proximity of the optimal target and the CNN-LSTM prediction with its uncertainty interval.

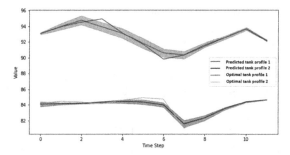

Fig. 1. Storage state profiles: prediction, credible interval, and optimal target.

Distance to a Feasible Solution. Since the optimal solutions of (P) are not unique, these metrics only show the ability of the deep learning model to predict the target one. We now estimate the distance of the predictions to feasible solutions by measuring the number of predictions that actually lead to feasible solutions through ADM (with penalty terms ρ initialized to 50). When starting from CNN-LSTM predictions, ADM finds a feasible solution for 49 out of the 50 instances in benchmark VZ12, but only half (27 instances out of 50) from FFW predictions. For the 26 instances where both algorithms reach a feasible solution, the associated objective function value derived with CNN-LSTM is smaller for all but two instances, with an average 4.8% decrease.

6.5   Performance of the Hybrid Algorithm

The previous experiment provides a first insight into the efficiency of algorithm HA. It hints that better predictions increase the chance of ending up with a feasible solution or a better upper bound. We now validate the capability of HA to compute good feasible solutions for the daily instances of the Van Zyl network with $T = 12, 24, 48$. We evaluate the hybrid algorithm with the penalty terms ρ in ADM initialized to 50 (HA50) or to 2 (HA2), and we compare it with the two variants of branch-and-check, with (BCpre) and without (BC) advanced preprocessing. Each algorithm stops when reaching one first feasible solution or the CPU time limit: 1800 seconds for $T = 12$, 3600 seconds for $T = 24$, and 7200 seconds for $T = 48$. Table 1 reports, for each benchmark set (column 1) and for each algorithm (column 2), the number of instances solved (column 3) among the 50 instances in the set, then, on the subset of instances solved by the algorithm: the average (column 4) and maximum (column 5) computational time in seconds, and the average (column 6) and maximum (column 7) estimated optimality gap, measured as the deviation in % of the solution cost to the best known lower bound computed with BCpre.

The results of the global solver BC illustrate well the computational complexity of this problem, particularly as the horizon length T increases. With a tailored preprocessing (running for a minimum of 100 seconds included in the time reported in the table), BCpre quickly reaches one first feasible solution

Table 1. Instances solved, computation time, and optimality gap.

bench	algo	#solved	avg time	max time	avg gap	max gap
VZ12	HA50	49	254 s	1570 s	6.6%	21.2%
	HA2	44	305 s	1577 s	4.6%	11.3%
	BC	48	121 s	681 s	5.4%	12.5%
	BCpre	50	124 s	137 s	4.3%	12.4%
VZ24	HA50	50	285 s	1257 s	9.5%	23.4%
	HA2	50	279 s	1711 s	8.4%	16.3%
	BC	5	1097 s	3117 s	11.1%	12.6%
	BCpre	50	809 s	2430 s	7.5%	39.6%
VZ48	HA50	50	776 s	7069 s	9.8%	21.0%
	HA2	49	1014 s	5548 s	10.3%	19.7%
	BC	1	–	–	–	–
	BCpre	32	2517 s	6404 s	6.4%	8.9%

of high quality for all instances with $T = 12$. The performance falls dramatically for the largest instances with $T = 24$ and $T = 48$. Without preprocessing, BC cannot even compute one feasible solution for most of the instances within the allowed time. Preprocessing strongly improves the performance, still, BCpre cannot reach feasible solutions in two hours for 18 instances with $T = 48$.

While the hybrid algorithm is globally outperformed on the instances with $T = 12$, it is well suited to the hardest instances with $T = 24$ and $T = 48$, as HA50 can compute a feasible solution for all these instances, within an average 10% estimated optimality gap (which is overestimated as the lower bound computed by BCpre is probably far below the optimum value). Because of the scaling mechanism, the prediction error is probably worse for these instances, compared to $T = 12$ where the deep learning predictions are directly used as starting points for ADM. However, the space of the feasible solutions is also less scarce, and the optimal storage state profiles for the coarse and fine discretization problems are probably not too dissimilar. This should explain the good performance of the hybrid algorithm on the hardest benchmark sets VZ24 and VZ48 and how this approach overcomes the difficulty of solving larger dimensional problems as the scheduling horizon increases.

Using a lower penalization of the time coupling constraints, HA2 is less robust than HA50, as it solves fewer instances, but the computed feasible solutions have a slightly better cost, at least when $T = 12$ and $T = 24$. On average, for these instances, the first feasible solutions computed by BCpre are still better, but the average computational time is also about three times larger. For $T = 48$, HA50 finds feasible solutions in 13 min on average for all instances and in 8 min on the 18 instances for which BCpre does not acquire any solution in 2 h.

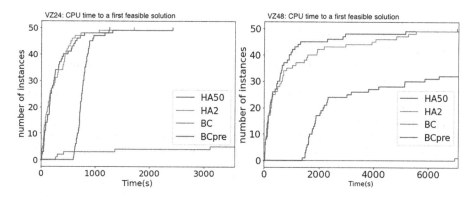

Fig. 2. Number of instances solved to feasibility over time on benchmarks VZ24 (left) and VZ48 (right): hybrid algorithm (HA2, HA50) vs branch-and-check (BC, BCpre).

Figure 2 depicts the number of instances solved in VZ24 and VZ48 as a function of time for each algorithm. It clearly illustrates the speed of the hybrid algorithm.

7   Conclusions

The proposed strategy is to address a certain broad class of control problems through the mapping between the discrete control decisions and the continuous storage state profiles. Searching in this continuous space allowed us to derive an original combination of deep learning and variable splitting, linked with scaling and multi-start mechanisms to overcome the lack of training data and the prediction errors, respectively. The approach proves to be effective in tackling the feasibility issue in hard combinatorial and nonconvex problems, such as pump scheduling. A major asset is its robustness to the lengthening of the scheduling horizon, which results from the storage/control decomposition strategy. A perspective of this work is to investigate different ways to drive the search, e.g., to learn also dual starting points (penalty terms) for ADM or to apply numerical algorithms (black-box, nonsmooth) to the augmented Lagrangian relaxation that could exploit the same decomposition and mapping.

Acknowledgement. This work is supported by the French government, through the 3IA Côte d'Azur Investments in the Future project managed by the National Research Agency (ANR) with the reference number ANR-19-P3IA-0002.

References

1. Bonvin, G., Demassey, S.: Extended linear formulation of the pump scheduling problem in water distribution networks. In: 9th International Network Optimization Conference, pp. 13–18 (2019)
2. Bonvin, G., Demassey, S., Lodi, A.: Pump scheduling in drinking water distribution networks with an LP/NLP-based B&B. Optim. Engin. **22**, 1275–1313 (2021)

3. Borovykh, A., Bohte, S., Oosterlee, C.W.: Conditional time series forecasting with convolutional neural networks. preprint: arXiv:1703.04691 (2017)
4. Demassey, S., Sessa, V., Tavakoli, A.: Strengthening mathematical models for pump scheduling in water distribution. In: 14th International Conference on Applied Energy (2022)
5. Ding, J.Y., et al.: Accelerating primal solution findings for mixed integer programs based on solution prediction. In: AAAI Conference on Artificial Intelligence, vol. 34, pp. 1452–1459 (2020)
6. Gal, Y., Ghahramani, Z.: Dropout as a Bayesian approximation: Representing model uncertainty in deep learning. In: 33rd International Conference Machine Learning, vol. 48, pp. 1050–1059 (2016)
7. Geißler, B., Morsi, A., Schewe, L., Schmidt, M.: Penalty alternating direction methods for mixed-integer optimization: a new view on feasibility pumps. SIAM J. Optim. **27**(3), 1611–1636 (2017)
8. Ghaddar, B., Naoum-Sawaya, J., Kishimoto, A., Taheri, N., Eck, B.: A Lagrangian decomposition approach for the pump scheduling problem in water networks. Eur. J. Oper. Res. **241**(2), 490–501 (2015)
9. Goodfellow, I., Bengio, Y., Courville, A.: Deep Learning. MIT Press, Cambridge (2016). http://www.deeplearningbook.org
10. Gorski, J., Pfeuffer, F., Klamroth, K.: Biconvex sets and optimization with biconvex functions: a survey and extensions. Math. Meth. Oper. Res. **66**, 373–407 (2007)
11. Kingma, D.P., Ba, J.: Adam: a method for stochastic optimization. preprint: arXiv:1412.6980 (2014)
12. Kleinert, T., Schmidt, M.: Computing feasible points of bilevel problems with a penalty alternating direction method. INFORMS J. Comput. **33**, 198–215 (2020)
13. Kool, W., Van Hoof, H., Welling, M.: Attention, learn to solve routing problems! preprint: arXiv:1803.08475 (2018)
14. Mackle, G.: Application of genetic algorithms to pump scheduling for water supply. In: International Conference on Genetic Algorithms in Engineering Systems: Innovations and Applications, pp. 400–405 (1995)
15. Naoum-Sawaya, J., Ghaddar, B., Arandia, E., Eck, B.: Simulation-optimization approaches for water pump scheduling and pipe replacement problems. Eur. J. Oper. Res. **246**, 293–306 (2015)
16. Rockafellar, R.T.: Network Flows and Monotropic Optimization. Athena Scientific, Nashua (1999)
17. Srivastava, N., Hinton, G., Krizhevsky, A., Sutskever, I., Salakhutdinov, R.: Dropout: a simple way to prevent neural networks from overfitting. J. Mach. Learn. Res. **15**(1), 1929–1958 (2014)
18. Todini, E., Pilati, S.: A gradient algorithm for the analysis of pipe networks. In: Computer Applications in Water Supply, vol. 1. Research Studies Press Ltd. (1988)
19. Van Zyl, J.E., Savic, D.A., Walters, G.A.: Operational optimization of water distribution systems using a hybrid genetic algorithm. J. Water Res. Plan. Manage. **130**(2), 160–170 (2004)
20. Wang, Y., Yin, W., Zeng, J.: Global convergence of ADMM in nonconvex Nonsmooth optimization. J. Sci. Comput. **78**, 29–63 (2019)

Branch and Cut for Partitioning a Graph into a Cycle of Clusters

Leon Eifler[1], Jakob Witzig[1,2], and Ambros Gleixner[1,3(✉)]

[1] Zuse Institute Berlin, Takustr. 7, 14195 Berlin, Germany
eifler@zib.de
[2] SAP SE, Dietmar-Hopp-Allee 17, 69190 Walldorf, Germany
jakob.witzig@sap.com
[3] HTW Berlin, 10313 Berlin, Germany
gleixner@htw-berlin.de

Abstract. In this paper we study formulations and algorithms for the cycle clustering problem, a partitioning problem over the vertex set of a directed graph with nonnegative arc weights that is used to identify cyclic behavior in simulation data generated from nonreversible Markov state models. Here, in addition to partitioning the vertices into a set of coherent clusters, the resulting clusters must be ordered into a cycle such as to maximize the total net flow in the forward direction of the cycle. We provide a problem-specific binary programming formulation and compare it to a formulation based on the reformulation-linearization technique (RLT). We present theoretical results on the polytope associated with our custom formulation and develop primal heuristics and separation routines for both formulations. In computational experiments on simulation data from biology we find that branch and cut based on the problem-specific formulation outperforms the one based on RLT.

Keywords: Branch and cut · cycle clustering · graph partitioning · valid inequalities · primal heuristics

1 Introduction

This paper is concerned with solving a curious combination of a graph partitioning problem with a cyclic ordering problem called *cycle clustering*. This combinatorial optimization problem was introduced in [16] to analyze simulation data from nonreversible stochastic processes, in particular Markov state models, where standard methods using spectral information are not easily applicable. One typical example for such processes are catalytic cycles in biochemistry, where a set of chemical reactions transforms educts to products in the presence of a catalyst.

Generally, we are given a complete undirected graph $G = (V, E)$ with vertex set $V = \{1, \ldots, n\}$, edge set $E = \binom{n}{2}$, weights $q_{i,j} \geq 0$ for all $i, j \in V$, and a pre-specified number of clusters $m \in \mathbb{N}$. It is essential to the problem that q is not assumed to be symmetric, i.e., we may have $q_{i,j} \neq q_{j,i}$. We refer to directed edges (i,j) also as arcs and let $A = \{(i,j) \in V \times V : i \neq j\}$. Our goal is then to partition V into an ordered set of clusters $C_1, \ldots, C_m$ such as to maximize an objective function based on the following definition.

Definition 1. *Let* $Q \in \mathbb{R}_{\geq 0}^{n \times n}$ *and* $S, T \subseteq V = \{1, \ldots, n\}$ *be two disjoint sets of vertices, then we define the net flow from* S *to* T *as*

$$f(S,T) := \sum_{i \in S, j \in T} q_{i,j}^- \quad \text{with } q_{i,j}^- := q_{i,j} - q_{j,i},$$

and the coherence *of* S *as*

$$g(S) := \sum_{i,j \in S} q_{i,j} = \sum_{i,j \in S, i \leq j} q_{i,j}^+ \quad \text{with } q_{i,j}^+ := \begin{cases} q_{i,i}, & \text{if } i = j, \\ q_{i,j} + q_{j,i}, & \text{otherwise.} \end{cases}$$

Given a scaling parameter α with $0 < \alpha < 1$, we want to compute a partitioning $C_1, \ldots, C_m$ of V that maximizes the combined objective function

$$\alpha \sum_{t=1}^{m} f(C_t, C_{t+1}) + (1 - \alpha) \sum_{t=1}^{m} g(C_t), \tag{1}$$

where we use the cyclic notation $C_{m+1} = C_1, C_{m+2} = C_2, \ldots$, when indexing clusters in order to improve readability. W.l.o.g. we assume in the following that $q_{i,i} = 0$ for all $i \in V$, since nonzero values only add a constant offset to (1).

In the application domain, the vertices may correspond to Markov states and $Q = (q_{i,j})_{i,j \in V}$ then constitutes the transition matrix of the unconditional probabilities that a transition from state i to state j is observed during the simulation. In this case, the coherence of a set of states gives the probability that a transition takes place inside this set, hence the second component of (1) aims to separate states that rarely interact with each other and cluster states into so-called metastabilities. The (forward) flow $\sum_{i \in S, j \in T} q_{i,j}$ between two disjoint sets of states S and T corresponds to the probability of observing a transition from S to T. Hence, the first component of (1) aims at arranging clusters into a cycle such that the probability of observing a transition from one cluster to the next minus the probability of observing a transition in the backwards direction is maximized. With α close to one, finding near-optimal cycle clusterings can be used to detect whether there exists such cyclic behavior in simulation data. For details on the motivation and interpretation of cycle clustering, we refer to [16].

Since the solution of cycle clustering problems to global optimality has proven to be challenging, we have developed a problem-specific branch-and-cut algorithm that is the focus of this paper. Our contributions are as follows. In Sect. 2, we first present two binary programming formulations for cycle clustering and give basic results on the underlying polytopes. In Sect. 3, we provide the basis for a branch-and-cut algorithm by introducing three classes of valid inequalities and discuss suitable separation algorithms for them; we show that the computationally most useful class of extended subtour and path inequalities can be separated in polynomial time. In Sect. 4, we describe four primal heuristics that aim at quickly constructing near-optimal solutions before or during branch and cut: a greedy construction heuristic, an LP-based rounding heuristic, a Lin-Kernighan [10] style improvement heuristic, and a sub-MIP heuristic based on sparsification of the arc set. We conclude the paper with Sect. 5, where we evaluate and compare the effectiveness of the primal and dual techniques numerically on simulation data that are generated to exhibit cyclic structures.

2   Formulations

We start with a straightforward formulation as a quadratic binary program. Let $\mathcal{K} := \{1, \dots, m\}$ denote the index set of clusters. For each vertex $i \in V$ and cluster C_s, $s \in \mathcal{K}$, we introduce a binary variable $x_{i,s}$ with

$$x_{i,s} = 1 \iff \text{vertex } i \text{ is assigned to cluster } C_s.$$

Then a first, nonlinear formulation for the cycle clustering problem is

$$\max \quad \alpha \sum_{s \in \mathcal{K}} \sum_{(i,j) \in A} q_{i,j}^- x_{i,s} x_{j,s+1} + (1-\alpha) \sum_{s \in \mathcal{K}} \sum_{\{i,j\} \in E} q_{i,j}^+ x_{i,s} x_{j,s} \tag{2a}$$

$$\text{s.t.} \quad \sum_{s \in \mathcal{K}} x_{i,s} = 1 \qquad \text{for all } i \in V, \tag{2b}$$

$$\sum_{i \in V} x_{i,s} \geq 1 \qquad \text{for all } s \in \mathcal{K}, \tag{2c}$$

$$x_{i,s} \in \{0,1\} \qquad \text{for all } i \in V, s \in \mathcal{K}. \tag{2d}$$

Again, note that we use the cyclic notation $x_{i,m+1} = x_{i,1}$ for convenience. The two parts of the objective function correspond to net flow and coherence, respectively. Constraints (2b) ensure that each vertex is assigned to exactly one cluster, while constraints (2c) ensure that no cluster is empty.

Following Padberg [14], this binary quadratic program could be linearized by introducing a binary variable and linearization constraints for each bilinear term, also called McCormick inequalities [13]. A more compact linearization proposed by Liberti [11] and refined by Mallach [12] exploits the set partitioning constraints (2b). Instead of McCormick inequalities, it relies on the reformulation-linearization technique (RLT) [2]. Each of equations in (2b) is multiplied by $x_{j,t}$ for all $j \in V$, $t \in \mathcal{K}$, and the bilinear terms $x_{i,s} x_{j,t}$ are replaced by newly introduced linearization variables $w_{i,j}^{s,t}$. This gives the mixed-binary program (RLT)

$$\max \quad \alpha \sum_{s \in \mathcal{K}} \sum_{(i,j) \in A} q_{i,j}^- w_{i,j}^{s,s+1} + (1-\alpha) \sum_{s \in \mathcal{K}} \sum_{\{i,j\} \in E} q_{i,j}^+ w_{i,j}^{s,s}$$

$$\text{s.t.} \quad (2b), (2c),$$

$$\sum_{s \in \mathcal{K}} w_{i,j}^{s,t} = x_{j,t} \qquad \text{for all } i,j \in V, t \in \mathcal{K},$$

$$x_{i,s} \in \{0,1\} \qquad \text{for all } i \in V, s \in \mathcal{K},$$

$$w_{i,j}^{s,t} \in [0,1] \qquad \text{for all } (i,j) \in A, s,t \in \mathcal{K},$$

where, by symmetry arguments, we can use the same variable for $w_{i,j}^{s,s} = w_{j,i}^{s,s}$.

Our second, problem-specific formulation exploits the fact that we only need to distinguish three cases for each edge $\{i,j\} \in E$ in order to model its contribution to the objective function. Either i and j are in the same cluster, in consecutive clusters, or more than one cluster apart. We capture this by introducing binary variables $y_{i,j}$ with

$$y_{i,j} = 1 \Leftrightarrow \text{vertices } i \text{ and } j \text{ are in the same cluster}$$

for all edges $\{i,j\} \in E$, $i < j$, and

$$z_{i,j} = 1 \Leftrightarrow \text{vertex } i \text{ is one cluster before } j \text{ along the cycle}$$

for all arcs $(i,j) \in A$. We will use the shorthand $y_{i,j} = y_{j,i}$ if $i > j$. With these variables, we obtain the significantly more compact binary program (CC)

$$\max \quad \alpha \sum_{(i,j) \in A} q_{i,j}^- z_{i,j} + (1-\alpha) \sum_{\{i,j\} \in E} q_{i,j}^+ y_{i,j} \tag{3a}$$

s.t.

$$(2b), (2c), \tag{}$$

$$y_{i,j} + z_{i,j} + z_{j,i} \leq 1 \quad \text{for all } \{i,j\} \in E, \tag{3b}$$

$$x_{i,s} + x_{j,s} - y_{i,j} + z_{i,j} - x_{j,s+1} - x_{i,s-1} \leq 1 \quad \text{for all } (i,j) \in A, s \in \mathcal{K}, \tag{3c}$$

$$x_{i,s} + x_{j,s+1} - z_{i,j} + y_{i,j} - x_{j,s} - x_{i,s+1} \leq 1 \quad \text{for all } (i,j) \in A, s \in \mathcal{K}, \tag{3d}$$

$$x_{i,s} \in \{0,1\} \quad \text{for all } i \in V, s \in \mathcal{K}, \tag{3e}$$

$$y_{i,j}, z_{i,j}, z_{j,i} \in \{0,1\} \quad \text{for all } \{i,j\} \in E, i < j. \tag{3f}$$

Again, the first sum in the objective function expresses the total net flow between consecutive clusters, while the second sum expresses the coherence within all clusters. Constraints (3b) ensure that two vertices cannot both be in the same cluster and in consecutive clusters. Constraints (3c) and (3d) are best explained by examining several weaker constraints first. For $(i,j) \in A$, $s \in \mathcal{K}$, consider

$$x_{i,s} + x_{j,s} - y_{i,j} \leq 1, \tag{4a}$$

$$x_{i,s} + z_{i,j} - x_{j,s+1} \leq 1, \tag{4b}$$

$$x_{j,s} + z_{i,j} - x_{i,s-1} \leq 1. \tag{4c}$$

The reasoning behind (4a) is that if i and j are in the same cluster, then $y_{i,j}$ has to be equal to one. If i is in some cluster s and $z_{i,j} = 1$, then (4b) forces j to be in the next cluster $s+1$. Analogously, if j is in cluster s and $z_{i,j} = 1$, then (4c) ensures that i is in the cluster preceding s. All of these three cases are covered by (3c), since $y_{i,j}$ and $z_{i,j}$ are binary and at most one of the two can be nonzero at the same time. Constraints (3d), in the same way, cover the functionality of the weaker constraints

$$x_{i,s} + x_{j,s+1} - z_{i,j} \leq 1,$$

$$x_{i,s} + y_{i,j} - x_{j,s} \leq 1,$$

$$x_{j,s+1} + y_{i,j} - x_{i,s+1} \leq 1.$$

Since (3c) and (3d) are defined for all $(i,j) \in A$ and $s \in \mathcal{K}$, the following implications hold:

- If i and j are in the same cluster, then $y_{i,j} = 1$ due to (3c).
- If i and j are in consecutive clusters, then $z_{i,j} = 1$ due to (3d).
- If $y_{i,j} = 1$, there exists $s \in \{1, \dots, m\}$ with $x_{i,s} = x_{j,s} = 1$ due to (3d).
- If $z_{i,j} = 1$, there exists $s \in \{1, \dots, m\}$ with $x_{i,s} = x_{j,s+1} = 1$ due to (3c).

To compare both formulations (RLT) and (CC), we first note that one can transform any feasible solution of (RLT) into a feasible solution of (CC) and vice versa, since the assignment of the x-variables in both cases uniquely implies the assignment of the remaining linearization variables. Moreover, it holds for all $(i, j) \in A, s \in \mathcal{K}$ that $y_{i,j} = \sum_{s \in \mathcal{K}} w_{i,j}^{s,s}$ and $z_{i,j} = \sum_{s \in \mathcal{K}} w_{i,j}^{s,s+1}$, hence also any valid inequality for (CC) can be transformed into a valid inequality for (RLT).

Second, we observe that the LP relaxations of both formulations yield the same dual bound at the root node. For both formulations an LP-optimal solution is induced by assigning $x_{i,s} = 1/m$ for all $i \in V$, $s \in \mathcal{K}$.

Regarding size, (RLT) has roughly $0.5m^2$ times as many variables as (CC), whereas (CC) has about twice the number of constraints. Note that in practice we can exploit in both formulations that linearization variables for pairs of vertices $i, j \in V$ are only needed if $q_{i,j} + q_{j,i} > 0$.

For the following section, define the *cycle clustering polytope CCP* as the convex hull of all feasible incidence vectors of (CC), i.e.,

$$CCP := \operatorname{conv}\left(\{(x, y, z) \in \mathbb{R}^{nm+1.5|A|} : (2b - 2c), (3b - 3f)\}\right)$$

One can show that the dimension of CCP is $(m - 1)n + |A|$ if $m = 3$ and $(m - 1)n + 1.5|A|$ if $4 \le m \le n - 2$. Furthermore, the lower bound constraints $x_{i,s} \ge 0$, $y_{i,j} \ge 0$, and $z_{i,j} \ge 0$ define facets of CCP, whereas the upper bound constraints $x_{i,s} \le 1$, $y_{i,j} \le 1$, and $z_{i,j} \le 1$ do not. For $m \ge 4$, one can show that also the model inequalities (3c) and (3d) define facets of CCP.

3   Valid Inequalities

As explained in Sect. 2, any valid inequality for (CC) can be transformed to a valid inequality for (RLT). Hence, the following valid inequalities for the cycle clustering polytope CCP apply equally to (RLT). The formal proofs that some of the inequalities are facet-defining follow standard patterns and are left for an extended version of the paper due to space limitations.

In this section we use the notation $\kappa(i) \in \{1, \dots, m\}$ to denote the (index of the) cluster that a vertex $i \in V$ is assigned to.

3.1   Triangle Inequalities

First, consider the special case of exactly three clusters $m = 3$ and three vertices $i, j, k \in V$. If j is in the successor cluster of $\kappa(i)$, and k is in the successor cluster of $\kappa(j)$, then also i has to be in the successor cluster of $\kappa(k)$. This is expressed by the valid inequality

$$z_{i,j} + z_{j,k} - z_{k,i} \le 1 \qquad \text{for all } (i, j), (j, k), (k, i) \in A.$$

Now assume $m \geq 4$. The next inequality builds only on y-variables and can be found in the graph partitioning literature, e.g., [5,8]. If i and j, and j and k are in the same cluster, then also i and k must be in the same cluster, i.e., we have

$$y_{i,j} + y_{j,k} - y_{i,k} \leq 1 \qquad \text{for all } (i,j),(j,k),(k,i) \in A. \qquad (5)$$

Similarly, if i and j are in the same cluster, and k is in the successor of $\kappa(i)$, then k also has to be in the successor of $\kappa(j)$. This gives the valid inequality

$$y_{i,j} + z_{i,k} - z_{j,k} \leq 1 \qquad \text{for all } (i,j),(j,k),(k,i) \in A, \qquad (6)$$

and if we choose k to be in the predecessor of $\kappa(i)$, then we obtain

$$y_{i,j} + z_{k,i} - z_{k,j} \leq 1 \qquad \text{for all } (i,j),(j,k),(k,i) \in A. \qquad (7)$$

With these observations, one can prove the following result.

Theorem 1. *Let $m \geq 4$ and $i,j,k \in V$ with $(i,j),(j,k),(i,k) \in A$, then*

$$y_{i,j} + y_{j,k} - y_{i,k} + \frac{1}{2}(z_{i,j} + z_{j,i} + z_{j,k} + z_{k,j} - z_{i,k} - z_{k,i}) \leq 1 \qquad (8)$$

is a valid, facet-defining inequality for CCP.

The final set of triangle inequalities is derived from the following observations. If vertex i is in the predecessor of $\kappa(j)$ as well as in the predecessor of $\kappa(k)$, then j and k must be assigned to the same cluster. This gives the valid inequality

$$z_{i,j} + z_{i,k} - y_{j,k} \leq 1 \qquad \text{for all } (i,j),(i,k),(j,k) \in A. \qquad (9)$$

Conversely, if i is in the successor of $\kappa(j)$ as well as in the successor of $\kappa(k)$, then j and k must be assigned to the same cluster, i.e., we have

$$z_{j,i} + z_{k,i} - y_{j,k} \leq 1 \qquad \text{for all } (i,j),(i,k),(j,k) \in A. \qquad (10)$$

In the special case that $m = 4$ we can prove that the stronger inequality

$$z_{i,j} + z_{i,k} - 2y_{j,k} - (z_{j,k} + z_{k,j} + z_{j,i} + z_{k,i}) \leq 0 \qquad (11)$$

must hold. The reason is that if j is assigned to the successor of $\kappa(i)$, but k is not assigned to the successor of $\kappa(i)$, then k has to be in one of the other three clusters. In each of those three cases, one of the variables $z_{j,k}$, $z_{k,j}$, $z_{k,i}$ has to be set to one: If k is in the same cluster as i, then $z_{k,j}$ has to be one; if k is in the predecessor of $\kappa(i)$, then $z_{k,i}$ has to be one; if k is in the successor of $\kappa(j)$, then $z_{j,k}$ has to be one. The same argument holds if $z_{i,k}$ is one, but $z_{i,j}$ is not. Regarding the strength of the above triangle inequalities, we can show the following.

Theorem 2. *Let $i,j,k \in V$ with $(i,j),(j,k),(i,k) \in A$. If $m = 4$, then (11) is facet-defining for CCP. If $m > 4$, then (9) and (10) are facet-defining for CCP.*

All types of triangle inequalities can be separated at once by complete enumeration in $\mathcal{O}(n^3)$ time.

3.2   Partition Inequalities

The following generalization of the triangle inequalities (9) is inspired by [8].

Theorem 3. *Let $S, T \subseteq V$ with $S \cap T = \emptyset$. The partition inequality*

$$\sum_{i \in S, j \in T} z_{i,j} - \sum_{i,j \in S, i<j} y_{i,j} - \sum_{i,j \in T, i<j} y_{i,j} \leq \min\{|S|, |T|\} \tag{12}$$

is valid for CCP. If $m > 4$ and $|S| - |T| = \pm 1$, it is facet-defining.

These inequalities can be separated heuristically by deriving S, T from almost violated triangle inequalities. To limit the computational effort, we only create partition inequalities with $|S| + |T| \leq 5$ in our implementation.

3.3   Subtour and Path Inequalitities

Since we consider a fixed number of clusters, if (i, j) is an arc between two clusters, then there must be exactly $m - 1$ further arcs along the cycle before vertex i is reached again. Formally, let $K = \{(i_1, i_2), (i_2, i_3), \ldots, (i_{\ell-1}, i_\ell), (i_\ell, i_1)\} \subseteq E$ be any cycle of length $1 < |K| < m$, then the *subtour (elimination) inequality*

$$\sum_{(i,j) \in K} z_{i,j} \leq |K| - 1 \tag{13}$$

is valid for CCP. This inequality can be extended by adding variables for arcs inside a cluster. Let $U \subset K$ be any strict subset of K, then the *extended subtour (elimination) inequality*

$$\sum_{(i,j) \in K} z_{i,j} + \sum_{(i,j) \in U} y_{i,j} \leq |K| - 1 \tag{14}$$

is valid for CCP, because in any cycle the number of forward transitions must be a multiple of the number of clusters m. The inequality (14) is stronger than (13) in the sense that it defines a higher-dimensional face of CCP. Figure 1 illustrates an extended subtour inequality. While extended subtour inequalities may not define facets, they prove effective in practice and can be separated efficiently.

Theorem 4. *Extended subtour inequalities can be separated in polynomial time.*

Proof. Let (x, y, z) be a given LP solution. By rotational symmetry, we may assume w.l.o.g. that $U = K \setminus \{(i_1, i_2)\}$. For a fixed start node i_1, define weights

$$c_{i,j} := \begin{cases} z_{i,j}, & \text{if } i = i_1, \\ z_{i,j} + y_{i,j}, & \text{otherwise,} \end{cases}$$

for all $(i, j) \in A$, and let $A' = \{(i, j) \in A : c_{i,j} > 0\}$. Then violated extended subtour inequalities correspond to cycles in the directed graph $D = (V, A')$ with length $\ell < m$ and weight greater than $\ell - 1$.

Hence, (14) can be separated by computing, for each start node, maximum weight walks between all pairs of nodes for $\ell = 2, \ldots, m - 1$. Since $c_{i,j} \geq 0$, this is possible in $\mathcal{O}(n^3 m)$ time by dynamic programming.

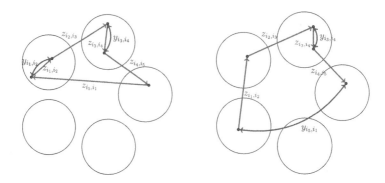

Fig. 1. Illustration of an extended subtour inequality (left) and a path inequality (right) in a 5-cluster problem. Not all of the displayed variables can be set to one.

Similarly, let $P = \{(i_1, i_2,), \ldots, (i_{m-1}, i_m)\}$ be a path from i_1 to $i_m \neq i_1$ of length $m - 1$, and let $U \subset P$. Then the *path inequality*

$$\sum_{(i,j)\in P} z_{i,j} + \sum_{(i,j)\in U} y_{i,j} + y_{i_1,i_m} \leq m - 1 \qquad (15)$$

is valid for CCP, for the following reason. Suppose $\sum_{(i,j)\in P} z_{i,j} + \sum_{(i,j)\in U} y_{i,j}$ equals $m - 1$, then there are between one and $m - 1$ forward transitions from i_1 to i_m. Therefore, i_1 and i_m cannot be in the same cluster and y_{i_1,i_m} has to be zero. These inequalities can be separated in the same way as the extended subtour inequalities. Figure 1 gives an example of a path inequality.

4   Primal Heuristics

The heuristics presented in this section exploit the fact that an integer solution for (RLT) and (CC) is determined completely by the assignment of x-variables. We present three start heuristics and one improvement heuristic.

Greedy. This heuristic constructs a feasible solution by iteratively assigning vertices to clusters in a greedy fashion. First, one vertex is assigned to each cluster, such that (2c) is satisfied. Second, we repeatedly compute for all unassigned vertices and each of the m clusters the updated objective value (in $\mathcal{O}(n)$ time), and choose the vertex-cluster assignment with largest improvement. In our implementation, the algorithm is called once as a start heuristic. It runs in $\mathcal{O}(n^2m)$ time and always generates a feasible solution since it ensures that each cluster contains at least one vertex and that all vertices are assigned.

Sparsify. For this sub-MIP heuristic, we remove 97% of the edges with smallest weight $q_{i,j} + q_{j,i}$ and call the branch-and-cut solver on the reduced instance with a node limit of one, i.e., we only process the root node in order to limit its effort.

Rounding. This heuristic assumes a fractional LP solution and rounds the x-variables. For each vertex $i \in V$, it selects a cluster $s^* \in \mathcal{K}$ with largest x_{i,s^*}

and fixes it to one; all other $x_{i,s}$, $s \neq s^*$, are fixed to zero. If $\arg \max_{s^* \in \mathcal{K}} x_{i,s^*}$ is not unique, we choose the smallest cluster. For each vertex, m clusters have to be considered, hence the heuristic runs in $\mathcal{O}(nm)$ time. If no cluster remains empty, the heuristic terminates with a feasible solution.

Exchange. This heuristic assumes an integer feasible solution and applies a series of exchange steps, where each time one vertex is transferred to a different cluster. It is inspired by the famous Lin-Kernighan graph partitioning heuristic [10]. Each vertex is reassigned once. We repeatedly compute for all unprocessed vertices and each of the $m-1$ alternative clusters the updated objective value (in $\mathcal{O}(n)$ time), and choose the swap with largest improvement, even if it is negative. We track the solution quality and return the clustering with best objective value.

The algorithm runs in $\mathcal{O}(n^3 m)$ time and is applied to each new incumbent. It can be executed repeatedly, with the solution of the previous run as the input, until no more improvement is found. In this case, it can be useful to continue from a significantly different solution. To this end, half of the vertices are selected randomly from each cluster and moved to the next cluster. In our implementation, we apply at most 5 perturbations.

5   Computational Experiments

We implemented both formulations (RLT) and (CC) within the branch-and-cut framework SCIP [7] and added the separation routines and primal heuristics described above. Except for the problem-specific changes, we used default settings for all test runs. The time limit was set to 7200 seconds, with only one job running per node at a time on a compute cluster of 2.7 GHz Intel Xeon E5-2670 v2 CPUs with 64 GB main memory each.

We consider 65 instances over graphs with 20 to 250 vertices that were created from four different types of simulations of non-reversible Markov state models such as to exhibit between 3 and 15 clusters. In the objective function, net flow and coherence are weighted 1 : 0.001, i.e., $\alpha = 1/1.001$. The first 40 instances stem from a model of artificial catalytic cycles, simulated using a hybrid Monte-Carlo method [4] as described in [16], with 3, 4, and 6 minima; for each of those, transition matrices over graphs with 20, 30, 50, and 100 vertices were created. For 12 instances we used repressilator simulations [6], a very prominent example of a synthetic genetic regulatory network; these instances feature 40, 80, and 200 vertices and between 3 and 6 clusters. Another 8 instances were created from simulations of the Hindmarsh-Rose model [9], which is used to study neuronal activity in the human heart; these instances feature 50 and 250 vertices with 3, 5, 7, and 9 clusters each. The final set of 5 instances stems from the dynamics of an Amyloid-β peptide [15] over 220 vertices, with 3 to 15 clusters.

We first compare the performance of (RLT) and (CC), both with and without problem-specific separation. In all cases, the problem-specific heuristics were enabled, since they apply equally to both models. Table 1 reports aggregated results, using shifted geometric means as defined in [1] with a shift of 10 seconds for time, 100 nodes for the size of the search tree, and 1000 for the primal and

Table 1. Performance of formulations CC and RLT with and without problem-specific separation: number of solved instances, shifted geometric means of solving time, number of branch-and-bound nodes, primal and dual integral, and arithmetic mean of final gap. Subset "1-solved" contains all instances that could be solved to optimality by at least one setting. Problem-specific heuristics are enabled, hence the lines "CC" and "CC+sepa" over all instances match the last two lines in Table 2.

Instances	setting	solved	time [s]	nodes	gap [%]	primal int.	dual int.
All (65)	RLT	29	1184.1	182.0	72.3	25.4	475.7
	RLT+sepa	28	1377.9	35.3	69.1	25.5	392.1
	CC	24	1550.9	676.4	35.9	11.6	815.6
	CC+sepa	31	755.3	125.0	25.2	10.8	347.6
1-solved (34)	RLT	29	221.8	206.1	3.2	6.2	63.1
	RLT+sepa	28	299.0	45.6	2.1	6.9	40.4
	CC	24	376.8	1425.2	5.3	2.5	104.5
	CC+sepa	31	89.0	137.6	0.7	2.2	13.1

dual integrals. The primal integral [3] is a useful metric to quantify how quickly good primal solutions are found and improved during the course of the solving process. The dual integral is defined analogously and can give insight into how quickly good dual bounds are established. It is computed using the difference between the current dual bound and the best known primal solution. The gap at primal bound p and dual bound d is computed as $(d - p)/\min\{p + \epsilon, d + \epsilon\}$ for $\epsilon = 10^{-6}$ and can exceed 100% [1].

While plain (CC) solves less instances than both variants of (RLT), we observe that (CC) with separation clearly outperforms all other settings overall. (CC) with separation solves 7 more instances than plain (CC) and reduces solving time by 51.3% and the number of nodes by 81.5%. By contrast, for (RLT) we observe that separation negatively impacts overall performance. While it reduces the number of nodes significantly by a similar factor than in (CC), the addition of cutting planes slows down the speed of LP solving and overall node processing time. This may be explained by the fact that (RLT) contains many more variables than constraints. The above analysis is confirmed on the subset "1-solved" of instances, where all instances not solved by any setting are excluded.

In order to quantify the performance impact of individual solving techniques, we continued to use (CC) as a baseline with all cutting planes and primal heuristics deactivated. In Table 2 we report the performance when a single separator is enabled, when all separators are enabled simultaneously, and similarly for the four primal heuristics. Here we can observe that each separation routine individually improves performance. Subtour separation seems to be the most useful single separator, reducing the solving time by 40.4% and the number of nodes by 67.1%. All separators combined deliver the best results, with 4 more instances solved, a speedup of 45%, and a reduction in the number of nodes by 81.7%.

Table 2. Performance impact of separation routines and primal heuristics over all 65 instances with formulation CC: on the number of solved instances, on the shifted geometric means of solving time, number of branch-and-bound nodes, primal and dual integral, and on the arithmetic mean of the final gap.

Setting	solved	time [s]	nodes	gap [%]	primal int.	dual int.
Default	27	1592.6	1240.6	7439.0	622.5	767.6
+triangle only	29	1147.4	639.0	7231.0	422.8	488.8
+partition only	27	1372.0	593.4	7652.5	581.5	608.4
+subtour only	30	949.9	408.1	4906.8	456.0	453.8
+all sepas	31	876.0	227.0	5050.1	408.0	388.1
Default	27	1592.6	1240.6	7439.0	622.5	767.6
+greedy only	25	1741.3	1084.3	1356.1	490.7	729.5
+sparsify only	23	1763.2	1109.6	113.5	475.4	751.9
+rounding only	27	1376.4	937.4	1724.3	345.8	502.7
+exchange only	28	1305.2	684.4	35.0	12.5	733.2
+all heurs	24	1550.9	676.4	35.9	11.6	815.6
All sepas+all heurs	31	755.3	125.0	25.2	10.8	347.6

However, the final gap still remains large because on hard instances that time out, good primal solutions cannot be found by the LP solutions alone.

Among the primal heuristics, the exchange heuristic proves to be the most important single technique. While also the other heuristics all reduce the primal integral and the final gap, only the exchange heuristic increases the number of solved instances (by one instance) and yields the smallest final gap of 35.0%. Still, applying all heuristics together gives the best results. The main benefit of the greedy and sparsify heuristics is to run once in order to provide the exchange heuristic with starting solutions to improve upon. The best performance in all metrics is obtained when all separators and heuristics are active.

To conclude, we observed the best out-of-the-box performance using an RLT formulation including improvements from [11,12]. However, this formulation did not profit from problem-specific separation due to an increased cost in solving the LP relaxations. In turn, our problem-specific formulation responded very well to the separation routines, and we observed that all solving techniques complement each other in improving performance. Still, only 52.3% of the instances could be solved to proven optimality, which underlines the importance of the primal heuristics when applying cycle clustering to real-world data. In this respect, we found that repeated application of a Lin-Kernighan style exchange heuristic with random perturbations was highly successful in improving initial solutions provided by simple construction heuristics.

Acknowledgments. We wish to thank Konstantin Fackeldey, Andreas Grever, and Marcus Weber for supplying us with simulation data for our experiments. This work has been supported by the Research Campus MODAL funded by the German Federal Ministry of Education and Research (BMBF grants 05M14ZAM, 05M20ZBM).

Disclosure of Interests. The authors have no competing interests to declare that are relevant to the content of this article.

References

1. Achterberg, T.: Constraint Integer Programming. Ph.D. thesis (2007). http://nbn-resolving.de/urn/resolver.pl?urn:nbn:de:0297-zib-11129
2. Adams, W.P., Sherali, H.D.: A reformulation-linearization technique for solving discrete and continuous nonconvex problems (1999). https://doi.org/10.1007/978-1-4757-4388-3
3. Berthold, T.: Measuring the impact of primal heuristics. OR Lett. **41**(6), 611–614 (2013). https://doi.org/10.1016/j.orl.2013.08.007
4. Brooks, S., Gelman, A., Jones, G., Meng, X.L.: Handbook of Markov Chain Monte Carlo. CRC Press, Boca Raton (2011)
5. Chopra, S., Rao, M.R.: The partition problem. Math. Prog. **59**(1), 87–115 (1993). https://doi.org/10.1007/BF01581239
6. Elowitz, M.B., Leibler, S.: A synthetic oscillatory network of transcriptional regulators. Nature **403**(6767), 335–338 (2000). https://doi.org/10.1038/35002125
7. Gleixner, A., Bastubbe, M., Eifler, L., et. al: The SCIP optimization suite 6.0. ZIB-Report 18-26 (2018). https://nbn-resolving.org/urn:nbn:de:0297-zib-69361
8. Grötschel, M., Wakabayashi, Y.: Facets of the clique partitioning polytope. Math. Prog. **47**(1), 367–387 (1990). https://doi.org/10.1007/BF01580870
9. Hindmarsh, J.L., Rose, R.: A model of neuronal bursting using three coupled first order differential equations. Proc. Roy. Soc. Lond. B. Biol. Sci. **221**(1222), 87–102 (1984). https://doi.org/10.1098/rspb.1984.0024
10. Kernighan, B.W., Lin, S.: An efficient heuristic procedure for partitioning graphs. Bell Syst. Tech. J. **49**(2), 291–307 (1970). https://doi.org/10.1002/j.1538-7305.1970.tb01770.x
11. Liberti, L.: Compact linearization for binary quadratic problems. 4OR **5**(3). 231–245 (2007). https://doi.org/10.1007/s10288-006-0015-3
12. Mallach, S.: Compact linearization for binary quadratic problems subject to assignment constraints. 4OR **16**(3), 295–309 (2018).https://doi.org/10.1007/s10288-017-0364-0
13. McCormick, G.P.: Computability of global solutions to factorable nonconvex programs: Part I - convex underestimating problems. Math. Prog. **10**(1), 147–175 (1976). https://doi.org/10.1007/BF01580665
14. Padberg, M.: The Boolean quadric polytope: some characteristics, facets and relatives. Math. Prog. **45**(1), 139–172 (1989). https://doi.org/10.1007/BF01589101
15. Reuter, B., et al.: Generalized Markov state modeling method for nonequilibrium biomolecular dynamics: exemplified on Amyloid β conformational dynamics driven by an oscillating electric field. J. Chem. Theory Comput. **14**(7), 3579–3594 (2018). https://doi.org/10.1021/acs.jctc.8b00079
16. Witzig, J., Beckenbach, I., Eifler, L., et al.: Mixed-integer programming for cycle detection in nonreversible Markov processes. Multiscale Model. Simul. **16**(1), 248–265 (2018). https://doi.org/10.1137/16M1091162

Graph Theory

Computing the Edge Expansion of a Graph Using Semidefinite Programming

Akshay Gupte[1] , Melanie Siebenhofer[2] , and Angelika Wiegele[2,3(✉)]

[1] University of Edinburgh, Edinburgh, UK
akshay.gupte@ed.ac.uk
[2] Alpen-Adria-Universität Klagenfurt, Klagenfurt, Austria
{melanie.siebenhofer,angelika.wiegele}@aau.at
[3] Universität zu Köln, Cologne, Germany

Abstract. Computing the edge expansion of a graph is a famously hard combinatorial problem for which there have been many approximation studies. We present two versions of an exact algorithm using semidefinite programming (SDP) to compute this constant for any graph. The SDP relaxation is used to first reduce the search space considerably. One version applies then an SDP-based branch-and-bound algorithm, along with heuristic search. The other version transforms the problem into an instance of a max-cut problem and solves this using a state-of-the-art solver. Numerical results demonstrate that we clearly outperform mixed-integer quadratic solvers as well as another SDP-based algorithm from the literature.

Keywords: Edge expansion · Cheeger constant · bisection problems · semidefinite programming

1 Introduction

Let G be a simple graph on n vertices. The *(unweighted) edge expansion*, also called the *Cheeger constant* or *conductance* or *sparstest cut*, of G is defined as

$$h(G) = \min_{\emptyset \neq S \subset V} \frac{|\partial S|}{\min\{|S|, |S'|\}} = \min_{S \subset V} \left\{ \frac{|\partial S|}{|S|} : 1 \leq |S| \leq \frac{n}{2} \right\}, \qquad (1)$$

where $\partial S = \{(i,j) \in E(G) : i \in S, j \in S'\}$ is the cut-set associated with S, and $S' = V \backslash S$. This constant is positive if and only if the graph is connected. A graph with $h(G)$ small is said to have a bottleneck. A threshold for good expansion properties is having $h(G) \geq 1$, which is desirable in many applications.

Edge expansions arise in the study of expander graphs, which have applications in network science, coding theory, cryptography, complexity theory, etc. [9,13]. The famous Mihail-Vazirani conjecture [7,20] in polyhedral combinatorics claims that the graph (1-skeleton) of any 0/1 polytope has edge expansion at least 1.

© The Author(s) 2024
A. Basu et al. (Eds.): ISCO 2024, LNCS 14594, pp. 111–124, 2024.
https://doi.org/10.1007/978-3-031-60924-4_9

Computing the edge expansion is related to the *uniform sparsest cut* problem which is defined as

$$\phi(G) = \min_{\emptyset \neq S \subset V} \frac{|\partial S|}{|S| \cdot |S'|} = \min_{S \subset V} \left\{ \frac{|\partial S|}{|S| \cdot |S'|} : 1 \leq |S| \leq \frac{n}{2} \right\}. \tag{2}$$

Since $\frac{n}{2} \leq |S'| \leq n$ it holds that $|S| \cdot |S'| \leq |S| \cdot n \leq 2|S| \cdot |S'|$ and hence

$$h(G) \leq n \cdot \phi(G) \leq 2h(G)$$

and a cut (S, S') that is α-approx for $\phi(G)$ is a 2α-approx for $h(G)$.

Both $h(G)$ and $\phi(G)$ are NP-hard to compute [15], and the latter has received considerable attention from the approximation algorithms community.

The classical bounds on $h(G)$ are from the spectral relation due to Alon and Milman [1] who showed that $\frac{\lambda_2}{2} \leq h(G) \leq \sqrt{2\lambda_2 \Delta}$, where λ_2 is the second smallest eigenvalue of the Laplace matrix of G (spectral gap) and Δ is the maximum degree of G. The best-known approximation for $\phi(G)$ is the famous $\mathcal{O}(\sqrt{\log n})$ factor by Arora et al. [2] which improved upon the earlier $\mathcal{O}(\log n)$-approximation [15]. Meira and Miyazawa [19] developed a branch-and-cut algorithm for $h(G)$ using LP relaxations and SDP-based heuristics. To the best of our knowledge, there is no other exact solver for the edge expansion.

Contribution and Outline. We develop an algorithm in two phases for computing the edge expansion of a graph. In the first phase, our algorithms splits the problem into subproblems and by computing lower and upper bounds for these subproblems, we can exclude a significant part of the search space. In the second phase, we either solve the remaining subproblems to optimality or until a subproblem can be pruned due to the bounds. For the second phase, we develop two versions. The first version implements a tailored branch-and-bound (B&B) algorithm, in the second version we transform the subproblem into an instance of a max-cut problem and solve this max-cut problem to optimality. We perform numerical experiments on different types of instances which demonstrate the effectiveness of our results. To the best of our knowledge, no other algorithms are capable of computing the edge expansion for graphs with a few hundred vertices.

In Sect. 2 we formulate the problem as a mixed-binary quadratic program and present an SDP relaxation. Section 3 investigates a related problem, namely the k-bisection problem. Our algorithm is introduced in Sect. 4, the performance of the algorithm is demonstrated in Sect. 5, followed by conclusions in Sect. 6.

Notation. The trace inner product for two real symmetric matrices X, Y is defined as $\langle X, Y \rangle = \text{tr}(XY)$ and the operator $\text{diag}(X)$ returns the main diagonal of matrix X as a vector. We denote by e the vector of all ones, and define $E = ee^\top$.

2   QP Formulation and a Semidefinite Relaxation

Consider a graph G with vertices $V = \{1, \ldots, n\}$ and its Laplacian matrix L, which is defined as $L = \text{Diag}(d) - A$, where A is the adjacency matrix of the graph and $d = (d_1, d_2, \ldots, d_n)$ is the vector of vertex degrees. Any binary vector $x \in \{0,1\}^n$ represents a cut in this graph and the value of this cut can be computed as $x^\top L x$. Hence, the expansion can be computed as

$$h(G) = \min_{x \in \{0,1\}^n} \left\{ \frac{x^\top L x}{e^\top x} : 1 \le e^\top x \le \frac{n}{2} \right\}$$

$$= \min_{\substack{x \in \{0,1\}^n \\ y \in \mathbb{R}}} \left\{ y : \frac{x^\top L x}{e^\top x} \le y, \ 1 \le e^\top x \le \frac{n}{2} \right\},$$

which can be equivalently written as the mixed-binary quadratic problem

$$h(G) = \min_{\substack{x \in \{0,1\}^n \\ y \in \mathbb{R}}} \left\{ y : x^\top L x - y e^\top x \le 0, 1 \le e^\top x \le \frac{n}{2} \right\}.$$

Standard solvers like Gurobi, CPLEX, Mosek, or CBC can handle this formulation but require a large computation time even for small instances. For example, Gurobi (version 11.0 with default parameter setting) terminated after $1.65\,\mathrm{h}/3\,548$ work units (resp. more than $24\,\mathrm{h}/59\,000$ work units) on a graph with 29 vertices and 119 edges (resp. 37 vertices and 176 edges) corresponding to the grevlex polytope in dimension 7 (resp. 8) [10].

A way to derive the spectral lower bound on $h(G)$ is via the SDP relaxation

$$
\begin{array}{llll}
h(G) \ge \min_{\tilde{X}, k} & \frac{1}{k} \langle L, \tilde{X} \rangle & = \min_X & \langle L, X \rangle & = \frac{\lambda_2(L)}{2}, \\
\text{s.t.} & \operatorname{tr}(\tilde{X}) = k & \text{s.t.} & \operatorname{tr}(X) = 1 & \\
& \langle E, \tilde{X} \rangle = k^2 & & 1 \le \langle E, X \rangle \le \frac{n}{2} & \quad (3) \\
& 1 \le k \le \frac{n}{2} & & X \succeq 0 & \\
& \tilde{X} \succeq 0 & & &
\end{array}
$$

where $\tilde{X}$ models $x x^\top$ and we scale $X = \frac{1}{k} \tilde{X}$ to eliminate the variable k. By considering the dual of the second SDP, it can be shown that the optimum is $\lambda_2(L)/2$. To strengthen (3) we round down the upper bound to $\lfloor \frac{n}{2} \rfloor$ and add the following inequalities.

Lemma 1. *The following are valid inequalities for X for all $1 \le i, j, \ell \le n$.*

$$0 \le X_{ij} \le X_{ii} \tag{4a}$$

$$X_{i\ell} + X_{j\ell} - X_{ij} \le X_{\ell\ell} \tag{4b}$$

$$X_{ii} + X_{jj} - X_{ij} \le 1/k \le 1 \tag{4c}$$

$$X_{ii} + X_{jj} + X_{\ell\ell} - X_{ij} - X_{i\ell} - X_{j\ell} \le 1/k \le 1 \tag{4d}$$

Proof. The inequalities result from scaling $\tilde{X}$ in the facet-inducing inequalities of the boolean quadric polytope for $\tilde{X}$ which are given as $0 \le \tilde{X}_{ij} \le \tilde{X}_{ii}$, $\tilde{X}_{i\ell} + \tilde{X}_{j\ell} - \tilde{X}_{ij} \le \tilde{X}_{\ell\ell}$, $\tilde{X}_{ii} + \tilde{X}_{jj} - \tilde{X}_{ij} \le 1$, $\tilde{X}_{ii} + \tilde{X}_{jj} + \tilde{X}_{\ell\ell} - \tilde{X}_{ij} - \tilde{X}_{i\ell} - \tilde{X}_{j\ell} \le 1$. $\square$

Note, that in (4c) and (4d) we have to replace $\frac{1}{k}$ by its upper bound 1 in order to obtain a formulation without k.

Table 1 compares different lower bounds on the example on graphs of the grlex polytope, which is described in [10]. The first three columns indicate the dimension of the polytope, the number of vertices in the associated graph, and the edge expansion that is known to be one for these graphs [10]. In the fourth and fifth columns the spectral bound and the strenghened SDP bound (3) are displayed. Column 6 displays a bound that is very easy to compute: the minimum cut of the graph divided by the largest possible size of the smaller set of the partition. In the last column, the minimum of the lower bounds ℓ_k for $1 \le k \le \lfloor \frac{n}{2} \rfloor$ is listed with ℓ_k being a bound related to the solution of (3) for k fixed. The definition of ℓ_k follows in Sect. 3.1.

The numbers in the table show that some of these bounds are very weak, in particular, if the number of vertices increases. Interestingly, if we divide the edge-expansion problem into $\lfloor \frac{n}{2} \rfloor$ many subproblems with fixed denominator (as we did to obtain the numbers in column 6), the lower bound we obtain by taking the minimum over all SDP relaxations for the subproblems seems to be stronger than the other lower bounds presented in Table 1. We will, therefore, take this direction of computing the edge expansion, namely, we will compute upper and lower bounds on the problem with fixed k.

Table 1. Lower bounds for graphs from the grlex polytope in dimension d.

d	n	$h(G)$	$\lambda_2/2$	(3) & (4)	min-cut$(G)/\lfloor \frac{n}{2} \rfloor$	$\min_k(\ell_k)$
4	11	1	0.6662	0.7095	0.8000	1
5	16	1	0.5811	0.6271	0.6250	1
6	22	1	0.5231	0.5743	0.5455	1
7	29	1	0.4820	0.5395	0.5000	1
8	37	1	0.4516	0.5164	0.4444	1

3   Fixing the Size k: Bisection Problem

If the size k of the smaller set of the partition of an optimum cut is known, the edge expansion problem would result in a scaled bisection problem. That is, we ask for a partition of the vertices into two parts, one of size k and one of size $n - k$, such that the number of edges joining these two sets is minimized. This problem is NP-hard [8] and has the following formulation,

$$h_k = \frac{1}{k} \min_{x \in \{0,1\}^n} \left\{ x^\top L x : e^\top x = k \right\} \tag{5}$$

which standard branch-cut solvers can solve in reasonable time only for small-sized graphs.

Since SDP-based bounds have been shown to be very strong for partitioning problems [cf 14, 18, 21, 22], we exploit these bounds by developing two kinds of solvers. We develop a tailored B&B algorithm based on semidefinite programming to solve the bisection problem. In the subsequent sections, we describe how to obtain lower and upper bounds on h_k (Sect. 3.1 and Sect. 3.2) as well as further ingredients of this exact solver (Sect. 3.3). An alternative to this B&B solver is presented in Sect. 3.4, where we transform the bisection problem into an instance of a max-cut problem which is then solved using the state-of-the-art solver BiqBin [11].

3.1 SDP Lower Bounds for the Bisection Problem

A computationally cheap SDP relaxation of the bisection problem is

$$\ell_{\text{bisect}}(k) = \min_{X,x} \left\{ \langle L, X \rangle : \text{tr}(X) = k, \langle E, X \rangle = k^2, \text{diag}(X) = x, X \succeq xx^\top \right\}. \quad (6)$$

Since the k-bisection of a graph has to be an integer, we get that

$$\ell_k = \frac{\lceil \ell_{\text{bisect}}(k) \rceil}{k}$$

is a lower bound on the scaled bisection h_k.

There are several ways to strengthen (6). In [22] a vector lifting SDP relaxation, tightened by non-negativity constraints, has been introduced. In our setting, this results in the following doubly non-negative programming (DNN) problem.

$$
\begin{aligned}
\min_X \quad & \frac{1}{2} \langle L, X^{11} + X^{22} \rangle \\
\text{s.t.} \quad & \text{tr}(X^{11}) = k, \ \langle E, X^{11} \rangle = k^2, \\
& \text{tr}(X^{22}) = n - k, \ \langle E, X^{22} \rangle = (n - k)^2, \\
& \text{diag}(X^{12}) = 0, \ \text{diag}(X^{21}) = 0, \ \langle E, X^{12} + X^{21} \rangle = 2k(n - k), \quad (7) \\
& X = \begin{pmatrix} 1 & (x^1)^\top & (x^2)^\top \\ x^1 & X^{11} & X^{12} \\ x^2 & X^{21} & X^{22} \end{pmatrix} \succeq 0, \ x^i = \text{diag}(X^{ii}), \ i = 1, 2, \\
& X \geq 0,
\end{aligned}
$$

where X is a matrix of size $(2n + 1) \times (2n + 1)$.

Meijer et al. [18] use this relaxation and strengthen it further by cutting planes from the boolean quadric polytope. Since this SDP cannot be solved by standard methods due to the large number of constraints, they present an alternating direction method of multipliers (ADMM) to (approximately) solve this relaxation even for graphs with up to 1000 vertices, and a post-processing algorithm is applied to guarantee a valid lower bound. Using this algorithm, we can compute strong lower bounds for each k with reasonable computational effort.

3.2   A Heuristic for the Bisection Problem

The graph bisection problem can be written as a quadratic assignment problem (QAP). To do so, we set the weight matrix W to be the Laplacian matrix L and the distance matrix D to the matrix with a top left block of size k with all ones and the rest zero. The resulting QAP for this weight and distance matrix is

$$\min_{\pi \in \Pi_n} \sum_{i=1}^{n} \sum_{j=1}^{n} W_{i,j} D_{\pi(i),\pi(j)} = \min_{\pi \in \Pi_n} \sum_{i=1}^{k} \sum_{j=1}^{k} L_{\pi^{-1}(i),\pi^{-1}(j)} = kh_k.$$

To compute an upper bound u_k on h_k, we use a simulated annealing heuristic for the QAP, as introduced in [6], and divide the solution by k.

3.3   A Branch-and-bound Algorithm for the Bisection Problem

We implement an open-source B&B solver to solve graph bisection problems of medium size to optimality using the ingredients described in the previous sections, namely the SDP bound as described in Sect. 3.1 and the upper bound as described in Sect. 3.2.

We base our branching decision on the solution of the relaxation of the subproblem. Namely, we branch on the node with corresponding value in x^1 being closest to 0.5. It turns out that we can write the subproblems again as problems of a similar form. In particular, if we set a variable x_i to be 0, we can write the problem as the minimization problem $\min\{\bar{x}^\top \bar{L}\bar{x} : e^\top \bar{x} = k\}$, where $\bar{x}$ is obtained from x by deleting x_i and $\bar{L}$ by deleting the i-th row and column of L. The subproblem where we set $x_i = 1$ can be written as $\min\{\bar{x}^\top \tilde{L}\bar{x} + c : e^\top \bar{x} = k - 1\}$, with $\bar{x}$ again resulting from x by deleting x_i and $\tilde{L}$ is obtained from L by adding the i-th row and column to the diagonal before deleting them and $c = L_{ii}$. Note that for both types of subproblems, although they are no bisection problems anymore, we can still use the methods discussed in Sect. 3.1 and Sect. 3.2 to compute bounds.

3.4   Transformation to a Max-Cut Problem

A different approach to solving the graph bisection problem is to transform it to a max-cut problem and use a max-cut solver, e.g. the open source parallel solver from [11]. To do so, we first need to transform the bisection problem into a quadratic unconstrained binary problem (QUBO).

Lemma 2. Let $\tilde{x} \in \{0,1\}^n$ such that $e^\top \tilde{x} = k$, and denote $\mu_k = \tilde{x}^\top L\tilde{x}$. Then

$$h_k = \frac{1}{k} \cdot \min_{x \in \{0,1\}^n} \left\{ x^\top (L + \mu_k E)x - 2\mu_k k e^\top x + \mu_k k^2 \right\}.$$

Proof. First note that $x^\top Lx + \mu_k \|e^\top x - k\|^2 = x^\top (L + \mu_k ee^\top)x - 2\mu_k k e^\top x + \mu_k k^2$. Let $x \in \{0,1\}^n$. For $e^\top x = k$ we have $x^\top Lx + \mu_k \|e^\top x - k\|^2 = x^\top Lx$. And if $e^\top x \neq k$, then $x^\top Lx + \mu_k \|e^\top x - k\|^2 > \mu_k$. Hence, for any infeasible $x \in \{0,1\}^n$, the objective is greater than the given upper bound μ_k and therefore the minimum can only be attained for $x \in \{0,1\}^n$ with $e^\top x = k$.   □

It is well known that a QUBO problem can be reduced to a dense max-cut problem with one additional binary variable [cf. 5].

4   Split and Bound

We now assemble the tools developed in the previous section to compute the edge expansion of a graph by splitting the problem into $\lfloor \frac{n}{2} \rfloor$ many bisection problems. Since the bisection problem is NP-hard as well, we want to reduce the number of bisection problems we have to solve exactly as much as possible. To do so, we start with a pre-elimination of the bisection problems.

4.1   Pre-elimination

The size k of the smaller set of the partition can theoretically be any value from 1 to $\lfloor \frac{n}{2} \rfloor$. However, it can be expected that for some candidates, one can quickly check that the optimal solution cannot be attained for that k. As a first quick check, we use the cheap lower bound ℓ_k obtained by solving the SDP (6) in combination with the upper bound introduced in Sect. 3.2. We do not need to further consider values of k where the lower bound ℓ_k of the scaled bisection problem is already above an upper bound u^* on the edge expansion. A pseudo-code of this pre-elimination step is given in Algorithm 1.

Algorithm 1: Pre-eliminate certain values of k

1 **for** $k \in \{1, \ldots, \lfloor \frac{n}{2} \rfloor\}$ **do**
2 $\quad$ Compute an upper bound u_k using a heuristic from Sect. 3.2;
3 $\quad$ Compute a lower bound ℓ_k by solving the cheap SDP (6);
4 Global upper bound $u^* := \min \{u_k : 1 \leq k \leq \lfloor \frac{n}{2} \rfloor\}$;
5 **if** $\min \ell_k = u^*$ **then**
6 $\quad$ $\mathcal{I} = \emptyset$, $h(G) = u^*$;
7 **else**
8 $\quad$ $\mathcal{I} := \{k \in \{1, \ldots, \lfloor \frac{n}{2} \rfloor\} : \ell_k < u^*\}$;

As it can be seen in Fig. 1, for a graph associated to a randomly generated 0/1 polytope and for a network graph, about 2/3 of the potential values of k can be excluded already by considering the cheap lower bound ℓ_k.

We can further reduce the number of candidates for k by computing a tighter lower bound $\tilde{\ell}_k$ by solving the DNN relaxation (7) with additional cutting planes. Note that in our implementation we do not compute the tighter bound $\tilde{\ell}_k$ as part of the pre-elimination, since this bound is computed in the root node of the B&B tree in the algorithm from Sect. 3.3.

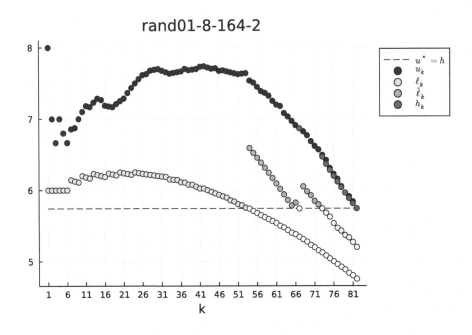

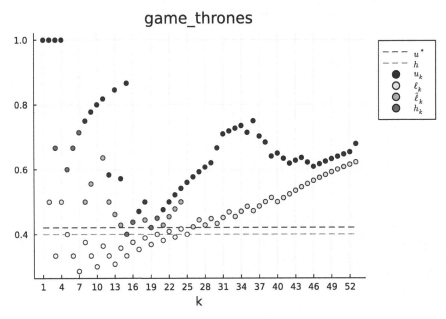

Fig. 1. Lower and upper bounds for each k.

4.2   Computational Aspects

Stopping Exact Computations Earlier. For all values of k that are not excluded in the pre-elimination step, we have to compute the scaled bisection h_k. For some values of k, however, the optimum h_k is greater than the threshold u^*. In that case, it makes no sense to compute h_k but stop as soon as it is clear that we will not find the optimum with this choice of k.

Order of Considering Sub-Problems. If we find a smaller upper bound while computing h_k, this is also a better upper bound on the edge expansion. This affects all other open bisection problems since a better upper bound means that we can stop the B&B algorithms even earlier. Therefore, we do another 30 trials of simulated annealing for each k in $\mathcal{I}$. We expect the best further improvement on the upper bound for k with the smallest upper bound u_k and therefore consider the subproblems in ascending order of their upper bound.

5   Numerical Results

All of our algorithms were written[1] in Julia[2]. All computations were carried out on an AMD EPYC 7532 with 32 cores with 3.30 GHz and 1024 GB RAM. All max-cut problems were solved with the max-cut solver of BiqBin[3]. The SDPs to compute our cheap lower bounds ℓ_k are solved with MOSEK[4] and MINLPs are solved with Gurobi[5] using JuMP[6].

5.1   Benchmark Instances

The first class of graphs are the *graphs of random 0/1-polytopes*. The polytopes are generated by randomly selecting n_d vertices of the polytope in dimension d according to the uniform model in [16]. For any pair (d, n_d) with $n_8 \in \{164, 189\}$, $n_9 \in \{153, 178, 203, 228, 253, 278\}$, and $n_{10} \in \{256, 281\}$, we generated 3 instances. Another class of polytopes we consider are the *grlex* and *grevlex* polytopes introduced and investigated in [10]. The last category of graphs originates from the graph partitioning and clustering application. The set of *DIMACS instances* are the graphs of the 10th DIMACS challenge on graph partitioning and graph clustering [4] with at most 500 vertices. Additionally, we consider some more *network graphs* obtained from an online network repository[7].

---

[1] The code is available as ancillary files from the arXiv page of this paper at https://arxiv.org/src/2403.04657/anc and on the GitHub repository https://github.com/melaniesi/EdgeExpansion.jl.

[2] Julia version 1.9.2, https://julialang.org/.

[3] Biqbin version 1.1.0, https://gitlab.aau.at/BiqBin/biqbin.

[4] MOSEK Optimizer API for C 10.0.47, https://docs.mosek.com/10.0/capi/index.html.

[5] Gurobi Optimizer version 11.0, https://www.gurobi.com.

[6] JuMP modeling language, https://jump.dev/.

[7] Tiago P. Peixoto, *The Netzschleuder network catalogue and repository*, https://networks.skewed.de/.

5.2   Discussion of the Experiments

Our numerical experiments indicate that the variant using BiqBin demonstrates superior performance compared to the B&B algorithm for bisection. For example, on the instance chesapeake from the DIMACS set, it took 2 s to compute the edge expansion compared to 9.8 s with the tailored B&B algorithm. Therefore, and due to space restrictions, we only report the results using BiqBin to solve the scaled bisection problems.

The detailed results are given in Table 2–3. In each of these tables, the first three columns give the name of the instance and the number of vertices and edges. In the columns 4–6 we report the optimal solution, i.e., the edge expansion of the graph, and the global lower and upper bound after the pre-elimination. The number of candidates for k after the pre-elimination is given in column 7. Column 8 lists the total number of B&B nodes in the max-cut algorithm for all values of k considered. The last two columns display the time spent in the pre-elimination and the total time (including pre-elimination) of the algorithm.

As reported in the tables, the pre-elimination phase only leaves a comparably small number of candidates for k to be further investigated. This indicates that already the cheap SDP bound is of good quality. We also observe that the SDP bound in the root-node of the B&B tree is of high quality: for many of the instances the gap is closed within the root node. This holds for all instances where the number of B&B nodes coincides with $|\mathcal{I}|$. As for comparing the run times of an instance for different values of k, no general statement can be derived. Typically, for k that is around the size where the optimum is attained and for k close to $\frac{n}{2}$ we experience the longest run time. The heuristic for computing upper bounds also performs extremely well: for almost all instances the upper bound found is the edge expansion of the graph, cf. columns titled $h(G)$ and u^*. Overall, we solve almost all of the considered instances within a few minutes, for very few instances the B&B tree grows rather large and therefore computation times exceed several hours.

6   Summary and Future Research

We developed a split & bound algorithm to compute the edge expansion of a graph. The splitting refers to separately considering different sizes k of the smaller partition. We used semidefinite programming in both phases of our algorithm: on the one hand, SDP-based bounds are used to eliminate several values for k and we use SDP-based bounds in a B&B algorithm that solves the problem for k fixed. Through numerical results on various classes of graphs, we demonstrate that our algorithm outperforms other existing methods like the exact solver of [19] reporting an average run time of 2.7 h for instances with 60 vertices.

In some applications, one wants to check whether a certain value is a lower bound on the edge expansion, e.g., the Mihail-Vazirani conjecture. This verification is a straightforward modification of our algorithm and we are currently working on an implementation that enables this option. Another line of research

Table 2. Results of split & bound for graphs of random 0/1, grlex and grevlex polytopes.

Instance	n	m	$h(G)$	$\min \ell_k$	u^*	$|\mathcal{I}|$	B&B nodes	Algorithm 1 time (s)	total time (s)
rand01-9-153-0	153	4081	18.7500	17.7763	18.7500	5	5	43.2	129.4
rand01-9-153-1	153	4044	18.4868	17.5789	18.4868	5	5	39.9	111.9
rand01-9-153-2	153	4107	19.0000	17.8421	19.0000	6	6	45.2	220.4
rand01-8-164-0	164	1868	5.7683	4.8659	5.7683	17	123	62.0	2037.7
rand01-8-164-1	164	1837	5.3537	4.7073	5.3537	15	27	56.9	774.7
rand01-8-164-2	164	1808	5.7439	4.7561	5.7439	29	251	85.3	5347.0
rand01-9-178-0	178	4590	17.0787	16.0899	17.0787	6	18	92.6	320.4
rand01-9-178-1	178	4467	16.7079	15.3933	16.7079	9	11	87.9	506.8
rand01-9-178-2	178	4537	16.7528	15.6517	16.7528	7	7	70.0	219.1
rand01-8-189-0	189	1768	4.2234	3.4681	4.2234	23	633	99.2	5581.6
rand01-8-189-1	189	1745	4.0426	3.3723	4.0426	26	128	103.8	2634.7
rand01-8-189-2	189	1719	4.0745	3.3511	4.0638	28	100	97.9	2669.6
rand01-9-203-0	203	4900	15.1386	14.0198	15.1386	9	41	109.7	892.1
rand01-9-203-1	203	4781	14.8416	13.5545	14.8416	12	388	117.2	3591.5
rand01-9-203-2	203	4720	14.3762	13.3861	14.3762	9	9	105.8	412.1
rand01-9-228-0	228	5065	13.2368	12.0439	13.2368	13	129	166.0	2083.8
rand01-9-228-1	228	4927	9.0000	9.0000	9.0000	0	0	135.6	135.6
rand01-9-228-2	228	4984	12.8246	11.8070	12.8246	11	11	174.3	619.9
rand01-9-253-0	253	5258	11.8730	10.6825	11.873	16	684	234.5	10547.7
rand01-9-253-1	253	5053	9.0000	9.0000	9.0000	0	0	186.9	186.9
rand01-9-253-2	253	5072	11.2222	10.1190	11.2222	16	402	232.7	8709.2
rand01-10-256-0	256	11056	30.4766	29.4219	30.4766	5	5	228.8	547.7
rand01-10-256-1	256	10611	28.8438	27.7031	28.8438	6	18	233.5	926.9
rand01-10-256-2	256	10746	29.375	28.1563	29.375	6	20	240.6	769.7
rand01-9-278-0	278	5224	10.0719	8.9065	10.0719	20	1292	326.8	17542.8
rand01-9-278-1	278	5007	9.0000	8.3237	9.0000	15	387	336.6	8153.3
rand01-9-278-2	278	5132	9.9209	8.6906	9.9209	22	2238	338.1	31125.4
rand01-10-281-0	281	11828	28.9000	27.7357	28.9000	7	75	311.7	1807.9
rand01-10-281-1	281	11490	27.7929	26.5214	27.7929	8	30	321.2	1776.4
rand01-10-281-2	281	11454	27.7500	26.4571	27.7500	8	66	316.9	2435.7
grlex-7	29	119	1.0000	1.0000	1.0000	0	0	0.3	0.3
grlex-8	37	176	1.0000	1.0000	1.0000	0	0	0.6	0.6
grlex-9	46	249	1.0000	1.0000	1.0000	0	0	1.5	1.5
grlex-10	56	340	1.0000	0.8571	1.0000	7	7	2.7	22.7
grlex-11	67	451	1.0000	0.8333	1.0000	12	12	3.6	148.0
grlex-12	79	584	1.0000	0.8000	1.0000	15	15	5.8	280.4
grlex-13	92	741	1.0000	0.8000	1.0000	18	1788	8.5	14037.2

(continued)

Table 2. (*continued*)

Instance	n	m	$h(G)$	$\min \ell_k$	u^*	$|\mathcal{I}|$	B&B nodes	Algorithm 1 time (s)	total time (s)
grevlex-7	29	119	2.4615	2.1429	2.4615	3	3	0.4	1.0
grevlex-8	37	176	2.8333	2.3889	2.8333	5	5	1.0	5.8
grevlex-9	46	249	2.9565	2.5652	2.9565	5	5	1.5	20.7
grevlex-10	56	340	3.2222	2.7857	3.2222	6	6	2.9	33.8
grevlex-11	67	451	3.6667	3.0909	3.6667	8	20	3.5	193.9
grevlex-12	79	584	3.9231	3.3333	3.9231	9	241	6.9	1315.5
grevlex-13	92	741	4.0000	3.5435	4.0000	7	475	9.4	2246.3

Table 3. Results of split & bound for network instances.

Instance	n	m	$h(G)$	$\min \ell_k$	u^*	$|\mathcal{I}|$	B&B nodes	Algorithm 1 time (s)	total time (s)
karate	34	78	0.5882	0.5000	0.5882	4	4	0.7	2.3
chesapeake	39	170	2.1667	2.000	2.1667	8	8	1	2.0
dolphins	62	159	0.2857	0.2000	0.2857	16	16	4	13.2
lesmis	77	254	0.3000	0.2500	0.3	2	2	4.7	14.7
polbooks	105	441	0.3654	0.3269	0.3654	37	37	18	540.0
adjnoun	112	425	1.0000	1.0000	1.0000	0	0	16.9	16.9
football	115	613	1.0702	0.9825	1.0702	5	55	15.2	399.9
jazz	198	2742	1.0000	1.0000	1.0000	0	0	118.4	118.4
celegansneural	297	2148	1.0000	1.0000	1.0000	0	0	389.3	389.3
celegans_metabolic	453	2025	0.4000	0.3333	0.5000	20	24	1475.6	2383.3
moviegalaxies-567	52	146	0.3810	0.3636	0.3810	3	3	2.3	3.5
moviegalaxies-52	59	119	0.5385	0.4000	0.5385	27	27	3.9	16.3
terrorists-911	62	152	0.2174	0.2000	0.2174	6	6	3.2	10.7
train_terrorists	64	243	0.6000	0.4000	0.6000	20	20	5.2	44.9
highschool	70	274	0.9143	0.7059	0.9143	26	26	5.5	131.2
blumenau_drug	75	181	0.5000	0.5000	0.5000	0	0	5.1	5.1
sp_office	92	755	3.3696	3.1739	3.3696	5	5	9.9	19.3
swingers	96	232	0.3333	0.3333	0.3333	0	0	10.2	10.2
game_thrones	107	352	0.4000	0.2857	0.4211	22	22	13	290.6
revolution	141	160	0.0962	0.0770	0.0962	33	111	39.4	1595.6
foodweb_little_rock	183	2434	1.0000	1.0000	1.0000	0	0	99.2	99.2
cintestinalis	205	2575	1.0000	1.0000	1.0000	0	0	117.9	117.9
malaria_genes_HVR_1	307	2812	0.2377	0.2105	0.2377	120	1890	503.1	62943.4

is to replace the simulated annealing approach by a more sophisticated heuristic, e.g., in the spirit of the Goemans-Williamson rounding. This is necessary if one wants to obtain high-quality solutions for larger instances. We will also investigate convexification techniques by using recent results on fractional programming [12,17] and on exploiting submodularity [3] of the cut function.

Acknowledgments. This research was funded in part by the Austrian Science Fund (FWF) [10.55776/DOC78]. For open access purposes, the authors have applied a CC BY public copyright license to any author-accepted manuscript version arising from this submission.

Disclosure of Interests. The authors have no competing interests to declare that are relevant to the content of this article.

References

1. Alon, N., Milman, V.D.: λ_1, isoperimetric inequalities for graphs, and superconcentrators. In: J. Comb. Theory. Ser. B **38**(1), 73–88 (1985)
2. Arora, S., Rao, S., Vazirani, U.: Expander flows, geometric embeddings and graph partitioning. J. ACM **56**(2), 1–37 (2009)
3. Atamtürk, A., Gómez, A.: Submodularity in conic quadratic mixed 0–1 optimization. Oper. Res. **68**(2), 609–630 (2020)
4. Bader, D.A., Meyerhenke, H., Sanders, P., Wagner, D., (eds.): Graph partitioning and graph clustering. In: 10th DIMACS Implementation Challenge Workshop, Georgia Institute of Technology, Atlanta, GA, USA, February 13-14, 2012. Proceedings. vol. 588. Contemporary Mathematics. American Mathematical Society (2013)
5. Barahona, F., Jünger, M., Reinelt, G.: Experiments in quadratic 0–1 programming. Math. Program. **44**(1–3), 127–137 (1989)
6. Burkard, R.E., Rendl, F.: A thermodynamically motivated simulation procedure for combinatorial optimization problems. EJOR **17**, 169–174 (1984)
7. Feder, T., Mihail, M.: Balanced matroids. In: Proceedings of the Twenty-Fourth Annual ACM Symposium on Theory of Computing. STOC 1992, pp. 26–38. Association for Computing Machinery. Victoria, British Columbia, Canada (1992)
8. Garey, M.R., Johnson, D.S., Stockmeyer, L.: Some simplified NPcomplete graph problems. Theoret. Comput. Sci. **1**(3), 237–267 (1976)
9. Goldreich, O.: Basic facts about expander graphs. In: Goldreich, O. (ed.) Studies in Complexity and Cryptography. Miscellanea on the Interplay between Randomness and Computation. LNCS, vol. 6650, pp. 451–464. Springer, Heidelberg (2011). https://doi.org/10.1007/978-3-642-22670-0_30
10. Gupte, A., Poznanović, S.: On dantzig figures from graded lexicographic orders. Discrete Math. **341**(6), 1534–1554 (2018)
11. Gusmeroli, N., Hrga, T., Lužzar, B., Povh, J., Siebenhofer, M., Wiegele, A.: BiqBin: a parallel branch-and-bound solver for binary quadratic problems with linear constraints. ACM Trans. Math. Softw. **48**(2) (2022)
12. He, T., Liu, S., Tawarmalani, M.: Convexification techniques for fractional programs (2023). http://arxiv.org/abs/2310.08424arXiv: 2310.08424 [math.OC]
13. Hoory, S., Linial, N., Wigderson, A.: Expander graphs and their applications. Bull. AMS **43**(4), 439–561 (2006)
14. Karisch, S.E., Rendl, F.: Semidefinite programming and graph equipartition. In: Topics in Semidefinite and Interior-point Methods, vol. 18, pp. 77–95. (Toronto, ON, 1996). Fields Inst. Commun. Amer. Math. Soc., Providence, RI (1998)

15. Leighton, T., Rao, S.: Multicommodity max-flow min-cut theorems and their use in designing approximation algorithms. J. ACM **46**(6), 787–832 (1999)

16. Leroux, B., Rademacher, L.: Expansion of random 0/1 polytopes. Random Struct. Algorithms, 1–11 (2023)

17. Mehmanchi, E., Gómez, A., Prokopyev, O.A.: Fractional 0–1 programs: links between mixed-integer linear and conic quadratic formulations. J. Glob. Optim. **75**, 273–339 (2019)

18. de Meijer, F., Sotirov, R., Wiegele, A., Zhao, S.: Partitioning through projections: strong SDP bounds for large graph partition problems. Comput. Oper. Res. 151, 20 (2023). Id/No 106088

19. Meira, L.A.A., Miyazawa, F.K.: Semidefinite programming based algorithms for the sparsest cut problem. RAIRO Oper. Res. **45**(2), 75–100 (2011)

20. Mihail, M.: On the expansion of combinatorial polytopes. In: Havel, I.M., Koubek, V. (eds.) MFCS 1992. LNCS, vol. 629, pp. 37–49. Springer, Heidelberg (1992). https://doi.org/10.1007/3-540-55808-X_4

21. Wiegele, A., Zhao, S.: SDP-based bounds for graph partition via extended ADMM. Comput. Optim. Appl. **82**(1), 251–291 (2022)

22. Wolkowicz, H., Zhao, Q.: Semidefinite programming relaxations for the graph partitioning problem. Discrete Appl. Math. 96/97, 461–479 (1999)

Minimizing External Vertices in Hypergraph Orientations

Alberto José Ferrari[1], Valeria Leoni[1,2], Graciela Nasini[1,2],
and Gabriel Valiente[3(✉)]

[1] Universidad Nacional de Rosario, Rosario, Argentina
{aferrari,valeoni,nasini}@fceia.unr.edu.ar
[2] CONICET, Rosario, Argentina
[3] Algorithms, Bioinformatics, Complexity and Formal Methods Research Group,
Technical University of Catalonia, Barcelona, Spain
gabriel.valiente@upc.edu

Abstract. We introduce the problem of assigning a direction to the hyperedges of a hypergraph such that the number of source and sink vertices is minimized. We consider hypergraphs whose hyperedges are partitioned in two disjoint subsets of vertices, which will become the tail and the head of the hyperedge when oriented. We prove that the problem is NP-hard even when restricted to hypergraphs where each vertex belongs to exactly two hyperedges, and that it becomes polynomial-time solvable on graphs. We give a compact ILP formulation for the general problem, and apply it to the biochemical reactions stored in the KEGG database by representing compounds as vertices, reactions as hyperedges, and metabolic pathways and networks as hypergraphs. We provide experimental results showing that metabolic pathways and networks with thousands of compounds and reactions can be oriented in a few seconds on a personal computer.

1 Introduction

A metabolic pathway or network can be modeled as a hypergraph $H = (V, E)$ and a partition $\{X_j, Y_j\}$ of each hyperedge $e_j \in E$. An orientation of H is an assignment of a direction (X_j, Y_j) or (Y_j, X_j) to every hyperedge $e_j \in E$. The computational complexity of the graph orientation problem, that is when $|X_j| = |Y_j| = 1$ for every edge $e_j \in E$, has been studied under various constraints, such as that the out-degree of the vertices satisfy given lower and upper bounds [1], that the number of vertices with out-degree at most one be minimized [14], or that the maximum in-degree of the vertices be minimized [20]. However, the orientation problem has only been studied for hypergraphs under the constraint that $|Y_j| = 1$ for every hyperedge $e_j \in E$ [6].

Supported by the Ministerio de Ciencia e Innovación (MCI), the Agencia Estatal de Investigación (AEI) and the European Regional Development Fund (ERDF) through project METACIRCLE PID2021-126114NB-C44, and by the Agency for Management of University and Research Grants (AGAUR) through grant 2021-SGR-01419 (ALB-COM).

A. Basu et al. (Eds.): ISCO 2024, LNCS 14594, pp. 125–136, 2024.
https://doi.org/10.1007/978-3-031-60924-4_10

We further study in this paper the hypergraph orientation problem, consisting in finding an orientation that minimizes the number of source and sink vertices. Our results imply that the problem of finding an orientation of the biochemical reactions in a metabolic pathway or network with the minimum number of substrates and products, is easy for reactions with one substrate and one product (where the metabolic pathway or network is represented as a directed graph) but it is hard when the reactions have more than one substrate or more than one product (and the metabolic pathway or network is represented as a directed hypergraph). We present experimental results using the biochemical reactions stored in the KEGG (Kyoto Encyclopedia of Genes and Genomes) database [12] to show the adequacy of the ILP model to determine an optimal orientation of a metabolic pathway or network. As a matter of fact, it allows for computing an optimal orientation of a metabolic network consisting of more than three thousand compounds and four thousand reactions in less than two seconds of solver time on a personal computer.

The rest of this paper is organized as follows. In the next section we introduce the basic notations and definitions on graphs and hypergraphs, and we present formally the problem addressed in this paper. Then, in Sect. 3, we prove that this problem is NP-hard, even when restricted to 2-regular hypergraphs without twin vertices, we prove in Sect. 4 that the problem can be solved in polynomial time on graphs, and we present a compact ILP model for solving the problem on hypergraphs in Sect. 5. Finally, in Sect. 6, we present experimental results on the orientation of metabolic pathways and networks, ranging from a few dozen to several thousands of biochemical compounds and reactions. The last section summarizes the main conclusions and presents some open lines for future research.

2 Notations, Definitions and Preliminaries

Given $n, m \in \mathbb{Z}$, $n \leqslant m$, we denote $[n, m] = \{z \in \mathbb{Z} : n \leqslant z \leqslant m\}$ and for $n \in \mathbb{Z}^+$, $[n] = [1, n]$. A *hypergraph* $H = (V, E)$ consists of a set of vertices $V = \{v_i : i \in [n]\}$ and a set of hyperedges $E = \{e_j \subseteq V : e_j \neq \emptyset, j \in [m]\}$. A vertex is a *pendant vertex* if it belongs to only one hyperedge. A graph is just a hypergraph whose hyperedges have all cardinality two. In this work, graphs and hypergraphs are undirected and simple (that is, hyperedges are pairwise distinct).

A *directed hypergraph* is a hypergraph $H = (V, E)$ with a bipartition of its hyperedges such that, for all $j \in [m]$, $e_j = X_j \cup Y_j$ with $X_j \cap Y_j = \emptyset$, $X_j \neq \emptyset$, $Y_j \neq \emptyset$. In a directed hypergraph, each hyperedge e_j, for $j \in [m]$, is given by an ordered pair of the form $e_j = (X_j, Y_j)$, meaning that it is oriented from X_j to Y_j. The first set in the pair is usually called the *tail* of the hyperedge and the second set, the *head* [7].

Figure 1 (left) shows a representation of the hypergraph $H = (V, E)$ with $V = \{v_1, \ldots, v_7\}$ and $E = \{e_1, \ldots, e_4\}$, where $e_1 = \{v_1, v_2, v_3\}$, $e_2 = \{v_2, v_3, v_4\}$, $e_3 = \{v_4, v_5, v_6, v_7\}$, $e_4 = \{v_5, v_7\}$. In Fig. 1 (right) a representation of the

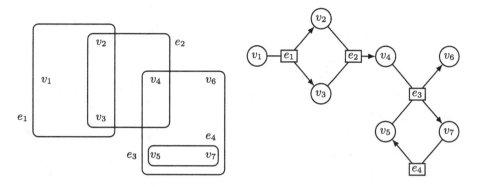

Fig. 1. A hypergraph (left) and a directed hypergraph (right).

same hypergraph is shown, with partitions of its hyperedges $X_1 = \{v_1\}, Y_1 = \{v_2, v_3\}, X_2 = \{v_2, v_3\}, Y_2 = \{v_4\}, X_3 = \{v_4, v_5\}, Y_3 = \{v_6, v_7\}, X_4 = \{v_7\}, Y_4 = \{v_5\}$, and $e_j = (X_j, Y_j)$ for all $j \in [4]$.

In a directed hypergraph $H = (V, E)$, where $e_j = (X_j, Y_j)$ for all $j \in [m]$, a vertex $v \in V$ is called a *source* if $v \in \cup_{j \in [m]} X_j$ and $v \notin \cup_{j \in [m]} Y_j$, and it is called a *sink* if $v \in \cup_{j \in [m]} Y_j$ and $v \notin \cup_{j \in [m]} X_j$. A vertex is called *external* if it is either a source or a sink. Otherwise, it is called an *internal* vertex.

Given a hypergraph $H = (V, E)$ and a partition $\{X_j, Y_j\}$ of each hyperedge $e_j \in E$, let $\Gamma_H(v) = \bigcup_{j \in [m]} (\{X_j : v \in X_j\} \cup \{Y_j : v \in Y_j\})$, for each vertex $v \in V$. Then, $d_H(v) = |\Gamma_H(v)|$ is the degree of v in H. A hypergraph $H = (V, E)$ is called *k-regular* if every vertex has degree k. For vertices v and w in $H = (V, E)$, we denote $N_X(v) = \{j : v \in X_j\}$ and $N_Y(v) = \{j : v \in Y_j\}$, and say that v and w are *twins* in H if $N_X(v) = N_X(w)$ and $N_Y(v) = N_Y(w)$.

An *orientation* of an hypergraph H with a partition $\{X_j, Y_j\}$ of each hyperedge $e_j \in E$ is a directed hypergraph $H' = (V, E')$ such that, for all $j \in [m]$, either $(X_j, Y_j) \in E'$ or $(Y_j, X_j) \in E'$. In this work, we address the following optimization problem:

Problem 1. **Hypergraph orientation problem (HOP)**

INSTANCE: A hypergraph $H = (V, E)$ and a bipartition $\{X_j, Y_j\}$ of every hyperedge $e_j \in E$, for $j \in [m]$.

OBJECTIVE: Find an orientation of H with the minimum number of external vertices.

In the next section we prove that HOP is NP-hard, even on instances corresponding to 2-regular hypergraphs without twin vertices.

3   Computational Complexity of HOP

Given a 2-regular hypergraph $H = (V, E)$, with $V = \{v_i : i \in [n]\}$, $E = \{e_j \subseteq V : j \in [m]\}$ and a bipartition $\{X_j, Y_j\}$ of hyperedge $e_j \in E$ for each $j \in [m]$, we

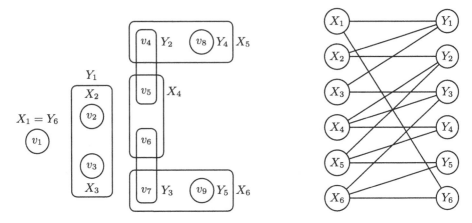

Fig. 2. A 2-regular hypergraph $H = (V, E)$ without twin vertices and its graph G^H.

define the graph $G^H = (V^H, E^H)$, where $V^H = \{X_j : j \in [m]\} \cup \{Y_j : j \in [m]\}$ and $E^H = \{\{X_j, Y_j\} : j \in [m]\} \cup \{\Gamma_H(v_i) : i \in [n]\}$. From the construction, it follows that G^H is a simple graph without pendant vertices, with $2m$ vertices and $n + m$ edges, and $M = \{\{X_j, Y_j\} : j \in [m]\}$ is a perfect matching of G^H. Figure 2 shows a representation of a 2-regular hypergraph $H = (V, E)$ with $V = \{v_1, \ldots, v_9\}$ and $E = \{e_1, \ldots, e_6\}$ (left), where $e_1 = \{v_1\} \cup \{v_2, v_3\}$, $e_2 = \{v_2\} \cup \{v_4, v_5\}$, $e_3 = \{v_3\} \cup \{v_6, v_7\}$, $e_4 = \{v_5, v_6\} \cup \{v_8\}$, $e_5 = \{v_4, v_8\} \cup \{v_9\}$, $e_6 = \{v_7, v_9\} \cup \{v_1\}$, and the graph G^H constructed above (right).

Next proposition shows that 2-regular hypergraphs without twin vertices can be characterized as graphs without pendant vertices that have a perfect matching.

Lemma 1. *Let $G = (V, E)$ be a simple graph without pendant vertices that has a perfect matching. Then, there exists a 2-regular hypergraph H without twin vertices such that $G = G^H$ up to isomorphism.*

Proof. Let $M = \{\{v_j^1, v_j^2\} : j \in [m]\}$ be a perfect matching in G. For $j \in [m]$, we define $X_j = \Gamma_G(v_j^1) \backslash M$ and $Y_j = \Gamma_G(v_j^2) \backslash M$ and the hypergraph $H = (E \backslash M, \{X_j \cup Y_j : j \in [m]\})$.

From the construction of H we have that, if $\{v_j^i, v_k^\ell\} \in E$ for some $j \neq k$ and $i, \ell \in \{1, 2\}$, then

$$\{v_j^i, v_k^\ell\} = \begin{cases} X_j \cap X_k & \text{if } i = \ell = 1 \\ Y_j \cap Y_k & \text{if } i = \ell = 2 \\ X_j \cap Y_k & \text{if } i = 1 \text{ and } \ell = 2 \\ Y_j \cap X_k & \text{if } i = 2 \text{ and } \ell = 1. \end{cases}$$

Thus, every vertex of H belongs to exactly two hyperedges and H is 2-regular. Moreover, it is easy to verify that, for $j \neq k$ and $j, k \in [m]$:

- $X_j \cup X_k \in E^H$ if and only if $\{v_j^1, v_k^1\} \in E$;

- $X_j \cup Y_k \in E^H$ if and only if $\{v_j^1, v_k^2\} \in E$;
- $Y_j \cup Y_k \in E^H$ if and only if $\{v_j^2, v_k^2\} \in E$.

Then, an isomorphism between G^H and G directly follows from the previous observations. $\qquad\square$

The following result can be exploited to restate HOP:

Theorem 1. *Let $H = (V, E)$ be a 2-regular hypergraph without twin vertices and $\{X_j, Y_j\}$ be a partition of each hyperedge $e_j \in E$, for $j \in [m]$. Then, H has an orientation with no external vertices if and only if the graph G^H (defined above) is bipartite.*

Proof. It is clear that there is a one-to-one correspondence between the orientations of H and functions $f : V^H \to \{0, 1\}$ with $f(X_j) \neq f(Y_j)$ for each $j \in [m]$. Moreover, from the construction of G^H follows that for a given orientation f of H, v is an external vertex in H if and only if $\sum_{w \in N_H(v)} f(w) = 1$.

Then, f is an orientation of H without external vertices if and only if f is a 2-coloring of G^H. It only remains to recall that a graph is 2-colorable if and only if it is bipartite. $\qquad\square$

In the light of Theorem 1 we can restate HOP as follows:

Problem 2.

INSTANCE: A graph $G = (V, E)$ without pendant vertices and a perfect matching M of G.

OBJECTIVE: Find a subset A of $E \backslash M$ of minimum cardinality such that the graph $G' = (V, E \backslash A)$ is bipartite.

Theorem 2. *HOP on 2-regular hypergraphs without twin vertices is equivalent to Problem 2 (defined above).*

Proof. Let H be a 2-regular hypergraph without twin vertices and with a partition $\{X_j, Y_j\}$ of each hyperedge $e_j \in E$, and consider the graph G^H defined above and a perfect matching M of G^H.

Let B denote a (possibly empty) subset of edges in $E^H \backslash M$ such that the graph $G = (V^H, E^H \backslash B)$ is bipartite, and let f be a $\{0, 1\}$-coloring of G. Then, the orientation of H induced by f is a solution to HOP on H with at most $|B|$ external vertices.

Conversely, let f be an orientation of H with minimum number of external vertices, and consider the graph G^H defined above. For each external vertex $v \in E(G^H)$, delete the edge in E^H corresponding to v. Then, the restriction of f to the remaining edges is a 2-coloring of G, that is, G is bipartite. $\qquad\square$

We will show that HOP is NP-hard by proving that Problem 2 is NP-hard. In order to do this, let us consider Problem 3 below, which we already know that is NP-hard from [21, Theorem 13].

Problem 3. **Edge deletion problem for bipartite subgraph**

INSTANCE: A graph $G = (V, E)$.
OBJECTIVE: Find a subset $B \subseteq E$ of minimum cardinality such that the graph $G' = (V, E \backslash B)$ is bipartite.

Given a graph G, since for every pendant vertex, the edge incident to it does not belong to any circuit in G, we can equivalently solve Problem 3 on the subgraph of G induced by the non-pendant vertices. Thus, Problem 3 is NP-hard even on graphs without pendant vertices.

Theorem 3. *HOP is NP-hard, even when restricted to 2-regular hypergraphs without twin vertices.*

Proof. We will reduce polynomially Problem 3 to Problem 2.

Given an instance $G = (V, E)$ of Problem 3 without pendant vertices, consider a paw (that is, a graph with vertices a, b, c, d and edges ab, ac, ad, bc) for each vertex $v \in V$ and identify v with the pendant vertex of the paw. Call G' the graph built in this way and consider a perfect matching M in G'. Observe that G' preserves all the edges in E. It is clear that M contains a perfect matching of each paw. A solution to Problem 2 on G' with M contains an edge of each triangle in the paws together with a subset of edges which is a solution of Problem 3 on G. Thus, the intersection with E of an optimal solution of Problem 2 on G' is an optimal solution of Problem 3 on G. And conversely, given an optimal solution of Problem 3 on G, an optimal solution of Problem 2 on G' is obtained by adding an edge of every paw, not in M. In all, the optimum of Problem 2 minus $|V|$ coincides with the minimum number of edges that must be deleted from G to obtain a bipartite graph. $\square$

In the next section, we prove that HOP is solvable in polynomial time on instances given by graphs.

4   Polynomial Instances of HOP

On the one hand, it is clear that if a given graph has an Eulerian cycle, we can orient its edges along this cycle obtaining an orientation where all the vertices are internal.

On the other hand, in any orientation of a graph, every pendant vertex is external. We can prove:

Lemma 2. *The minimum number of external vertices in an orientation of a graph is equal to the number of pendant vertices in the graph.*

Proof. Let $G = (V, E)$ be a graph. If G is Eulerian, it has an orientation with no external vertices. Otherwise, we only need to prove that there exists an orientation of G such that the set of external vertices coincides with the set of pendant vertices.

Let us index with v_i, with $i \in [2k]$ and $k \geq 1$, the vertices of G of odd degree. We consider a set M of new edges of the form $\{v_i, v_{i+k}\}$ for $i = 1, \ldots, k$, and denote by G' the graph with vertex set V and edge set $E \cup M$. Clearly, G' is Eulerian.

Consider an orientation of G' along an Euler cycle of G'. With this orientation, its vertices have as many incoming as outgoing arcs in G'. Now, each vertex of even degree in G has the same number of incoming and outgoing arcs in G, while the number of incoming and outgoing arcs in G for each vertex of odd degree in G differ by 1. If the degree in G of a vertex is an odd number greater than or equal to 3, there is at least one incoming and one outgoing arc, that is, the vertex is internal. Thus, the only external vertices in G with this orientation are its pendant vertices.   □

From [3,4] we derive:

Corollary 1. *Given a graph $G = (V, E)$, an orientation of G with the minimum number of external vertices can be obtained in $O(|V| + |E|)$ time.*

5   An ILP Model for HOP

Let $H = (V, E)$ be a hypergraph with $V = \{v_i : i \in [n]\}$ and $E = \{e_j \subseteq V : j \in [m]\}$, with partition $\{X_j, Y_j\}$ of each hyperedge $e_j \in E$.

Observe that the orientations of H can be characterized by vectors $z \in \{0,1\}^m$ such that, for each $j \in [m]$, $z_j = 1$ if and only if e_j is oriented from Y_j to X_j. Moreover, given an orientation defined by $z \in \{0,1\}^m$, the number of hyperedges coming into a vertex v_i and the number of hyperedges going out of a vertex v_i are, respectively,

$$\sum_{j \in N_X(v_i)} z_j + \sum_{j \in N_Y(v_i)} (1 - z_j) \quad \text{and} \quad \sum_{j \in N_X(v_i)} (1 - z_j) + \sum_{j \in N_Y(v_i)} z_j.$$

We present the following ILP, which only has $n + m$ binary variables and $2n$ inequalities, and prove that it models HOP on a hypergraph H.

Model 1. *ILP model for HOP*

Maximize

$$\sum_{v_i \in V} t_i \tag{1}$$

subject to

$$t_i \leqslant \sum_{j \in N_X(v_i)} z_j + \sum_{j \in N_Y(v_i)} (1 - z_j) \qquad (v_i \in V) \tag{2}$$

$$t_i \leqslant \sum_{j \in N_X(v_i)} (1 - z_j) + \sum_{j \in N_Y(v_i)} z_j \qquad (v_i \in V) \tag{3}$$

$$t_i \in \{0, 1\} \qquad (v_i \in V) \tag{4}$$

$$z_j \in \{0, 1\} \qquad (e_j \in E) \tag{5}$$

```
R00959 C00103 <=> C00668
R13199 C00668 <=> C00085
R00756 C00002 + C00085 <=> C00008 + C00354
R01068 C00354 <=> C00111 + C00118
R01015 C00118 <=> C00111
R01061 C00118 + C00009 + C00003 <=> C00236 + C00004 + C00080
R01512 C00002 + C00197 <=> C00008 + C00236
R01518 C00631 <=> C00197
R00658 C00631 <=> C00074 + C00001
R00200 C00002 + C00022 <=> C00008 + C00074
```

Fig. 3. Biochemical reactions for the glycolysis pathway in KEGG, from the glycolysis and gluconeogenesis metabolic pathway (www.genome.jp/pathway/map00010).

Lemma 3. *The ILP in Model 1 solves HOP on a hypergraph $H = (V, E)$.*

Proof. It is clear that minimizing the number of external vertices is equivalent to maximizing the number of internal vertices. If the model ensures that, in any optimal solution, $t_i = 1$ if and only if v_i is an internal vertex, then the objective function (1) models this goal.

Next, we show that this is indeed what the restrictions impose. That is, we show that, if $z \in \{0,1\}^m$ corresponds to an optimal orientation of H, then v_i is an internal vertex in this orientation if and only if $t_i = 1$.

From the previous observations, inequations (2) and (3) ensure that, if no hyperedge is coming into v_i or no hyperedge is going out of v_i (that is, if v_i is a source or a sink, repectively), then $t_i = 0$.

Conversely, if there is at least one hyperedge coming into v_i and at least one hyperedge going out of v_i, these inequalities do not impose any restriction to the binary variable t_i. Then, since the objective function is to maximize the sum of these variables, t_i will take value 1. □

6 Computational Experiments

Several models have been proposed for representing biochemical reactions in metabolic pathways and networks, including compound graphs, reaction graphs, bipartite graphs, and directed hypergraphs [2,19]. When biochemical reactions are represented as hypergraphs, the problem of orienting a hypergraph with the minimum number of external vertices becomes the problem of finding an orientation of the biochemical reactions in a metabolic pathway or network with the minimum number of substrates and products for the overall biochemical transformation.

In fact, most biochemical reactions are reversible, and the orientation found in databases such as KEGG [12] is often arbitrary [9]. However, in the context of a particular metabolic pathway or network, reactions tend to adopt an orientation because of biochemical reasons such as stoichiometry and energy consumption and production in the cell [11].

In the KEGG database, chemical compounds and biochemical reactions have unique identifiers, such as C00001 for H_2O and R00200 for the transformation of C00008 (adenosine diphosphate) and C00074 (phosphoenolpyruvate) into

Table 1. Number of compounds, number of biochemical reactions, and number of source and sink vertices in an optimal orientation of 15 metabolic pathways and 11 metabolic networks from the KEGG database.

Pathway	Compounds	Reactions	External	Network	Compounds	Reactions	External
00010	62	56	17	01100	3,506	4,565	600
00020	43	29	6	01110	2,418	2,582	533
00030	68	59	25	01120	1,184	1,444	187
00040	79	83	18	01200	160	201	21
00051	80	85	14	01210	165	151	27
00052	70	58	22	01212	123	111	22
00053	83	74	18	01220	294	253	77
00500	59	77	15	01230	171	166	27
00520	153	148	40	01232	87	134	8
00562	65	62	16	01240	391	331	80
00620	85	84	30	01250	238	209	78
00630	97	85	34				
00640	71	65	20				
00650	77	72	20				
00660	52	38	14				

C00002 (adenosine triphosphate) and C00022 (pyruvate). The biochemical reactions in the glycolysis pathway are shown in Fig. 3, where + indicates multiple substrates or products, and <=> stands for reversibility of the biochemical reaction. All of the 10 reactions involved in the gylcolysis pathway are reversible.

We downloaded from the KEGG database the biochemical reactions in 15 reference metabolic pathways (for carbohydrate metabolism) and 11 reference metabolic networks (see Table 1). These are just a few of the many metabolic pathways in KEGG, and there are also 8 pathways for energy metabolism, 16 for lipid metabolism, 2 for nucleotide metabolism, 14 for amino acid metabolism, 7 for metabolism of other amino acids, 22 for glycan biosynthesis and metabolism, 12 for metabolism of cofactors and vitamins, 21 for metabolism of terpenoids and polyketides, 31 for biosynthesis of other secondary metabolites, 21 for xenobiotics biodegradation and metabolism, and 9 for chemical structure transformation maps, making a total of 178 pathways and 11 networks.

Figure 4 shows the directed hypergraph for an optimal orientation of the compounds and reactions involved in the glycolysis pathway [8, Chapter 18], contained in KEGG reference metabolic pathway 00010 (glycolysis and gluconeogenesis). Figure 4 (left) shows the phosphorylation of C00103 (glucose) and conversion to C00118 (glyceraldehyde-3-phosphate), while Fig. 4 (right) illustrates the subsequent conversion of C00118 to C00022 (pyruvate), where some of the *currency metabolites* C00002 (adenosine triphosphate, ATP), C00003 (nicotinamide adenine dinucleotide, NAD$^+$), C00004 (reduced nicotinamide adenine dinucleotide, NADH), C00008 (adenosine diphosphate, ADP), C00009 (phosphate), and C00080 (hydron, H$^+$) are repeated, for clarity. Currency metabo-

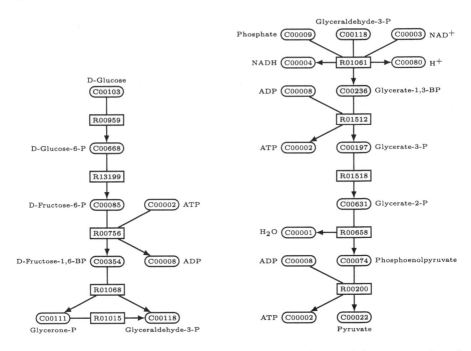

Fig. 4. Directed hypergraph for an optimal orientation of some of the compounds and reactions in KEGG metabolic pathway 00010 (glycolysis and gluconeogenesis).

lites [10,16] are ubiquitous compounds, assumed to be plentiful in the cell, and they are often duplicated in order to make the graphical representation of metabolic pathways simpler.

An optimal orientation of the largest reference metabolic network, with KEGG identifier 01100, was computed in less than 2 s on a computer with a 6-core Intel Core i5-10600 processor running at 3.30 GHz and 16 GB RAM using the AMPL modelling language [5] release 2023/04/30 and the COIN-OR Branch-and-Cut solver [15] release 2.10.7. The hypergraph representation of this metabolic network consists of 3,506 vertices and 4,565 hyperedges, and an optimal orientation has 600 external vertices.

The optimal orientations obtained for the metabolic pathways and networks in Table 1 are consistent with current knowledge in the biochemical literature [8,13,18]. For example, in the optimal orientation of the 10 biochemical reactions in the glycolysis pathway (see Fig. 4), C00103 (glucose) and C00022 (pyruvate) become external vertices, although in the context of the 56 biochemical reactions in the glycolysis and gluconeogenesis pathway (KEGG reference pathway 00010) they are indeed internal vertices, because of the presence of alternative paths for the phosphorylation of glucose to produce C00668 (glucose 6-phosphate) and since pyruvate is converted into C00186 (lactate) for the propanoate metabolism (KEGG reference pathway 00640). Furthermore, in the context of the 4,565 biochemical reactions in the whole metabolic network

(KEGG reference pathway 01100), C00103 and C00022 also become internal vertices because glucose is produced by the starch and sucrose metabolism (KEGG reference pathway 00500), and pyruvate is consumed in the pyruvate metabolism (KEGG reference pathway 00620). In a similar vein, the currency metabolites C00001 (H_2O), C00003 (NAD^+), C00004 (NADH), C00009 (phosphate), and C00080 (H^+) also become external vertices in the optimal orientation of these 10 reactions, although they are internal vertices in the context of a whole metabolic pathway or network.

7   Conclusion

We have addressed in this paper the problem of, given a hypergraph and a partition of its hyperedges in two disjoint subsets of vertices, assigning a direction to the hyperedges in order to minimize the number of source and sink vertices. This constitutes a novel addition to the criteria found in the literature for the orientation of hypergraphs.

We have proved that the problem is NP-hard even when restricted to hypergraphs where each vertex belongs to exactly two hyperedges, and that it can be solved in polynomial time on graphs. We have also presented a compact ILP formulation for the general problem, and have applied it to the metabolic pathways and networks stored in the KEGG database. The ILP optimal solution is consistent with current knowledge in the biochemical literature, and it only takes a few seconds of solver time on a personal computer to compute an optimal orientation of a metabolic network with thousands of compounds and reactions.

Open lines for future research include the study of hypergraph orientation with additional constraints (such as fixing some internal, external, source or sink vertices, or fixing the orientation of some hyperedges) and the application of hypergraph orientation to the reconstruction of metabolic pathways [17].

Acknowledgment. We thank Flavia Bonomo, Mercè Llabrés, Dora Tilli and Zephyr Verbist for detailed comments on a preliminary version of this paper.

References

1. Asahiro, Y., Jansson, J., Miyano, E., Ono, H.: Degree-constrained graph orientation: maximum satisfaction and minimum violation. Theory Comput. Syst. **58**(1), 60–93 (2016). https://doi.org/10.1007/s00224-014-9565-5
2. Deville, Y., Gilbert, D., van Helden, J., Wodak, S.J.: An overview of data models for the analysis of biochemical pathways. Brief. Bioinform. **4**(3), 246–259 (2003). https://doi.org/10.1093/bib/4.3.246
3. Ebert, J.: Computing Eulerian trails. Inf. Process. Lett. **28**(2), 93–97 (1988). https://doi.org/10.1016/0020-0190(88)90170-6
4. Edmonds, J., Johnson, E.L.: Matching, Euler tours and the Chinese postman. Math. Program. **5**(1), 88–124 (1973). https://doi.org/10.1007/BF01580113
5. Fourer, R., Gay, D.M., Kernighan, B.W.: AMPL: A Modeling Language for Mathematical Programming, 2nd edn. Cengage Learning, Boston, Massachusetts (2002)

6. Frank, A., Király, T., Király, Z.: On the orientation of graphs and hypergraphs. Discret. Appl. Math. **131**(2), 385–400 (2003). https://doi.org/10.1016/S0166-218X(02)00462-6

7. Gallo, G., Longo, G., Pallottino, S., Nguyen, S.: Directed hypergraphs and applications. Discret. Appl. Math. **42**(2–3), 177–201 (1993). https://doi.org/10.1016/0166-218X(93)90045-P

8. Garrett, R.H., Grisham, C.M.: Biochemistry, 7th edn. Cengage Learning, Boston, Massachusetts (2017)

9. van Helden, J., Wernisch, L., Gilbert, D., Wodak, S.J.: Graph-based analysis of metabolic networks. In: Mewes, H.W., Seidel, H., Weiss, B. (eds.) Bioinformatics and Genome Analysis. Ernst Schering Research Foundation Workshop, vol. 38, pp. 245–274. Springer, Heidelberg (2002). https://doi.org/10.1007/978-3-662-04747-7_12

10. Huss, M., Holme, P.: Currency and commodity metabolites: their identification and relation to the modularity of metabolic networks. IET Syst. Biol. **1**(5), 280–285 (2007). https://doi.org/10.1049/iet-syb:20060077

11. Judge, A., Dodd, M.S.: Metabolism. Essays Biochem. **64**(4), 607–647 (2020). https://doi.org/10.1042/EBC20190041

12. Kanehisa, M., Furumichi, M., Sato, Y., Kawashima, M., Ishiguro-Watanabe, M.: KEGG for taxonomy-based analysis of pathways and genomes. Nucleic Acids Res. **51**(D1), D587–D592 (2023). https://doi.org/10.1093/nar/gkac963

13. Kennelly, P.J., Botham, K.M., McGuinness, O.P., Rodwell, V.W., Weil, P.A.: Harper's Illustrated Biochemistry, 32nd edn. McGraw Hill, New York, NY (2023)

14. Khoshkhah, K.: On finding orientations with the fewest number of vertices with small out-degree. Discret. Appl. Math. **194**(1), 163–166 (2015). https://doi.org/10.1016/j.dam.2015.05.007

15. Lougee-Heimer, R.: The Common Optimization INterface for operations research: promoting open-source software in the operations research community. IBM J. Res. Dev. **47**(1), 57–66 (2003). https://doi.org/10.1147/rd.471.0057

16. Ma, H.W., Zeng, A.P.: Reconstruction of metabolic network from genome information and its structural and functional analysis. In: Kriete, A., Eils, R. (eds.) Computational Systems Biology: From Molecular Mechanisms to Disease, chap. 7, pp. 113–131. Academic Press, San Diego, CA, 2nd edn. (2014). https://doi.org/10.1016/B978-0-12-405926-9.00007-1

17. Mendoza, S.N., Olivier, B.G., Molenaar, D., Teusink, B.: A systematic assessment of current genome-scale metabolic reconstruction tools. Genome Biol. **20**, 158 (2019). https://doi.org/10.1186/s13059-019-1769-1

18. Michal, G., Schomburg, D. (eds.): Biochemical Pathways: An Atlas of Biochemistry and Molecular Biology, 2nd edn. Wiley, Hoboken (2012)

19. Pearcy, N., Crofts, J.J., Chuzhanova, N.: Hypergraph models of metabolism. Int. J. Bioeng. Life Sci. **8**(8), 829–833 (2014). https://doi.org/10.5281/zenodo.1094247

20. Venkateswaran, V.: Minimizing maximum indegree. Discret. Appl. Math. **143**(1–3), 374–378 (2004). https://doi.org/10.1016/j.dam.2003.07.007

21. Yannakakis, M.: Node-and edge-deletion NP-complete problems. In: Lipton, R.J., Burkhard, W., Savitch, W., Friedman, E.P., Aho, A. (eds.) Proc. 10th Annual ACM Symposium on Theory of Computing, pp. 253–264. Association for Computing Machinery, New York, NY (1978). https://doi.org/10.1145/800133.804355

Open-Separating Dominating Codes in Graphs

Dipayan Chakraborty[1,2]([envelope]) [ORCID] and Annegret K. Wagler[1]

[1] Université Clermont-Auvergne, CNRS, Mines de Saint-Étienne,
Clermont-Auvergne-INP, LIMOS, 63000 Clermont-Ferrand, France
{dipayan.chakraborty,annegret.wagler}@uca.fr
[2] Department of Mathematics and Applied Mathematics,
University of Johannesburg, Auckland Park 2006, South Africa

Abstract. Using dominating sets to separate vertices of graphs is a well-studied problem in the larger domain of identification problems. In such problems, the objective is to choose a suitable dominating set C of a graph G such that the neighbourhoods of all vertices of G have distinct intersections with C. Such a dominating and separating set C is often referred to as a *code* in the literature. Depending on the types of dominating and separating sets used, various problems arise under various names in the literature. In this paper, we introduce a new problem in the same realm of identification problems whereby the code, called *open-separating dominating code*, or OSD-*code* for short, is a dominating set and uses open neighbourhoods for separating vertices. The paper studies the fundamental properties concerning the existence, hardness and minimality of OSD-codes. Due to the emergence of a close and yet difficult to establish relation of the OSD-codes with another well-studied code in the literature called open locating dominating codes, or OLD-codes for short, we compare the two on various graph families. Finally, we also provide an equivalent reformulation of the problem of finding OSD-codes of a graph as a covering problem in a suitable hypergraph and discuss the polyhedra associated with OSD-codes, again in relation to OLD-codes of some graph families already studied in this context.

Keywords: open-separating codes · dominating sets · open locating-dominating codes · NP-completeness · hypergraphs · polyhedra

1 Introduction

The problem of placing surveillance devices in buildings to locate an intruder (like a fire, a thief or a saboteur) leads naturally to different location-domination type problems in the graph modeling the building (where rooms are represented as vertices and connections between rooms as edges). Depending on the characteristics of the detection devices (to detect an intruder only if it is present in the room where the detector is installed and/or to detect one in any neighbouring room), different kinds of dominating sets can be used to detect the existence of

A. Basu et al. (Eds.): ISCO 2024, LNCS 14594, pp. 137–151, 2024.
https://doi.org/10.1007/978-3-031-60924-4_11

an intruder, whereas different locating-type properties are considered to exactly locate the position of an intruder in the building.

More precisely, let $G = (V(G), E(G)) = (V, E)$ be a graph and let $N(v) = \{u \in V : uv \in E\}$ (respectively, $N[v] = N(v) \cup \{v\}$) denote the *open neighbourhood* (respectively, *closed neighborhood*) of a vertex $v \in V$. A subset $C \subseteq V$ is *dominating* (respectively, *total-dominating*) if $N[v] \cap C$ (respectively, $N(v) \cap C$) is a non-empty set for each $v \in V$. Moreover, $C \subseteq V$ is *closed-separating* (respectively, *open-separating*) if $N[v] \cap C$ (respectively, $N(v) \cap C$) is a unique set for each $v \in V$. Furthermore, the set C is *locating* if $N(v) \cap C$ is a unique set for each $v \in V \backslash C$.

So far, the following combinations of location/separation and domination properties have been studied in the literature over the last decades:

- closed-separation with domination and total-domination leading to *identifying codes* (ID-*codes* for short) [11] and *differentiating total-dominating codes* (DTD-*codes* for short) [12], respectively;
- location with domination and total-domination leading to *locating-dominating codes* (LD-*codes* for short) [16] and *locating total-dominating codes* (LTD-*codes* for short) [12], respectively;
- open-separation with total-domination leading to *open locating-dominating codes* (OLD-*codes* for short) [17].

Such problems have several applications, e.g. in fault-detection in multiprocessor networks [11], locating threats/intruders in facilities using sensor networks [18], logical definability of graphs [15] and canonical labeling of graphs for the graph isomorphism problem [4], to name a few. An extensive internet bibliography containing over 500 articles around these topics is maintained by Jean and Lobstein [13].

In this paper, we aim at studying open-separation combined with domination. We call a subset $C \subseteq V$ of a graph $G = (V, E)$ an *open-separating dominating code* (OSD-*code* for short) if

- $N[v] \cap C$ is a non-empty set for each $v \in V$; and
- $N(v) \cap C$ is a unique set for each $v \in V$.

Note that not all graphs admit codes of all the studied types. Accordingly, in Sect. 2 of this paper, we address the conditions for the existence of the OSD-codes and their relations to codes of other types. It turns out that the OSD-codes possess a particularly close relationship with the OLD-codes as the minimum cardinalities of the two differ by at most one. Moreover, for any $X \in \{ID, DTD, LD, LTD, OLD\}$, the problem of determining an X-code of minimum cardinality $\gamma^X(G)$ of a graph G, called the X-*number* of G, has been shown to be NP-hard [7,8,17]. In Sect. 2, we show that NP-hardness holds for OSD-codes as well. Furthermore, in view of the close relationship between the OSD- and the OLD-numbers of a graph, we show that deciding whether the two numbers differ is NP-complete. This motivates us to compare the OSD- and the OLD-codes of graphs of different families in Sect. 3 and to study their related polyhedra in Sect. 4. We close with some concluding remarks and lines of future research.

2 Existence, Bounds and Hardness

We next address fundamental questions concerning the existence of OSD-codes, bounds for OSD-numbers and the hardness of the OSD-problem.

Existence of OSD-*Codes.* It has been observed in the literature that the studied domination and separation properties may not apply to all graphs, see for example [11,12,17]. More precisely,

- total-domination excludes the occurrence of *isolated vertices* in graphs, that is, vertices v with $N(v) = \emptyset$;
- closed-separation (respectively, open-separation) excludes the occurrence of *closed twins* (respectively, *open twins*), that is, distinct vertices u, v with $N[u] = N[v]$ (respectively, $N(u) = N(v)$).

Calling a graph G to be X-*admissible* if G has an X-code, we see that while, for example, every graph G is LD-admissible, a graph G is OLD-admissible if and only if G has neither isolated vertices nor open twins. Accordingly, we conclude the following regarding the existence of OSD-codes of graphs.

Corollary 1. *A graph G is* OSD-*admissible if and only if G has no open twins.*

Since any two distinct isolated vertices of a graph are open twins with the empty set as both their open neighbourhoods, Corollary 1 further implies that an OSD-admissible graph has at most one isolated vertex.

Bounds on OSD-*Numbers and their Relations to Other X-numbers.* We prove the following general bounds on the OSD-number of a graph in terms of the number of vertices of the graph. The upper bound is based on results in [6].

Theorem 1. *For a graph G on $n \geq 2$ vertices without open twins and any isolated vertices, we have $\log n \leq \gamma^{\mathrm{OSD}}(G) \leq n - 1$.*

The following are bounds on the OSD-numbers in relation to other X-numbers.

Theorem 2. *Let $G = (V, E)$ be an* OSD-*admissible graph.*

(a) We have $\gamma^{\mathrm{LD}}(G) \leq \gamma^{\mathrm{OSD}}(G)$.
(b) If G is a disjoint union of a graph G' and an isolated vertex, then we have $\gamma^{\mathrm{OSD}}(G) = \gamma^{\mathrm{OLD}}(G') + 1$; otherwise, $\gamma^{\mathrm{OLD}}(G) - 1 \leq \gamma^{\mathrm{OSD}}(G) \leq \gamma^{\mathrm{OLD}}(G)$.

Proof (sketch). (a) It can be verified that any minimum OSD-code C of G is also a locating set of G, and hence, the result holds.

(b) Let G be a disjoint union of a graph G' and an isolated vertex v. Moreover, let C be any minimum OSD-code of G. Then, we must have $N(v) \cap C = \emptyset$. Hence, the result follows from the fact that the set $C - \{v\}$ must total-dominate all vertices in $V - \{v\}$, that is, $C - \{v\}$ is an OLD-code of G'.

Let us now assume that G has no isolated vertices. Then, G is also OLD-admissible. The inequality $\gamma^{\mathrm{OSD}}(G) \leq \gamma^{\mathrm{OLD}}(G)$ follows from the fact that any OLD-code of G is also an OSD-code of G. To prove the other inequality, consider

Table 1. Comparison of OSD-numbers and other X-numbers of some graphs, where $X \in \{ID, DTD, LTD\}$. The black vertices constitute the respective codes.

X	X-code	OSD-code	Graph name	$\gamma^{OSD}(G)$	$\gamma^X(G)$
X=ID			Gem	3	4
			$\overline{\text{Gem}}$	5	4
X=DTD			Bull	3	4
			Bow	5	3
X=LTD			$2P_2$	3	4
			P_4	3	2

a minimum OSD-code C of G. If C is also an OLD-code of G, then the result holds trivially. Otherwise, C is not a total-dominating set of G which implies that there exists a vertex v of G such that $C \cap N(v) = \emptyset$. One can show that v is the only vertex with this property. Thus, $C \cup \{u\}$ for some neighbour u of v (which exists as v is not an isolated vertex of G) is a total-dominating set of G and hence, is also an OLD-code of G. Therefore, we have $\gamma^{OLD}(G) \leq |C| + 1 = \gamma^{OSD}(G) + 1$. □

The tightness and extremal examples of graphs whose OSD-numbers attain the above bounds are discussed in the next section. Apart from the relations in Theorem 2, the OSD-numbers of graphs are not generally comparable to the other X-numbers with $X \in \{ID, DTD, LTD\}$, see Table 1 for an illustration.

Hardness of the OSD-*Problem.* It has been established in the literature that all the previously studied X-problems are NP-hard [7,8,17]. We next address the hardness of the OSD-problem.

> OSD
> *Instance.* (G, ℓ): An OSD-admissible graph G and a non-negative integer ℓ.
> *Question.* Is $\gamma^{OSD}(G) \leq \ell$?

Theorem 3. OSD *is NP-complete.*

As exhibited in Theorem 2(b), given an OLD-admissible graph G, we have either $\gamma^{OLD}(G) = \gamma^{OSD}(G)$ or $\gamma^{OLD}(G) = \gamma^{OSD}(G) + 1$. This poses the question how hard it is, for a given graph, to decide which of the two relations hold.

> OLD= OSD+1
> *Instance.* (G, ℓ): An OLD-admissible graph G and a non-negative integer ℓ.
> *Question.* Is $\gamma^{OSD}(G) = \ell$ and $\gamma^{OLD}(G) = \ell + 1$?

As the next theorem shows, despite the closeness, it is hard to decide if the OSD- and the OLD-numbers on a graph differ.

Theorem 4. OLD= OSD+1 *is NP-complete.*

To prove Theorems 3 and 4, we reduce in polynomial-time another well known NP-complete problem of Linear Satisfiability (LSAT) to the above two problems. The problem of LSAT is stated formally as the following.

LSAT

Instance. $I = (X, Y)$: A set X of variables and a set Y of clauses over X such that

1. each clause contains exactly 3 variables of the form x or $\bar{x}$ (each referred to as a *literal*), for some $x \in X$; and
2. each literal is contained in at most two clauses.
3. any two distinct clauses have at most one variable in common.

   Question. Is there a satisfactory truth assignment for the instance?

The proofs of Theorems 3 and 4 are based on the following construction of a suitable LSAT instance and several technical lemmatas.

Construction 1. *Let $I = (X, Y)$ be an instance of LSAT such that $|X| = k$ and $|Y| = l$. For each variable $x \in X$, construct a (variable gadget) graph G_x on 5 vertices, namely, $v^x, v^{\bar{x}}, a_1^x, a_2^x, a_3^x$, as shown in Fig. 1a. Next, for each clause $y \in Y$, take a (clause gadget) 3-path P_y on vertices u_1^y, u_2^y, u_3^y as in Fig. 1b. Finally construct the graph G_I on $5k + 3l$ vertices by adding edges between all pairs (u_1^y, v^x) (respectively, $(u_1^y, v^{\bar{x}})$) of vertices if and only if x (respectively, $\bar{x}$) is a literal in y. See Fig. 1c.*

3   OSD-numbers of Some Graph Families

In this section, we study the OSD-numbers of graphs belonging to some well-known graph families. Moreover, motivated by the hardness of deciding for which graphs the OSD- and the OLD-numbers differ, we compare in the following the two numbers on some chosen graph families. This comparison also exhibits extremal cases for the upper bounds in Theorem 1.

Cliques and their Disjoint Unions. Cliques K_n are clearly open twin-free so that for $n \geq 2$ both OSD- and OLD-codes exist. It is known that the following holds.

$$\gamma^{\text{OLD}}(K_n) = \begin{cases} 2, & \text{if } n = 2; \\ n - 1, & \text{if } n \geq 3. \end{cases}$$

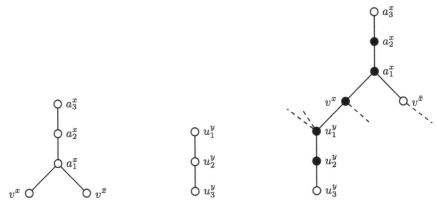

(a) G_x: Variable gadget for the variable $x \in X$.

(b) P_y: Clause gadget for the variable $y \in Y$.

(c) G_I: Instance of OLD. The black vertices represent those in a code.

Fig. 1. Polynomial-time construction of the graph G_I from an LSAT instance $I = (X, Y)$.

Lemma 1. *For a clique K_n with $n \geq 2$, we have $\gamma^{\text{OSD}}(K_n) = n - 1$.*

Hence, the OSD- and the OLD-numbers of cliques K_n differ only for $n = 2$ and are equal for all $n \geq 3$. Moreover, the upper bound in Theorem 1 is attained for all $n \geq 3$. Consider now a graph $G = K_{n_1} + \ldots + K_{n_k}$ that is the disjoint union of $k \geq 2$ cliques with $1 < n_1 \leq \ldots \leq n_k$. It is well-known that the OLD-number of the disjoint union of two or more graphs is the sum of their OLD-numbers. Hence, we have

$$\gamma^{\text{OLD}}(K_{n_1} + \ldots + K_{n_k}) = \sum_{n_i = 2} 2 + \sum_{n_i \geq 3} (n_i - 1).$$

To compare this with the corresponding OSD-numbers, we have the following.

Lemma 2. *Let $G = K_{n_1} + \ldots + K_{n_k}$ be a disjoint union of $k \geq 2$ cliques with $1 < n_1 \leq \ldots \leq n_k$.*

(a) If $n_1 = 2$, then $\gamma^{\text{OSD}}(G) = -1 + \sum_{n_i = 2} 2 + \sum_{n_i \geq 3} (n_i - 1)$,
(b) If $n_1 \geq 3$, then $\gamma^{\text{OSD}}(G) = \sum_{1 \leq i \leq k} (n_i - 1)$.

Hence, for graphs G that are disjoint unions of cliques, the OSD- and the OLD-numbers are equal if all components are cliques of order ≥ 3, but differ otherwise. In particular, if G is a *matching* (i.e. if $n_i = 2$ for all $1 \leq i \leq k$) and $k \geq 2$, the OLD-number of G is strictly greater than its OSD-number, and the upper bound of $\gamma^{\text{OSD}}(G) = |V(G)| - 1$ from Theorem 1 is attained.

Bipartite Graphs. A graph $G = (U \cup W, E)$ is *bipartite* if its vertex set can be partitioned into two stable sets U and W so that every edge of G has one endpoint in U and the other in W. We next exhibit families of bipartite graphs where the OSD- and the OLD-numbers differ.

For any integer $k \geq 1$, the *half-graph* $B_k = (U \cup W, E)$ is the bipartite graph with its stable vertex sets $U = \{u_1, \ldots, u_k\}$ and $W = \{w_1, \ldots, w_k\}$ and edges $u_i w_j$ if and only if $i \leq j$ (see Fig. 2a). In particular, we have $B_1 = K_2$, $B_2 = P_4$. Moreover, we clearly see that half-graphs are connected and open-twin-free and hence, are both OSD- and OLD-admissible.

In [10] it was shown that the only graphs whose OLD-numbers equal the order of the graph are the disjoint unions of half-graphs. In particular, we have $\gamma^{\mathrm{OLD}}(B_k) = |V(B_k)| = 2k$. Now, let $G = (V, E)$ be a graph that is a disjoint union of half-graphs. By Theorem 2(b), therefore, we have $\gamma^{\mathrm{OSD}}(G) \geq \gamma^{\mathrm{OLD}}(G) - 1 = |V| - 1$. Moreover, by Theorem 1, we have $\gamma^{\mathrm{OSD}}(G) \leq |V| - 1$. Hence, combining the two inequalities, we get the following corollary.

Corollary 2. *For a graph $G = (V, E)$ being the disjoint union of half-graphs, we have $\gamma^{\mathrm{OSD}}(G) = |V| - 1$. In particular, for a half-graph B_k, we have $\gamma^{\mathrm{OSD}}(B_k) = 2k - 1$.*

Corollary 2 shows in particular that half-graphs and their disjoint unions are extremal examples of graphs whose OSD-numbers also attain the general upper bound in Theorem 1. We further note that the upper bound from Theorem 1 does not apply to OSD-admissible graphs having an isolated vertex. To see this, consider the graph $G = B_k + K_1$ for some $k \geq 1$. By Theorem 2(b), we have $\gamma^{\mathrm{OSD}}(G) = \gamma^{\mathrm{OLD}}(B_k) + 1 = 2k + 1 = |V(G)|$. As half-graphs and their disjoint unions are the only graphs whose OLD-numbers equal the order of the graph by [10], adding an isolated vertex to them yields the only graphs whose OSD-numbers equal the order of the graph.

A *k-double star* $D_k = (U \cup W, E)$ is the bipartite graph with its stable vertex sets $U = \{u_0, u_1, \ldots, u_k\}$ and $W = \{w_1, \ldots, w_k\}$ and edges $u_i w_i$ and $u_0 w_i$ for all $w_i \in W$ (see Fig. 2b). Then, we have $D_1 = P_3$ and $D_2 = P_5$. Moreover, we clearly see that k-double stars with $k \geq 2$ are connected and open-twin-free and hence, are both OSD- and OLD-admissible. As the next Lemma shows, k-double stars also provide examples of bipartite graphs where the OLD- and the OSD-numbers disagree.

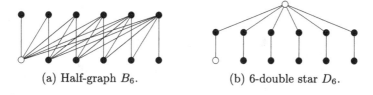

(a) Half-graph B_6.   (b) 6-double star D_6.

Fig. 2. The black vertices depict an OSD-code of the respective graph.

Lemma 3. *For a k-double star D_k with $k \geq 2$, we have $\gamma^{OSD}(D_k) = 2k - 1$ and $\gamma^{OLD}(D_k) = 2k$.*

Split Graphs. A graph $G = (Q \cup S, E)$ is a *split graph* if its vertex set can be partitioned into a clique Q and a stable set S. In order to study OSD-codes of split graphs and compare them with the OLD-codes, we restrict ourselves to split graphs G without open twins and isolated vertices. This further implies that G is connected and Q non-empty (as, otherwise, every component not containing the clique Q needs to be an isolated vertex from S, contradicting our assumptions). Figure 3 shows some small OLD-admissible graphs. It is easy to see that $\gamma^{OSD}(G)$ and $\gamma^{OLD}(G)$ differ for $G \in \{P_4, \text{gem}\}$ and are equal for $G \in \{\text{net, sun}\}$.

We next examine OSD-codes in two families of split graphs for which the exact OLD-numbers are known from [3]. A *headless spider* is a split graph with $Q = \{q_1, \ldots, q_k\}$ and $S = \{s_1, \ldots, s_k\}$. In addition, a headless spider is *thin* (respectively, *thick*) if s_i is adjacent to q_j if and only if $i = j$ (respectively, $i \neq j$). By definition, it is clear that the complement of a thin headless spider H_k is a thick headless spider $\overline{H}_k$, and vice-versa. We have $H_2 = \overline{H}_2 = P_4$, the two headless spiders $H_3 = net$ and $\overline{H}_3 = sun$ are depicted in Figs. 3(c) and 3(d), respectively. Moreover, it is easy to check that the thin and the thick headless spiders have no twins.

In [3], it was shown that $\gamma^{OLD}(H_k) = k$ for $k \geq 3$ and $\gamma^{OLD}(\overline{H}_k) = k + 1$ for $k \geq 3$. We next analyse the OSD-numbers of the thin and the thick headless spiders.

Lemma 4. *For any integer $k \geq 3$ and thin and thick headless spiders H_k and $\overline{H}_k$, respectively, we have $\gamma^{OSD}(H_k) = k$ and $\gamma^{OSD}(\overline{H}_k) = k + 1$.*

Hence, Lemma 4 combined with the results from [3] show that for the thin and the thick headless spiders H_k and $\overline{H}_k$, respectively, the OSD- and the OLD-numbers are equal for all $k \geq 3$. It would be interesting to study whether there exist families of open twin-free split graphs where the OSD- and the OLD-numbers differ.

Thin Suns. The latter result on thin headless spiders can be further generalized to thin suns. A *sun* is a graph $G = (C \cup S, E)$ whose vertex set can be partitioned into S and C, where, for an integer $k \geq 3$, the set $S = \{s_1, \ldots, s_k\}$ is a stable set and $C = \{c_1, \ldots, c_k\}$ is a (not necessarily chordless) cycle. A *thin sun* $T_k =$

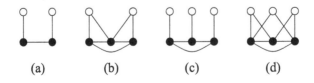

   (a)   (b)   (c)   (d)

Fig. 3. Split graphs (the black vertices belong to Q and the white vertices to S), where (a) is the P_4, (b) the gem, (c) the net, (d) the sun.

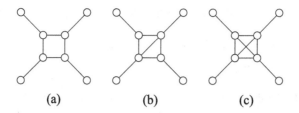

Fig. 4. The three thin suns T_4 where (a) is a sunlet and (c) a thin headless spider.

$(C \cup S, E)$ is a sun where s_i is adjacent to c_j if and only if $i = j$. Therefore, thin headless spiders are special thin suns where all chords of the cycle C are present (such that C induces a clique). Other special cases of thin suns are *sunlets* where no chords of the cycle C are present (such that C induces a hole). For illustration, for $k = 3$, the (only) thin sun T_3 equals the thin headless spider H_3 (see Fig. 3(c)); for $k = 4$, the three possible thin suns T_4 are depicted in Fig. 4.

We call two vertices c_i and c_j of a thin sun $T_k = (C \cup S, E)$ open C-twins if c_i and c_j are non-adjacent and $N_C(c_i) = N_C(c_j)$, where $N_C(v) = N(v) \cap C$. For instance, the sunlet in Fig. 4(a) and the thin sun in Fig. 4(b) have open C-twins, whereas the thin headless spider in Fig. 4(c) does not.

In [3], it was shown that for a thin sun T_k with $k \geq 4$ and without open C-twins, the set C is the unique minimum OLD-code of T_k and thus, we have $\gamma^{\mathrm{OLD}}(T_k) = k$. Now, with regards to OSD-numbers of thin suns, we show the following.

Lemma 5. *For a thin sun $T_k = (C \cup S, E)$ with $k \geq 4$ and without open C-twins, C is a minimum OSD-code and hence, we have $\gamma^{\mathrm{OSD}}(T_k) = |C| = k$.*

Therefore, thin suns without open C-twins are examples of graphs where the OSD- and the OLD-number are equal. This applies in particular to sunlets T_k with $k \geq 5$ and to thin headless spiders. However, for thin suns T_k with open C-twins, $\gamma^{\mathrm{OSD}}(T_k)$ and $\gamma^{\mathrm{OLD}}(T_k)$ may differ. For instance, for the thin sun T_4 depicted in Fig. 4(b), it can be checked that $\gamma^{\mathrm{OSD}}(T_4) = 4 < 5 = \gamma^{\mathrm{OLD}}(T_4)$. We call a thin sun $T_k = (C \cup S, E)$ *almost complete* if $k = 2\ell$ and c_i is non-adjacent to $c_{i+\ell}$ but is adjacent to all other $c_j \in C$. We can show:

Lemma 6. *For an almost complete thin sun $T_{2\ell}$ with $\ell \geq 3$, we have $\gamma^{\mathrm{OSD}}(T_{2\ell}) = 3\ell - 1$ and $\gamma^{\mathrm{OLD}}(T_{2\ell}) = 3\ell$.*

Hence, there exist infinitely many thin suns with open C-twins for which the OSD- and the OLD-numbers differ.

4   Polyhedra Associated with OSD-codes

As polyhedral methods turned out to be successful for many NP-hard combinatorial optimization problems in the literature, it was suggested in [2,3] to apply such techniques to locating-dominating type problems. For that, a reformulation of the studied X-problem in a graph G as a covering problem in a suitable hypergraph $\mathcal{H}_X(G)$ is in order. The incidence matrix of $\mathcal{H}_X(G)$ then defines the constraint system of the resulting covering problem. We next study the OSD-codes in this context.

Hypergraph Representation of the OSD-*Problem.* Given a graph $G = (V, E)$ and a problem X, we look for a hypergraph $\mathcal{H}_X(G) = (V, \mathcal{F}_X)$ so that $C \subset V$ is an X-code of G if and only if C is a *cover* of $\mathcal{H}_X(G)$ satisfying $C \cap F \neq \emptyset$ for all $F \in \mathcal{F}_X$. Then the *covering number* $\tau(\mathcal{H}_X(G))$, defined as the minimum cardinality of a cover of $\mathcal{H}_X(G)$, equals by construction the X-number $\gamma^X(G)$ of G. The hypergraph $\mathcal{H}_X(G)$ is called the X-*hypergraph* of the graph G.

It is a simple observation that for an X-problem involving domination (respectively, total-domination), $\mathcal{F}_X$ needs to contain the closed (respectively, open) neighborhoods of all vertices of G. In order to encode the separation of vertices, that is, the fact that the intersections of an X-code C with the neighborhood of each vertex is *unique*, it was suggested in [2,3] to use the symmetric differences of the neighborhoods. Here, given two sets A and B, their *symmetric difference* is defined by $A \Delta B = (A \backslash B) \cup (B \backslash A)$. In fact, it has been shown in [2,3] that a code C of a graph G is closed-separating (respectively, open-separating) if and only if $(N[u] \Delta N[v]) \cap C \neq \emptyset$ (respectively, $(N(u) \Delta N(v)) \cap C \neq \emptyset$) for all pairs of distinct vertices u, v of G. This implies for OSD-codes:

Corollary 3. *The* OSD-*hypergraph* $\mathcal{H}_{\mathrm{OSD}}(G) = (V, \mathcal{F}_{\mathrm{OSD}})$ *of a graph* $G = (V, E)$ *is composed of*

- *the closed neighborhoods* $N[v]$ *of all vertices* $v \in V$ *and*
- *the symmetric differences* $N(u) \Delta N(v)$ *of open neighborhoods of distinct vertices* $u, v \in V$

as hyperedges in $\mathcal{F}_{\mathrm{OSD}}$ *and* $\gamma^{\mathrm{OSD}}(G) = \tau(\mathcal{H}_{\mathrm{OSD}}(G))$ *holds.*

Note that a graph $G = (V, E)$ is not X-admissible if there is $\emptyset \in \mathcal{F}_X$ as then there is no $C \subset V$ satisfying $F \cap C \neq \emptyset$ for all $F \in \mathcal{F}_X$. For OSD-codes, we see that $N(u) \Delta N(v) = \emptyset$ whenever u, v are open twins, again showing that only open twin-free graphs are OSD-admissible.

It was observed in [2,3] that $\mathcal{H}_X(G) = (V, \mathcal{F}_X)$ may contain redundant hyperedges. In fact, if there are two hyperedges $F, F' \in \mathcal{F}_X$ with $F \subset F'$, then $F \cap C \neq \emptyset$ also implies $F' \cap C \neq \emptyset$ for every $C \subset V$. Thus, F' is *redundant* as $(V, \mathcal{F}_X - \{F'\})$ suffices to encode the domination and separation properties of the X-codes of G. This motivates to consider the X-*clutter* $\mathcal{C}_X(G)$ of the graph G obtained from $\mathcal{H}_X(G)$ by removing all redundant hyperedges of the latter. We note that clearly $\tau(\mathcal{H}_{\mathrm{OSD}}(G)) = \tau(\mathcal{C}_{\mathrm{OSD}}(G))$ holds.

Moreover, a special interest lies in hyperedges of $\mathcal{C}_X(G)$ consisting of a single vertex, called the *forced vertex*, as each forced vertex needs to belong to every X-code of G. For OSD-codes, we denote the set of forced vertices of G by $\mathcal{F}^1_{OSD}(G)$ and can characterise them in the following manner.

Lemma 7. *For an* OSD-*admissible graph* G, *we have* $\mathcal{F}^1_{OSD}(G) = V_0 \cup V_1$, *where*

$$V_0 = \{x \in V : N(x) = \emptyset\}; \text{ and}$$
$$V_1 = \{y \in V : \exists \text{ non-adjacent } u, v \text{ with } \{y\} = N(u)\Delta N(v)\}.$$

Note that for an OSD-admissible graph G having an isolated vertex v, V_1 contains all vertices u of degree 1 (as $N(u)\Delta N(v) = N(u)\Delta\emptyset = N(u)$ holds). Accordingly, we express the OSD-clutter of a graph $G = (V, E)$ by $\mathcal{C}_{OSD}(G) = (V, \mathcal{F}^1_{OSD}(G) \cup \mathcal{F}^2_{OSD}(G))$, where $\mathcal{F}^2_{OSD}(G)$ is composed of all non-redundant hyperedges of $\mathcal{H}_{OSD}(G)$ with size at least 2. For illustration, we construct $\mathcal{H}_{OSD}(P_4)$ and $\mathcal{C}_{OSD}(P_4)$. The OSD-hypergraph $\mathcal{H}_{OSD}(P_4)$ is composed of

$$
\begin{aligned}
N[1] &= \{1,2\} & N(1)\Delta N(2) &= \{1,2,3\} & N(1)\Delta N(3) &= \{4\} \\
N[2] &= \{1,2,3\} & N(2)\Delta N(3) &= \{1,2,3,4\} & N(1)\Delta N(4) &= \{2,3\} \\
N[3] &= \{2,3,4\} & N(3)\Delta N(4) &= \{2,3,4\} & N(2)\Delta N(4) &= \{1\} \\
N[4] &= \{3,4\}
\end{aligned}
$$

Clearly, the OSD-clutter $\mathcal{C}_{OSD}(P_4)$ only contains the symmetric differences of open neighborhoods of non-adjacent vertices, namely, the sets $\{1\}, \{2,3\}, \{4\}$. Moreover, we have $\mathcal{F}^1_{OSD}(P_4) = \{1,4\}$ and $\mathcal{F}^2_{OSD}(P_4) = \{\{2,3\}\}$. Note that for the previously studied X-problems, it has been shown in [2,3] that $\mathcal{C}_X(G)$ does not contain symmetric differences of neighborhoods of non-adjacent vertices without common neighbor. This does not apply to OSD-clutters, as $N(1)\Delta N(4) = \{2,3\}$ from the above example demonstrates.

Polyhedra Associated with OSD-*Codes.* Due to $\gamma^{OSD}(G) = \tau(\mathcal{C}_{OSD}(G))$, we can determine a minimum OSD-code in a graph $G = (V, E)$ by solving the following covering problem

$$\min \mathbf{1}^T \mathbf{x}$$
$$M_{OSD}(G)\, \mathbf{x} \geq \mathbf{1}$$
$$\mathbf{x} \in \{0, 1\}^{|V|}$$

where $\mathbf{1}$ is the vector having 1-entries only and $M_{OSD}(G)$ is the incidence matrix of the OSD-clutter $\mathcal{C}_{OSD}(G)$ encoding row-wise its hyperedges F (that is, the row of $M_{OSD}(G)$ corresponding to F is a 0/1-vector of length $|V|$ having a 1-entry if $v \in F$ and a 0-entry otherwise). For any 0/1-matrix M with n columns, the associated covering polyhedron is $P(M) = \text{conv}\{\mathbf{x} \in \mathbf{Z}^n_+ : M\mathbf{x} \geq \mathbf{1}\}$. Accordingly, the OSD-*polyhedron* of $G = (V, E)$ is defined by

$$P_{OSD}(G) = P(M_{OSD}(G)) = \text{conv}\{\mathbf{x} \in \mathbf{Z}^{|V|}_+ : M_{OSD}(G)\, \mathbf{x} \geq \mathbf{1}\}.$$

Based on results from [5] on general covering polyhedra, we prove the following.

Theorem 5. *Let $G = (V, E)$ be an OSD-admissible graph. $P_{OSD}(G)$ has*

(a) *the equation $x_v = 1$ for all forced vertices $v \in \mathcal{F}^1_{OSD}(G)$;*
(b) *a nonnegativity constraint $x_v \geq 0$ for all vertices $v \notin \mathcal{F}^1_{OSD}(G)$ and*
(c) *$\sum_{v \in F} x_v \geq 1$ for all hyperedges F of $\mathcal{C}_{OSD}(G)$ with $F \in \mathcal{F}^2_{OSD}(G)$.*

For any covering polyhedron $P(M)$ associated with a 0/1-matrix M with n columns, $Q(M) = \{ \mathbf{x} \in \mathbf{R}^n_+ : M\mathbf{x} \geq \mathbf{1} \}$ is its linear relaxation. We have $P(M) \subseteq Q(M)$ in general and further constraints have to be added to the system $M\mathbf{x} \geq \mathbf{1}$ in order to describe $P(M)$ using real variables instead of integral ones.

We next study the OSD-polyhedra for some special graphs related to hypergraph $\mathcal{R}^q_n = (V, \mathcal{E})$ called *complete q-rose of order n*, where $V = \{1, \ldots, n\}$ and $\mathcal{E}$ contains all q-element subsets of V for $2 \leq q < n$. In [2] it was proved that the covering polyhedron of $\mathcal{R}^q_n$ is given by the nonnegativity constraints and

$$x(V') = \sum_{v \in V'} x_v \geq |V'| - q + 1$$

for all subsets $V' \subseteq \{1, \ldots, n\}$ with $|V'| \in \{q, \ldots, n\}$. Moreover, we have $\tau(\mathcal{R}^q_n) = n - q + 1$. Note that, for $q = 2$, $\mathcal{R}^q_n$ is in fact the complete graph K_n.

Determining the OSD-clutters of the graph families studied below showed their relation to different complete q-roses. Relying on the results from [2] on polyhedra associated to complete q-roses enabled us to prove the following.

Theorem 6. *Let $G = (V, E)$ be either a clique K_n with $n \geq 2$ or a matching kK_2 with $k \geq 1$ and $n = 2k$. Then, we have $\mathcal{C}_{OSD}(G) = \mathcal{R}^2_n = K_n$ and $P_{OSD}(G)$ is given by*

(a) *a nonnegativity constraint $x_v \geq 0$ for all vertices $v \in V$ and*
(b) *$x(V') = \sum_{v \in V'} x_v \geq |V'| - 1$ for all subsets $V' \subseteq V$ with $|V'| \geq 2$.*

Note that two graphs with equal OSD-clutters have the same set of OSD-codes and thus also the same OSD-numbers and the same OSD-polyhedra. Theorem 6 shows that this applies to cliques and matchings. The following two theorems show that the OSD-numbers of thin and thick headless spiders, as calculated in Lemma 4, can also be arrived by the use of polyhedral techniques.

Theorem 7. *Let $\overline{H}_k = (Q \cup S, E)$ be a thick headless spider with $k \geq 4$. Then, we have $\mathcal{C}_{OSD}(\overline{H}_k) = \mathcal{R}^{|S|-1}_{|S|} \cup \mathcal{R}^2_{|Q|}$ and $P_{OSD}(\overline{H}_k)$ is given by the constraints*

(a) *$x_v \geq 0$ for all vertices $v \in Q \cup S$,*
(b) *$x(V') = \sum_{v \in V'} x_v \geq |V'| - k + 2$ for all $V' \subseteq S$ with $|V'| \geq k - 1$,*
(c) *$x(V') = \sum_{v \in V'} x_v \geq |V'| - 1$ for all $V' \subseteq Q$ with $|V'| \geq 2$.*

Comparing this result with the result from [3] on OLD-codes of thick headless spiders, we observe that $\mathcal{C}_{OSD}(\overline{H}_k) = \mathcal{C}_{OLD}(\overline{H}_k)$. Hence, a vertex subset is an OSD-code of $\overline{H}_k$ if and only if it is an OLD-code of $\overline{H}_k$. Accordingly, the OSD- and the OLD-numbers and as well as the OSD- and the OLD-polyhedra are equal for thick headless spiders.

Theorem 8. *Consider a thin headless spider $H_k = (Q \cup S, E)$ with $k \geq 4$. Then, we have $\mathcal{C}_{\mathrm{OSD}}(H_k) = H_k$ and $P_{\mathrm{OSD}}(H_k)$ is given by the constraints*

(a) $x_v \geq 0$ for all vertices $v \in Q \cup S$,
(b) $x_{q_i} + x_{s_i} \geq 1$ for $1 \leq i \leq k$,
(c) $x(V') = \sum_{v \in V'} x_v \geq |V'| - 1$ for all $V' \subseteq Q$ with $|V'| \geq 2$.

Combining all the constraints in (b) yields $x(Q) + x(S) \geq k$ and this implies $\gamma^{\mathrm{OSD}}(H_k) \geq k$. It is also easy to see that Q is a cover of $\mathcal{C}_{\mathrm{OSD}}(H_k)$ and hence, $\gamma^{\mathrm{OSD}}(H_k) = k$. This illustrates how, on the one hand, polyhedral arguments can be used to determine lower bounds for OSD-numbers and, on the other hand, an analysis of the OSD-clutter provides OSD-codes. Moreover, if the order of the latter meets the lower bound, the OSD-number of the studied graph is determined.

We note further that manifold hypergraphs have been already studied in the covering context, see e.g. [1,9,14] to mention just a few. The same techniques as illustrated above with the help of complete q-roses can be applied whenever the OSD-clutter of some graph equals such a hypergraph or contains such a hypergraph as substructure, which gives an interesting perspective of studying OSD-polyhedra further.

5   Concluding Remarks

In this paper, we introduced and studied open-separating dominating codes in graphs. We showed that such codes exist in graphs without open twins and that finding minimum OSD-codes is NP-hard. Moreover, we provided bounds on the OSD-number of a graph both in terms of its number of vertices and in relation to other X-numbers, notably showing that OSD- and OLD-number of a graph differ by at most one. Despite this closeness between the OSD- and the OLD-numbers, we proved that it is NP-complete to decide if the two said parameters of a graph actually differ. This further motivated us to compare the two numbers on several graph families. This study revealed that they

- are equal, for example, for cliques K_n with $n \geq 3$, thin and thick headless spiders H_k and $\overline{H}_k$, respectively, with $k \geq 3$, and thin suns $T_k = (C \cup S, E)$ with $k \geq 4$ and without open C-twins;
- differ for example, for matchings kK_2 with $k \geq 1$, half-graphs B_k with $k \geq 1$ and their disjoint unions, k-double stars D_k with $n \geq 2$, and almost complete thin suns $T_{2\ell}$ with $\ell \geq 3$.

In particular, this showed that the OSD-numbers of cliques, half-graphs and their disjoint unions attain the upper bound in Theorem 1. Moreover, we provided an equivalent reformulation of the OSD-problem as a covering problem in a suitable hypergraph composed of the closed neighborhoods and the symmetric differences of open neighborhoods of vertices. We also discussed the polyhedra associated with the OSD-codes, particularly, in relation to the OLD-codes of some graph families already studied in this context. The latter illustrated how polyhedral arguments can be used to determine lower bounds for OSD-numbers, how an analysis of the OSD-clutter can provide the OSD-codes, and that combining both arguments can yield the OSD-numbers of the studied graphs.

The future lines of our research include studying the OSD-problem on more graph families and also searching for extremal cases concerning the lower bounds for OSD-numbers (that is, the logarithmic bound in Theorem 1 and the LD-number in Theorem 2).

Even though the problem of deciding if the OSD- and the OLD-numbers differ is NP-complete in general, it would be interesting to see if for some particular graph families, this problem becomes polynomial-time solvable. In that case, it would be further interesting to provide a complete dichotomy as to for which graphs of that graph family the two code-numbers differ and for which they are equal. Finally, it would be interesting to address the question of whether or not similar relations as for OSD- and OLD-numbers of a graph (who differ by at most one) also hold for ID- and DTD-numbers (combining closed-separation with domination and total domination, respectively) and for LD- and LTD-numbers (combining location with domination and total domination, respectively) on connected graphs (e.g. by bounding their possible differences).

Acknowledgement. This research was financed by a public grant overseen by the French National Research Agency as part of the "Investissements d'Avenir" through the IMobS3 Laboratory of Excellence (ANR-10-LABX-0016), by the French government IDEX-ISITE initiative 16-IDEX-0001 (CAP 20-25) and the International Research Center "Innovation Transportation and Production Systems" of the I-SITE CAP 20-25.

References

1. Argiroffo, G., Bianchi, S.: On the set covering polyhedron of circulant matrices. Discret. Optim. **6**, 162–173 (2009)
2. Argiroffo, G., Bianchi, S., Lucarini, Y., Wagler, A.K.: Polyhedra associated with identifying codes in graphs. Discret. Appl. Math. **245**, 16–27 (2018)
3. Argiroffo, G., Bianchi, S., Lucarini, Y., Wagler, A.K.: Polyhedra associated with locating-dominating, open locating-dominating and locating total-dominating sets in graphs. Discret. Appl. Math. **322**, 465–480 (2022)
4. Babai, L.: On the complexity of canonical labeling of strongly regular graphs. SIAM J. Comput. **9**(1), 212–216 (1980)
5. Balas, E., Ng, S.M.: On the set covering polytope: I. All the facets with coefficients in 0,1,2. Math. Program. **43**, 57–69 (1989)
6. Bondy, J.A.: Induced subsets. J. Comb. Theory **12**(B), 201–202 (1972)
7. Charon, I., Hudry, O., Lobstein, A.: Minimizing the size of an identifying or locating-dominating code in a graph is NP-hard. Theoret. Comput. Sci. **290**, 2109–2120 (2003)
8. Colburn, C., Slater, P.J., Stewart, L.K.: Locating-dominating sets in series-parallel networks. Congr. Numer. **56**, 135–162 (1987)
9. Cornuéjols, G., Combinatorial Optimization: Packing and Covering, SIAM, CBMS **74** (2001)

10. Foucaud, F., Ghareghani, N., Roshani-Tabrizi, A., Sharifani, P.: Characterizing extremal graphs for open neighbourhood location-domination. Discret. Appl. Math. **302**, 76–79 (2021)

11. Karpovsky, M.G., Chakrabarty, K., Levitin, L.B.: On a new class of codes for identifying vertices in graphs. IEEE Trans. Inf. Theory **44**(2), 599–611 (1998)

12. Haynes, T.W., Henning, M.A., Howard, J.: Locating and total-dominating sets in trees. Discret. Appl. Math. **154**, 1293–1300 (2006)

13. Jean, J., Lobstein, A.: Watching systems, identifying, locating-dominating and discriminating codes in graphs. https://dragazo.github.io/bibdom/main.pdf

14. Nobili, P., Sassano, A.: Facets and lifting procedures for the set covering polytope. Math. Program. **45**, 111–137 (1989)

15. Pikhurko, O., Veith, H., Verbitsky, O.: The first order definability of graphs: upper bounds for quantifier depth. Discret. Appl. Math. **154**(17), 2511–2529 (2006)

16. Slater, P.J.: Dominating and reference sets in a graph. J. Math. Phys. Sci. **22**, 445–455 (1988)

17. Seo, S.J., Slater, P.J.: Open neighborhood locating dominating sets. Australas. J. Comb. **46**, 109–119 (2010)

18. Ungrangsi, R., Trachtenberg, A., Starobinski, D.: An implementation of indoor location detection systems based on identifying codes. In: Aagesen, F.A., Anutariya, C., Wuwongse, V. (eds.) INTELLCOMM 2004. LNCS, vol. 3283, pp. 175–189. Springer, Heidelberg (2004). https://doi.org/10.1007/978-3-540-30179-0_16

On the Complexity of the Minimum Chromatic Violation Problem

Diego Delle Donne[1], Mariana Escalante[2,3], and María Elisa Ugarte[3(✉)]

[1] ESSEC Business School, Cergy, France
[2] CONICET, Rosario, Argentina
[3] FCEIA, Universidad Nacional de Rosario, Rosario, Argentina
ugarte@fceia.unr.edu.ar

Abstract. In this paper, we consider a generalization of the classical vertex coloring problem of a graph, where the edge set of the graph is partitioned into *strong* and *weak edges*; the endpoints of a weak edge can be assigned to the same color and the *minimum chromatic violation problem* (MCVP) asks for a coloring of the graph minimizing the number of weak edges having its endpoints assigned to the same color. Previous works in the literature on MCVP focus on defining integer programming formulations and performing polyhedral studies on the associated polytopes but, to the best of our knowledge, very few computational complexity studies exist for MCVP. In this work, we focus on the computational complexity of this problem over several graph families such as interval and unit interval graphs, among others. We show that MCVP is NP-hard for general graphs and it remains NP-hard when the graph induced by the strong edges is unit interval or distance hereditary. On the other side, we provide a polynomial algorithm that properly solves MCVP when the graph is a unit interval graph without triangles with two or more weak edges.

Keywords: chromatic violation · graph coloring · complexity

1 Introduction

A k-coloring of a graph $G = (V, E)$ with a given number of colors $k \in \mathbb{N}$ is an assignment $c : V \to \{1, \ldots, k\}$ such that $c(v) \neq c(w)$ for each edge $vw \in E$. As a shortcut, throughout this paper, we use $[k]$ to denote the set $\{1, \ldots, k\}$. Given a graph G, the classical vertex coloring problem (VCP) aims to determine the smallest number, $k \in \mathbb{N}$, required to provide a k-coloring of the given graph G. This value is known as the chromatic number of G and is denoted as $\chi(G)$.

VCP has diverse applications in scheduling [16], course/event timetabling [7,11], register allocation [9], frequency assignment [3,13], and communication networks [18]. In practical scenarios, colors often represent available resources, such as rooms in a classroom distribution setup, registers in allocation problems, or frequencies/channels in network communication planning.

A. Basu et al. (Eds.): ISCO 2024, LNCS 14594, pp. 152–162, 2024.
https://doi.org/10.1007/978-3-031-60924-4_12

Vertices of the graph usually represent activities that need to use one of the available resources. In this setting, an edge between two vertices states that the associated activities cannot use the same resource, thus representing a conflict between them. This type of graph is usually known as *conflict graph*. If the fixed number k of available colors is not sufficient, the situation may become infeasible, requiring alternative approaches. For instance, in class allocation problems with limited rooms but numerous lectures, relaxing constraints to allow certain lectures to share a room (like small groups, lab sessions, or office hours) becomes necessary. Similarly, in frequency allocation scenarios, where channel availability is restricted due to technology specs, antennae with overlapping coverage areas must not use the same channel. However, if the available channels are fewer than the conflict graph's chromatic number, a feasible assignment becomes impossible. In this case, allowing some degree of interference but minimizing this might be an option. The *minimum chromatic violation problem* (MCVP), which we describe below, is a variant of the VCP that may assess these cases.

Given a graph $G = (V, E)$ and a subset of edges $F \subseteq E$, we note $G - F$ to the graph having V and $E \backslash F$ as vertex and edge sets, i.e., $G - F = (V, E \backslash F)$. A *weak k-coloring* of G w.r.t. F is a k-coloring of $G - F$. In other words, a weak k-coloring must assign different colors to endpoints of edges that are not in F. When the number of available colors is clear from the context we simply refer to it as a weak coloring of the graph. Edges in $E \backslash F$ (resp. on F) are called to be *strong* (resp. *weak*) edges.

Given an integer k, and a weak k-coloring $c : V \rightarrow [k]$, the *chromatic violation of c* is

$$\nu_{G,F}(c) = |\{vw \in F : c(v) = c(w)\}|,$$

i.e., the number of edges in F whose endpoints share the same color. Given a graph G and a set of weak edges F, the *minimum chromatic violation problem* (MCVP) consists of finding a weak k-coloring $c : V \rightarrow [k]$ with minimum $\nu_{G,F}(c)$. The *chromatic violation number* of G w.r.t. F is

$$\nu(G, F, k) = \min \{\nu_{G,F}(c) : c \text{ is a weak } k\text{-coloring of } G \text{ w.r.t. } F\}.$$

Previous works in the literature on MCVP focus on defining integer programming formulations and performing polyhedral studies on the associated polytopes but, to the best of our knowledge, very few computational complexity studies exist for MCVP. In [6], an integer programming formulation for MCVP is introduced and an initial polyhedral study of the polytope arising from this formulation is provided, along with partial characterizations of facet-inducing inequalities and facet-generating procedures. Separation procedures for some families of valid inequalities are presented in [10], and a branch-and-cut algorithm using these inequalities is presented for solving the MCVP.

In [5] the authors define a new framework for locally checkable problems, called r-locally checkable problems. By resorting to this setting, they show that MCVP can be solved in linear time when the graph G has a bounded treewidth.

Unfortunately, graph classes satisfying this property are not too many.

In this work, we further study the computational complexity of MCVP over several graph families in which the treewidth is not bounded, such as distance hereditary, interval, and unit interval graphs, among others. We show that MCVP is NP-hard for general graphs and it remains NP-hard when the graph induced by the strong edges is unit interval or distance hereditary. On the other side, we provide a polynomial algorithm that properly solves MCVP when the graph is a unit interval graph without triangles with two or more weak edges.

1.1 Basic Definitions

In this paper, we consider simple undirected graphs $G = (V, E)$ where V stands for its vertex set and E is its edge set. An induced subgraph of G is a graph G' with vertex set $V' \subset V$ and edge set $E' = \{ij \in E : i, j \in V'\}$.

The complete graph with n vertices, denoted by K_n, is a graph in which every pair of distinct vertices is connected by an edge. A *chordal* graph is a graph without cycles of length at least 4, as induced subgraphs.

A graph $G = (V, E)$ is an *intersection graph* for a finite family F of subsets of a non-empty set if there is a one-to-one correspondence between F and V such that two sets in F have non-empty intersection if and only if their corresponding vertices in V are adjacent. A graph is an *interval* graph if it is the intersection graph of a set of intervals over the real line. A *unit interval* graph is the intersection graph of a set of intervals of length one.

The distance between two vertices of G is the number of edges in a path of minimum length in G between them. A graph is *distance-hereditary* if any two vertices are equidistant in every connected induced subgraph containing them.

2 Complexity of MCVP

Given a graph G and an integer number k, the decision problem associated with the classical VCP consists of determining whether there exists a k-coloring of G or not. This decision problem is NP-complete for general graphs [14].

We consider the decision problem associated with MCVP: given a graph G, a set of weak edges F, an integer k and an integer t, determine whether $\nu(G, F, k) \leq t$ or not. Note that when $F = \emptyset$ and $t = 0$, this problem corresponds to the classical k-coloring decision problem, thus it is NP-complete in general graphs. We then focus on the study of the complexity of MCVP in different classes of graphs intending to identify classes in which the problem can be solved in polynomial time or in which we can prove that it remains NP-complete.

In this section, we present a result on the complexity of MCVP, by resorting to a polynomial reduction from a known NP-complete problem, namely *the precoloring extension problem* (PrExt) [2]. Given a graph $G = (V, E)$, an integer k, and a function $f : A \to [k]$ for some $A \subset V$, which pre-assigns colors

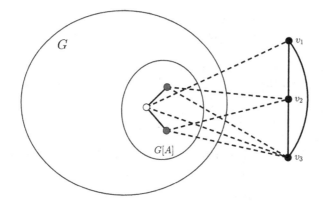

Fig. 1. An example, with $k = 3$, of the graph G' in the proof of Theorem 1. Vertices v_1, v_2 and v_3 on the right are the additional vertices representing the colors. Dotted lines correspond to weak edges in F connecting each original vertex v from the set A with every additional vertex except for the one representing the color pre-assigned to vertex v.

to some of the vertices of G, the precoloring extension problem asks whether there exists a coloring $c : V \to [k]$ of G such that $c(i) = f(i)$ for every $i \in A$. Precoloring extension is NP-complete for general graphs, and it remains NP-complete when restricted to interval graphs [2], unit interval graphs [17], and distance-hereditary graphs [4], among others. It is worth noting that the classical vertex coloring problem is polynomial in the mentioned classes [15], thus stating a qualitative difference between the computational complexity of these two combinatorial optimization problems.

Theorem 1. *Let $\mathcal{G}$ and $\mathcal{H}$ be two families of graphs such that $\{G \cup K_r : G \in \mathcal{G}, r \in \mathbb{N}\} \subset \mathcal{H}$. If PrExt is NP-complete on $\mathcal{G}$, then the decision problem associated with MCVP is NP-complete on the family of graphs H with weak edges F such that $H - F \in \mathcal{H}$.*

Proof. Given an instance (G, f, k) of PrExt with $G = (V, E) \in \mathcal{G}$, we define an instance (G', F, k) of MCVP as follows. Add one vertex to V for each color in $[k]$, $V' = V \cup \{v_1, v_2, \ldots, v_k\}$. Let $G' = (V', E')$ where $E' = E \cup F \cup \{v_i v_j : i \neq j\}$ for $F = \{vv_i : v \in A, f(v) \neq i\}$. Figure 1 depicts this construction in an example using $k = 3$.

In this way, the graph G' is such that $G' - F = G \cup K_k \in \mathcal{H}$. Let us prove that, MCVP on (G', F, k) has a solution with $\nu(G', F, k) = 0$ if and only if f can be extended to a coloring of G.

Let $g : V' \to [k]$ be a coloring of G' with $\nu_{G',F}(g) = 0$. Hence, g is a coloring of G' and $v_1, v_2, \ldots, v_k$ receive different colors, since they induce a complete subgraph of G'. W.l.o.g. assume that $g(v_i) = i$, for each $i \in [k]$ (otherwise, we perform a permutation of color classes). Furthermore, by definition of E', a vertex $v \in A$ is adjacent to every vertex v_j, with $j \neq f(v)$. As there are only

k available colors, v can only receive the color assigned to $v_{f(v)} = v_i$, i.e., color $g(v_i) = i = f(v)$. Therefore, the restriction of g to V is a coloring of G, that extends f.

Conversely, if f admits an extension $\tilde{f} : V \rightarrow [k]$ that is a coloring of G, let $g : V' \rightarrow [k]$ be such that

$$g(v) = \begin{cases} \tilde{f}(v) & \text{if } v \in V, \\ i & \text{if } v = v_i \text{ for } i = 1, \dots, k. \end{cases}$$

Clearly, this function defines a weak coloring of G' with chromatic violation equal to zero, thus completing the proof. □

Note that this reduction does not preserve the graph. In particular, many graph classes are not closed under this kind of operation. According to the previous theorem, we are interested in those classes of graphs G such that $G \cup K_k$ also belongs to that class. Some examples of that kind of graph classes are: interval, unit interval, distance-hereditary, and chordal graphs, where, precisely, PrExt is NP-complete. Therefore, we have the following result.

Corollary 1. *The decision problem associated with MCVP remains NP-complete in the family of graphs G with weak edges F, whenever $G - F$ is a unit interval or distance-hereditary graph.*

Proof. Let G be a unit interval (resp. distance hereditary) graph. Clearly, $G \cup K_r$ is also a unit interval (resp. distance hereditary) graph, for every $r \in \mathbb{N}$. Therefore, the results follow from Theorem 1, when $\mathcal{G}$ and $\mathcal{H}$ are both the family of unit interval (resp. distance hereditary) graphs. □

3   A Polynomial Algorithm for MCVP on a Subclass of Unit Interval Graphs

In this section, we present a polynomial time algorithm for determining the chromatic violation number of unit interval graphs without some particular substructures, which we define next. Our approach, inspired by the ideas in [2,8], is based on solving a minimum cost flow problem in a particular network associated with the MCVP instance.

We recall that in the *minimum cost flow of size k problem*, we are given an integer k and a directed graph (a network) $D = (V, A, \text{cost}, \text{cap}, \text{dem})$, where V is a set of vertices, A is a set of (directed) arcs, and associated with each arc e is a nonnegative cost $\text{cost}(e)$, a positive capacity $\text{cap}(e)$ and a nonnegative demand $\text{dem}(e)$. Also, there are designated vertices s and t. A flow from s to t is a mapping f defined on A assigning a real number to each arc satisfying the following conditions:

1. for each vertex u, except s and t, the flow into u is equal to the flow out of u, and

2. for each edge e, $\text{dem}(e) \leq f(e) \leq \text{cap}(e)$.

The *size* of flow f is the net flow into t. The *cost* of flow f is $\sum_{e \in E} f(e)\text{cost}(e)$. The objective in the minimum cost flow of size k problem is to find a flow of size k from s to t that is of minimum cost among all such flows. A polynomial time algorithm for finding such a flow is given in [1,12].

Given an instance (G, F) of MCVP, a *weak* triangle is an induced K_3 in G with all its edges in F. When only two of the three edges belong to F, we refer to this subgraph as a *semi-weak* triangle. Left images of Figs. 4 and 5 depict these structures (dotted lines represent weak edges).

Theorem 2. *The chromatic violation number of unit interval graphs without weak and semi-weak triangles can be found in polynomial time.*

Proof. Consider a unit interval graph $G = (V, E)$ with $V = \{v_1, \ldots, v_n\}$, a set of colors $[k]$ and a set of weak edges $F \subseteq E$ not inducing any weak or semi-weak triangles. Let $\{(a_i, b_i), i \in [n]\}$ be the interval representation of G, with $b_i = a_i + 1$. W.l.o.g., we assume that $a_i < a_{i+1}$, for all $i \in [n-1]$. Figure 2 gives an example of an MCVP instance (left) and its interval representation (right). The dotted lines in the graph represent the weak edges.

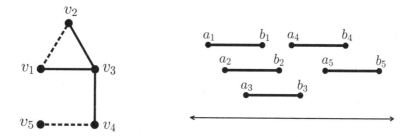

Fig. 2. An interval graph (left) and its interval representation (right). The dotted lines in the graph represent the weak edges.

We construct a network D with nodes $s_1, s_2, a_1, b_1, a_2, b_2, \ldots, a_n, b_n, t$ and directed arcs defined as follows:

- a *coloring* arc $s_1 s_2$,
- a *starting* arc $s_2 a_i$, for each $i \in [n]$,
- an *interval* arc $a_i b_i$, for each $i \in [n]$,
- a *split* arc $b_i a_j$, for each $i, j \in [n]$ with $b_i < a_j$,
- a *weak* arc $b_i a_j$, for each $e = v_i v_j \in F$, $i < j$, and
- a *finishing* arc $b_i t$, for each $i \in [n]$.

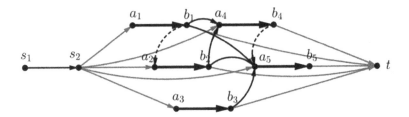

Fig. 3. Network resulting from the example graph depicted in Fig. 2. Gray arrows represent starting and finishing arcs, thick black arrows represent interval arcs, thin black arrows represent split arcs (except for the arrow from s_1 to s_2, which is the coloring arc), and dotted arrows represent weak arcs.

Figure 3 illustrates the network resulting from the example graph depicted in Fig. 2. Every arc of the constructed network has a capacity of 1, except for the coloring arc $s_1 s_2$, whose capacity is k. The cost of a weak arc is 1, and the cost of all other arcs is 0. The only arcs with positive demands are the interval arcs, with demand equal to 1.

For proving this result we state the following claims.

Claim (1). For every integer flow on network D with size $r \in \mathbb{N}$ and cost $t \in \mathbb{N}$, there exists a weak coloring c of G w.r.t. F with r colors and chromatic violation $\nu_{G,F}(c) = t$.

Claim (2). For every weak r-coloring c of G w.r.t. F and chromatic violation $\nu_{G,F}(c) = t$, there exists an integer flow on the network D of size r and cost t.

If the previous claims hold, then an optimal solution for MCVP can be found by solving the minimum cost flow problem on D, which may be done in polynomial time [1]. We recall that an integer optimal flow in D can always be found, as all costs, demands, and capacities are integer values [1].

Proof. of Claim (1)

Consider an integer flow in D of size $r \leq k$ and cost $t \in \mathbb{N}$. Since the flow is integer, we can interpret that each unit of flow follows a single path from s_1 to t. In fact, the r resulting paths from s_2 to t are arc-disjoint, due to the capacity of 1 of every arc in D. Moreover, these paths are also node-disjoint (except for nodes s_2 and t); this can be seen because nodes a_i have just one leaving arc, and nodes b_i have just one entering arc, hence at most one unit of flow can pass through these nodes (again, due to arc capacities). Let $\mathcal{P} = \{P_1, \ldots, P_r\}$ be the set of such disjoint paths and define a coloring $c : V \to [r]$, such that $c(v_i) = j$ if and only if arc $a_i b_i \in P_j$. This is, a path P_j represents a color which is assigned to every *interval arc* $a_i b_i$ traversed by P_j. We shall prove that this coloring is a weak coloring of G w.r.t. F and its chromatic violation is $\nu_{G,F}(c) = t$.

Assume that $c(i) = c(j)$ for some strong edge $v_i v_j \in E \backslash F$ and, w.l.o.g., assume $i < j$. Then, there exists a path $P \in \mathcal{P}$ which first traverses arc $a_i b_i$ and then arc $a_j b_j$. Since these two intervals represent a strong edge, they intersect in the representation (i.e., $a_i < a_j < b_i$) and so there is no *split* arc from b_i

to a_j in D. Hence, there should exist another interval arc $a_h b_h$ traversed by P between $a_i b_i$ and $a_j b_j$ (i.e., with $a_i < a_h < a_j$). Also, given that the model is a *unit* interval representation, it follows that this third arc intersects with both $a_i b_i$ and $a_j b_j$, meaning that there exist edges $v_i v_h$ and $v_h v_j$ in E. If these two are weak edges, then they form a semi-weak triangle with the strong edge $v_i v_j$. Figure 4 illustrates the representation of a semi-weak triangle. If $v_i v_h$ is a strong edge, we can follow the same reasoning but using $v_i v_h$ instead of $v_i v_j$ (as we know that $c(v_i) = c(v_h)$) and finally finding a semi-weak triangle. Analogously, we can do the same in case $v_h v_j$ is a strong edge. Hence, we got a contradiction from the assumption that $c(i) = c(j)$ for some strong edge $v_i v_j \in E \backslash F$, thus c is a valid weak coloring.

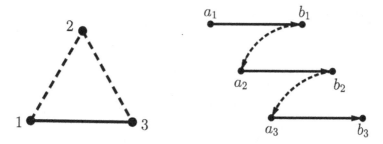

Fig. 4. A semi-weak triangle in G (left) and its representation in the network D (right). Dotted lines correspond to weak edges and weak arcs, respectively.

We now prove that $\nu_{G,F}(c) = t$. Note that if a weak arc $b_i a_j$ in D is traversed by some path $P \in \mathcal{P}$, then $c(v_i) = c(v_j)$, hence it corresponds to a weak edge whose endpoints receive the same color. Assume now that there exists a weak edge $v_i v_j \in F$, for which $c(v_i) = c(v_j) = q$ but for which the weak arc $b_i a_j$ in D is not traversed by the corresponding path (i.e., by path P_q). Assume also, w.l.o.g., $i < j$. In this case, since P_q traverses both interval arcs, first $a_i b_i$ and then $a_j b_j$, there should exist another interval arc $a_h b_h$ traversed by P_q between $a_i b_i$ and $a_j b_j$ (i.e., with $a_i < a_h < a_j$). Also, given that the model is a *unit* interval representation, it follows that this third arc intersects both with $a_i b_i$ and $a_j b_j$, meaning that there exist edges $v_i v_h$ and $v_h v_j$ in E. These two edges should be weak since the intervals representing vertices v_i, v_h and v_j are traversed by the same path and we already proved that the resulting coloring is valid for the strong edges. Then, these edges define a weak triangle with the weak edge $v_i v_j$, hence a contradiction. Figure 5 illustrates the representation of a weak triangle.

Therefore, we have that a weak arc is traversed by some path from $\mathcal{P}$ if and only if it represents a weak edge whose endpoints receive the same color. Since the only arcs with positive cost are the weak arcs, the flow traverses exactly t weak arcs. This reasoning proves that $\nu_{G,F}(c) = t.\diamond$

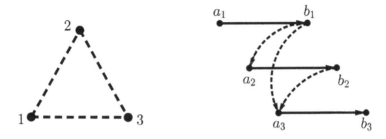

Fig. 5. Weak triangle in G and its representation in the network D. Dotted lines correspond to weak edges and weak arcs, respectively.

Proof. of Claim (2)

Consider a weak r-coloring c of G w.r.t. F and chromatic violation $\nu_{G,F}(c) = t$. For each color $q \in [r]$, let $\{v_{i_1}, \ldots, v_{i_h}\}$ be the set of vertices receiving color q, such that $i_1 < i_2 < \cdots < i_h$. Define path P_q in the network D as $P_q = [s_1, s_2, a_{i_1}, b_{i_1}, \ldots, a_{i_h}, b_{i_h}, t]$. This is a feasible path in D since each arc $b_{i_j} a_{i_{j+1}}$ in this path is either a weak arc (if $v_{i_j} v_{i_{j+1}} \in F$) or a split arc (if $v_{i_j} v_{i_{j+1}} \notin E$). This follows from the fact that $v_{i_j} v_{i_{j+1}}$ cannot be a strong edge since $c(v_{i_j}) = c(v_{i_{j+1}})$.

Consider now a flow that sends one unit through each of the paths $P_1, \ldots, P_r$. It is easy to check that the resulting flow is feasible, as the paths are arc-disjoint (except for the arc $s_1 s_2$ which has capacity k). Then we just need to prove that this flow has cost $\nu_{G,F}(c) = t$, i.e., that it traverses exactly t weak arcs.

First, note that the flow only traverses weak arcs associated with weak edges in G whose endpoints receive the same color in coloring c. Then, we just need to prove that all such weak arcs are traversed. Assume then that there exists a weak edge $v_i v_j \in F$, for which $c(v_i) = c(v_j) = q$ but for which the weak arc $b_i a_j$ in D is not traversed by the corresponding path (i.e., by path P_q). W.l.o.g. assume that $i < j$. In this case, P_q traverses both interval arcs, first $a_i b_i$ and then $a_j b_j$, then there should exist another interval arc $a_h b_h$ traversed by P_q between $a_i b_i$ and $a_j b_j$ (i.e., with $a_i < a_h < a_j$). Also, given that the model is a *unit* interval representation, the reasoning followed in the proof of Claim (1) shows that there exists a weak triangle, a contradiction. Therefore, a weak arc is traversed by some path in $\mathcal{P}$ if and only if the arc represents a weak edge whose endpoints receive the same color. Since the number of weak edges whose endpoints receive the same color is $\nu_{G,F}(c) = t$, the cost of this flow is exactly t. ◇

As previously explained, these two claims complete the proof of Theorem 2. □

4   Concluding Remarks

We considered a generalization of the classical vertex coloring problem of a graph, the *minimum chromatic violation problem* (MCVP), which asks for a coloring of a graph in which some edges are identified as *weak edges*, minimizing the number of weak edges having its endpoints assigned to the same color. We focus on the computational complexity of MCVP over several graph families and prove that MCVP is NP-hard even when the graph induced by the strong edges is unit interval or distance hereditary. This result marks a particular difference in the complexity of MVCP when compared with the classical vertex coloring problem (VCP), which is polynomial when the graph being colored belongs to either of these classes. As we mention in Sect. 2, a similar situation exists when considering the precoloring extension problem PrEXt.

On the other hand, we provide a polynomial algorithm for MCVP which finds the optimal solution when the graph is a unit interval graph without triangles with two or more weak edges. It is not clear if the same can be achieved for PrExt in this class of graphs (recall that PrExt is NP-complete for unit-interval graphs). This may pose a qualitative difference in the computational complexity of these two problems, namely MCVP and PrExt.

Our results show that the complexity boundary of MVCP considering interval graphs is not yet properly delimited, thus leaving some straightforward open questions. For example, it is not clear if MCVP is NP-hard when the graph is a unit interval graph with no other extra conditions. Also, it is uncertain yet if there exists a polynomial algorithm for MCVP when the original graph is an interval graph. We plan to investigate these open aspects in the future.

References

1. Ahuja, R.K., Magnanti, T.L., Orlin, J.B.: Network Flows: Theory, Algorithms, and Applications. Prentice Hall, Hoboken (1993)
2. Biró, M., Hujter, M., Tuza, Z.S.: Precoloring extension. I. Interval graphs. Discrete Math. **100**, 267–279 (1992)
3. Borndörfer, R., Eisenblätter, A., Grötschel, M., Martin, A.: The Orientation Model for Frequency Assignment Problems, ZIB-Berlin TR **98**–01 (1998)
4. Bonomo, F., Durán, G., Marenco, J.: Exploring the complexity boundary between coloring and list-coloring. Ann. Oper. Res. **169**, 3–16 (2009)
5. Bonomo-Braberman, F., Gonzalez, C.: A new approach on locally checkable problems. Discret. Appl. Math. **314**, 53–80 (2022)
6. Braga, M., Delle Donne, D., Escalante, M.S., Marenco, J., Ugarte, M.E., Varaldo, M.C.: The minimum chromatic violation problem: a polyhedral approach. Discret. Appl. Math. **281**, 69–80 (2020)
7. Burke, E., Marecek, J., Parkes, A., Rudová, H.: A supernodal formulation of vertex colouring with applications in course timetabling. Ann. Oper. Res. **179**–1, 105–130 (2010)
8. Carlisle, M.C., Lloyd, E.L.: On the k-coloring of intervals. Discret. Appl. Math. **59**, 225–235 (1995)
9. Chow, F.C., Hennessy, J.L.: The priority-based coloring approach to register allocation. ACM Trans. Program. Lang. Syst. **12**–4, 501–536 (1990)
10. Delle Donne, D., Escalante, M.S., Ugarte, M.E.: Implementing cutting planes for the chromatic violation problem. In: Proceedings of the Joint ALIO/EURO International Conference 2021–2022 on Applied Combinatorial Optimization, OpenProceedings.org, pp. 17–22 (2022)

11. de Werra, D.: An introduction to timetabling. Eur. J. Oper. Res. **19**, 151–162 (1985)
12. Edmonds, J., Karp, R.M.: Theoretical improvements in algorithmic efficiency for network flow problems. J. Assoc. Comput. Mach. **19**, 248–264 (1972)
13. Gamst, A.: Some lower bounds for a class of frequency assignment problems. IEEE Trans. Veh. Technol. **35–1**, 8–14 (1986)
14. Garey, M., Johnson, D.: Computers and Intractability: A Guide to the Theory of NP-Completeness (1979). Freeman, W. H
15. Grötschel, M., Lovász, L., Schrijver, A.: The ellipsoid method and its consequences in combinatorial optimization. Combinatorica **1**, 169–197 (1981)
16. Leighton, F.T.: A graph coloring algorithm for large scheduling problems. J. Res. Natl. Bur. Stand. **84–6**, 489–503 (1979)
17. Marx, D.: Precoloring extension on unit interval graphs. Discret. Appl. Math. **154**, 995–1002 (2006)
18. Woo, T.K., Su, S.Y.W., Newman Wolfe, R.: Resource allocation in a dynamically partitionable bus network using a graph coloring algorithm. IEEE Trans. Commun. **39–12**, 1794–1801 (2002)

Crystal Trees

Rafael Castro de Andrade[(✉)] [ID]

Department of Statistics and Applied Mathematics, Federal University of Ceará,
Campus do Pici, BL 910, CEP 60.455-760 Fortaleza, Ceará, Brazil
rafael.andrade@ufc.br

Abstract. This work introduces the class of crystal trees and their
mathematical modeling. Consider a simple undirected weighted graph
$G = (V, E)$ of edge weights $c_e > 0$, for all $e \in E$. Let T_k be a spanning
tree of G rooted at vertex $k \in V$, and $\mathcal{P}_{ij}^{T_k}$ denote the edge set of the path
between vertices i and j in T_k. Associate with every vertex $v \in V$ of T_k
a potential $\Phi_v^{T_k} = \{c_e \mid e \in \mathcal{P}_{kv}^{T_k}\}$, with $\Phi_k^{T_k} = \emptyset$. Let $\Delta_{uv} = \Phi_v^{T_k} \Delta \Phi_u^{T_k}$ be
the multiset symmetric difference between the potentials of the extrem-
ities of an edge $\{u, v\} \in E$. We say that these potentials are: (i) in
equilibrium if $max\{c_e \mid e \in \Phi_v^{T_k} \Delta \Phi_u^{T_k}\} \leq c_{uv}$; or (ii) in non-equilibrium,
otherwise. If, for all edges in E, the potentials of their extremities are
in equilibrium, then T_k is a crystal tree. We show that this new class of
trees generalizes the Minimum Spanning Trees (MSTs) of a graph. We
present theoretical results for crystal trees and an algebraic representation allowing us to describe MSTs by a system of linear inequalities.
This opens up new possibilities for solving optimization problems with
optimal tree structure in the set of constraints.

Keywords: Crystal tree · MST system · Algebraic multiset modeling

1   Introduction

In this work, we introduce the new class of Crystal Trees (CTs). They are rooted
spanning trees of a simple and undirected weighted graph satisfying a certain
'blind' optimality condition. The discovery of this new concept is a byproduct of
the effort to develop a system of linear inequalities whose solutions correspond
to the graph's Minimum Spanning Trees (MSTs). A set of constraints describing
MSTs of a graph is unprecedented in the literature and has potential application
for solving problems such as the robust spanning tree [2], in which the mathematical formulation requires defining an MST in the set of constraints of the
problem.

What would be our initial motivation, solving the robust tree problem, made
it possible to discover the new tree structure and novel theoretical results that
distinguish it from classical types of trees in graphs (such as MSTs and Shortest
Path Trees). In addition to making it possible to classify MSTs as a subclass
of CTs, we develop a characterization of CTs through multiset operations and
create an algebraic representation for them.

A. Basu et al. (Eds.): ISCO 2024, LNCS 14594, pp. 163–172, 2024.
https://doi.org/10.1007/978-3-031-60924-4_13

As a consequence, we propose a compact mixed integer programming model for CTs. We show that the intersection of CTs for different vertices as their roots can result in MSTs. This leads to open questions, one of which is knowing how many of these CT intersections are sufficient to determine an MST.

The reader will be faced with other interesting questions still unanswered that depend on the types of CTs (defined later), such as knowing the computational complexity to determine whether (or not) a given graph contains a certain type of CT.

Exploring uncharted territory in the realm of crystal trees, delving into their potential for resolving optimization problems with MST structures embedded within their constraints, alongside unveiling a novel class of spanning trees presented for the first time in the literature in this work, underscores the significance and pertinence of delving into the study of this new class of trees.

To grasp the mathematical intricacies of crystal trees and appreciate the theoretical contributions of this endeavor, it necessitates an introduction to a repertoire of unconventional concepts, ideas, and novel propositions introduced in the subsequent sections.

The proofs of the proposed propositions are omitted in this brief paper, as well as the utilization of CTs to model the robust spanning tree problem. Interested readers will find these details in the extended version of this work.

2   Preliminary Concepts

Let $G = (V, E)$ be a graph with node set V and edge set E, where each edge $\{u, v\} \in E$ has a positive edge weight denoted by c_{uv}. We will adopt the following notation.

- T_k is a spanning tree of G rooted at a given node $k \in V$.
- $E(T_k)$ denotes the set of edges of a tree T_k.
- $\mathcal{P}_{uv}^{T_k}$ is the edge set in the path (i.e., a path for short) from node u to node v in T_k.
- $\Phi_v^{T_k} = \{c_e \mid e \in \mathcal{P}_{kv}^{T_k}\}$ is the potential of node v in T_k, with the root node k having null potential, i.e., $\Phi_k^{T_k} = \emptyset$.
- $M \Delta M'$ is the symmetric difference between multisets M and M'.[1]

Definition 1 (Potentials' equilibrium). *We say that the potentials of nodes u and v in a k-rooted spanning tree T_k w.r.t. an edge $\{u, v\} \in E$ are*

$$in\ equilibrium\ if\ \max_{e \in \Phi_v^{T_k} \Delta \Phi_u^{T_k}} \{c_e\} \leq c_{uv},$$

or in non-equilibrium, otherwise.

Definition 2 (Type of equilibrium). *If the potentials of nodes u and v in T_k are in equilibrium w.r.t. an edge $\{u, v\} \in E$, it can be*

---

[1] If p_1 (resp. p_2) copies of an element belong to M (resp. M'), then exactly $|p_1 - p_2|$ copies of this element belong to the resulting multiset. For example, $\{2, 2, 3, 4, 4, 4, 5\} \Delta \{2, 4, 4, 6, 7, 7\} = \{2, 3, 4, 5, 6, 7, 7\}$.

– *stable if $c_{uv} \geq c_e$, for all $e \in \mathcal{P}_{uv}^{T_k}$;*
– *or unstable, otherwise; i.e., if $c_{uv} < c_e$, for some $e \in \mathcal{P}_{uv}^{T_k}$;*

To examplify the equilibrium notion, consider the spanning tree T_1 of the graph G rooted at node 1 in Fig. 1 and the edges $\{2,3\}$ and $\{2,5\}$. The potentials of nodes 2 and 3 in T_1 are $\Phi_2^{T_1} = \{1\}$ and $\Phi_3^{T_1} = \{5,1\}$, respectively. Because $\Phi_2^{T_1} \Delta \Phi_3^{T_1} = \{1\}\Delta\{5,1\} = \{5\}$ and $\max\{5\} = 5 \geq c_{23} = 5$, then for these nodes their potentials are in stable equilibrium. On the other hand, the potential of node 5 is $\Phi_5^{T_1} = \{5,1,1\}$. With respect to nodes 2 and 5, the difference of their potentials is $\Phi_2^{T_1} \Delta \Phi_5^{T_1} = \{1\}\Delta\{5,1,1\} = \{5,1\}$. Because $\max\{5,1\} = 5 > c_{25} = 4$, then their potentials are not in equilibrium.

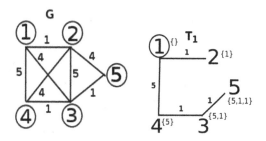

Fig. 1. On the left, a graph G with its edges and respective weights. On the right, a spanning tree T_1 of G, rooted at node 1, with node potentials indicated into braces.

Definition 3 (Crystal tree). *A crystal tree is a k-rooted spanning tree T_k of a weighted graph $G = (V, E)$ whose node potentials w.r.t. the extremities u and v of every edge $\{u, v\} \in E$ are in equilibrium in T_k.*

If the potentials of the extremities of all edges of a graph G present stable equilibrium in a k-rooted spanning tree T_k of G, then we say that T_k is a stable crystal tree. A crystal tree presenting at least a pair of potentials in unstable equilibrium w.r.t. the extremities of an edge of E is an unstable crystal tree.

In Fig. 2, we give examples of two crystal trees referring to the graph G in Fig. 1. We can check by inspecting the potentials' equilibrium w.r.t. the extremities of all the edges of G that the spanning trees T_1 and T_2 are crystal trees of stable and unstable equilibrium, respectively.

3 Crystal Trees Are Distinct from Well Known Classes of Trees in Graphs

In this section, we distinguish crystal trees from existing classes of trees. More specifically, from Minimum Spanning Trees (MSTs) and Shortest Path Trees (SPTs) of weighted graphs. Initially, consider the known property of MSTs.

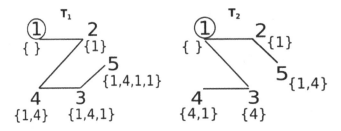

Fig. 2. Example of crystal trees of the graph G in Fig. 1 rooted at node 1 with distinct equilibrium.

Property 1. *Let T^* be a minimum spanning tree of a weighted graph $G = (V, E)$. The weight of every edge $\{u, v\} \in E \setminus E(T^*)$ is larger than or equal to the largest weight of an edge in the path $\mathcal{P}_{uv}^{T^*}$ from u to v in T^*.*

Property 1 allows us to conclude that the potentials of the extremities of all edges of a graph G w.r.t. an MST of this graph are in stable equilibrium, resulting in the following proposition.

Proposition 1. *Every minimum spanning tree T^* of a weighted graph G is a crystal tree independently of its root node.*

An example of Proposition 1 is the tree T_1 rooted at node 1 in Fig. 2 w.r.t. the graph G of Fig. 1. Changing the root node of this tree will reconfigure the node potentials. However, the resulting tree remains a crystal tree. This is not the case if, e.g., we change the root node of tree T_2 in the same figure. Indeed, keeping the same tree topology and fixing its root at node 5, we obtain a tree that is in non-equilibrium, for instance, w.r.t. edge $\{3, 5\}$. Consequently, we introduce a new proposition.

Proposition 2. *A crystal tree T_k is not necessarily a minimum spanning tree of a weighted graph G.*

Finally, we distinguish crystal trees from shortest-path trees. For instance, concerning the crystal tree T_2 in Fig. 2, it is one of the shortest-path trees rooted at node 1. But this is not always the case.

Proposition 3. *A shortest-path tree T_k of a weighted graph G rooted at a node k is not necessarily a crystal tree rooted at this node.*

Figure 3 shows an example of a shortest-path tree T_3 of the graph G in Fig. 1 rooted at node 1. This tree is not crystal because, for instance, the potentials of the extremities of the edge $\{2, 4\}$ are not in equilibrium.

We give a final distinction between CTs, SPTs, and MSTs in the next proposition, for which the tree T_2 in Fig. 2 is an example.

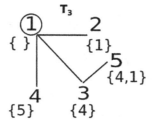

Fig. 3. Example of a shortest-path tree rooted at node 1. It is not a crystal tree.

Proposition 4. *If T_k is simultaneously a k-rooted shortest-path tree and a k-rooted crystal tree of a graph G, then T_k is not necessarily a minimum spanning tree of G.*

Since every minimum spanning tree is a crystal tree, we can efficiently obtain a stable CT in polynomial time. However, the computational complexity of determining whether a given graph contains an unstable CT remains unclear, and it is uncertain if this problem can be solved in polynomial time.

To assess the stability of a crystal tree T_k, the process involves checking if its equilibrium is unstable, and this can be accomplished in at most $O(|E| \times |V|)$. This entails performing a depth-first search in $O(|V|)$ time to identify, within the tree, the largest edge weight in the path between the two extremities of the edges not present in T_k. Subsequently, this weight is compared with the corresponding edge weight in question.

As a result, the decision problem concerning the existence of unstable Crystal Trees is classified within the computational complexity class NP. As discussed further below, resolving this question is of significant importance in obtaining a reduced system of MSTs.

4   Algebraic Definition of Crystal Trees

The main objective of this section is to explore the concept of crystal trees in mathematical optimization models. Considering their multiset definition is not convenient. To overcome this matter, we propose an algebraic definition for them. The idea is to represent a multiset algebraically. To avoid creating a new symbol for the algebraic representation of a multiset, we adopt the same notation as before.

Initially, let $\mathcal{O} = \{o_1, o_2, \cdots, o_s\}$, with $s \leq |E|$, be the set of s distinct values of the edge costs in E, and consider the following definitions.

Definition 4 (Expanded edge weight). *The expanded weight of $w \in \mathcal{O}$ is*

$$\phi(w) = (\gamma b)^{|\{o \in \mathcal{O} \mid o < w\}|} \tag{1}$$

where γb, with $b = 1 + \min\{|V| - 1, M\}$, denotes the base of expansion and M is the number of occurrences of the most frequent edge weight. $\gamma \geq 2$ is a positive integer constant.

Definition 5 (Expanded node potential). *Let T_k be a k-rooted spanning tree of G. The expanded node potential $\Phi_v^{T_k}$ of a node v in T_k, with $\Phi_k^{T_k} = 0$, is*

$$\Phi_v^{T_k} = \sum_{e \in \mathcal{P}_{kv}^{T_k}} \phi(c_e) \tag{2}$$

As an example, consider the graph G in Fig. 1 and the tree T_2 in Fig. 2. For this graph, $M = 3$ and $b = 1 + min\{5 - 1, 3\} = 4$. Considering $\gamma = 2$, the base of expansion γb is equal to 8. In T_2, the root node has expanded potential $\Phi_1^{T_2} = 0$. The remaining nodes have expanded potentials given by $\Phi_2^{T_2} = 8^0$, $\Phi_3^{T_2} = 8^1$, and $\Phi_4^{T_2} = \Phi_5^{T_2} = 8^0 + 8^1$. Note that each multiset (potential) is written uniquely in the numerical base $\gamma b = 8$, as ensured by the next propositions.

Proposition 5. *There is a unique representation of $\Phi_v^{T_k}$ in base γb.*

Proposition 6. *Let w be any edge weight in the path $\mathcal{P}_{kv}^{T_k}$ from k to v in T_k. The expanded weight $\phi(w)$ cannot be obtained by the sum of expanded potentials of edge weights smaller than w in $\mathcal{P}_{kv}^{T_k}$.*

Proposition 7. *The following statements are equivalent*

1. *The difference of potentials of the two extremities of an edge is empty;*
2. *The difference of their expanded potentials is zero;*

The propositions above allow characterizing crystal trees in the following theorem.

Theorem 1 (Algebraic definition of crystal tree). *A k-rooted spanning tree T_k of a graph G is crystal if and only if*

$$|\Phi_u^{T_k} - \Phi_v^{T_k}| \leq (b-1)\phi(c_{uv}), \quad \forall \{u, v\} \in E \tag{3}$$

with Φ, ϕ, and b from Definitions 4 and 5.

The formal proof of Theorem 1 is extensive, thus omitted. The idea is to show that the value of γ must be at least 2 in Definition 4. Under this assumption: (i) inequalities (3) avoid difference of potentials that are not in equilibrium, i.e., if there exists an edge $e' \in \mathcal{P}_{uv}^{T_k}$ of weight larger than the one of an edge $\{u, v\}$ not in T_k in the resulting difference of the potentials of u and v, then $|\Phi_u^{T_k} - \Phi_v^{T_k}|$ is of the order $O((\gamma b)^{|\{o \in \mathcal{O} | o < c_{e'}\}|})$ which is larger than $(b-1)(\gamma b)^{|\{o \in \mathcal{O} | o < c_{uv}\}|}$, thus violating the corresponding inequality (3), because $c_{e'} > c_{uv}$ and, by Proposition 6, the expanded weight of $c_{e'}$ cannot be obtained as a linear combination of the one of c_{uv}; and (ii) inequalities (3) lead to stable or unstable difference of potentials of the extremities of the edges not in T_k and, consequently, it is crystal.

To show the importance of the γ's value in the proof of Theorem 1, we introduce some intermediary results. Suppose that the most frequent edge weight of G is w with M occurrences and that $b = M + 1$ or, equivalently, $M = b - 1$.

Consider that w' is the weight immediately larger than w in $\mathcal{O}$, i.e., w' has order one unit larger than the order of the weight w. In Fig. 4 let T_a in (i) and T_b in (ii) be two spanning trees of G rooted at nodes a and b, respectively. Assume, for T_a, that there exists $(b-1)$ edges of weight w between nodes a and v and that there exists an edge $\{a, v\} \in E \setminus E(T_a)$. For this tree, the difference of potentials between the extremities of this edge is exactly $(b-1)\phi(c_{av})$. Also assume, for T_b, that there exists $(b-2)$ edges of weight w between nodes b and v, an edge $\{b, u\}$ of weight w', and that there exists an edge $\{u, v\} \in E \setminus E(T_b)$ of weight w. In this tree, the difference of potentials $|\Phi_u^{T_b} - \Phi_v^{T_b}|$ is $(\gamma b - b + 2)\phi(w)$.

Proposition 8. *The bound in the hight-hand side of inequalities (3) is tight.*

Proposition 9. *Given a path $\mathcal{P}_{uv}^{T_k}$ in a spanning tree T_k of G between the two extremities u and v of an edge $\{u, v\} \in E \setminus E(T_k)$, if this path contains only two consecutive weights in $\mathcal{O}$, say w and w', with $w < w'$, then the minimum difference of potentials for these nodes is $(\gamma b - b + 2)\phi(w)$ and this bound is tight.*

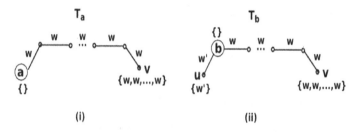

(i)   (ii)

Fig. 4. Bounding situations for the difference of potentials between the extremities of edges $\{a, v\}$ (for Proposition 8) and $\{u, v\}$ (for Proposition 9) in distinct rooted trees T_a and T_b, respectively.

In Fig. 4, the potentials of nodes u and v in T_b are not in equilibrium and, consequently, it is not a crystal tree. Theorem 1 fails for $\gamma = 1$ in the cited situation of edge weights in the path $\mathcal{P}_{uv}^{T_b}$ between these nodes. We observed, in this case, that inequalities (3) being satisfied is not sufficient for the tree to be crystal if the restriction $\lambda \geq 2$ is not used.

For example, consider the graph with three nodes in Fig. 5 and the spanning tree rooted at node b represented by edges $\{a, b\}$ and $\{b, c\}$. Let $b = 1 + \min\{3 - 1, 2\} = 3$ and consider $\gamma = 1$. For this tree, $|\Phi_a - \Phi_c| = |3^0 - 3^1| \leq (3-1)\phi(1) = (2)3^0$. This inequality is true, but the tree is not crystal. Adopting, e.g., $\gamma = 2$ resolves the problem.

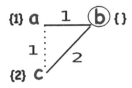

Fig. 5. A graph with a spanning tree. Weights are indicated near the edges and potentials near the nodes.

To show that $\gamma \geq 2$, we consider the situation from Proposition 9. By observing that T_b is not crystal in Fig. 4, we obtain the following result.

Proposition 10. *Any integer value for γ in the Definition 4 must satisfy*

$$\gamma > 2 - \frac{3}{b}, \quad \forall\, b \geq 2 \tag{4}$$

We close this section by highlighting the contributions discussed above. They refer to the two alternative (multiset and algebraic) definitions of the novel concept of crystal trees, and the original results concerning the ten propositions. Having presented the main ingredients to define the node potentials of a crystal tree by algebraic expressions, now we show how to obtain all the crystal trees of a graph G as solutions of a linear system.

4.1   A System of Crystal Trees

A crystal tree T_k of a graph G can be represented by an arborescence rooted at node k. To model T_k as an arborescence of G, every edge $\{u, v\} \in E$ gives place to two arcs (u, v) and (v, u) of weights equal to the one of the corresponding edge. Let A be the set of resulting arcs. Arbitrarily, we also let $\mathcal{P}_{kv}^{T_k}$ be the set of arcs in the path between the nodes k and v in T_k, with the potential of a node v being determined as above by the weights of the arcs in $\mathcal{P}_{kv}^{T_k}$.

We represent T_k by a characteristic vector $x^k \in \{0, 1\}^{|A|}$, where $x_{uv}^k = 1$ if the arc (u, v) belongs to T_k; and $x_{uv}^k = 0$, otherwise. To capture the expanded potentials of the nodes in T_k in function of x^k, we use non-negative continuous variables $\Phi^k \in \mathbb{R}_+^{|V|}$, with $\Phi_k^k = 0$. Consider the mathematical system (S_k)

$$x^k \in \{0, 1\}^{|A|} \text{ defines an arborescence of } G \tag{5}$$

$$\Phi_v^k - \Phi_u^k \leq \phi(c_{uv}) + \mathcal{M}(1 - x_{uv}^k),\ \forall\, (u, v) \in A \tag{6}$$

$$(S_k) \qquad \Phi_v^k - \Phi_u^k \geq \phi(c_{uv}) - \mathcal{M}(1 - x_{uv}^k),\ \forall\, (u, v) \in A \tag{7}$$

$$\Phi_v^k - \Phi_u^k \leq (b - 1)\phi(c_{uv}),\ \forall\, (u, v) \in A \tag{8}$$

$$\Phi_k^k = 0, \quad \Phi^k \geq \mathbf{0}, \tag{9}$$

where $\mathcal{M}$ is a very large positive value (e.g., $\mathcal{M} = (2b)^{|V|}$), with ϕ defined in equality (1). In constraint (5), we define an arborescence of G represented by x^k with any existing model [1]. Constraints (6) and (7) impose that if an arc (u, v)

is in the arborescence x^k, then we must have $\Phi_v^k - \Phi_u^k = \phi(c_{uv})$; otherwise, both constraints become redundant. Constraints (8) impose that inequality (3) of the Theorem 1 must be satisfied. Constraints (9) are the domain of the potential Φ^k variables, with the expanded potential of the root node k being null.

Theorem 2. *The set of feasible solutions for the model (S_k) defined by constraints (5)-(9) corresponds to the complete set of k-rooted crystal trees of G.*

Model (S_k) obtains crystal trees rooted at a given node k. In case we are interested in determining an arborescence that is crystal for two or more roots, we intersect the corresponding blocks of constraints (5)-(9), one for each required root. Additionally, we must impose that the CTs, for the different roots, share the same edges. For example, if we want a given spanning tree to be crystal for all nodes, we must consider the following constraints.

$$x_{uv}^1 + x_{vu}^1 = x_{uv}^k + x_{vu}^k, \quad \forall \{u,v\} \in E, \ \forall k \in V - \{1\}. \tag{10}$$

Corollary 1. *Consider the intersection $(S_1) \bigcap (S_2) \cdots \bigcap (S_{|V|}) \bigcap (10)$, where the $|V|$ blocks of constraints (S_k), $k = 1, \cdots, |V|$, represent arborescences x^k defining k-rooted crystal trees of G. The solutions of the resulting system are minimum spanning trees of G.*

Corollary 1 tell us that to obtain MSTs of a graph G, we can intersect $|V|$ blocks of constraints defining (S_k), $k = 1, \cdots, |V|$, and intersect with (10). We can show that intersecting $|V| - 1$ of these blocks is sufficient to obtain MSTs. An interesting open question is to determine the minimum number of blocks necessary to obtain all the MSTs of a graph. Using less than this minimum value, what would be the probability that the partial system solution is not an MST of G? Is it possible to estimate such a probability? The only certainty we have is that if for a given root node there is no unstable crystal tree, then the respective block for that node does not need to be considered in the system. However, answering this question appears to be an NP-complete decision problem.

Finally, as we are dealing with a mathematical model, we must remember that the use of a big-M value is a weakness of model (S_k) when using a mathematical solver to handle it. However, it seems we can overcome this difficulty if we implement a branch-and-bound enumeration tree to solve (S_k) algorithmically without employing commercial solvers.

5   Conclusion

This article introduces the basic theory of the new class of k-rooted crystal trees. We propose original theoretical results for them. We also distinguish these trees from existing ones such as k-rooted shortest-path trees and minimum spanning trees of a graph. We develop a linear system with a polynomial number of variables and constraints whose feasible solutions are k-rooted crystal trees. This system can be used to obtain the complete set of minimum spanning trees of a graph. Therefore, opening new possibilities, as future work, for solving optimization problems constrained to have such an optimal tree structure in their set of constraints. Finally, some interesting open questions concerning the crystal trees need to be answered and also constitute a nice direction for future research.

Acknowledgments. Research funded by 'Coordenação de Aperfeiçoamento de Pessoal de Nível Superior - Brasil (CAPES)' (Finance Code 001), Cearense Foundation for Scientific and Technological Development Support (FUNCAP)' (Grant PGP-0192-00040.01.00/22), and the Brazilian National Council for Scientific and Technological Development (Grant 303988/2021-5). The author thanks the four anonymous referees for their constructive and valuable comments and suggestions.

Disclosure of Interests. The author has no competing interests to declare that are relevant to the content of this article.

References

1. Magnanti, T., Wolsey, L.: Optimal trees. CORE Discussion Paper 1994026, Université Catholique de Louvain, Center for Operations Research and Econometrics (1994)
2. Yaman, H., Karasan, O.E., Pinar, M.: The robust spanning tree problem with interval data. Oper. Res. Lett. **29**(1), 31–40 (2001)

Parameterized Algorithms

Reducing Treewidth for SAT-Related Problems Using Simple Liftings

Ernst Althaus[(✉)][iD] and Daniela Schnurbusch[(✉)]

Johannes Gutenberg-Universität Mainz, Mainz, Germany
{ernst.althaus,daschnur}@uni-mainz.de

Abstract. Tree decompositions are a powerful tool to obtain parameterized algorithms, in particular to solve different variants of the satisfiability problem. Most algorithms are based on a tree decomposition of the so called primal graph. Variants of the satisfiability problem that allow parameterized algorithms in the treewidth of the primal graph are for example Model Counting, MaxSat or QBF.

To obtain efficient algorithms in practice, reducing the size of the instance by preprocessing is a very important technique and hence is highly investigated. In this paper, we investigate how preprocessing techniques can be used to reduce the parameter of a parameterized algorithm other than the size of the instance. In particular, we look at satisfiability and related problems and try to preprocess the formula in order to reduce the treewidth of the resulting primal graph. To the best of our knowledge, this is the first such approach.

We show how to compute a set of auxiliary variables and an equisatisfiable (w.r.t. the original variables) formula using those such that the treewidth of the resulting primal graph is minimal under all sets of auxiliary variables. To reach this goal, we restrict our attention to auxiliary variables such that their value has to be the value of a subclause of the formula for each satisfying truth assignment.

We implemented our approach and evaluated it on standard benchmark instances. While our approach is able to reduce the treewidth of around 10% of the instances, there is no clear improvement in the running time when solving the formula, due to the dependence of the practical efficiency of the solver on the structure of the formula.

Keywords: Tree Decomposition · Preprocessing · Satisfiability

2012 ACM Subject Classification Theory of computation · Fixed parameter tractability

1 Introduction

In this paper, we investigate how preprocessing techniques can be used to reduce the parameter of a parameterized algorithm other than the size of the instance. In particular, we look at satisfiability and related problems and try to preprocess

A. Basu et al. (Eds.): ISCO 2024, LNCS 14594, pp. 175–191, 2024.
https://doi.org/10.1007/978-3-031-60924-4_14

the formula in order to reduce the treewidth of the resulting primal graph. To the best of our knowledge, this is the first such approach.

Tree decompositions are a powerful tool to obtain parameterized algorithms, in particular to solve different variants of the satisfiability problem, e.g. given a Boolean formula in conjunctive normal form (CNF), is there a truth assignment such that the formula is satisfied. Variants of the satisfiability problem that allow for parameterized algorithms in the treewidth of the primal graph are for example:

- Model Counting [19], i.e. counting the number of satisfying assignments,
- MaxSat [18], i.e. find an assignment such that as many clauses as possible are satisfied,
- Quantified Boolean Formula (QBF) [3], i.e. the variables are additionally either existentially or universally quantified. This algorithm takes the number of quantifier alternations as an additional parameter.

Most algorithms for solving satisfiability problems are based on a tree decomposition of the so called primal graph, i.e. the graph consisting of a vertex for each variable of the formula and an edge between two vertices, if there is a clause containing the two corresponding variables (see e.g. [7] for the classical satisfiability problem, [5] or [11] and references therein for model counting or [3] for QBF). The running time of parameterized algorithms for solving QBFs depends exponentially on the treewidth, where the exponential function is a tower, whose height is the number of quantifier alternations. This high running time is justified by almost matching lower bounds by Fichte, Hecher and Pfandler [6].

To obtain efficient algorithms in practice, reducing the size of the instance by preprocessing is a very important technique and hence is highly investigated (see e.g. [1,10] and references therein). If the preprocessing technique guarantees that the size of the instance becomes bounded in a parameter, this is called a kernelization (see [4]).

Samer and Szeider [19] showed that the model counting problem also admits a parameterized algorithm in the width of the following graphs.

- Dual Graph: the graph consisting of a vertex for each clause, in which two vertices are adjacent if the corresponding clauses have a joint variable.
- Incidence Graph: the graph consisting of a vertex for each variable and a vertex for each clause, where a vertex for a variable has an edge with a vertex for a clause if the clause contains the variable.

The dependency of the running time of their algorithm solving the model counting problem on the width is worse for these two graphs than for the primal graph. For the incidence graph, the algorithm was later improved to have the same running time dependency as the best algorithm for the primal graph [20]. It is asymptotically faster than the algorithm proposed in this paper: with the approach proposed here, we can not reduce the treewidth of a primal graph below the treewidth of the incidence graph. To the best our knowledge, for the QBF-Problem, only an algorithm for the primal graph is known.

The algorithm of Samer and Szeider [19] for the incidence graph can also be interpreted as first transforming the formula to an equivalent one and then using the known algorithm parameterized on the primal graph on the modified formula. This transformation is done by introducing new variables and forcing them to take on the value of a specific subclause (defined more formally later). We call such a construction a simple lifting as we lift the set of satisfying assignments to a higher dimensional space.

The construction of first computing a tree decomposition and then transforming the formula does not necessarily result in the optimal lifting, i.e. a lifting such that the resulting tree decomposition of the primal graph has smallest width. In this paper, we investigate whether it is possible to compute a lifting such that the treewidth of the resulting primal graph is minimal under all liftings. We restrict our attention to liftings that correspond to the value of a subclause of the formula.

The paper is organized as follows. In the next section, we state some related work. Afterwards, we introduce the formal notation and definitions and define the question under consideration. In Sect. 3, we give a solution where the treewidth of the resulting primal graph of the lifted formula only depends on local properties of a tree decomposition of the incidence graph. We call this the modified width of the tree decomposition. In Sect. 4, we show that this approach is essentially optimal, i.e. we show that for each lifted formula, there is a tree decomposition of the incidence graph such that its modified width is at most one more than the treewidth of the primal graph. In Sect. 5, we argue that current approaches to compute the treewidth exactly cannot directly be generalized to optimally compute the modified width and adopt heuristics that are commonly used. Finally, we give some experimental results in Sect. 6 and conclude the paper.

1.1   Related Work

Notice that the problem of representing a Boolean function by a formula in CNF with additional variables has been studied before, even for more general semantics of the additional variables [22], but these techniques are not equipped with an automatic method to transform a given CNF to one with minimal or at least smaller treewidth. In particular, if the CNF is unsatisfiable, the representation with smallest treewidth is a single empty clause. Hence, finding the optimal representation in general is NP-hard.

Markov and Shi [17] show how to reduce the number of literals of each clause to at most three without increasing the treewidth by more than one (more generally for arbitrary circuits, not only for formulas in CNF). Notice that our approach can easily be modified to construct only clauses with at most three literals without changing the treewidth obtained. Hence, our approach achieves at least the same treewidth as the approach of Markov and Shi, but potentially reduces the treewidth significantly as in the example of Sect. 2.4.

Notice that there is plenty of work for preprocessing a Boolean formula to reduce its size, specialized for different versions of the SAT problem. For model counting, we refer to Lagniez and Marquis [12] and the recent work of Soos and Meel [21]. For QBF, Lonsing and Egly recently implement a preprocessor [14, 15]. In all our experiments, we first simplified all formulas with the appropriate preprocessor to overcome possible trivial reductions of the treewidth by removing variables and/or clauses.

Hoder and Voronkov made experiments with different heuristics for splitting clauses with many literals into smaller ones and handling the splitted clauses differently in a first order reasoner [9]. One of the approaches corresponds to our approach of simple liftings. In contrary to their heuristic approach, we decide on which clauses to split by the treewidth of the resulting graph.

Lokshtanov, Panolan, Fahad and Ramanujan [13] consider the problem of finding a partial assignment to the variables that allows a satisfying completion (i.e. an satisfying assignment of the remaining variables) and such that the resulting simplified formula has at most a fixed treewidth. These partial assignments are called backdoors. The problem of finding backdoor is W[2]-hard in general and the authors make a detailed investigation on its complexity for special kinds of formula. Recently, Mählmann, Siebertz and Vigny [16] extended this idea to recursive backdoors.

Hecher ([8] and earlier works) tries to keep the treewidth small when reducing another problem, namely answer set programming, to the satisfiability problem.

2   Definitions and Formalization of the Question

2.1   Graphs and Tree Decompositions

Given an (undirected) graph $G = (V(G), E(G))$ a tree decomposition of G is a tuple $\mathcal{T} = (T, (X_t)_{t \in V(T)})$, where $T = (V(T), E(T))$ is a tree and $X_t \subseteq V(G)$ for all $t \in V(T)$ with the following properties:

- Node coverage: for each $u \in V(G)$, there is $t \in V(T)$ with $u \in X_t$
- Edge coverage: for each $uv \in E(G)$, there is $t \in V(T)$ with $u, v \in X_t$
- Consistency: for all $t_1, t_2, t_3 \in V(T)$ such that t_2 lies on the path between t_1 and t_3, $X_{t_1} \cap X_{t_3} \subseteq X_{t_2}$.

The sets X_t are called the bags of the tree decomposition. The width of a tree decomposition is the maximal cardinality of a bag minus one and denoted by tw. The treewidth of a graph G is the minimal width of any tree decomposition of G.

To distinguish between the vertices of the graph and the tree, from here on out we call the elements of $V(G)$ vertices and the elements of $V(T)$ nodes. For an introduction to tree decompositions and their usage, we refer to Cygan et al. [4]. There is a notion of canonical tree decompositions, which we will not define here. As e.g. shown in [2], for each graph G there is a canonical tree decomposition whose width is the treewidth of G. For this paper, it suffices that for a canonical tree decomposition $\mathcal{T} = (T, (X_t)_{t \in V(T)})$, the bags are pairwise different and it is not possible to choose $t^* \in V(T)$ and construct a tree decomposition $\mathcal{T}' =$

$(T'(X'_t)_{t \in V(T')})$, such that for all $t' \in V(T')$ either $X'_{t'} = X_t$ for some $t \in V(T)$ or $X_{t'} \subsetneq X_{t^*}$. Informally, each bag of the alternative tree decomposition is either equal to a bag different from t^* or a strict subset of the bag of t^*.

2.2 Formulas and Satisfiability Problems

A Boolean formula in conjunctive normal form is defined over a finite set of variables, typically denoted as $\{x_1, \ldots, x_n\}$. A literal is a variable x_i or its negation $\neg x_i$, a clause is a disjunction of literals, denoted by $\ell_1 \vee \cdots \vee \ell_k$ for literals ℓ_i and a Boolean formula is a conjunction of clauses, denoted as $\phi = c_1 \wedge \cdots \wedge c_m$ for clauses c_j.

A truth-assignment for a set of variables $\{x_1, \ldots, x_n\}$ is a function $\pi : \{x_1, \ldots, x_n\} \to \{\text{TRUE}, \text{FALSE}\}$. Given a Boolean formula ϕ, its truth value $\phi(\pi)$ for π is either TRUE or FALSE and determined as follows. A literal x_i has truth value $\pi(x_i)$ and a literal $\neg x_i$ has truth value TRUE, iff $\pi(x_i) = \text{FALSE}$. A clause has truth value FALSE, iff all its literals have truth value FALSE. A formula has truth value TRUE, iff all its clauses have truth value TRUE. Notice that the empty formula evaluates to TRUE for all truth-assignments and an empty clause evaluates to FALSE.

Given a Boolean formula ϕ, the satisfiability problem asks whether there is a truth assignment such that ϕ evaluates to TRUE and the model counting problem asks how many truth assignments there are such that ϕ evaluates to true.

For a subset $X \subseteq \{x_1, \ldots, x_n\}$, we use $\pi|_X$ to denote the restriction of π to the variables X. For all assignments π with $\pi(x_1) = \text{TRUE}$, given a Boolean formula ϕ over $\{x_1, \ldots, x_n\}$, we can easily find a Boolean formula $\phi^{x_1 = \text{TRUE}}$ over $\{x_2, \ldots, x_n\}$ such that $\phi^{x_1 = \text{TRUE}}(\pi|_{x_2, \ldots, x_n}) = \phi(\pi)$ by removing all clauses containing x_1 and removing $\neg x_1$ from all clauses containing it. Similarly, we can easily find a Boolean formula $\phi^{x_1 = \text{FALSE}}$ over $\{x_2, \ldots, x_n\}$ such that $\phi^{x_1 = \text{FALSE}}(\pi|_{x_2, \ldots, x_n}) = \phi(\pi)$ for all assignments π with $\pi(x_1) = \text{FALSE}$ by removing all clauses containing $\neg x_1$ and removing x_1 from all clauses containing it.

A quantified Boolean formula in prenex normal form over a set $\{x_1, \ldots, x_n\}$ of variables is of the form $Q_1 x_1 Q_2 x_2 \ldots Q_n x_n \phi$, where ϕ is a Boolean formula over $\{x_1, \ldots, x_n\}$ and $Q_i \in \{\exists, \forall\}$ for $1 \leq i \leq n$. The truth value is recursively defined as follows:

- If $Q_1 = \exists$, the truth value of an assignment of $\{x_2, \ldots, x_n\}$ is FALSE, iff $\phi^{x_1 = \text{TRUE}}$ and $\phi^{x_1 = \text{FALSE}}$ have truth value FALSE for this assignment.
- If $Q_1 = \forall$, the truth value of an assignment of $\{x_2, \ldots, x_n\}$ is TRUE, iff $\phi^{x_1 = \text{TRUE}}$ and $\phi^{x_1 = \text{FALSE}}$ have truth value TRUE for this assignment.

Given a quantified Boolean formula in prenex normal form, the QBF problem asks whether the value of the formula is TRUE.

2.3 Graphs Associated with Formula

The primal graph G_ϕ^P of a Boolean formula ϕ over the variables $\{x_1, \ldots, x_n\}$ is the graph with vertices $\{\tilde{x}_1, \ldots, \tilde{x}_n\}$ (one vertex for each variable), where two vertices $\tilde{x}_i$ and $\tilde{x}_j$ are connected by an edge iff there is a clause containing a literal of x_i and a literal of x_j.

The incidence graph G_ϕ^I of a Boolean formula ϕ over the variables $\{x_1, \ldots, x_n\}$ and clauses $c_1, \ldots, c_m$ is the graph with vertices $\{\tilde{x}_1, \ldots, \tilde{x}_n\} \cup \{\tilde{c}_1, \ldots, \tilde{c}_m\}$ (a vertex for each variable and a vertex for each clause), where $\tilde{x}_i$ and $\tilde{c}_j$ are connected by an edge iff the clause c_j contains a literal of x_i. We denote the vertices $\{\tilde{x}_1, \ldots, \tilde{x}_n\}$ by V_ϕ^X and the vertices $\{\tilde{c}_1, \ldots, \tilde{c}_m\}$ by V_ϕ^C.

For a variable y and a clause $c \equiv \bigvee_{k=1}^m \ell_k$, we write $y \equiv c$ as a short notation for the set of clauses $\{\neg y \vee c\} \cup \{y \vee \neg \ell_k \mid 1 \le k \le m\}$, i.e. a set of clauses forcing y to have the same value as c. For a subset $X \subseteq \{x_1, \ldots, x_n\}$ and a clause c, we use $c|_X$ to denote the restriction of c to the clauses in X, i.e. the clause obtained from c by removing all literals of variables not contained in X.

Let $\phi = \bigwedge_{j=1}^m c_j$ be a formula over the variables $x_1, \ldots, x_n$ with clauses $c_j = \bigvee_{i=1}^{m_j} \ell_i^j$. A subclause of c_j is the clause $\bigvee_{i \in I} \ell_i^j$ for $I \subseteq \{1, \ldots, m_j\}$. A subclause of ϕ is a subclause of one of the clauses of ϕ.

2.4   Simple Lifting and Problem Statement

Next, we will define simple liftings as basis of our research.

For a formula ϕ over $\{x_1, \ldots, x_n\}$, we call a formula ϕ' over an extended set of variables $\{x_1, \ldots, x_n\} \cup \{y_1, \ldots, y_k\}$ a lifting of ϕ, if

- for each satisfying assignment π of ϕ, there is exactly one satisfying assignment π' of ϕ' with $\pi'(x_i) = \pi(x_i)$ for all $1 \le i \le n$ and
- for each non-satisfying assignment π of ϕ there is no satisfying assignment of ϕ' of $\pi'(x_i) = \pi(x_i)$ for all $1 \le i \le n$.

We call $\{x_1, \ldots, x_n\}$ the original variables and $\{y_1, \ldots, y_k\}$ the added variables in the following and the assignment π' the lifted assignment of π.

We call a formula a simple lifting, if it is a lifting and the following condition holds: for each satisfying assignment π of ϕ and each added variable y_j, the value of y_i in the lifted assignment of π is equal to the value of a subclause of ϕ in π.

Note that we can construct a lifting of ϕ with an additional variable y forced to have the same value as a clause $c = \ell_1 \vee \cdots \vee \ell_m$ (where c is not necessarily part of ϕ) for each satisfying assignment by adding the clauses $\neg y \vee c$ and $y \vee \neg l_i$ for $1 \le i \le m$ to ϕ. We call this set of clauses the clauses enforcing $y \equiv c$. Furthermore, if we have variables $z_1, \ldots, z_k$ who are forced to have the values of the clauses $c_1, \ldots, c_k$ then the clause $c_1 \vee \cdots \vee c_j$ can be replaced in a formula ϕ by the clause $z_1 \vee \cdots \vee z_k$ without changing the set of satisfying assignments. Note that the literals in the clauses $c_1, \ldots, c_k$ are not necessarily disjoint and that the variables z_j can be original or added variables.

Note that a tree decomposition of the primal graph of a formula $c_1 \wedge \cdots \wedge c_m$ has to contain a bag X_t containing all variables of a clause c_j for all $1 \le j \le m$. For ease of notation, we say that if a bag of a tree decomposition contains a vertex $\tilde{x}_i$, it contains the literals x_i and $\neg x_i$. Hence, in order to enforce $y_i \equiv c$, its corresponding vertex $\tilde{y}_i$ has to be in a common bag with vertices $\tilde{z}_j$ ($1 \le j \le m$)

with an enforced $z_j \equiv c_j$ for some clauses c_j such that $c = c_1 \lor \cdots \lor c_m$ (each literal of y_i has to be a literal in at least one such subclause c_j). We say that the subclauses $c_1, \ldots, c_j$ cover y_i in this case.

As an example, we can lift the formula consisting of the single clause $x_1 \lor x_2 \lor \cdots \lor x_n$ by adding variables h_j for $2 \le j \le n$ where the value of h_j is forced to the value of the subclause $x_1 \lor \cdots \lor x_j$. With these auxiliary variables, we can replace the clause $x_1 \lor x_2 \lor \ldots x_n$ by clauses implying $h_2 \equiv x_1 \lor x_2$, $h_j \equiv h_{j-1} \lor x_j$ for $2 \le j \le n$ and $h_n \equiv$ TRUE. The treewidth of the resulting primal graph drops from n to 2.

Note that clauses may have subclauses in common. When constructing a lifting for such a subclause, we only need one new variable, but it is possible that our approach constructs copies for each appearance in a single clause. Hence, it would be possible to reduce the number of added variables in this case. In the following, we restrict our interest to simple liftings that do not use this improvement to reduce the number of added variables and hence the treewidth of the corresponding primal graph. In this paper, we will focus on the computation of such a simple lifting of ϕ resulting in the graph with the smallest treewidth.

3   A Solution Using the Incidence Graph

Consider a formula ϕ and the corresponding incidence graph G_ϕ^I.

We may assume that each variable appears at most once in each clause. (If a variable appears twice positive or twice negative, we can remove one appearance without changing the value. If it appears positive and negative, the clause is satisfied for every truth-assignment and hence can be removed). For a clause c_j and a variable x_i, let $\ell^j(x_i)$ be the (unique) literal of c_j corresponding to x_i and FALSE if x_i does not appear in c_j.

Definition 1. *Let $\mathcal{T} = (T, (X(t) \cup C(t))_{t \in V(T)})$ be a (canonical) tree decomposition of G_ϕ^I where we split the bags of the nodes $t \in V(T)$ into the vertices $X(t)$ in V_ϕ^X and the vertices $C(t)$ in V_ϕ^C. For a node $t \in V(T)$ and vertex $\tilde{c}_j \in C(t)$, let $w(t, j) = |\{t' \in N(t) \mid \tilde{c}_j \in C(t')\}|$, where $N(t)$ is the set of neighbors of t. In other words, $w(t, j)$ is the number of neighbors of t having $\tilde{c}_j$ in their bag. Let $w'(t, j) = w(t, j)$ if $w(t, j) \le 1$ and $w'(t, j) = w(t, j) - 1$ otherwise. For a node $t \in V(T)$, we define its modified width $\operatorname{tw}'(t) = |X(t)| - 1 + \sum_{j|\tilde{c}_j \in C(t)} w'(t, j)$. The modified width of a tree decomposition $\mathcal{T}$ is $\operatorname{tw}'(\mathcal{T}) := \max_{t \in V(T)} \operatorname{tw}'(t)$.*

Intuitively, a clause c_j in a bag contributes nothing to the size of it, if c_j is only contained in that bag. If it is contained in one neighbor, it contributes one. If it is contained in several neighbors, it contributes the number of neighbors minus one. The modified width of the bag is then the number of variables in the bag plus the contribution of all clauses.

Theorem 2. *Given a Boolean formula ϕ and a tree decomposition $\mathcal{T}$ of G_ϕ^I, we can efficiently construct a simple lifting ϕ' of ϕ and a tree decomposition $\mathcal{T}'$ of $G_{\phi'}^P$ of width $\operatorname{tw}(\mathcal{T}') \le \operatorname{tw}'(\mathcal{T}) + 1$.*

Proof. Let $\phi = c_1 \wedge \cdots \wedge c_m$ be a Boolean formula and $\mathcal{T}$ be a tree decomposition of the incidence graph G_ϕ^I.

For the construction, we iteratively replace each vertex $\tilde{c}_j$ by variables enforced to subclauses of c_j, which we call the simple liftings of c_j, and clauses enforcing $\tilde{c}_j \equiv c_j$, i.e. the bags of the tree decomposition contain vertices and clauses. The formula ϕ' is the conjunction of all clauses added to the bags. At the end, ϕ' is a simple lifting of ϕ and the constructed tree decomposition (after removing the clauses assigned to the bags, as the vertices of the primal graph correspond only to the variables) is a tree decomposition for the primal graph of ϕ'. Being a tree decomposition of the primal graph is guaranteed, as all clauses constructed are assigned to a node of the tree decomposition whose bag contains all variables of the clause. During that construction, we sometimes have to modify the tree decomposition. The construction is as follows:

- If a vertex $\tilde{c}_j$ is in only one bag of a node $t \in V(T)$, we simply remove it from the bag of t, add the clause c_j and assign it to t, which is possible since all literals of c_j have to be in $X(t)$. Hence, in the following, we will assume that each vertex $\tilde{c}_j$ is contained in at least two bags.
- The construction will guarantee that each node $t \in V(T)$ contains at most $w'(t,j)+1$ simple liftings of c_j for each clause c_j. Furthermore, we will ensure that for each node $t \in V(T)$ there is at most one clause c_j such that the bag of t contains $w'(t,j)+1$ lifted variables of c_j. In this case, we say the node t is marked with c_j. This will ensure that the number of vertices of a bag is at most $1 + \sum_{j=1}^{m} w'(t,j)$ in the final tree decomposition and hence prove the theorem.
- A split of node t marked with c_j towards a neighbor t' is defined as follows (see Fig. 1). We add a new node t'' on the edge (t,t') whose bag is the bag of t. Then we remove all lifted variables of c_j from t which are not in a neighbor of t other than t'' (to ensure that the constructed tree decomposition is consistent). From the bag of t'' we remove all vertices $\tilde{c}_i, i \neq j$ which are not in the bag of t'. A split is feasible, if the set of simple liftings of c_j in the bag of t is different from that in t' and if the bag of t contains at most $w'(t,j)$ simple liftings of c_j. Hence t is not marked after a feasible split. All clauses assigned to t are reassigned to t''.
- A split does not increase the modified width of the tree-decomposition, as
  - the modified width of t does not increase: the set of vertices in the bag of t is a subset of the vertices originally contained in the bag of t and each clause-variable $\tilde{c}_i$ that is in t'' is also in t' and hence the number of neighbors of t containing $\tilde{c}_i$ does not increase and
  - the modified width of t'' is at most that of t before the split as the set of vertices of the bag of t'' is a subset of the vertices originally contained in the bag of t and the number of neighbors containing a clause vertex $\tilde{c}_i$ is at most two as the degree of t'' is two.
- If t is marked with c_j, we will guarantee that t has at least two neighbors allowing a feasible split. Notice that after a feasible split, the new node t'' is marked with c_j and the splits of t'' towards t' and towards t are feasible.

- For each vertex $\tilde{c}_j \in V_\phi^C$, consider the subtree T_j of T of nodes whose bags contain $\tilde{c}_j$. We will root the tree T_j at an arbitrary node r_j of degree one w.r.t. T_j. Let $\mathrm{ch}_j(t)$ be the children of t w.r.t. this rooting and $\mathrm{ch}_j^+(t)$ be the set of all descendants of t (including t). We will replace $\tilde{c}_j$ in all nodes of T_j bottom-up. The clauses created during the construction for t are assigned to t.

- For each node t in T_j, we will create one additional simple lifting for c_j, which we call y_j^t. If we do a split of a node t towards t' creating t'', we do not create additional simple liftings, but give the simple lifting $y_j^{t'}$ the additional name $y_j^{t''}$. Hence, for each node t in T_j, there is exactly one simple lifting of c_j with name y_j^t.

- For each leaf t in the rooted T_j, we replace $\tilde{c}_j$ with a new variable y_j^t and add clauses for $y_j^t \equiv c_j \mid_{X(t)}$, i.e. the subclause containing the variables of the bag t. Notice that t is not marked with c_j.

- For the root r_j with child ch, we replace $\tilde{c}_j$ by y_j^{ch} and add the clause $y_j^{\mathrm{ch}} \vee c_j \mid_{X(t)}$. Notice that r_j is not marked with c_j.

- For an internal node $t \in V(T)$ that is marked with some clause c_i, we first do a split of t towards a node t' different from the parent w.r.t. T_j (the node is not necessarily in T_j) such that the split is feasible. If t' is in T_j, the new node t'' will be in T_j and we replace $\tilde{c}_j$ by $y_j^{t'}$ and give this variable the name $y_j^{t''}$ too. Notice that now t is not marked anymore and we can continue the construction for t as a non-marked internal node.

- For a non-marked internal node $t \in V(T)$ we do the following. We replace c_j by $\cup_{t' \in \mathrm{ch}_j(t)} y_j^{t'}$ and add the additional variable y_j^t. Furthermore, we add clauses to ensure $y_j^t \equiv c_j \mid_{X(t)} \vee \bigvee_{t' \in \mathrm{ch}_j(t)} y_j^{t'}$. Notice that now t is marked with c_j and each split towards a child ch is feasible as y_j^{ch} is neither contained in the bag of any other child of t nor will it be in the bag of the parent (no matter whether the parent will be split or not). Furthermore, the split towards the parent will be feasible, as y_j^t will be contained in the bag of the parent (no matter whether it will be split or not), but is not contained in the bag of any child of t. Hence, there are at least two feasible splits.

- Notice that $y_j^t \equiv c_j \mid_{\cup_{t \in \mathrm{ch}_j^+(t)} X(t)}$, i.e. the value of the lifted variable y_j^t is the same as the value of the clause c_j restricted to the variables below t w.r.t. to the rooted T_j. Hence, for the root r_j, c_j is ensured by the added clauses. □

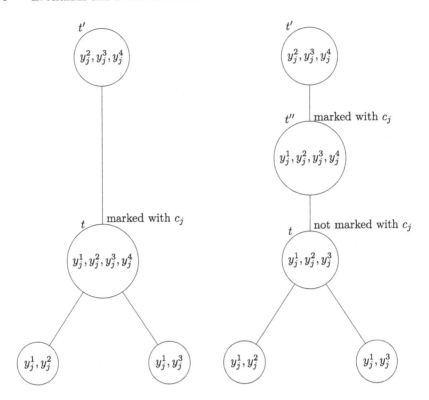

Fig. 1. Example for a split of t at t'. In the left figure, a tree decomposition where the node t that is marked with c_j is shown. In the bags, we only give the lifted variables of c_j. The right figure shows the tree decomposition after the split of t towards t'. It is feasible, as t' contains a different set of lifted variables for c_j and after the split, one lifted variable of c_j is removed from t. After the split, t'' is marked with c_j, but the split towards both its neighbors t and t' is feasible.

4   Near-Optimality of the Approach

In this section, we show that the construction above is optimal up to an additional error of one.

Theorem 3. *Let ϕ' be an arbitrary simple lifting of ϕ. There is a tree decomposition of G_ϕ^I whose modified width is at most the treewidth of $G_{\phi'}^P$.*

Proof. Let ϕ' be a feasible simple lifting of ϕ with additional variables Y and let $\mathcal{T} = (T, (X_t \cup Y_t)_{t \in V(T)})$ be a canonical tree decomposition of $G_{\phi'}^P$, where the vertices of the bags of t are split into vertices in $X(t)$ corresponding to original variables and vertices in $Y(t)$ corresponding to additional variables. Note that we can assume each variable in $Y(t)$ to be contained in at least two bags. Otherwise we could remove it from the single bag in which it is contained (the value is not propagated into the tree decomposition and hence not needed).

We consider an arbitrary tree decomposition T' of G_ϕ^I. In T' we replace each occurrence of a subclause of c_j by c_j itself. Furthermore, if T does not contain any subclause of a clause c_j, there has to be at least one bag containing all variables of c_j. We can assign c_j to one of these bags without changing the modified width, as the variable $\tilde{c}_j$ does not increase the modified width of the bag if it is not contained in any of its neighbors. We argue that the modified width of T' is the (ordinary) width of T and that T' is a tree decomposition of G_B.

For each clause c_j there has to be a bag, say r_j, that covers c_j. In the following, we consider T rooted at r_j. Consider a node $t \in V(T)$. If for a child t', the bag $X(t')$ contains a variable x_i of c_j not in $X(t)$, there has to be a simple lifting in $Y(t)$ and $Y(t')$ carrying the information of x_i to r_j. If there are several such children, all of them have to be different and contained in $Y(t)$. If there is a variable x_i of c_j in $X(t)$ but not in the bag of its parent, there has to be at least one simple lifting carrying this information in $X(t)$. Hence, the number of simple lifting variables of c_j in a bag which are replaced by c_j is at least $w'(t,j)$.

Coverage should be clear, as for each variable x_i contained in c_j either x_i is in a bag covering c_j or there has to be a simple lifting of c_j in a bag containing x_i. Consistency follows from the discussion above (if the set of nodes whose bags contain a variable of a subclause of c_j do not form a subtree, we can remove all variables of subclauses of c_j that are not connected to r_j).

Notice that this theorem shows that the width of the tree decomposition of the constructed formula is at most one larger than the width of the original formula and can be arbitrarily smaller. Compared to the treewidth of the incidence graph, we are at most a factor two larger and can be equally small.

5   Computation of a Lifting

With the results of the previous section, the computation of an almost optimal lifting can be done by computing a tree decomposition of minimal modified width. Unfortunately, almost all algorithms and heuristics restrict the search to canonical tree decompositions. However, there are formulas ϕ such that no canonical tree decomposition of G_ϕ^I has minimal modified width, e.g. a tree decomposition for which a clause variable is contained in the bags of four neighbors of a node can never be optimal (see below). To resolve this issue, we will define the width measure w'' which gives the width after splitting all bags of degree larger than three. But even with this width measure, it remains unclear whether a there is always a canonical tree decomposition that is optimal.

Nevertheless, we implemented variants of the most common heuristics, which compute a canonical tree decomposition and roughly work as follows. It is known that the bag X_t of a leaf t of a canonical tree decomposition T has the form $\{u\} \cup \{v \mid uv \in E(G)\}$ for some vertex $u \in V(G)$. Removing t from T leaves us with a tree decomposition of the graph G^u, obtained from G by removing all vertices in X_t for which all neighbors are in X_t (and hence at least u) and adding edges between all other pairs of vertices in X_t. The heuristics guess a vertex u, construct the graph G^u and recursively compute a tree decomposition of G^u. There are several rules to select u, namely:

- Min-Degree: Chose a vertex u with minimal degree in the current graph
- Min-Fill: Chose a vertex u such that the number of edges added to the graph is minimal
- Random: Chose a vertex u (uniformly) at random

To adopt these heuristics to the modified width, we make the following observations:

- In the resulting tree decomposition, we can reduce the degree of any node to at most 3 by making copies of nodes of higher degree. This does not increase the modified width, if we add variables corresponding to clauses only when necessary, i.e. on all nodes of paths that connect two neighbors containing the clause variable in their bag. Hence $w'(t, j)$ is at most two in the resulting tree decomposition. Instead of restricting to tree decompositions of degree at most three, we can modify w' to w'' with $w''(t, j) := \min(w'(t, j), 2)$.
- Notice that the modified width of G_ϕ^I is at most one less than its width. This can be seen as follows. The modified width of a bag X_t can only be less than its width by more than one if it contains more than one clause vertex $\tilde{c}_j$ with $w(t, j) = 0$. In this case, we can remove all such clauses from X_t and attach new leaf nodes with bags $X_t \cap V_\phi^X \cup \{c_j\}$ for each such clause vertex $\tilde{c}_j$ to t.
- Consider the incidence graph $G_\phi''^I$, which contains a copy $\tilde{c}_j'$ of each vertex in $\tilde{c}_j \in V_\phi^C$, i.e. for each clause c_j, we have two vertices $\tilde{c}_j$ and $\tilde{c}_j'$, both having the vertices corresponding to the variables of the clause as neighbors. There is an optimal tree decomposition w.r.t. the treewidth such that for each clause c_j either both copies of the corresponding vertices are in the same set of bags or both copies appear in exactly one bag, both of which are leaves sharing the same parent. In the second case, there is exactly one clause variable in the bag. This can be seen as follows. Consider a tree decomposition in which neither copy can be removed from any bag. If one copy, say $\tilde{c}_j$ is contained in only one bag, it can be made a leaf by removing $\tilde{c}_j$ from the bag and attaching a new leaf to the bag containing $\tilde{c}_j$ and its neighbors. We make a similar copy for $\tilde{c}_j'$ and remove $\tilde{c}_j'$ from all other bags to obtain a tree decomposition of the desired kind. If both copies are in at least two bags, we argue that we can remove all occurrences in bags containing only one copy. First notice, that there has to be at least one bag containing both copies, as otherwise any cut between the subtree containing $\tilde{c}_j$ and $\tilde{c}_j'$ has to contain all neighbors of $\tilde{c}_j$ and we can remove $\tilde{c}_j$ from all but one bag. Hence, the remaining bags form a subtree. Furthermore, each edge, say $\tilde{c}_j u$, is covered in the modified tree decomposition, as u has to be in the bag covering $\tilde{c}_j u$ and $\tilde{c}_j' u$ in the given tree decomposition and hence on a bag containing both $\tilde{c}_j$ and $\tilde{c}_j'$.

We compare the modified width of such a tree decomposition $\mathcal{T}'$ with the width of the tree decomposition $\mathcal{T}$ of G_ϕ^I, where we remove all vertices $\tilde{c}_j'$ from the bags. For each bag t and each clause variable $\tilde{c}_j$, it holds that

- if $w(t, j) = 0$, the modified width of t in $\mathcal{T}'$ is at most one less than the width of t in $\mathcal{T}$
- if $w(t, j) = 1$, the modified width of t in $\mathcal{T}'$ is at most twice the width of t in $\mathcal{T}$ and
- if $w(t, j) = 2$, the modified width of t in $\mathcal{T}'$ is equal to the width of t in $\mathcal{T}$.

  Hence, the width of a tree decomposition of G''^I_ϕ is a better approximation of the modified width of G^I_ϕ than the treewidth of G^I_ϕ.

- Instead of constructing G''^I_ϕ, we modify the choice of the vertex in G^I_ϕ such that the choice corresponds to min-degree or min-fill in G''^I_ϕ.

Furthermore, we tried to modify the random choice such that the probability of choosing a vertex decreases with its degree.

6 Experiments

We implemented our approach and tested it on standard benchmarks for the different variants

- Model Counting: https://mccompetition.org/2022/mc_description.html
- QBFLib http://www.qbflib.org/qbfeval22.php

The experiments were conducted in the following environment: Intel Core i5-6600K@3.5 GHz CPU, 16 GB DDR4 RAM, Windows 10 (64 bit), Java version jre1.8.0_271, .NET version 4.8. The time measured is the CPU time, including the time for garbage collection.

We wanted to experimentally test, whether our approach is capable to reduce the treewidth of the primal graph of a formula in a significant number of cases and whether such a reduction can decrease the solution time of a given solver for Model Counting or QBF.

In this paper, we give detailed results for the QBFLib22 instances, as the dependence of the running time on the treewidth is worst and hence our improvement should be most significant in the total running time. Furthermore, dynQBF [3], the only solver based on a tree decomposition we are aware of, is open source and has be shown to be efficient even without our technique. All instances are first preprocessed with qratpre+ [15] and then solved with dynQBF, both with and without our lifting approach. We set a lime limit of half an hour and no memory limit for each instance. As dynQBF does not use randomization and internally determines a suitable order of the variables, we solve each instance only once and did not try to reorder the formula.

Figure 2 shows all instances except for the instances in the subfolder "Organic_Synthesis", as no such instance could be solved by either approach. Furthermore, we exclude all instances for which dynQBF took less than 1s with the standard settings with and without our approach. This leaves us with 260 instances. Interestingly, for the instances of the 2022 benchmark sets of Model Counting and QBF, we reduce the treewidth significantly (a reduction of the

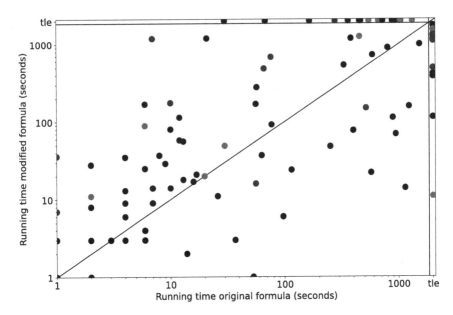

Fig. 2. A scatter plot of our experimental results. The running time is given in seconds, where tle denotes that the time limed was exceeded. Dots over the line indicate a higher running time when using the modified formula. Dots on the top line are instances that are not solved when using the modified formula, whereas dots on the right are instances that are not solved with the original formula. Instances which are not solved with either formula are not shown. Most dots are black, meaning that the treewidth of the clique graphs found by the heuristics of the two formula differ by at most 1. Blue dots indicate that the treewidth for the modified formula is at least two smaller, whereas red dots indicate that the treewidth is at most two higher. (Color figure online)

treewidth of at least five) for only around 10% of the instances, whereas for the 2020 edition of the Model Counting and QBF benchmarks, we can improve the treewidth significantly for around 30% of the instances. Notice that all instances of the model counting competitions were preprocessed with Arjun [21] before applying our approach.

For 27 of the instances, the treewidth increases with our approach by more than 1, showing that the heuristic for computing a tree decomposition of small modified width should be improved. The effect of restructuring the formula seems to be bigger than the effect of different treewidths, as Fig. 2 indicates there is no correlation. Our approach is currently not able to reduce the average running time, even for instances where we significantly reduce the treewidth.

With our approach, we can solve 15 instances that cannot be solved using the original formula, and here we reduce the treewidth of 3 instances and increase the treewidth of 1 instance. On the downside, 17 instances were not solved using the modified formula but with the original formula.

The worst case running time of the algorithm of Charwat et. al [3] depends, besides the treewidth, on the number of quantifier alternations. Notice that the lifted variables have to be existentially quantified at the inner quantification block. As all instances have an existential quantifier as inner quantification before the modification, there is no increase of quantifier alternations.

The heuristics to compute the tree decompositions are quite fast (≤ 1.5 s for primal graphs and ≤ 2 s for incidence graphs in this study). As the instances in Organic_Synthesis are much larger, the running times for computing tree decompositions are much higher (up to 5 min).

7   Conclusion

To the best of our knowledge, we gave the first preprocessing technique that does not aim at reducing the size of an instance, but the parameter of a parameterized algorithm. More specifically, we gave a method to reduce the treewidth of the primal graph of a Boolean formula in conjunctive normal form by introducing new variables corresponding to subformulas, allowing an increase in the size of the formula. Our experimental results show that this method can sometimes significantly reduce the treewidth of the primal graph, but the effect on the running time is not as significant as the effect of restructuring the formula.

For future work on this specific approach, we want to investigate how we can incorporate shared subformulas for computing lifted formulas such that the resulting primal graph has smaller treewidth. Furthermore, we want to investigate whether it is possible to compute the optimal tree decomposition w.r.t. the modified width in reasonable time and find better heuristics. Furthermore, we are developing an (practical) algorithm for the Model Counting problem based on a tree decomposition, where the impact of restructuring the formula on the practical efficiency of the known solver should not be as big as in the case of QBF.

More generally, we want to derive preprocessing methods to reduce other parameters for important parameterized algorithms.

References

1. Biere, A., Järvisalo, M., Kiesl, B.: Preprocessing in SAT solving. In: Biere, A., Heule, M., van Maaren, H., Walsh, T. (eds.) Handbook of Satisfiability, 2nd edn., vol. 336 of Frontiers in Artificial Intelligence and Applications, pp. 391–435. IOS Press (2021)
2. Bodlaender, H.L., Koster, A.M.: Treewidth computations I. Upper bounds. Inf. Comput. **208**(3), 259–275 (2010)
3. Charwat, G., Woltran, S.: Expansion-based QBF solving on tree decompositions. Fundam. Informaticae **167**(1–2), 59–92 (2019)
4. Cygan, M., et al.: Parameterized Algorithms. Springer, Heidelberg (2015). https://doi.org/10.1007/978-3-319-21275-3

5. Fichte, J.K., Hecher, M., Morak, M., Thier, P., Woltran, S.: Solving projected model counting by utilizing treewidth and its limits. Artif. Intell. **314**, 103810 (2023)
6. Fichte, J.K., Hecher, M., Pfandler, A.: Lower bounds for qbfs of bounded treewidth. In: Hermanns, H., Zhang, L., Kobayashi, N., Miller, D. (eds.) LICS 2020: 35th Annual ACM/IEEE Symposium on Logic in Computer Science, Saarbrücken, Germany, 8–11 July 2020, pp. 410–424. ACM (2020)
7. Habet, D., Paris, L., Terrioux, C.: A tree decomposition based approach to solve structured SAT instances. In: ICTAI 2009, 21st IEEE International Conference on Tools with Artificial Intelligence, Newark, New Jersey, USA, 2–4 November 2009, pp. 115–122. IEEE Computer Society (2009)
8. Hecher, M.: Treewidth-aware reductions of normal ASP to SAT - is normal ASP harder than SAT after all? Artif. Intell. **304**, 103651 (2022)
9. Hoder, K., Voronkov, A.: The 481 ways to split a clause and deal with propositional variables. In: Bonacina, M.P. (ed.) Automated Deduction - CADE-24 - 24th International Conference on Automated Deduction, Lake Placid, NY, USA, 9–14 June 2013. Proceedings, vol. 7898 of Lecture Notes in Computer Science, pp. 450–464. Springer, Heidelberg (2013). https://doi.org/10.1007/978-3-642-38574-2_33
10. Ihalainen, H., Berg, J., Järvisalo, M.: Clause redundancy and preprocessing in maximum satisfiability. In: Blanchette, J., Kovács, L., Pattinson, D. (eds.) Automated Reasoning - 11th International Joint Conference, IJCAR 2022, Haifa, Israel, 8–10 August 2022, Proceedings, vol. 13385 of Lecture Notes in Computer Science, pp. 75–94. Springer, Heidelberg (2022). https://doi.org/10.1007/978-3-031-10769-6_6
11. Korhonen, T., Järvisalo, M.: Integrating tree decompositions into decision heuristics of propositional model counters (short paper). In: Michel, L.D. (ed.) 27th International Conference on Principles and Practice of Constraint Programming, CP 2021, Montpellier, France (Virtual Conference), 25–29 October 2021, vol. 210 of LIPIcs, pp. 8:1–8:11. Schloss Dagstuhl - Leibniz-Zentrum für Informatik (2021)
12. Lagniez, J., Marquis, P.: On preprocessing techniques and their impact on propositional model counting. J. Autom. Reason. **58**(4), 413–481 (2017)
13. Lokshtanov, D., Panolan, F., Ramanujan, M.S.: Backdoor sets on nowhere dense SAT. In: Bojańczyk, M., Merelli, E., Woodruff, D.P. (eds.) 49th International Colloquium on Automata, Languages, and Programming (ICALP 2022), vol. 229 of Leibniz International Proceedings in Informatics (LIPIcs), Dagstuhl, Germany, pp. 91:1–91:20. Schloss Dagstuhl – Leibniz-Zentrum für Informatik (2022)
14. Lonsing, F.: Qbfrelay, qratpre+, and depqbf: incremental preprocessing meets search-based QBF solving. J. Satisf. Boolean Model. Comput. **11**(1), 211–220 (2019)
15. Lonsing, F., Egly, U.: Qratpre+: effective QBF preprocessing via strong redundancy properties. In: Janota, M., Lynce, I. (eds.) Theory and Applications of Satisfiability Testing - SAT 2019 - 22nd International Conference, SAT 2019, Lisbon, Portugal, July 9-12, 2019, Proceedings, vol. 11628 of Lecture Notes in Computer Science, pp. 203–210. Springer, Heidelberg (2019). https://doi.org/10.1007/978-3-030-24258-9_14
16. Mählmann, N., Siebertz, S., Vigny, A.: Recursive backdoors for SAT. CoRR arxiv:2102.04707 (2021)
17. Markov, I.L., Shi, Y.: Constant-degree graph expansions that preserve treewidth. Algorithmica **59**(4), 461–470 (2011)
18. Sæther, S.H., Telle, J.A., Vatshelle, M.: Solving #SAT and MAXSAT by dynamic programming. J. Artif. Intell. Res. **54**, 59–82 (2015)

19. Samer, M., Szeider, S.: Algorithms for propositional model counting. J. Disc. Algor. **8**(1), 50–64 (2010)
20. Slivovsky, F., Szeider, S.: A faster algorithm for propositional model counting parameterized by incidence treewidth. In: Pulina, L., Seidl, M. (eds.) Theory and Applications of Satisfiability Testing - SAT 2020 - 23rd International Conference, Alghero, Italy, July 3-10, 2020, Proceedings, vol. 12178 of Lecture Notes in Computer Science, pp. 267–276. Springer, Heidelberg (2020). https://doi.org/10.1007/978-3-030-51825-7_19
21. Soos, M., Meel, K.S.: Arjun: an efficient independent support computation technique and its applications to counting and sampling. In: ICCAD 2022, New York, NY, USA. Association for Computing Machinery (2022)
22. Wallon, R., Mengel, S.: Revisiting graph width measures for CNF-encodings. J. Artif. Intell. Res. **67**, 409–436 (2020)

Total Matching and Subdeterminants

Luca Ferrarini[1] , Samuel Fiorini[2] , Stefan Kober[2(✉)] ,
and Yelena Yuditsky[2]

[1] École des Ponts ParisTech, Champs-sur-Marne, France
luca.ferrarini@enpc.fr
[2] Université libre de Bruxelles, Brussels, Belgium
{Samuel.Fiorini,Stefan.Kober,Yelena.Yuditsky}@ulb.be

Abstract. In the total matching problem, one is given a graph G with weights on the vertices and edges. The goal is to find a maximum weight set of vertices and edges that is the non-incident union of a stable set and a matching. We consider the natural formulation of the problem as an integer program (IP), with variables corresponding to vertices and edges. Let $M = M(G)$ denote the constraint matrix of this IP. We define $\Delta(G)$ as the maximum absolute value of the determinant of a square submatrix of M. We show that the total matching problem can be solved in strongly polynomial time provided $\Delta(G) \leq \Delta$ for some constant $\Delta \in \mathbb{Z}_{\geq 1}$. We also show that the problem of computing $\Delta(G)$ admits an FPT algorithm. We also establish further results on $\Delta(G)$ when G is a forest.

1 Introduction

Let $G = (V(G), E(G))$ be a graph. We say that two elements (vertices or edges) $\alpha, \beta \in V(G) \cup E(G)$ are *incident* if $\alpha = \beta$ or if one is an edge and the other is an endpoint of that edge. Given weights $w(v) \in \mathbb{R}$ for each vertex $v \in V(G)$ and $w(e) \in \mathbb{R}$ for each edge $e \in E(G)$, the *total matching problem* asks for the maximum weight union of a stable set and a matching, such that the elements are pairwise non-incident. Note that we can assume that G is simple: if G contains any loop or parallel edge, we can replace them with a copy that attains the maximum weight. The natural integer programming formulation of the total matching problem is as follows[1]:

$$
\begin{aligned}
\max \ & \textstyle\sum_{v \in V(G)} w(v)x_v + \sum_{e \in E(G)} w(e)y_e \\
\text{s.t.} \ & x_v + \textstyle\sum_{e \in \delta(v)} y_e \leq 1 \quad \forall v \in V(G) \\
& x_v + x_w + y_e \leq 1 \quad \forall e = vw \in E(G) \\
& x \geq 0, \ y \geq 0 \\
& x, y \text{ integer.}
\end{aligned}
$$

We denote by $M(G)$ the constraint matrix of the above integer program, ignoring the nonnegativity constraints. That is, we let $M(G) := \begin{bmatrix} I & B \\ B^\intercal & I \end{bmatrix}$, where $B =$

---

[1] We use $\delta(v)$ to denote the set of edges incident to vertex v.

A. Basu et al. (Eds.): ISCO 2024, LNCS 14594, pp. 192–204, 2024.
https://doi.org/10.1007/978-3-031-60924-4_15

$B(G) \in \{0,1\}^{V(G) \times E(G)}$ denotes the (vertex-edge) incidence matrix of G and I denotes the identity matrix of appropriate size. Notice that $M(G)$ is simply the matrix of that incidence relation on $V(G) \cup E(G)$.

The study of the total matching problem has been initiated by Alavi, Bezhad, Lesniak-Foster, and Nordhaus [1]. As a generalization of the stable set problem, it is clearly NP-complete. In fact, the problem is NP-hard already for bipartite and planar graphs [11]. Efficient algorithms are only known for very structured classes of graphs, such as trees, cycles, or complete graphs [10,11]. Recently the first polyhedral study of the total matching problem was introduced [4].

We let $\Delta(G)$ denote the maximum subdeterminant of $M(G)$, that is, the maximum of $|\det M'|$ taken over all square submatrices M' of $M(G)$.

Before exploring the total matching problem further, we discuss our motivations to study this problem through the prism of subdeterminants. It is conjectured that the IP $\max\{w^\mathsf{T} x : Ax \leq b,\ x \in \mathbb{Z}^n\}$ can be solved in strongly polynomial time whenever A and b are integer and the maximum subdeterminant of the constraint matrix A is bounded by a constant $\Delta \in \mathbb{Z}_{\geq 1}$.

If $\Delta = 1$ then A is *totally unimodular* and solving the IP is equivalent to solving its LP relaxation $\max\{w^\mathsf{T} x : Ax \leq b\}$, and can hence be done in strongly polynomial time. This is a foundational result in discrete optimization, with countless applications. Artmann, Weismantel and Zenklusen [3] proved that this generalizes to $\Delta = 2$. To this day, the general conjecture remains open for all $\Delta \geq 3$, despite recent efforts [2,6–8,12–14]. However, it is known to hold in some special cases, for instance when A has at most two nonzeroes per row or per column [5].

While the conjecture remains out of reach, it is worthwhile to *test* it on interesting problems, and the total matching is one of them. One of the expected outcomes of such research efforts is to better understand how the structure of the constraint matrix A relates to the maximum subdeterminant, and more specifically what a constant bound on the maximum subdeterminant entails in terms of forbidden submatrices of A.

For instance, incidence matrices of odd cycles have determinant ± 2, and unsurprisingly they are relevant to our work. However, it turns out that incidence matrices of *near-pencils* play an even more important role. These are defined as follows. For $k \in \mathbb{Z}_{\geq 1}$, let N_k denote the $(1 + k) \times (1 + k)$ matrix defined as $N_k := \begin{bmatrix} 1 & \mathbf{1}^\mathsf{T} \\ \mathbf{1} & I \end{bmatrix}$. This matrix is the incidence matrix of a *near-pencil* of order k. Notice that $\det N_k = 1 - k$.

We conclude this introduction with an overview of the paper. We first consider the case of general graphs in Sect. 2 and prove our main algorithmic results, which are a $O(2^{O(\Delta \log \Delta)}(|V(G)| + |E(G)|))$ time algorithm to solve the total matching problem on graphs G with $\Delta(G) \leq \Delta$ and a $O(|V(G)| + |E(G)| + 2^{O(\Delta \log \Delta)})$ time algorithm to either correctly report that $\Delta(G) > \Delta$ or compute $\Delta(G)$ exactly. In Sect. 3, we study further properties of $\Delta(G)$ when G is a forest. In Sect. 4, we give an outlook on the questions solved in this paper and also on future research.

2 General Graphs

In this section, we first prove some structural results that relate $\Delta(G)$ to the structure of G (in Sect. 2.1), and then use these results to obtain our main two algorithmic results (in Sect. 2.2).

2.1 Maximum Subdeterminants and Graph Structure

Lemma 1. *If $G_1, \ldots, G_c$ denote the components of graph G, then*

$$\Delta(G) = \prod_{i=1}^{c} \Delta(G_i).$$

Proof. Consider any ordering of the elements (vertices and edges) of G such that the elements of each component G_i are consecutive. With respect to such an ordering, $M(G)$ has a block-diagonal structure, with one block per component. The result follows. □

The next two lemmas implicitly use the fact that $\Delta(G)$ is monotone under taking subgraphs: if $G \subseteq H$ then $\Delta(G) \leq \Delta(H)$.

Lemma 2. *Let G be a graph and let $d_{\max}(G)$ denote the maximum degree of G. Then*
$$\Delta(G) \geq d_{\max}(G) - 1.$$

Proof. If G contains a vertex of degree $k \geq 1$, then $M(G)$ contains N_k as a submatrix. Hence, $\Delta(G) \geq |\det N_k| = k - 1$. □

Below, we use the notation $C = v_1 e_1 v_2 e_2 \cdots v_n e_n v_1$ to denote the cycle with vertex set $V(C) = \{v_1, v_2, \ldots, v_n\}$ and edge set $E(C) = \{e_1, e_2, \ldots, e_n\}$ where $e_i = v_i v_{i+1}$ for $i \in [n-1] = \{1, 2, \ldots, n-1\}$ and $e_n = v_n v_1$. Later, we use a similar notation for paths.

Lemma 3. *Let $k \in \mathbb{N}$. If a graph G contains k disjoint cycles, then $\Delta(G) \geq 2^k$.*

Proof. Firstly we consider the case where $k = 1$. We may assume without loss of generality that G is a cycle. If G is an odd cycle, then its incidence matrix $B = B(G)$ has determinant ± 2 and hence $\Delta(G) \geq |\det B| = 2$.

Now assume that $G := v_1 e_1 v_2 e_2 \cdots v_n e_n v_1$ is an even cycle. Consider the submatrix M' of $M(G)$ whose rows are indexed by $V(G) \cup \{e_n\}$ and whose columns are indexed by $E(G) \setminus \{e_n\} \cup \{v_1, v_n\}$. It is easy to see that M' is the incidence matrix of an odd cycle, hence we get again $\Delta(G) \geq |\det M'| = 2$.

The bound for $k \geq 2$ follows from the above and Lemma 1. □

Lemma 4. *$M(G)$ is totally unimodular if and only if every component of G is a path.*

Proof. First, assume that every component of G is a path. By Lemma 1, we may assume that G is a path, say $G = v_1 e_1 v_2 e_2 \cdots e_{n-1} v_n$. Using this ordering of its elements, we see that $M(G)$ has the consecutive ones property, and is hence totally unimodular.

Conversely, assume that $M(G)$ is totally unimodular. By Lemma 2, the maximum degree of G is at most 2. By Lemma 3, every component of G is a path. □

Let G be any graph. Consider the corresponding constraint matrix $M = M(G)$, and a *witness*, i.e., a square submatrix M' of M whose determinant is maximum in absolute value. It is natural to assume that every proper square submatrix M'' of witness M' has $|\det M''| < |\det M'|$, in which case we say that M' is *minimal*. In particular, if M' is a minimal witness then every row and column of M' has at least two ones.

In order to facilitate the analysis, we use the following terminology for representing the witness M'. We say that an element of G (vertex or edge) is *red* if the corresponding row is selected as a row of M', and *cyan* if the corresponding column is selected as a column of M'. Notice that an element can be red and cyan at the same time, in which case we say that it is *bichromatic* (or *bicolored*). The elements that belong to a single color are said to be *monochromatic*. We allow some elements to be *uncolored*. Clearly the square submatrices of M correspond to the colorings of the elements of G with k red elements and k cyan elements, where k is a positive integer.

We say that witness $M' \in \{0,1\}^{k \times k}$ has a *fault* if M' has two rows (resp. columns) such that, regarded as sets, one is contained in the other and the two differ by at most one element. A minimal witness M' cannot have a fault, since either $\det M' = 0$ or one can remove one row and column of M' to obtain a submatrix $M'' \in \{0,1\}^{(k-1) \times (k-1)}$ such that $|\det M''| = |\det M'|$. This can easily be seen by performing a row (resp. column) operation on M'.

Lemma 5. *Let G be a graph. Let x_0, x_1, x_2, x_3, x_4 be elements of G such that each x_i, $i \in [3]$ is incident to x_{i-1}, x_i, x_{i+1} and to no other element of G (possibly $x_0 = x_4$, but the elements are otherwise distinct). Consider a minimal witness M' for G and the corresponding coloring. The following hold:*

(i) x_1 is uncolored if and only if x_2 is uncolored;
(ii) if x_2 is bicolored and x_1, x_2, x_3 are all colored in the same color, then each of x_0, x_1, x_2, x_3, x_4 is colored in the opposite color.

Proof. (i) Toward a contradiction, suppose that x_1 is uncolored, but x_2 is colored, say red. Hence, x_2 and x_3 are both cyan. Looking at the corresponding columns of M', we see that M' has a fault, a contradiction. This shows that x_2 is uncolored whenever x_1 is. The same argument shows that x_1 is uncolored whenever x_2 is.

(ii) Assume for instance that x_1, x_2 and x_3 are all red, and x_2 is also cyan. In order to avoid a fault in the rows for x_1 and x_2, elements x_0 and x_3 need to be cyan. Similarly, by the considering the rows for x_2 and x_3 we conclude that x_1 and x_4 are cyan. □

Now consider any sequence $x_0, x_1, x_2, \ldots, x_k$ ($k \geq 4$) of elements of G such that for each $i \in [k-1]$, elements x_{i-1}, x_i, x_{i+1} are distinct and the only elements of G incident to x_i. Notice that such sequences essentially correspond to induced paths in G.

More precisely, assume that x_0 and x_k are vertices of G (hence k is even). Then, $P := x_2 x_3 \cdots x_{k-2}$ is an induced path in G.

Applying Lemma 5 several times, we see that if one element of P is uncolored, then they all are. Moreover, if some element x_i of P is bicolored with x_{i-1}, x_i, x_{i+1} of the same color then all elements of P are bicolored. These observations form the basis of the next two lemmas.

Lemma 6. *Let G be a graph containing a path P of length 7 all whose vertices have degree 2 in G. If G' denotes the graph obtained from G by contracting three edges of P, then $\Delta(G) = \Delta(G')$.*

Proof. Let $P = v_1 e_1 v_2 e_2 \ldots e_6 v_7$, where v_i, $i \in [7]$ denote the vertices of P and e_i, $i \in [6]$ denote the edges of P. Let e_0 (resp. e_7) denote the other edge of G incident to v_1 (resp. v_7).

Again, consider a minimal witness M' for G and the corresponding coloring. By Lemma 5, there are three cases to consider.

Case 1. All the elements of P are uncolored. Then M' is a submatrix of $M(G')$, hence $\Delta(G) = \Delta(G')$.

Case 2. All the elements of P are bicolored. Then also e_0 and e_7 are bicolored. We can add rows (or columns) to other rows (or columns) and remove rows (or columns) with exactly one non-zero entry of absolute value 1 and the corresponding column (or row) in order to obtain a square submatrix M'' of $M(G')$. In the determinant sense, these operations correspond to the linear combination of rows and columns and the Laplace expansion, which preserve the absolute value of the determinant, that is, $|\det M''| = |\det M'|$.

More formally, given an element $\alpha \in V(G) \cup E(G)$, we denote by r_α and c_α the row and column corresponding to α. We replace r_{e_1} by $r_{e_1} + r_{v_3} - r_{v_1} - r_{e_3}$, and c_{v_4} by $c_{v_1} + c_{v_4}$. Observe that now, r_{e_1} has its only non-zero entry in c_{v_1}. Thus we remove r_{e_1} and c_{v_1}. Further, we replace r_{v_1} by $r_{v_1} + r_{e_2} - r_{v_2} - r_{e_3}$, and c_{e_0} by $c_{e_0} + c_{e_3}$. Now, by considering the elements in the following order, $r_{e_3}, c_{e_1}, c_{v_2}, c_{e_2}$, and c_{v_3}, each element has one non-zero entry and can be removed with their counterpart, and we obtain the claimed submatrix.

We have once more that $\Delta(G) = \Delta(G')$.

Case 3. Every row and column of M' corresponding to an element of P has exactly 2 ones. Then we can contract 4 consecutive elements on P in the following way (we describe a process for rows that works in the same way for columns): Choose a row r_1 that is not the first or last row on P. Then there is two distinct columns c_1 and c_2 adjacent to r_1. Again, there is two distinct rows r_0 and r_2, where r_0 is adjacent to c_1 and r_2 is adjacent to c_2. We can apply the operations from the previous case in order to combine r_0 and r_2 into a single row and remove c_1, c_2, and r_1.

More formally, we replace c_1 by $c_1 - c_2$, and r_2 by $r_2 + r_0$. Now, r_1 and c_1 contain exactly one non-zero and can be removed along with their counterparts.

This corresponds to contracting the corresponding elements in the graph. We repeat this process as long as there is more than ten rows and columns corresponding to elements of P. Finally, we map the remaining elements to the remaining subpath. Again, there exists a square submatrix M'' of $M(G')$ with $|\det M''| = |\det M'|$, which implies $\Delta(G) = \Delta(G')$. □

Lemma 7. *Let G be a n-cycle. Then,*

$$\Delta(G) = \begin{cases} 3 \text{ if } n \equiv 1, 2 \pmod 3, \\ 2 \text{ if } n \equiv 0 \pmod 3. \end{cases}$$

Proof. Consider a minimal witness $M' \in \{0,1\}^{k \times k}$ for G and the corresponding coloring. By Lemma 5, either $M' = M(G)$ or M' is a submatrix of $M(G)$ having exactly two ones in each row and column (see also the proof of Lemma 3). In the first case, $k = 2n$ and $n \equiv 1, 2 \pmod 3$ and $|\det M'| = 3$ (if $n \equiv 0 \pmod 3$ then $|\det M'| = 0$, contradicting the fact that M' is a witness). In the second case, $k \equiv 1 \pmod 2$ and $|\det M'| = 2$ (if $k \equiv 0 \pmod 2$ then again $|\det M'| = 0$, a contradiction). □

Lemma 8. *Let G be a graph and let $D \subseteq V(G)$ be a set of vertices which have at least 2 neighbours in $\overline{D} := V(G) \backslash D$. Let H denote the bipartite graph obtained from G by deleting all edges contained in D or $\overline{D}$. Then*

$$\Delta(G) \geq \prod_{v \in D} (d_H(v) - 1) .$$

Proof. Let M' denote the principal submatrix of $M(G)$ whose rows and columns are indexed by vertices $v \in D$ and edges in $\delta_H(v)$ for $v \in D$. After reordering the rows (resp. columns) of M' so that the row (resp. column) of each vertex $v \in D$ is immediately followed by the rows (resp. columns) for the edges in $\delta_H(v)$ (in any order), we see that M' has a block-diagonal structure, with one block per vertex of D. Moreover, the block for $v \in D$ in M' is the near-pencil matrix N_k with $k := d_H(v)$. It follows that $\Delta(G) \geq |\det M'| = \prod_{v \in D} (d_H(v) - 1) .$ □

Lemma 9. *Every graph G has at most $7 \log \Delta(G)$ vertices of degree at least 3.*

Proof. Consider any inclusion-wise maximal packing of vertex-disjoint $K_{1,3}$ subgraphs in G, and let X denote the set of vertices covered by the packing. By Lemma 8, we have that $|X| \leq 4 \log \Delta(G)$.

By maximality of our packing, all degrees in $G - X$ are at most 2. Hence, $G - X$ has a proper 3-coloring. Let $Y \subseteq V(G - X)$ denote the set of vertices of $G - X$ whose degree in G is at least 3. Considering the color class that contains the largest number of vertices of Y, and using again Lemma 8, we see that $|Y| \leq 3 \log \Delta(G)$.

We conclude that the total number of vertices of degree at least 3 in G is at most $|X| + |Y| \leq 4 \log \Delta(G) + 3 \log \Delta(G) = 7 \log \Delta(G)$. □

Lemma 10. *Let $\Delta \in \mathbb{Z}_{\geq 1}$ be a constant, and let G be any graph such that $\Delta(G) \leq \Delta$. There exists a vertex subset Z such that $G - Z$ is a disjoint union of paths, $|Z| \leq 8 \log \Delta$ and $|\delta(Z)| \leq 8(\Delta + 1) \log \Delta$. Moreover, Z can be computed in time $O(|V(G)| + |E(G)|)$.*

Proof. Let $X := \{v \in V(G) \mid d(v) \geq 3\}$ denote the set of vertices of degree at least 3. By Lemma 9, we have $|X| \leq 7 \log \Delta$. Obviously, $G - X$ is a disjoint union of paths and cycles. By Lemma 3, there are at most $\log \Delta$ cycles in $G - X$ (all disjoint). Hence there exists a set $Y \subseteq V(G - X)$ of size $|Y| \leq \log \Delta$ intersecting every cycle of $G - X$. Letting $Z := X \cup Y$, we see that $G - Z$ is a disjoint union of paths. By Lemma 2, the degree of any vertex in G is at most $\Delta + 1$, hence the number of edges in the cut $\delta(Z)$ is at most $(\Delta + 1)|Z| \leq 8(\Delta + 1) \log \Delta$.

It is easy to see that the set Z can be found in time linear in the size of the graph.   □

2.2   Algorithmic Results

Theorem 1. *Let $\Delta \in \mathbb{Z}_{\geq 1}$ be a constant. There is an algorithm to find a total matching of maximum weight in a graph G such that $\Delta(G) \leq \Delta$ in time $O(2^{O(\Delta \log \Delta)}(|V(G)| + |E(G)|))$.*

Proof. Let $Z \subseteq V(G)$ be as in Lemma 10. Let $I = I(Z)$ denote the set of elements of G that are incident to some vertex of Z, including the edges with both ends in Z. Notice that $|I| = O(\log \Delta) + O(\Delta \log \Delta) + O(\log^2 \Delta) = O(\Delta \log \Delta)$.

We find a total matching of maximum weight by considering, one after the other, each subset total matching $M_1 \subseteq I$. For a fixed M_1, let Z' denote the set of vertices of $G - Z$ that are incident to some edge in M_1. Notice that by removing from $G - Z$ all the vertices of Z' (as well as the incident edges), we delete all the elements of $G - Z$ that are incident to some element of M_1 (and hence cannot be included in any total matching of G extending M_1). Let M_2 be a maximum weight total matching in $G - Z - Z'$. Such a total matching can be computed in linear time via dynamic programming, since $G - Z - Z'$ is a disjoint union of paths.

The optimal solution returned by the algorithm is the best total matching of the form $M_1 \cup M_2$ over all choices of a total matching $M_1 \subseteq I$. The number of choices for M_1 is bounded by $2^{O(\Delta \log \Delta)}$. Hence the running time of the above algorithm is $O(2^{O(\Delta \log \Delta)}(|V(G)| + |E(G)|))$.   □

Theorem 2. *Let $\Delta \in \mathbb{Z}_{\geq 1}$ be a constant. There is an algorithm that, given any graph G, either computes $\Delta(G)$ or correctly reports that $\Delta(G) > \Delta$ in time $O(|V(G)| + |E(G)| + 2^{O(\Delta \log \Delta)})$.*

Proof. By Lemma 10 in time $O(|V(G)| + |E(G)|)$ we either find a set Z as in the lemma or report correctly that $\Delta(G) > \Delta$.

Assume that we are in the former case. By Lemmas 1 and 4, every component of $G - Z$ that is not connected to Z does not contribute to $\Delta(G)$, and can be deleted from G without changing $\Delta(G)$. There are at most $|\delta(Z)| = O(\Delta \log \Delta)$

components of $G - Z$ that send an edge to Z. From the proof of Lemma 10, we see that each such component is a path whose vertices have all degree at most 2 in G.

By Lemma 6 we can shrink each component of $G - Z$ to a path of bounded length without changing $\Delta(G)$. This reduces G to a graph with $O(\Delta \log \Delta)$ vertices and edges. The maximum subdeterminant of $M(G)$ can now be computed in time $O(2^{O(\Delta \log \Delta)})$, by brute force. The result follows.   □

3   Forests

In this section, we explore several properties of $\Delta(G)$ for forests. Let G be a forest, and let M' be a minimal witness for G. As before, consider the corresponding coloring of the elements of G. Our first lemma states that M' is a principal submatrix of M.

Lemma 11. *No element of G is monochromatic.*

Proof. Toward a contradiction, assume that some element α is monochromatic, say red.

We claim that, for some $k \geq 2$, we can partition the red elements into $k + 1$ sets, $\{\alpha\}, R_1, \ldots, R_k$, and the cyan elements into k sets $C_1, \ldots, C_k$ such that (i) no element of R_i is incident to an element of C_j for $i \neq j$, and (ii) α is incident to at most one element in each C_j. In other words, we claim that, after permuting the rows and columns, M' takes the form

$$M' = \begin{bmatrix} a_1^\mathsf{T} & a_2^\mathsf{T} & \cdots & a_k^\mathsf{T} \\ M_1' & 0 & \cdots & 0 \\ 0 & M_2' & \cdots & 0 \\ \vdots & \vdots & \ddots & \vdots \\ 0 & 0 & \cdots & M_k' \end{bmatrix}$$

where each a_j is a 0/1 vector of the appropriate dimension, with at most one 1.

To prove the claim, we consider separately the two cases that can occur.

First, assume that $\alpha = e = v_1 v_2$ is an edge. By minimality of M', both v_1 and v_2 are cyan. In this case, we take $k = 2$ and consider an arbitrary decomposition of forest G as the union of $(\{v_1, v_2\}, \{e\})$ and two vertex-disjoint forests G_1 and G_2, the first containing v_1 and the second v_2. Then, for $i = 1, 2$, we let R_i (resp. C_i) be the set of red (resp. cyan) elements in G_i. It remains to check that (i) and (ii) are satisfied. Clearly, no element of R_i is incident to an element of C_j for $i \neq j$. Moreover, e is incident to exactly one element of C_1 (namely, v_1) and exactly one element of C_2 (namely, v_2).

Second, assume that $\alpha = v$ is a vertex. Let $e_1, \ldots, e_k$ denote the edges incident to v (by minimality of M', at least two of these edges are cyan, which implies in particular that $k \geq 2$). In this case, we decompose forest G as the union of k forests $G_1, \ldots, G_k$ all containing v but otherwise vertex-disjoint, in such a way that each G_i contains edge e_i, for $i = 1, \ldots, k$. Then, for $i = 1, \ldots, k$, we let R_i denote the red elements distinct from v in G_i and C_i denote the cyan elements in G_i. Again, properties (i) and (ii) are satisfied. This concludes the proof of the claim.

Now, for $i = 1, \ldots, k$, let $r_i := |R_i|$ and $c_i := |C_i|$. We have $r_i \leq c_i$ for each i since the r_i rows of M' for the elements of R_i are linearly independent vectors spanning a space of dimension at most c_i. Also, we have $c_i \leq r_i + 1$ for each i since the c_i columns of M' for the elements of C_i are linearly independent vectors spanning a space of dimension at most $r_i + 1$. Hence, $c_i - r_i \in \{0, 1\}$ for $i = 1, \ldots, k$. Since M' is square, we have

$$1 + \sum_{i=1}^{k} r_i = \sum_{i=1}^{k} c_i \implies \sum_{i=1}^{k} \underbrace{(c_i - r_i)}_{\in \{0,1\}} = 1 \,.$$

After permuting the indices if necessary, we may assume that $c_1 = 1 + r_1$ and $c_i = r_i$ for $i \geq 2$.

We find

$$\det M' = \det \begin{bmatrix} a_1^\mathsf{T} \\ M_1' \end{bmatrix} \det M_2' \cdots \det M_k'$$

By choice of M', none of these determinants is zero. In particular, α is incident to exactly one element of C_1, say β. Notice that the first determinant equals (in absolute value) the determinant of the square submatrix associated to R_1 and $C_1 \setminus \{\beta\}$. This means that we can uncolor element α and remove the color cyan from β and find the same determinant, which contradicts the minimality of M'. ☐

Resuming our discussion of the minimal witness M', by Lemma 11 we have

$$M' = \begin{bmatrix} I & B' \\ (B')^\mathsf{T} & I \end{bmatrix} ,$$

where B' is the submatrix[2] of the incidence matrix B with rows indexed by bichromatic edges and columns indexed by bichromatic vertices.

We use a Schur complement to express $\det M'$ in terms of a smaller matrix whose rows and columns are indexed by the bichromatic vertices. We find

$$\det M' = \det \begin{bmatrix} I - B'(B')^\mathsf{T} & 0 \\ (B')^\mathsf{T} & I \end{bmatrix} = \det(I - B'(B')^\mathsf{T}) \,.$$

Let G' denote the subgraph of G whose edges are the bichromatic edges. Let G'' denote the subgraph of G' induced by the bichromatic vertices. We denote by $\tilde{L}(G', G'')$ the matrix $B'(B')^\mathsf{T} - I$. Note that this matrix coincides with the matrix obtained from the adjacency matrix $A(G'')$ of G'' by changing the diagonal coefficient for each bichromatic vertex v to $d_{G'}(v) - 1$. In a formula,

$$\tilde{L}(G', G'') = B'(B')^\mathsf{T} - I = A(G'') + \mathrm{diag}(d_{G'}(v_1) - 1, \ldots, d_{G'}(v_k) - 1)$$

where $v_1, \ldots, v_k$ denote the bichromatic vertices.

---

[2] Notice that B' is not always a true incidence matrix since there might be bichromatic edges e such that only one endpoint of e is bichromatic.

From the discussion above, we directly obtain the following result. The final part of the theorem follows from the minimality of M'.

Theorem 3. *Let G be any forest (with at least one vertex). Then*

$$\Delta(G) = \max\{|\det \tilde{L}(G', G'')| : G'' \overset{\text{ind}}{\subseteq} G' \subseteq G\}$$

where G' is a subforest of G and G'' is an induced subforest of G'. Moreover, in the formula above we can restrict to choices of subgraphs such that every edge of G' has at least one endpoint in G'', and $d_{G'}(v) \geq 2$ for all $v \in V(G'')$ (provided that every component of G has at least three vertices).

We conclude this section with a lemma that bounds $\Delta(G)$ in terms of the degree sequence of G. The lower bound is similar to Lemma 8 in spirit, except that here we can be more explicit using the fact that G is bipartite.

Lemma 12. *Let G be a forest, and let $d_1 \geq d_2 \geq \cdots \geq d_n$ denote the degrees of the vertices of G. For $d \in \mathbb{N}$, let n_d denote the number of vertices of degree at least d in G. Then we have*

$$\sqrt[n_2]{\prod_{i=1}^{n_2}(d_i - 1)} \leq \Delta(G) \leq \left(\frac{\sum_{i=1}^{n_2} d_i}{n_2}\right)^{n_2}.$$

Proof. To prove the lower bound, consider any partition of the set of vertices of degree at least 2 into two stable sets S_1 and S_2 (that is, any bicoloring of the corresponding induced subgraph of G). Notice that $\det \tilde{L}(G, G[S_1]) \cdot \det \tilde{L}(G, G[S_2]) = \prod_{i=1}^{n_2}(d_i - 1)$. Hence, there exists some index $i \in [2]$ such that $\det \tilde{L}(G, G[S_i]) \geq \sqrt{\prod_{i=1}^{n_2}(d_i - 1)}$. We conclude by applying Theorem 3 with $G' := G$ and $G'' := G[S_i]$.

To prove the upper bound, let G' and G'' be any choice of subgraphs of G that achieve the maximum subdeterminant, see Theorem 3. Thus, $\Delta(G) = |\det \tilde{L}(G', G'')|$. Again, let $v_1, \ldots, v_k$ denote the vertices of G''. Since $d_{G'}(v_i) \geq d_{G''}(v_i)$ for all $i \in [k]$, we have

$$\tilde{L}(G', G'') + I \succcurlyeq A(G'') + \text{diag}(d_{G''}(v_1), \ldots, d_{G''}(v_k)) \succcurlyeq 0.$$

This implies that the eigenvalues $\lambda_1 \geq \lambda_2 \geq \cdots \geq \lambda_k$ of $\tilde{L}(G', G'')$ are all larger or equal to -1. Let $\ell \in [k]$ be the last index for which $\lambda_\ell \geq 0$. We have

$$\Delta(G) = |\det \tilde{L}(G', G'')| = \left|\prod_{i=1}^{k} \lambda_i\right| \leq \prod_{i=1}^{\ell} \lambda_i \leq \left(\frac{\sum_{i=1}^{\ell} \lambda_i}{\ell}\right)^{\ell}$$

$$\leq \left(\frac{(k - \ell) + \sum_{i=1}^{k} \lambda_i}{\ell}\right)^{\ell} = \left(\frac{\sum_{i=1}^{k} d_{G'}(v_i)}{\ell} - 1\right)^{\ell}$$

$$\overset{(\star)}{\leq} \left(\frac{\sum_{i=1}^{k} d_{G'}(v_i)}{k}\right)^{k} \leq \max\left\{\left(\frac{\sum_{i=1}^{p} d_i}{p}\right)^{p} \,\middle|\, p \in [n], \, p \leq n_2\right\}.$$

Above, we use the AM-GM-inequality, the fact that the sum of the eigenvalues of $\tilde{L}(G', G'')$ equals its trace, $G'' \subseteq G' \subseteq G$, and $d_{G'}(v_i) \geq 2$ for all $i \in [k]$. It remains to show that $(\star)$ holds and that $p = n_2$ achieves the maximum in our final upper bound.

In order to prove $(\star)$, let $\bar{d} := \frac{\sum_{i=1}^{k} d_{G'}(v_i)}{k}$ denote the average number of edges of G' incident to a vertex of G'' and $k := (1+\varepsilon)\ell$ for some $\varepsilon \geq 0$. Without loss of generality, we may assume $k > \ell$, i.e., $\varepsilon > 0$. By minimality of M', we have $\bar{d} \geq 2$, see Theorem 3.

Observe that $(\star)$ holds if and only if $(1 + \varepsilon - \bar{d}^{-1})^{\frac{1}{\varepsilon}} \leq \bar{d}$. In case $\bar{d} \geq$ e we get $(1 + \varepsilon - \bar{d}^{-1})^{\frac{1}{\varepsilon}} \leq (1+\varepsilon)^{\frac{1}{\varepsilon}} \leq$ e $\leq \bar{d}$. In case $2 \leq \bar{d} <$ e, we get $(1 + \varepsilon - \bar{d}^{-1})^{\frac{1}{\varepsilon}} \leq (1 + \varepsilon - e^{-1})^{\frac{1}{\varepsilon}} \leq 2 \leq \bar{d}$, where the second to last inequality holds since $2^{\varepsilon} - \varepsilon \geq \frac{1 + \ln \ln 2}{\ln 2} \geq 1 - e^{-1}$ for all $\varepsilon > 0$.

Finally let $1 \leq p < n_2$. Define $x := \sum_{i=1}^{p} d_i$ and observe that $x \geq 2p$. Then,

$$\frac{\left(\frac{x+d_{p+1}}{p+1}\right)^{p+1}}{\left(\frac{x}{p}\right)^p} = \left(\frac{p}{p+1}\right)^p \cdot \left(\frac{(x + d_{p+1})^{p+1}}{(p+1)x^p}\right) \geq e^{-1}\left(\frac{(x+2)^{p+1}}{(p+1)x^p}\right)$$

$$\geq e^{-1}\left(\frac{x^{p+1} + 2(p+1)x^p}{(p+1)x^p}\right) \geq e^{-1}\left(\frac{4p+2}{p+1}\right) \geq 1.$$

We conclude that $p = n_2$ maximizes the upper bound.   □

We conclude this section by comparing the bounds we just proved on $\Delta(G)$. Taking logarithms and assuming that $n_2 = O(n_3)$ (this is without loss of generality by Lemma 6), we see that

$$\log\left(\frac{\sum_{i=1}^{n_2} d_i}{n_2}\right)^{n_2} \leq O\left(\frac{\log d_{\max}}{\log d_{\min}}\right) \log \sqrt{\prod_{i=1}^{n_2}(d_i - 1)},$$

where $d_{\max}$ is the maximum degree and $d_{\min}$ is the minimum degree among all vertices of degree at least 3. Hence, in general, the bounds are quasipolynomially related. If the degree distribution is not too spread out, they are even polynomially related.

4   Conclusions and Future Work

The total matching problem has a natural formulation as an IP. In this paper we investigate instances such that all subdeterminants of the constraint matrix are bounded (in absolute value) by a constant $\Delta \in \mathbb{Z}_{\geq 1}$. Our main contributions include linear time algorithms for solving, as well as recognizing, such instances. Our algorithms rely on results on the structure of graphs G with bounded $\Delta(G)$, which we establish here. Moreover, we prove further structural results about $\Delta(G)$ where G is a forest.

It is not known in general how to check in polynomial time whether an integer matrix has all its subdeterminants bounded by a constant. In particular, no polynomial time algorithm appears to be known to test whether a given integer matrix is *totally bimodular*, that is, has all its subdeterminants bounded by 2 (see for instance [3] for more information).

The main question that we leave for future work is that of determining the complexity of computing $\Delta(G)$. We believe that this is as difficult as computing the maximum subdeterminant of the incidence matrix of a graph. This last problem amounts to computing the odd cycle packing number $\mathrm{ocp}(G)$ graph G. This last invariant is NP-hard to compute, and even NP-hard to approximate [9].

The constraint matrix $M(G)$ of the total matching IP contains the incidence matrix of G as a submatrix, hence $\Delta(G)$ is at least $2^{\mathrm{ocp}(G)}$. However, for most graphs $\Delta(G)$ seems to be dominated by other structural aspects of the graph, in particular its degree sequence.

Note that determining the complexity of computing $\Delta(G)$ remains open, even in the case where G is a forest. Even though the formula given in Theorem 3 slightly simplifies the problem, it remains not trivial to understand precisely which edges and vertices should be selected in the corresponding bicoloring. For instance, there are examples showing that insisting to bicolor certain (high-degree) vertices can be detrimental for the subdeterminant, see Fig. 1.

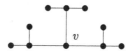

Fig. 1. A graph G with $\Delta(G) = 8$. In colorings where v is bicolored, the maximum subdeterminant is 4. Hence it is optimal to leave v uncolored.

When we consider general graphs, the picture becomes even less clear. Already for small graphs on up to six vertices, there is a large number of examples where certain elements have to be monochromatic in order to maximize the subdeterminant. We note that there appears to be a strong connection between principal subdeterminants and spectral graph theory (as used in the proof of Lemma 12), in particular for forests and regular graphs.

References

1. Alavi, Y., Behzad, M., Lesniak-Foster, L.M., Nordhaus, E.: Total matchings and total coverings of graphs. J. Graph Theory 1(2), 135–140 (1977)
2. Artmann, S., Eisenbrand, F., Glanzer, C., Oertel, T., Vempala, S., Weismantel, R.: A note on non-degenerate integer programs with small sub-determinants. Oper. Res. Lett. 44(5), 635–639 (2016)
3. Artmann, S., Weismantel, R., Zenklusen, R.: A strongly polynomial algorithm for bimodular integer linear programming. In: Proceedings of the 49th Annual ACM SIGACT Symposium on Theory of Computing, STOC 2017, pp. 1206–1219. Association for Computing Machinery (2017)

4. Ferrarini, L., Gualandi, S.: Total coloring and total matching: polyhedra and facets. Eur. J. Oper. Res. **303**(1), 129–142 (2022)
5. Fiorini, S., Joret, G., Weltge, S., Yuditsky, Y.: Integer programs with bounded sub-determinants and two nonzeros per row. In: 2021 IEEE 62nd Annual Symposium on Foundations of Computer Science (FOCS), pp. 13–24 (2022)
6. Glanzer, C., Stallknecht, I., Weismantel, R.: Notes on {a, b, c}-modular matrices. Vietnam J. Math. **50**(2), 469–485 (2022)
7. Gribanov, D., Shumilov, I., Malyshev, D., Pardalos, P.: On Δ-modular integer linear problems in the canonical form and equivalent problems. J. Glob. Optim. 1–61 (2022)
8. Gribanov, D.V., Veselov, S.I.: On integer programming with bounded determinants. Optim. Lett. **10**, 1169–1177 (2016)
9. Kawarabayashi, K.I., Reed, B.: Odd cycle packing. In: Proceedings of the Forty-Second ACM Symposium on Theory of Computing, pp. 695–704 (2010)
10. Leidner, M.E.: A study of the total coloring of graphs. Ph.D. thesis, University of Louisville (2012)
11. Manlove, D.F.: On the algorithmic complexity of twelve covering and independence parameters of graphs. Disc. Appl. Math. **91**(1–3), 155–175 (1999)
12. Nägele, M., Nöbel, C., Santiago, R., Zenklusen, R.: Advances on strictly Δ-modular IPs. In: International Conference on Integer Programming and Combinatorial Optimization, pp. 393–407. Springer, Heidelberg (2023). https://doi.org/10.1007/978-3-031-32726-1_28
13. Nägele, M., Santiago, R., Zenklusen, R.: Congruency-constrained TU problems beyond the bimodular case. In: Proceedings of the 2022 Annual ACM-SIAM Symposium on Discrete Algorithms (SODA), pp. 2743–2790. SIAM (2022)
14. Nägele, M., Sudakov, B., Zenklusen, R.: Submodular minimization under congruency constraints. Combinatorica **39**(6), 1351–1386 (2019)

A New Structural Parameter on Single Machine Scheduling with Release Dates and Deadlines

Maher Mallem[1](✉)[ID], Claire Hanen[1,2][ID], and Alix Munier-Kordon[1][ID]

[1] Sorbonne Université, CNRS, LIP6, 75005 Paris, France
{Maher.Mallem,Claire.Hanen,Alix.Munier}@lip6.fr
[2] UPL, Université Paris Nanterre, 92000 Nanterre, France
http://www.lip6.fr

Abstract. In this paper we study the single machine scheduling problem with release dates, deadlines and precedence relations where the objective is to minimize the makespan. This is a well-known strongly NP-hard scheduling problem [18]. We analyze the problem from the parameterized complexity point of view. We propose parameter q which is the maximum number of time windows $[r_j, d_j)$ that can strictly include a time window $[r_i, d_i)$ on both ends. We show that problems $1|prec, r_j, d_j|C_{max}$ and $1|prec, r_j|L_{max}$ are fixed-parameter tractable parameterized by q. We use a dynamic programming approach and define a new dominance rule, which we call the weak earliest deadline rule. This rule narrows down the number of relevant scheduling prefixes enough to complete the search via a fixed-parameter tractable number of dynamic programming states.

Keywords: parameterized complexity · scheduling · single machine

1 Introduction

Many scheduling problems, even seemingly basic ones such as $1|r_j|L_{max}$, have been proved strongly NP-hard [18] over the years. Upon reaching NP-hardness this quickly, it becomes difficult to establish anything deeper about these scheduling problems within the scope of classical complexity theory. To answer this, parameterized complexity theory gives additional tools for a refined analysis of such hard scheduling problems. Given a parameter k and denoting n the input size, a problem is called $fixed-parameter\ tractable\ (FPT)$ parameterized by k if it can be solved in time $\mathcal{O}(f(k) \cdot poly(n))$ with f an arbitrary computable function [6]. The idea is to identify k as the limiting property and give a polynomial time algorithm for all instances with a bounded value of k. When the studied problem is believed to not be FPT, many complexity classes are available as parameterized analogues to NP like para-NP or the W-hierarchy [7,10].

While parameterized complexity theory has been successfully applied to a lot of computer science areas, it has started to be studied on scheduling only quite recently [4,5,15,20]. One of the few older results was from [3] and showed that

A. Basu et al. (Eds.): ISCO 2024, LNCS 14594, pp. 205–219, 2024.
https://doi.org/10.1007/978-3-031-60924-4_16

$P|prec, p_j = 1|C_{max}$ is W[2]-hard parameterized by the number m of machines. When time windows are involved, two parameters have been considered with great success. First slack σ - which is defined as the maximum number of starting times allowed for any job - was successfully used in the context of single machine scheduling with setup times, and rejection [1]. Second pathwidth μ is the maximum number of overlapping job time windows at any given time. Several problems with arbitrary processing times and/or precedence constraints were proved FPT parameterized by μ [1,14,16,19], for some of them even in the context of parallel machines.

Despite their success, such parameters can overestimate the instance difficulty at times. Indeed a single job with large time window length is enough to get an unbounded slack. In the case of pathwidth μ we give an example in Fig. 1 with two instances of decision problem $1|r_j, d_j|\star$. In the left instance all jobs have the same release date and deadline, so this is an easy particular case which can be solved in linear time by adding up the processing times. Yet this instance has the same pathwidth as the right one, which simulates a PARTITION problem by using a fill job right in the middle. Note that both instances yield the same time window overlap graph - the complete graph K_n. So in order to distinguish them one would have to track time window interactions beyond their overlaps.

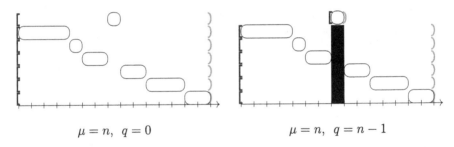

$$\mu = n, \quad q = 0 \qquad\qquad \mu = n, \quad q = n - 1$$

Fig. 1. Two instances of $1|r_j, d_j|\star$ with the same pathwidth μ but different q-proper levels q.

Instead of time window overlaps like with pathwidth μ, we propose to track whenever a time window $[r_j, d_j)$ strictly includes another time window $[r_i, d_i)$ on both ends. When this is the case we say that job j *surrounds* job i. Such time window interactions were considered in the past [8,9,11], albeit never in the context of parameterized complexity. We define parameter q as the maximum number of jobs j which can surround a job i. This definition of q is inspired by Proskurowski and Telle in [21] where the q-proper level of an interval graph was defined. This is applied to the interval graph given by the job time windows.

Going back to Fig. 1 the q-proper level effectively distinguishes between both instances: a low value for the easy instance on the left and a high value for the hard instance on the right. As the right instance suggests parameter q can be interpreted as the maximum size of any PARTITION subproblem which is encoded in the scheduling instance.

Now it is worth noting that scheduling rules have already been successfully used in some subproblems of $1|prec, r_j, d_j|C_{max}$ [17]. One of the most well-known examples is the *earliest deadline rule* [23] - which will be denoted (ED). Given an instance of problem $1|d_j|C_{max}$ the (ED) rule says the following: schedule jobs actively in nondecreasing order of their deadline. This gives the minimum makespan in time $\mathcal{O}(n \cdot log(n))$. The same can be achieved on problem $1|r_j|C_{max}$ with the *earliest release date rule* [2] - which will be denoted (ERD).

However given an instance of $1|prec, r_j, d_j|C_{max}$ such strategies become more difficult to pull off [12,13,22]. Indeed both (ED) and (ERD) rules can still work but only when release dates and deadlines follow the same order. If there is at least one job couple (i, j) such that $r_j < r_i$ and $d_i < d_j$ - i.e. job j *surrounds* job i - then both rules become wrong. Fittingly when this happens we have $q \geq 1$.

In response we propose a variant of the (ED) rule. We call it the *weak earliest deadline rule* and denote it (wED). It says the following: schedule jobs in nondecreasing order of their deadline, except when some job surrounds other jobs. Given each job $i \in [n]$ the only allowed exceptions are the jobs j which either surround i or surround a job k with a deadline $d_k \leq d_i$ and scheduled between j and i. For each job i we show that under the (wED) rule the number of such jobs j is bounded by our parameter q. This will be enough to get a number of dynamic programming states of the form $f(q) \cdot poly(n)$.

The paper is organized as follows: in Sect. 2 we give the necessary preliminaries to our approach. In Sect. 3 we define the (wED) rule on problem $1|prec, r_j, d_j|C_{max}$ and characterize the properties of the schedules which follow this dominance rule. This leads to the summit-based scheduling dominance originally proposed in [9] on problem $1|r_j, d_j|C_{max}$, which we prove to work with problem $1|prec, r_j, d_j|C_{max}$ too. Then in Sect. 4 we present a dynamic programming algorithm which computes the minimum makespan by only exploring the dominating scheduling prefixes. We show that this can be achieved in *FPT* time when parameterized by q. Finally we discuss how this approach can be adapted to criterion L_{max}.

2 Preliminaries

We consider a set of n jobs $\mathcal{J} = \{1, \ldots, n\}$ to be scheduled on a single machine. The machine can only process one job at a time. Each job j has a processing time $p_j \in \mathbb{N}^*$ and a time window $[r_j, d_j)$ with $r_j, d_j \in \mathbb{N}$ and $r_j < d_j$. Job j cannot start earlier than its release date r_j and must be completed no later than its deadline d_j. Moreover we consider precedence constraints given by a partial order relation $\rightarrow$ on $\mathcal{J}$: "$i \rightarrow j$" implies that j cannot start earlier than the completion of i. A feasible schedule $\sigma : \mathcal{J} \rightarrow \mathbb{N}$ assigns a starting time $\sigma(i)$ to each job i, following the previous constraints. Our optimization criterion is the makespan of the schedule, i.e. $C_{max}(\sigma) = \max_{i \in \mathcal{J}}[\sigma(i) + p_i]$. This problem is usually denoted $1|prec, r_j, d_j|C_{max}$.

We denote $[n]$ the integer interval $[1, n]$ and the set of jobs $\mathcal{J}$. Since we have both time windows constraints and precedence relations in our problem, we must ensure that the time constraints are compatible with the partial order $\rightarrow$.

Definition 1. *An instance* $I = \langle \mathcal{J}, prec, p_j, r_j, d_j \rangle$ *of* $1|prec, r_j, d_j|C_{max}$ *is called prec-consistent when:*
$$\forall (i,j) \in [\![n]\!]^2, \ (i \to j) \implies [(r_i + p_i \leq r_j) \wedge (d_i \leq d_j - p_j)].$$

Given an instance of $1|prec, r_j, d_j|C_{max}$ which is not *prec*-consistent, its release times and deadlines can be adjusted in time $\mathcal{O}(n^2)$ to fulfill this property by using path algorithms. Note that by doing so we only remove time window sections which are inaccessible according to precedence relations. So feasibility and the optimal makespan value are unchanged. While *prec*-consistency might look inconspicuous it is crucial to the dominance rule given in the next section. Without it one can build artificial counterexamples like the one in Fig. 2.

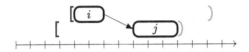

Fig. 2. A non *prec*-consistent counterexample to the (wED) rule.

Consider a *prec*-consistent instance. For the remainder of the paper we suppose that the jobs have been renamed in $[\![n]\!]$ in lexicographic nondecreasing order of (deadline, release date). So when we write "$i < j$" it means that: $(d_i < d_j) \vee [(d_i = d_j) \wedge (r_i \leq r_j)]$. We define proper sets and parameter q the following way:

Definition 2. *(Proper set Π_i) Let $i \in [\![n]\!]$. We say that a job $j \in [\![n]\!]$ surrounds job i when $r_j < r_i$ and $d_i < d_j$. In other words j surrounds i when time window $[r_i, d_i)$ is strictly included by time window $[r_j, d_j)$ on both ends. We define $\Pi_i = \{j \in [\![n]\!], j$ surrounds $i\}$.*
(q-proper level) We define parameter $q = max_{i \in [\![n]\!]}|\Pi_i|$. In other words q is the maximum number of jobs j which can surround a job i.

Finally like in [8] we highlight the following subset of jobs:

Definition 3. *We say that a job $j \in [\![n]\!]$ is a summit when j surrounds no job. In other words: for all jobs i in $[\![n]\!]$ j is not in Π_i. The summits are denoted $s_1 < \dots < s_N$ with N the number of summits.*

An example of a *prec*-consistent instance is given in Fig. 3. Jobs 1,2 and 3 surround no other job, so they are the summits of this instance. We conclude this section with the following properties:

Lemma 1. *(i) Every job either is a summit or surrounds a summit.*
(ii) If $s_i < s_k < j$ and $j \in \Pi_{s_i}$ then $j \in \Pi_{s_k}$.

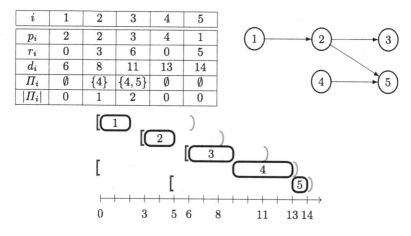

i	1	2	3	4	5
p_i	2	2	3	4	1
r_i	0	3	6	0	5
d_i	6	8	11	13	14
Π_i	$\emptyset$	$\{4\}$	$\{4,5\}$	$\emptyset$	$\emptyset$
$|\Pi_i|$	0	1	2	0	0

Fig. 3. An example with five jobs and q-proper level two.

Proof. (*i*) Let j be a job which is not a summit. Then j must surround at least one job j_1. Then either j_1 is a summit or it surrounds some job j_2. After k non-summit steps we would have $r_j < r_{j_1} < \ldots < r_{j_k}$ and $d_{j_k} < \cdots < d_{j_1} < d_j$. Thus all these jobs are distinct from each other and it eventually ends with some job j_K which is a summit. Then we also have $r_j < r_{j_K}$ and $d_{j_K} < d_j$, so j surrounds summit j_K.

(*ii*) Let $s_i < s_k < j$ such that $j \in \Pi_{s_i}$. s_k is a summit so $s_k \notin \Pi_{s_i}$. And since $s_i < s_k$ we necessarily have $r_{s_i} \le r_{s_k}$. So $r_j < r_{s_i} \le r_{s_k}$. Now $s_k < j$ implies that $d_{s_k} \le d_j$. If $d_{s_k} = d_j$ then we would have $j < s_k$ which leads to a contradiction. So $d_{s_k} < d_j$ and thus $j \in \Pi_{s_k}$. $\square$

3   The (wED) Rule

In this section we define the weak earliest deadline (wED) rule and establish the dominance of schedules following this rule. Then we show that such optimal schedules can be further modified to get optimal schedules with a stronger structure based on the summits of the instance.

3.1   Definition and Dominance

First we define the (wED) rule formally:

Definition 4. *A feasible schedule σ on an instance I follows the (wED) rule when:* $\forall (i,j) \in [\![n]\!]^2$, *if $i < j$ then either i is scheduled before j (i.e. $\sigma(i) < \sigma(j)$) or j surrounds i (i.e. $j \in \Pi_i$) or there exists $h < i$ such that $\sigma(j) < \sigma(h) < \sigma(i)$.*

On problem $1|prec, r_j, d_j|C_{max}$ we show that the schedules following the (wED) rule are dominant.

Lemma 2. *Given a feasible schedule σ on a prec-consistent instance I of problem $1|prec, r_j, d_j|C_{max}$, one can build a feasible schedule with a makespan lower than or equal to σ and which follows the (wED) rule.*

Proof. Let σ be a feasible schedule. If σ follows the (wED) rule then we are done. If not then let (i,j) be a job couple such that $i < j$, $j \notin \Pi_i$ and all jobs between j and i in σ are higher than i. Then in σ we have "jSi" where j and all jobs in sequence S are higher than i. Within the same time interval we propose job reordering "ijS" the following way: push sequence "jS" to the right by p_i time units then insert i right before. Indeed $i < j$ and $j \notin \Pi_i$ so necessarily we have $r_i \leq r_j$. Plus j and all jobs in S are greater than i. Since our instance is *prec*-consistent it means that there is no precedence relation from any of these jobs to i. Thus i can be inserted as desired. Now again j and all jobs in sequence S are greater than i. So their deadlines are non lower than d_i and they can be pushed as desired. Note that job couple (i,j) and all couples (i,s) with s in sequence S are (wED)-valid, now that i is scheduled before them. Thus the resulting schedule is feasible, has the same makespan as σ, and has at least one less (wED)-invalidating job couple with (i,j) becoming (wED)-valid and no job couple becoming (wED)-invalidating. This procedure can be repeated a maximum of $\frac{n(n-1)}{2}$ times until we get a feasible schedule which follows the (wED) rule.   □

3.2   A Summit-Based Decomposition

Upon closer inspection the (wED)-following schedules have these properties, which will serve as the basis of our dynamic programming approach in the next section:

Lemma 3. *Any schedule on a prec-consistent instance I of $1|prec, r_j, d_j|C_{max}$ which is feasible and follows the (wED) rule is of the form $T_1 s_1 \ldots T_N s_N T_{N+1}$ where:*

(i) $1 = s_1 < \ldots < s_N$ *are the summits of the instance,*
(ii) for all k in $[1, N]$ and $i < s_k$ we have $\sigma(i) < \sigma(s_k)$,
(iii) for all k in $[1, N]$ if $j > s_k$ and $\sigma(j) < \sigma(s_k)$ then $j \in \Pi_{s_k}$,
(iv)   a. every job in sequence T_1 is in set Π_{s_1},
       b. for all k in $[2, N]$ every job in sequence T_k is in set $\Pi_{s_{k-1}} \cup \Pi_{s_k}$, and
       c. every job in sequence T_{N+1} is in set Π_{s_N}.

Proof. Let σ be a feasible schedule which follows the (wED) rule. We start by proving point *(ii)*. Let $k \in [1, N]$ and $i < s_k$. We show that $\sigma(i) < \sigma(s_k)$. By the (wED) rule either $\sigma(i) < \sigma(s_k)$, or $s_k \in \Pi_i$, or there is some $h < i$ such that $\sigma(s_k) < \sigma(h) < \sigma(i)$. s_k is a summit so by its definition the second case is not possible. Now by contradiction suppose that the third case is true. Take h as the lowest job such that $\sigma(s_k) < \sigma(h) < \sigma(i)$. Then couple (h, s_k) must follow the (wED) rule while $\sigma(s_k) < \sigma(h)$ and there is no $h' < h$ such that $\sigma(s_k) < \sigma(h') < \sigma(h)$ by minimality of h. Then it means that s_k is in Π_h, which contradicts that s_k is a summit. Thus by the (wED) rule this only leaves us with $\sigma(i) < \sigma(s_k)$.

Next we prove point (iii). Let $k \in [1, N]$ and $j > s_k$ such that $\sigma(j) < \sigma(s_k)$. Then by the (wED) rule either $j \in \Pi_{s_k}$ or there is $h < s_k$ such that $\sigma(j) < \sigma(h) < \sigma(s_k)$. In the first case we are done, so suppose we are in the second phase. Take h minimum. Then by the (wED) rule $j \in \Pi_h$. But according to Lemma 1 (i) either h is a summit or it surrounds a summit. In the first case Lemma 1 (ii) would show that $j \in \Pi_{s_k}$ and we are done. In the second case we have a summit $s_i < h$ such that $r_j < r_h < r_{s_i}$ and $d_{s_i} < d_h < d_j$. Then $j \in \Pi_{s_i}$ and again Lemma 1 (ii) can be used to show that $j \in \Pi_{s_k}$.

Now consider two consecutive summits s_k and s_{k+1}. s_k is lower than s_{k+1}. So by point (ii) we know that s_k is scheduled before s_{k+1} in σ. This means that σ is of the form $T_1 s_1 \ldots T_N s_N T_{N+1}$ where all the jobs in sequences T_k are non-summits. Finally note job 1 has the lowest deadline and the lowest release date among the jobs with the same deadline. So job 1 is always a summit and $s_1 = 1$. This proves point (i).

Finally we prove point (iv). Let $k \in [1, N+1]$ and j be a job in sequence T_k.

a. If $k = 1$ then $s_1 = 1$, so $j > s_1$ and $\sigma(j) < \sigma(s_1)$. Thus by point (iii) $j \in \Pi_{s_1}$.
c. If $k = N + 1$: job j is not a summit so by Lemma 1 (i) it surrounds one summit s_ℓ. But necessarily $s_\ell \leq s_N$. If $\ell = N$ then we are done. Else we know that $\sigma(s_N) < \sigma(j)$ so by point (ii) $j > s_N$. Then $s_\ell < s_N < j$ with j surrounding s_ℓ. By Lemma 1 (ii) j also surrounds s_N.
b. If $k \in [2, N]$: again job j is not a summit so by Lemma 1 (i) it surrounds one summit s_ℓ. If $\ell \leq k-1$ then the same reasoning as in point c. can be applied to show that $j \in \Pi_{s_{k-1}}$. Now suppose $\ell \geq k$. Then we have $s_k \leq s_\ell < j$ and $\sigma(j) < \sigma(s_k)$. Thus by point (iii) $j \in \Pi_{s_k}$.   □

Then in any (wED)-following schedule every job lower than a summit s_k must be scheduled before s_k. And among the jobs higher than s_k only those in Π_{s_k} can be scheduled before s_k. With the next section methodology it would already lead to an FPT dynamic programming algorithm with $\mathcal{O}((2q)! \cdot 4^q \cdot N)$ states. However this number can be decreased to $\mathcal{O}(2^q \cdot N)$ by restricting further the set of dominant schedules. We do so by fixing the job order in each T_k. As a result we get the dominance proposed in [9] for problem $1|r_j, d_j|C_{max}$, except we must deal with precedence constraints and restrict ourselves to *prec*-consistent instances. Such schedules will be called *summit-ordered*.

Definition 5. *A schedule on a prec-consistent instance I of $1|prec, r_j, d_j|C_{max}$ is called summit-ordered when it is feasible and of the form $T_1 s_1 \ldots T_N s_N T_{N+1}$ where properties $(i), (ii), (iii)$ of Lemma 3 are satisfied and moreover the following property holds:*

(iv) a. *all jobs in sequence T_1 are in Π_{s_1} and scheduled in nondecreasing order of their release date.*
    b. *for all k in $[1, N-1]$ $T_{k+1} = C_k A_{k+1} B_{k+1}$ where:*
       * *all jobs in sequence C_k are in Π_{s_k}, not in $\Pi_{s_{k+1}}$ and scheduled in nondecreasing order of their deadline,*

> $*$ all jobs in sequence A_{k+1} are in $\Pi_{s_k} \cap \Pi_{s_{k+1}}$ and scheduled in any order (say in their lexicographic order),
> $*$ all jobs in sequence B_{k+1} are in $\Pi_{s_{k+1}}$, not in Π_{s_k} and scheduled in nondecreasing order of their release date.
> c. all jobs in sequence T_{N+1} are in Π_{s_N} and scheduled in nondecreasing order of their deadline.

For example consider the instance given in Fig. 3. A feasible schedule with job order "12453" (if it exists) would be *summit-ordered* with $C_2 = \emptyset$, $A_3 = \{4\}$ and $B_3 = \{5\}$, while one with job order "12543" would not be. We conclude this section with one final dominance result:

Proposition 1. On problem $1|prec, r_j, d_j|C_{max}$ the summit-ordered schedules are dominant.

Proof. Let σ be a feasible schedule. From Lemmas 2 and 3 one can build $\bar{\sigma}$ of the form $T_1 s_1 \ldots T_N s_N T_{N+1}$ with a makespan lower or equal to σ which follows the (wED) rule and verify points (i), (ii) and (iii). In order to prove point (iv) we propose a reordering of each sequence T_k with $k \in [1, N+1]$.

Let $k \in [1, N-1]$. Consider sequence "$s_k T_{k+1} s_{k+1}$". First we want to reorder each sequence T_{k+1} into a form $C_k A_{k+1} B_{k+1}$. Given the jobs in T_{k+1}, C_k groups those in Π_{s_k} and not in $\Pi_{s_{k+1}}$, A_{k+1} groups those common to both proper sets, and B_{k+1} groups those in $\Pi_{s_{k+1}}$ and not in Π_{s_k}. If T_{k+1} is not of this form then we first check the jobs from C_k. If there is a job from C_k scheduled after a job of $A_{k+1} \cup B_{k+1}$, let c be the leftmost of them. Then we have "$s_k CSc$" in schedule $\bar{\sigma}$ where the jobs in C are from C_k and the jobs in S are from $A_{k+1} \cup B_{k+1}$. We propose to reorder into "$s_k CcS$" the following way: push sequence S to the right by p_c units then insert c right before S. Indeed all jobs from S are in $\Pi_{s_{k+1}}$ so their deadlines are higher than $\sigma(s_{k+1})$ and they could be pushed to the right as desired. Plus c is in Π_{s_k} so its release date is lower than $\sigma(s_k)$ and it could be inserted as desired. So the only obstacle would be a precedence relation between some job α in S to job c. On the one hand α is in $\Pi_{s_{k+1}}$ so it is greater than s_{k+1}. On the other hand c is not in $\Pi_{s_{k+1}}$ and is scheduled before s_{k+1} in $\bar{\sigma}$. By Lemma 3 (iii) this means that c is lower than s_{k+1} and thus $c < \alpha$. Recall that our instance is *prec*-consistent, so it means that there cannot be a precedence relation from α to c. Thus the proposed reordering "$s_k CcS$" is allowed and does not increase the makespan of our schedule. Repeating this at most $|C_k|$ times allows us to get all jobs from C_k on the left side of interval $[\sigma(s_k) + p_{s_k}, \sigma(s_{k+1}))$. Then one can show that all jobs from B_{k+1} can be grouped on the right side of interval $[\sigma(s_k) + p_{s_k}, \sigma(s_{k+1}))$ with symmetric arguments.

Now let $k \in [1, N]$. We show that the jobs in C_k can be reordered in nondecreasing order of their deadline. Notice that jobs in $C_k \subset \Pi_{s_k}$ have release time lower than r_{s_k}. So they can be scheduled without idle time after s_k. If two consecutive jobs i, j of C_k satisfy $d_i > d_j$, we cannot have $i \to j$ as our instance is prec-consistent. So i, j can be swapped without modifying the feasibility nor the makespan of the schedule. Iterating such swaps leads to the desired order. Similarly one can show that all jobs from B_k can be reordered in nondecreasing order of their release dates with symmetric arguments. $\square$

4 The Algorithm

In this section we propose a dynamic programming algorithm which explores active *summit-ordered* schedules - "active" meaning that there is no unnecessary waiting time.

4.1 Core Concept

Assume that we have built a partial *summit-ordered* schedule until s_k. To extend this schedule to other jobs, we need to keep track of the information which ensures that each job is scheduled exactly once. By the definition of *summit-ordered* schedules we only need to recall the jobs greater than each summit s_k and scheduled before them. To this purpose we define the following set:

Definition 6. *(Lingering set Λ_k)* We define $\Lambda_k = (\bigcup_{1 \leq h \leq k} \Pi_{s_h}) \cap [s_k + 1, n]$.

Then a *summit-ordered* schedule $T_1 s_1 \ldots T_N s_N T_{N+1}$ can be represented as a state path $(s_1, L_1) \to (\ldots) \to (s_N, L_N)$ where every L_k is a subset of Λ_k. Such a path decomposition is motivated by the following lemma:

Lemma 4. *Let $\sigma = T_1 s_1 \ldots T_N s_N T_{N+1}$ be an active summit-ordered schedule on an instance I of $1|prec, r_j, d_j|C_{max}$. Then for every k in $[1, N]$ schedule $T_1 s_1 \ldots T_k s_k$ is a feasible active schedule on job subset $[1, s_k] \cup L_k$ where $L_k = (\bigcup_{1 \leq h \leq k} T_h) \cap [s_k + 1, n]$.*

Proof. Feasibility and activeness are stable properties by prefix operation. What is left to show is that the set of jobs featured in partial schedule $T_1 s_1 \ldots T_k s_k$ is indeed $[1, s_k] \cup L_k$.

($\subseteq$) Since we have a *summit-ordered* schedule, by Definition 5 (i) $s_1, \ldots, s_k$ are all lower or equal to s_k, so they are all included in $[1, s_k]$. Now given $\ell \in [1, k]$ and a job $j \in T_\ell$: if $j \leq s_k$ then $j \in [1, s_k]$, else we have $j \in [s_k + 1, n]$ and $j \in \bigcup_{1 \leq h \leq k} T_h$, so $j \in L_k$.

($\supseteq$) By definition L_k is included in $\bigcup_{1 \leq h \leq k} T_h$. Now let $j \in [1, s_k]$. Since we have a *summit-ordered* schedule, by Definition 5 (ii) all jobs lower than s_k are scheduled before s_k in σ. Thus either $j = s_\ell$ or $j \in T_\ell$ for some ℓ in $[1, k]$. □

Then the corresponding schedule can be retrieved solely from the sets L_k:

Lemma 5. *Let $\sigma = T_1 s_1 \ldots T_N s_N T_{N+1}$ be an active summit-ordered schedule on an instance I of $1|prec, r_j, d_j|C_{max}$. Let $(i_1, L_1) \to (\ldots) \to (i_N, L_N)$ be the corresponding state path. Then for every k in $[1, N]$:*

a. $B_1 = L_1$,
b. *for every k in $[2, N-1]$:*
   - $C_k = \{j \in \Pi_{s_k}, j \notin L_k \cup \Pi_{s_{k+1}}\}$,
   - $A_{k+1} = \{j \in \Pi_{s_k} \cap \Pi_{s_{k+1}} \cap L_{k+1}, j \notin L_k\}$,
   - $B_{k+1} = \{j \in \Pi_{s_{k+1}} \cap L_{k+1}, j \notin \Pi_{s_k} \cup L_k\}$,
c. $C_N = \{j \in \Pi_{s_N}, j \notin L_N\}$.

214   M. Mallem et al.

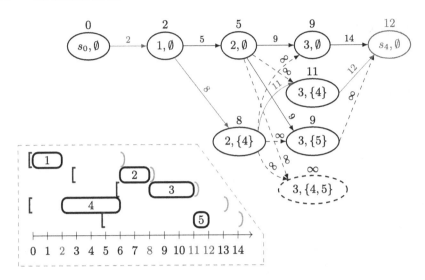

Fig. 4. State graph associated to the example given in Fig. 3. The value of each arc is the makespan of the scheduling candidate computed by extend(k, L_k, L_{k+1}). The red path corresponds to active schedule "14235", which gives the optimal makespan.

Proof. This directly comes from Definition 5 (*iv*) and Lemma 4.   □

Finally in order to find the optimal makespan of our instance, we show that we only need to consider *summit-ordered* schedules which are optimal on every prefix $T_1 s_1 \ldots T_k s_k$:

Lemma 6. *Let $\sigma = T_1 s_1 \ldots T_N s_N T_{N+1}$ be an active summit-ordered schedule with optimal makespan on an instance I of $1|prec, r_j, d_j|C_{max}$. Then one can build an active summit-ordered schedule $\sigma' = T_1' s_1 \ldots T_N' s_N T_{N+1}'$ also with optimal makespan and such that for all $k \in [1, N]$ schedule $T_1' s_1 \ldots T_k' s_k$ has optimal makespan among the schedules of state $(s_k, \bigcup_{1 \le h \le k} T_h' \cap [s_k + 1, n])$.*

Proof. This is proved by downward induction on k. Suppose that for all $j > k$ schedule $T_1 s_1 \ldots T_j s_j$ is optimal among the schedules associated to the same state - which is true in base case $k = N$. Let τ be a schedule associated to the same state as schedule $T_1 s_1 \ldots T_k s_k$ and with optimal makespan. Then by Proposition 1 one can build $\tau' = T_1' s_1 \ldots T_k' s_k$ with the corresponding properties and the same makespan as τ. We propose schedule $\sigma' = \tau' \cdot T_{k+1} s_{k+1} \ldots T_N s_N T_{N+1}$, for which the result holds for all $j \ge k$. This concludes the induction.   □

This motivates the procedure given in Algorithm 1. Set dummy summits s_0, s_{N+1} with processing time 0, an initial state $(s_0, \emptyset)$ and a final state $(s_{N+1}, \emptyset)$. For $k = 0$ to N use the optimal makespan from states (s_k, L_k) and extend them to the states (s_{k+1}, L_{k+1}) accordingly to Lemma 5. For each state pair either the resulting schedule is invalid or it is an optimal makespan candidate for the successor. At the end of the procedure the optimal makespan is given by the value of state $(s_{N+1}, \emptyset)$.

Algorithm 1. main()

1: preprocessing() ▷ Check *prec*-consistency. Compute proper sets and lingering sets.
2: $r_{s_0}, p_{s_0}, d_{s_0}, p_{s_{N+1}} \leftarrow 0$; $r_{s_{N+1}} \leftarrow max_{i \in [\![n]\!]}(r_i)$; $d_{s_{N+1}} \leftarrow \infty$
3: $\Pi_{s_0}, \Lambda_0, \Pi_{s_{N+1}}, \Lambda_{N+1} \leftarrow \emptyset$
4: Create table T with $N+2$ columns and $2^{|\Lambda_k|}$ slots in each column k.
5: Initialize T with ∞ everywhere.
6: $T[0, \emptyset] \leftarrow 0$
7: **for** $k = 0$ to N **do**
8:     **for each** $(L_k, L_{k+1}) \subseteq \Lambda_k \times \Lambda_{k+1}$ **do**
9:         $T[k+1, L_{k+1}] \leftarrow \min(T[k+1, L_{k+1}], \text{extend}(k, L_k, L_{k+1}))$
10: **return** $T[N+1, \emptyset]$

Algorithm 2. extend(k, L_k, L_{k+1})

▷ Input: $k \in [0, N], L_k \subseteq \Lambda_k, L_{k+1} \subseteq \Lambda_{k+1}$.
▷ Goal: start from an optimal schedule on job subset $[1, s_k] \cup L_k$ and attempt to extend it to job subset $[1, s_{k+1}] \cup L_{k+1}$.
1: $Cmax \leftarrow T[k, L_k]$
2: **if** $[1, s_k] \cup L_k \not\subseteq [1, s_{k+1}] \cup L_{k+1}$ or $Cmax = \infty$ **then return** ∞
    ▷ Add jobs between summits s_k and s_{k+1} accordingly to Lemma 5.
3: $C \leftarrow list(\{j \in \Pi_{s_k}, j \notin L_k \cup \Pi_{s_{k+1}}\})$ ▷ C_k
4: sort(C, nondecreasing d_j)
5: $A \leftarrow list(\{j \in \Pi_{s_k} \cap \Pi_{s_{k+1}} \cap L_{k+1}, j \notin L_k\})$ ▷ A_{k+1}
6: $B \leftarrow list(\{j \in \Pi_{s_{k+1}} \cap L_{k+1}, j \notin \Pi_{s_k} \cup L_k\})$ ▷ B_{k+1}
7: sort(B, nondecreasing r_j)
8: $add \leftarrow C \cdot A \cdot B \cdot [s_{k+1}]$
9: **if** prec_invalid(L_k, add) **then return** ∞
10: **for** j in add (in order) **do**
11:     $Cmax \leftarrow \max(Cmax, r_j) + p_j$
12:     **if** $Cmax > d_j$ **then return** ∞
13: **return** $Cmax$

The state graph corresponding to the Fig. 3 example is given in Fig. 4. We have $\Lambda_1 = \emptyset, \Lambda_2 = \{4\}$ and $\Lambda_3 = \{4,5\}$. The optimal makespan is obtained with *summit-ordered* active schedule "14235", which corresponds to state path $(s_0, \emptyset) \rightarrow (1, \emptyset) \rightarrow (2, \{4\}) \rightarrow (3, \{4\}) \rightarrow (s_4, \emptyset)$.

4.2 State Bounding

In this subsection we bound the size of lingering sets Λ_k. It turns out that only the proper set of summit s_k is needed to obtain Λ_k:

Lemma 7. $\forall k \in [1, N], \Lambda_k = \Pi_{s_k}$.

Proof. Let $k \in [\![n]\!]$. By definition every job j in Π_{s_k} is greater than s_k. So j is in Λ_k. Now let $\ell \in \Lambda_k$. We show that $\ell \in \Pi_{s_k}$. By the definition of Λ_k there is $j \in [1, k]$ such that $\ell \in \Pi_{s_j}$. If $j = k$ then we are done. Else we have $s_j < s_k < \ell$ with $\ell \in \Pi_{s_j}$. Then by Lemma 1 (*ii*) $\ell \in \Pi_{s_k}$. □

This bounds the size of Λ_k and the number of successors (s_{k+1}, L_{k+1}) for each state by parameter q:

Corollary 1. $\forall k \in [\![n]\!]$, $|\Lambda_k| \leq q$. Plus every state has at most 2^q successors.

4.3 Complexity

Now that the number of states and successors have been bounded we compute the time and space complexity of the proposed dynamic programming algorithm. In the preprocessing phase *prec*-consistency is checked then proper sets and lingering sets are computed. This takes time $\mathcal{O}(n^2)$ and space $\mathcal{O}(q \cdot N)$. Now by Corollary 1 each column has at most 2^q slots and for each one at most 2^q candidate successors are explored. For each couple $((s_k, L_k), (s_{k+1}, L_{k+1}))$ at most $2q + 1$ jobs are added to the schedule. By Lemma 5 they can be retrieved from sets L_k and L_{k+1} and appended accordingly to Definition 5 in time $\mathcal{O}(q \cdot log(q))$.

Now only the precedence checks related to the added jobs in $T_{k+1} \cup \{s_{k+1}\}$ are left. Since we build a *summit-ordered* schedule, following Definition 5 *(iv)*, there is no invalidating precedence constraint from one job in $T_{k+1} \cup \{s_{k+1}\}$ to another. So an invalidating one would necessarily go from a job i in $T_{k+1} \cup \{s_{k+1}\}$ to a job j in $\bigcup_{1 \leq h \leq k} T_h \cup \{s_h\}$. Since our instance is *prec*-consistent j must be greater than i, which is itself greater than or equal to s_{k+1} and thus greater than s_k. So $j \in L_k$. By Corollary 1 this leaves at most $2q + 1$ possible jobs at the start of the potentially invalidating precedence constraint and at most q possible jobs at the end of it. So the precedence checks take time $\mathcal{O}(q^2)$ for any couple $((s_k, L_k), (s_{k+1}, L_{k+1}))$.

Then either some deadline/precedence constraint is invalidated or the jobs are successfully added and a new candidate makespan value is proposed to the successor. So the main loop of the algorithm takes time $\mathcal{O}(q^2 \cdot 4^q \cdot N)$. Now we consider space complexity. For each state (i, L) we must remember index i, some encoding of set L and the minimum makespan currently found. By Corollary 1 this requires space $\mathcal{O}(q + log(D))$ where $D = max_{i \in [\![n]\!]}(d_i)$. Finally once the whole state graph has been computed, a feasible schedule with optimal makespan can be retrieved by starting from the final state and going backwards in the state graph.

Theorem 1. $1|prec, r_j, d_j|C_{max}$ can be solved in time $\mathcal{O}(max(1, q^2 \cdot 4^q) \cdot N + n^2)$ and space $\mathcal{O}((q + log(D)) \cdot 2^q \cdot N)$.

Note that precedence relations only intervene during the precedence checks between each couple $((s_k, L_k), (s_{k+1}, L_{k+1}))$. So if our instance has no precedence relations then each of these couples take time $\mathcal{O}(q \cdot log(q))$ to process instead of time $\mathcal{O}(q^2)$.

Corollary 2. $1|r_j, d_j|C_{max}$ can be solved in time $\mathcal{O}(max(1, q \cdot log(q) \cdot 4^q) \cdot N + n^2)$ and space $\mathcal{O}((q + log(D)) \cdot 2^q \cdot N)$.

Finally note that if only the optimal value is wanted and not a schedule, it is possible to reduce the space complexity to $\mathcal{O}((q + log(D)) \cdot 2^q)$. Indeed only the optimal makespan of the states associated to summit s_h must be remembered at any time in order to find the optimal makespan of all states associated to summit s_{h+1}. And when all the states associated to s_h have interacted with their successors, the space they take can be freed and used for the states associated to s_{h+2}. Finally instead of a preprocessing phase proper sets and lingering sets are only computed whenever needed and forgotten once all the states associated to the corresponding job have interacted with all their successors.

4.4   Minimizing L_{max}

In this subsection we consider instances of problem $1|prec, r_j|L_{max}$. Deadlines are replaced with due dates and we aim to minimize the maximum lateness of the instance. We run the dynamic programming algorithm for the makespan in order to get an initial value for L_{max}. This could be as far as $D' = [max_{i \in [\![n]\!]}(r_i) + \sum_{i \in [\![n]\!]} p_i]$ from the optimal value. Then we perform a binary search by solving a series of decision problems. We use deadlines instead of due dates and update them accordingly to the current step of the binary search. Each of these steps is solved by one run of our makespan algorithm.

Note that the value of parameter q is the same in every step of the search. Indeed when changing the L_{max} threshold all deadlines are shifted by the same amount of the time, so no relation of the form "$d_i < d_j$" is ever changed. This also means that $prec$-consistency, proper sets and lingering sets remain unchanged and thus do not need to be computed again at every step.

Corollary 3. *Any instance I of problem $1|prec, r_j|L_{max}$ can be solved in time $\mathcal{O}(max(1, q^2 \cdot 4^q) \cdot N \cdot log(D') + n^2)$ and space $\mathcal{O}((q + log(D')) \cdot 2^q \cdot N)$.*

5   Conclusion

In this paper we showed that problems $1|prec, r_j, d_j|C_{max}$ and $1|prec, r_j|L_{max}$ are FPT parameterized by the q-proper level of the interval graph given by the job time windows. We used a new dominance rule, which we called the weak earliest deadline rule (wED), to reduce the space of candidate schedules. This led to the same summit-based dominance as in [9], which we proved to still work in the presence of precedence constraints. Pairing this with a dynamic programming algorithm on a restricted selection of scheduling prefixes allowed us to solve both problems in FPT time.

Further research would use a similar approach beyond the single machine case, like a generalization to parallel machines. Allowing arbitrary processing times on two machines already leads to para-NP-hardness with pathwidth μ [14], which is a stronger parameter than q. So the parallel machine problem with unit-time processing times, precedence constraints and time windows could be the next step with parameter q.

References

1. Baart, R., de Weerdt, M., He, L.: Single-machine scheduling with release times, deadlines, setup times, and rejection. Eur. J. Oper. Res. **291**(2), 629–639 (2021)
2. Baker, K.R.: The effects of input control in a simple scheduling model. J. Oper. Manag. **4**(2), 99–112 (1984)
3. Bodlaender, H.L., Fellows, M.R.: W[2]-hardness of precedence constrained k-processor scheduling. Oper. Res. Lett. **18**(2), 93–97 (1995)
4. Bodlaender, H.L., van der Wegen, M.: Parameterized complexity of scheduling chains of jobs with delays. In: Cao, Y., Pilipczuk, M. (eds.) 15th International Symposium on Parameterized and Exact Computation (IPEC), December 14-18 2020, Hong Kong, China (Virtual Conference), LIPIcs, vol. 180, pp. 1–15 (2020)
5. Bredereck, R., Bulteau, L., Komusiewicz, C., Talmon, N., van Bevern, R., Woeginger, G.J.: Precedence-constrained scheduling problems parameterized by partial order width. In: Kochetov, Y., Khachay, M., Beresnev, V., Nurminski, E., Pardalos, P. (eds.) Discrete Optimization and Operations Research. Lecture Notes in Computer Science(), vol. 9869, pp. 105–120. Springer, Cham (2016). https://doi.org/10.1007/978-3-319-44914-2_9
6. Downey, R., Fellows, M.: Parameterized Complexity. Springer, Cham (1999)
7. Downey, R.G., Fellows, M.R., Regan, K.W.: Descriptive complexity and the W hierarchy. In: Proof Complexity and Feasible Arithmetics, pp. 119–134 (1996)
8. Erschler, J., Fontan, G., Merce, C.: Un nouveau concept de dominance pour l'ordonnancement de travaux sur une machine. RAIRO-Oper. Res. **19**(1), 15–26 (1985)
9. Erschler, J., Fontan, G., Mercé, C., Roubellat, F.: A new dominance concept in scheduling n jobs on a single machine with ready times and due dates. Oper. Res. **31**(1), 114–127 (1983). https://doi.org/10.1287/OPRE.31.1.114
10. Flum, J., Grohe, M.: Parameterized Complexity Theory. Springer, Cham (1998)
11. Gordon, V.S., Werner, F., Yanushkevich, O.: Single machine preemptive scheduling to minimize the weighted number of late jobs with deadlines and nested release/due date intervals. RAIRO-Oper. Res. **35**(1), 71–83 (2001)
12. Grigoreva, N.: Single machine scheduling with precedence constrains, release and delivery times. In: Wilimowska, Z., Borzemski, L., Swiatek, J. (eds.) Information Systems Architecture and Technology: Proceedings of 40th Anniversary International Conference on Information Systems Architecture and Technology – ISAT 2019. Advances in Intelligent Systems and Computing, vol. 1052, pp. 188–198. Springer, Cham (2020). https://doi.org/10.1007/978-3-030-30443-0_17
13. Hall, L.A., Shmoys, D.B.: Jackson's rule for single-machine scheduling: Making a good heuristic better. Math. Oper. Res. **17**(1), 22–35 (1992)
14. Hanen, C., Munier Kordon, A.: Fixed-parameter tractability of scheduling dependent typed tasks subject to release times and deadlines. J. Sched., 1–15 (2023)
15. Hermelin, D., Karhi, S., Pinedo, M., Shabtay, D.: New algorithms for minimizing the weighted number of tardy jobs on a single machine. Ann. Oper. Res. **298**(1), 271–287 (2021)
16. Kordon, A.M., Tang, N.: A fixed-parameter algorithm for scheduling unit dependent tasks with unit communication delays. In: Sousa, L., Roma, N., Tomas, P. (eds.) Euro-Par 2021: Parallel Processing. Lecture Notes in Computer Science(), vol. 12820, pp. 105–119. Springer, Cham (2021). https://doi.org/10.1007/978-3-030-85665-6_7

17. Lawler, E.L.: Optimal sequencing of a single machine subject to precedence constraints. Manage. Sci. **19**(5), 544–546 (1973)
18. Lenstra, J., Rinnooy Kan, A., Brucker, P.: Complexity of machine scheduling problems. Ann. Discrete Math. **1**, 343–362 (1977)
19. Mallem, M., Hanen, C., Munier-Kordon, A.: Parameterized complexity of a parallel machine scheduling problem. In: 17th International Symposium on Parameterized and Exact Computation (IPEC) (2022)
20. Mnich, M., Wiese, A.: Scheduling and fixed-parameter tractability. Math. Program. **154**(1), 533–562 (2015)
21. Proskurowski, A., Telle, J.A.: Classes of graphs with restricted interval models. Discrete Math. Theor. Comput. Sci. **3** (1999)
22. Sourd, F., Nuijten, W.: Scheduling with tails and deadlines. J. Sched. **4**(2), 105–121 (2001)
23. Ware, E.B.: Job shop simulation on the IBM 704. In: Preprints of Papers Presented at the 14th National Meeting of the Association for Computing Machinery (1959)

Fixed-Parameter Algorithms for Cardinality-Constrained Graph Partitioning Problems on Sparse Graphs

Suguru Yamada and Tesshu Hanaka[(✉)] [iD]

Kyushu University, 744, Motooka, Nishi-ku, Fukuoka 819-0395, Japan
yamada.suguru.896@s.kyushu-u.ac.jp, hanaka@inf.kyushu-u.ac.jp

Abstract. For an undirected and edge-weighted graph $G = (V, E)$ and a vertex subset $S \subseteq V$, we define a function $\varphi_G(S) := (1 - \alpha) \cdot w(S) + \alpha \cdot w(S, V \setminus S)$, where $\alpha \in [0, 1]$ is a real number, $w(S)$ is the sum of weights of edges having two endpoints in S, and $w(S, V \setminus S)$ is the sum of weights of edges having one endpoint in S and the other in $V \setminus S$. Then, given a graph $G = (V, E)$ and a positive integer k, Max (Min) α-Fixed Cardinality Graph Partitioning (Max (Min) α-FCGP) is the problem to find a vertex subset $S \subseteq V$ of size k that maximizes (minimizes) $\varphi_G(S)$. In this paper, we first show that Max α-FCGP with $\alpha \in [1/3, 1]$ and Min α-FCGP with $\alpha \in [0, 1/3]$ can be solved in time $2^{o(kd+k)}(e + ed)^k n^{O(1)}$ where k is the solution size, d is the degeneracy of an input graph, and e is Napier's constant. Then we consider Max (Min) Connected α-FCGP, which additionally requires the connectivity of a solution. For Max (Min) Connected α-FCGP, we give an $(e(\Delta - 1))^{k-1} n^{O(1)}$-time algorithm on general graphs and a $2^{O(\sqrt{k} \log^2 k)} n^{O(1)}$-time randomized algorithm on apex-minor-free graphs. Moreover, for Max α-FCGP with $\alpha \in [1/3, 1]$ and Min α-FCGP with $\alpha \in [0, 1/3]$, we propose an $(1 + d)^k 2^{o(kd) + O(k)} n^{O(1)}$-time algorithm. Finally, we show that they admit FPT-ASs when edge weights are constant.

Keywords: Graph partitioning · Fixed-parameter tractability · Degeneracy · Parameterized approximation

1 Introduction

Graph partitioning is a fundamental graph optimization problem. Several natural problems can be regarded as graph partitioning problems.

In an undirected and edge-weighted graph $G = (V, E)$, define the function $\varphi_G(S) := (1 - \alpha) \cdot w(S) + \alpha \cdot w(S, V(G) \setminus S)$ determined by a vertex subset $S \subseteq V$, where α is a real number satisfying $\alpha \in [0, 1]$, $w(S)$ is the sum of the weights

---

This work is partially supported by JSPS KAKENHI Grant Numbers JP21K17707, JP21H05852, JP22H00513, and JP23H04388.

of the edges whose both endpoints are contained in S, and $w(S, V(G) \setminus S)$ is the sum of the weights of the edges whose only one endpoint is contained in S. Then, given an undirected and edge-weighted graph $G = (V, E)$ and a positive integer k, MAX (MIN) α-FIXED CARDINALITY GRAPH PARTITIONING (MAX (MIN) α-FCGP) is the problem to find a set of vertices S of size k such that $\varphi_G(S)$ is maximized (minimized).

The MAX (MIN) α-FIXED CARDINALITY GRAPH PARTITIONING problem is a generalization of various important graph problems such as DENSEST (SPARSEST) k-SUBGRAPH ($\alpha = 0$), MAX (MIN) k-DEGREE SUM ($\alpha = 1/3$), MAX (MIN) k-VERTEX COVER ($\alpha = 1/2$), and MAX (MIN) $(k, n - k)$-CUT ($\alpha = 1$). Depending on α, the MAX (MIN) α-FCGP problem defines these problems.

1.1   Related Work

Bonnet et al. show that the MAX (MIN) α-FCGP problem can be solved in time $(2k\sqrt{\Delta})^{2k}n^{O(1)}$ where Δ is the maximum degree of an input graph [3]. They further design improved algorithms for MAX α-FCGP satisfying $\alpha \in (1/3, 1]$ and for MIN α-FCGP satisfying $\alpha \in [0, 1/3)$ that runs in time $\Delta^k n^{O(1)}$. These cases when $\alpha \in (1/3, 1]$ in MAX α-FCGP and $\alpha \in [0, 1/3)$ in MIN α-FCGP are called *degrading* [3]. The remaining cases are called *non-degrading*. Shachnai and Zehavi give a $4^{k+o(k)}\Delta^k n^{O(1)}$-time algorithm for MAX (MIN) α-FCGP [19], which improves the result in [3].

Koana et al. show that MAX α-FCGP is W[1]-hard if $\alpha \in [0, 1/3)$ while there exists a $(kd)^{O(k)}n^{O(1)}$-time algorithm if $\alpha \in [1/3, 1]$ when parameterized by the solution size k plus degeneracy d [14]. For MIN α-FCGP on unweighted graphs, they also show a $kd^{O(k)}n^{O(1)}$-time algorithm with respect to $\alpha \in [0, 1/3)$ and a $2^{O(dk)}n^{O(1)}$-time algorithm with respect to $\alpha \in (0, 1]$.

Very recently, Fomin et al. show that MAX (MIN) α-FCGP admits an FPT-AS on unweighted graphs if $\alpha \in (0, 1]$ [8]. Moreover, they show that there is a $2^{O(\sqrt{k})}n^{O(1)}$-time algorithm for the degrading cases of MAX (MIN) α-FCGP on apex-minor-free graphs.

For the structural parameterization of MAX (MIN) α-FCGP, Bonnet et al. show that it is fixed-parameter tractable when parameterized by treewidth for $\alpha \in [0, 1]$ [3].

1.2   Our Contribution

In this paper, we design fixed-parameter algorithms for MAX (MIN) (CONNECTED) α-FCGP on sparse graphs such as bounded degeneracy graphs, bounded degree graphs, and apex-minor-free graphs.

We first give an $(e+ed)^k 2^{o(kd)}n^{O(1)}$-time algorithm for MAX (MIN) α-FCGP where k is the solution size, d is the degeneracy of an input graph, and e is Napier's constant. Our algorithm works for $\alpha \in [1/3, 1]$ in the maximization case and $\alpha \in [0, 1/3]$ in the minimization case, i.e. the *degrading* cases [3]. Our result generalizes the algorithm for MAX k-VERTEX COVER proposed by Panolan and

Yaghoubizade [18]. Note that MAX α-FCGP with $\alpha \in [0, 1/3)$ is W[1]-hard when parameterized by $k + d$ [14]. Since the degeneracy of a planar graph is at most 5, it also implies that there exists a $2^{O(k)}n^{O(1)}$-time algorithm for MAX (MIN) α-FCGP on planar graphs, which improves a $2^{O(k \log k)}n^{O(1)}$-time algorithm proposed in [14]. Table 1 summarizes previous studies and our results. Note that the case of $\alpha = 1/3$ is equivalent to MAX (MIN) k-DEGREE SUM, which is the problem to find a vertex subset of size k that maximizes (minimizes) the sum of weighted degrees of vertices in S. This problem can be solved in polynomial time since the degree of each vertex can be checked in polynomial time and we only have to choose k vertices in descending order of the degrees.

Table 1. Previous studies and our result on MAX (MIN)α-FIXED CARDINALITY GRAPH PARTITIONING. Note that the results in the shaded area hold for edge-weighted graphs.

		α	0	$(0, \frac{1}{3})$	$\frac{1}{3}$	$(\frac{1}{3}, \frac{1}{2})$	$\frac{1}{2}$	$(\frac{1}{2}, 1)$	1
max	k		W[1]-hard [14]				W[1]-hard [4]	W[1]-hard [4]	
	$k + d$		W[1]-hard [14]		P	$(kd)^{O(k)}n^{O(1)}$ [14]	$(e+ed)^k 2^{o(kd)}n^{O(1)}$ [18] $(e+ed)^k 2^{o(kd)}n^{O(1)}$	$(kd)^{O(k)}n^{O(1)}$ [14]	
min	k		W[1]-hard [4]				W[1]-hard [14]		
	$k + d$		$(kd)^{O(k)}n^{O(1)}$ [14] $(e+ed)^k 2^{o(kd)}n^{O(1)}$		P		$2^{O(kd)}n^{O(1)}$ [14]		

Subsequently, we delve into the connected version of MAX (MIN) α-FCGP, considering it a natural direction, given the extensive exploration of various graph problems with connectivity (e.g., CONNECTED DENSEST k-SUBGRAPH [8, 13], CONNECTED VERTEX COVER [7,11], CONNECTED MAX CUT [6,12]). In MAX (MIN) CONNECTED α-FCGP, the goal is to find a set of vertices S of size k such that $\varphi_G(S)$ is maximized (minimized) and $G[S]$ is connected. In this paper, we give an $(e(\Delta - 1))^{k-1}n^{O(1)}$-time algorithm on general graphs and a $2^{O(\sqrt{k}\log^2 k)}n^{O(1)}$-time randomized algorithm on apex-minor-free graphs for every fixed $\alpha \in [0, 1]$. Moreover, for MAX α-FCGP with $\alpha \in [1/3, 1]$ and MIN α-FCGP with $\alpha \in [0, 1/3]$, we propose an $(1 + d)^k 2^{o(kd)+O(k)}n^{O(1)}$-time algorithm. Finally, we show that they admit FPT-ASs when edge weights are constant.

2   Preliminaries

Let $G = (V, E)$ be an undirected and edge-weighted graph, and let $n = |V|$ and $m = |E|$. We sometimes use $V(G)$ or $E(G)$ instead of V or E, respectively, to specify the graph G. The weight of an edge $\{u, v\}$ is denoted by w_{uv}. We suppose that the weight of an edge is a positive integer. Let $w_{\max}$ and $w_{\min}$ denote the maximum edge weight and the minimum edge weight, respectively. The subgraph induced by a vertex subset $S \subseteq V$ is denoted by $G[S]$. For a vertex $v \in V$, $N_G(v)$

denotes the set of neighbors of v. The degree of v is defined as $\deg(v) := |N_G(v)|$. Also, the weighted degree of v is defined by $\deg_w(v) := \sum_{u \in N_G(v)} w_{vu}$. The maximum degree of G is defined by $\Delta := \max_{v \in V} \deg(v)$. Similarly, the weighted maximum degree of G is defined by $\Delta_w := \max_{v \in V} \deg_w(v)$.

Let $[n] = \{1, \ldots, n\}$. Then a bijection $\lambda : V \to [n]$ denotes an ordering of vertices in V. For a vertex $v \in V$, $\lambda(v)$ denotes the order of v in a sequence λ. Let $\mathrm{PN}_\lambda(v) := \{u \in N_G(v) \mid \lambda(v) < \lambda(u)\}$. An ordering λ is called a d-posterior if $|\mathrm{PN}_\lambda(v)| \leq d$ for any vertex $v \in V$. A graph with a d-posterior ordering is called a d-degenerate graph. The degeneracy d of G is defined as the minimum d such that G is a d-degenerate graph. Note that $d \leq \Delta$ holds for any graph G. The degeneracy of a graph can be computed in linear time [17].

The degeneracy d of G is defined as the minimum d such that G is a d-degenerate graph. The degeneracy is one graph parameter of how sparse a graph is, and $m \leq dn$ holds for any d-degenerate graph. Also, $d \leq \Delta$ holds for any graph G. The degeneracy of a planar graph is at most 5 [16]. One can compute the degeneracy of a graph in linear time [17].

2.1   Fixed-Parameter Tractability

Given an instance I of a problem and a parameter k, an algorithm that solves the problem in time $f(k)|I|^{O(1)}$ is called an *fixed-parameter algorithm (FPT algorithm)* when parameterized by k. Problems that admit fixed-parameter algorithms are said to be *fixed-parameter tractable (FPT)* when parameterized by k.

For an instance I of a maximization (minimization) problem and parameter k, we say that the problem admits an *FPT Approximation Scheme (FPT-AS)* parameterized by k if there exists a $(1-\epsilon)$-approximation $((1+\epsilon)$-approximation) algorithm that runs in time $f(k, \epsilon)|I|^{O(1)}$ for any $\epsilon > 0$.

For further notions of parameterized complexity and fixed-parameter tractability, we refer the reader to the standard textbook [5].

2.2 (n, p, q)-Lopsided Universal Set

Definition 1 ((n, p, q)-lopsided universal set [10]). *A family $\mathcal{U}$ of subsets of $[n]$ is called an (n, p, q)-lopsided universal set if for every pair of disjoint subsets $A, B \subseteq [n]$ with $|A| = p$ and $|B| = q$, there exists a set U in $\mathcal{U}$ such that A is a subset of U and B is disjoint from U.*

Theorem 1 ([10]). *Given $n, p, q \geq 1$, there exists an algorithm that constructs an (n, p, q)-lopsided universal set of size $\binom{p+q}{p} \cdot 2^{o(p+q)} \log n$ in time $O(\binom{p+q}{p} \cdot 2^{o(p+q)} n \log n)$.*

2.3   Vertex-Weighted Max (Min) Connected k-Subgraph

Given a vertex-weighted graph $G = (V, E)$, VERTEX-WEIGHTED MAX (MIN) CONNECTED k-SUBGRAPH is the problem to find a set S of vertices such that

$|S| = k$, $G[S]$ is connected, and $\sum_{v \in S} w_v$ is maximized (minimized) where w_v is the weight of a vertex v.[1] VERTEX-WEIGHTED MAX (MIN) CONNECTED k-SUBGRAPH can be solved in time $2^{O(k)} n^{O(1)}$ [1].

3   Fixed-Parameter Algorithms for Max (Min) α-FCGP Parameterized by $k + d$

In this section, we design a fixed-parameter algorithm for the MAX (MIN) α-FIXED CARDINALITY GRAPH PARTITIONING with respect to the solution size k and the degeneracy d of an input graph. Our algorithm is based on the algorithm for MAX k-VERTEX COVER proposed by Panolan and Yaghoubizade [18], but generalizes it. In the following, we design a fixed-parameter algorithm by $k + d$ for MAX α-FIXED CARDINALITY GRAPH PARTITIONING. The algorithm for the minimization version can be obtained with minor modifications.

Theorem 2. *There exists an* $(e + ed)^k 2^{o(kd)} n^{O(1)}$*-time algorithm for* MAX α-FCGP *when* $\alpha \in [1/3, 1]$.

Proof. First, if $n \le k + kd$, the brute-force algorithm works in time $\binom{k+kd}{k} n^{O(1)} = (e + ed)^k n^{O(1)}$. Note that $\binom{n}{r} \le \left(\frac{en}{r}\right)^r$ holds for any r and n such that $1 \le r \le n$ [18]. Thus, we are done.

Otherwise, we construct an (n, p, q)-lopsided universal set $\mathcal{U}$ as $p := k$ and $q := \min\{n - k, kd\}$ by Theorem 1. For each $U \in \mathcal{U}$ of size k or larger, we define the value $\mathrm{val}_U(v)$ of v as follows.

$$\mathrm{val}_U(v) := \alpha \sum_{u \in N_G(v)} w_{uv} + (1 - 3\alpha) \sum_{u \in \mathrm{PN}_\lambda(v) \cap U} w_{uv}$$

Also, for $U \in \mathcal{U}$, we define $\mathrm{sol}(U)$ as the set of k vertices of which the sum of values is maximized, that is,

$$\mathrm{sol}(U) := \arg \max_{S \subseteq U, |S| = k} \sum_{s \in S} \mathrm{val}_U(s).$$

We further define $\mathrm{val}(U) := \sum_{v \in \mathrm{sol}(U)} \mathrm{val}_U(v)$.

Lemma 1. *If* $1/3 \le \alpha \le 1$, *it holds that* $\sum_{a \in A} \mathrm{val}_U(a) \le \varphi_G(A)$ *for any* $U \subseteq V$ *and* $A \subseteq U$.

Proof. We consider how the value of an edge $e = \{u, v\}$ is counted in $\sum_{a \in A} \mathrm{val}_U(a)$. If $u, v \notin A$, the value of e is clearly not counted in $\sum_{a \in A} \mathrm{val}_U(a)$.

Next, we consider the case that $u \notin A$ and $v \in A$. If $u \notin U \cap \mathrm{PN}_\lambda(v)$, then $\alpha \cdot w_{uv}$ is counted in $\sum_{a \in A} \mathrm{val}_U(a)$ by v, while it is never counted by u because

---

[1] VERTEX-WEIGHTED MAX (MIN) CONNECTED k-SUBGRAPH is also called GENERALIZED VERTEX WEIGHTED STEINER TREE in [1].

$u \notin A$. Therefore, $\alpha \cdot w_{uv}$ is counted by $\sum_{a \in A} \mathrm{val}_U(a)$ exactly once. On the other hand, if $u \in U \cap \mathrm{PN}_\lambda(v)$, $\mathrm{val}_U(v)$ counts $\alpha \cdot w_{uv} + (1 - 3\alpha) \cdot w_{uv} = (1 - 2\alpha) \cdot w_{uv}$, and u never contributes to $\sum_{a \in A} \mathrm{val}_U(a)$ because $u \notin A$. Therefore, $(1 - 2\alpha) \cdot w_{uv}$ is counted exactly once in $\sum_{a \in A} \mathrm{val}_U(a)$.

Then we consider the case that both endpoints of e are in A. Without loss of generality, u is on the right side of v in d-posterior ordering λ, i.e. $\lambda(v) < \lambda(u)$. Since $u \in A \subseteq U$, u is in $\mathrm{PN}_\lambda(v) \cap U$. Therefore, $\alpha \cdot w_{uv} + (1 - 3\alpha) \cdot w_{uv} = (1 - 2\alpha) \cdot w_{uv}$ is counted in $\mathrm{val}_U(v)$. On the other hand, since $v \notin \mathrm{PN}_\lambda(u)$, $\alpha \cdot w_{uv}$ is counted in $\mathrm{val}_U(u)$. Therefore, $(1 - 2\alpha) \cdot w_{uv} + \alpha \cdot w_{uv} = (1 - \alpha) \cdot w_{uv}$ is counted in $\sum_{a \in A} \mathrm{val}_U(a)$.

Thus, for an edge $e = \{u, v\}$ where $u, v \in A$, exactly $(1 - \alpha) w_{uv}$ is counted in $\sum_{a \in A} \mathrm{val}_U(a)$, and for an edge $e = \{u, v\}$ where $u \notin A$ and $v \in A$, at most $\max\{\alpha, 1 - 2\alpha\} w_{uv} = \alpha w_{uv}$ is counted. Note that $\alpha \geq 1 - 2\alpha$ if $\alpha \geq 1/3$. Therefore, we have:

$$\sum_{a \in A} \mathrm{val}_U(a) \leq (1 - \alpha) \cdot w(A) + \alpha \cdot w(A, V(G) \setminus A) = \varphi_G(A).$$

$\square$

We consider a subset $T \subseteq V \setminus S^*$ of size kd such that $\bigcup_{s \in S^*} \mathrm{PN}_\lambda(s) \setminus S^* \subseteq T$ for an optimal solution S^*. Note that such T always exists since $|\mathrm{PN}_\lambda(s)| \leq d$ holds for every vertex $s \in S^*$ on a d-degenerate graph with at least $k + kd$ vertices. From the definition of an (n, p, q)-lopsided universal set, there exists a set $\tilde{U} \in \mathcal{U}$ such that $S^* \subseteq \tilde{U}$ and $T \cap \tilde{U} = \emptyset$.

Lemma 2. *For an optimal solution S^* of* MAX (MIN) *α-FCGP and $\tilde{U} \in \mathcal{U}$ defined above, $\varphi_G(S^*) = \sum_{s \in S^*} \mathrm{val}_{\tilde{U}}(s)$ holds.*

Proof. It is sufficient to show that $\alpha \cdot w_{uv}$ is counted in $\sum_{s \in S^*} \mathrm{val}_{\tilde{U}}(s)$ for an edge $\{u, v\}$ such that exactly one endpoint is in S^* and $(1 - \alpha) \cdot w_{uv}$ is counted in $\sum_{s \in S^*} \mathrm{val}_{\tilde{U}}(s)$ for an edge $\{u, v\}$ such that both endpoints of $\{u, v\}$ are in S^*.

Consider an edge $\{u, v\}$ such that only one endpoint v is contained in S^*. Note that, from the definition of $\tilde{U}$, $\mathrm{PN}_\lambda(v) \setminus S^* \subseteq T$ for a vertex $v \in S^*$. Since $T \cap \tilde{U} = \emptyset$, it holds that $u \notin \mathrm{PN}_\lambda(v) \cap \tilde{U}$. Therefore, by the definition of $\mathrm{val}_U(v)$, $\alpha \cdot w_{uv}$ is counted exactly once in $\mathrm{val}_{\tilde{U}}(v)$.

For an edge $\{u, v\}$ whose endpoints are in S^*, as in Lemma 1, $(1 - \alpha) \cdot w_{uv}$ is counted exactly once. Therefore, $\varphi_G(S^*) = \sum_{s \in S^*} \mathrm{val}_{\tilde{U}}(s)$ holds. $\square$

For $U \in \mathcal{U}$ of size at least k, by Lemma 1, we have:

$$\mathrm{val}(U) = \sum_{v \in \mathrm{sol}(U)} \mathrm{val}_U(v) \leq \varphi_G(\mathrm{sol}(U)) \leq \varphi_G(S^*).$$

Note that the last inequality holds because an optimum solution S^* maximizes the value of φ_G. Thus, we have $\mathrm{val}(U) \leq \varphi_G(S^*)$.

On the other hand, for $\tilde{U} \in \mathcal{U}$, $\mathrm{sol}(\tilde{U})$ is the set S of k vertices in $\tilde{U}$ such that $\sum_{v \in S \subseteq \tilde{U}, |S| = k} \mathrm{val}_{\tilde{U}}(v)$ is maximized. Thus, from $S^* \subseteq \tilde{U}$ and Lemma 2, we have:

$$\mathrm{val}(\tilde{U}) = \sum_{v \in \mathrm{sol}(\tilde{U})} \mathrm{val}_{\tilde{U}}(v) \geq \sum_{s \in S^*} \mathrm{val}_{\tilde{U}}(s) = \varphi_G(S^*)$$

Therefore, $\mathrm{val}(\tilde{U}) = \varphi_G(S^*)$ holds for $\tilde{U} \in \mathcal{U}$, which implies that for the most valuable $U^* \in \mathcal{U}$, $\mathrm{val}(U^*) = \varphi_G(S^*)$ holds. Moreover, from $\mathrm{val}(U) \leq \varphi_G(\mathrm{sol}(U)) \leq \varphi_G(S^*)$, $\mathrm{sol}(U^*)$ is an optimal solution and $\varphi_G(\mathrm{sol}(U^*)) = \mathrm{val}(U^*)$. Therefore, we can return $\mathrm{sol}(U^*)$ for the most valuable $U^* \in \mathcal{U}$ as an optimal solution.

Finally, we analyze the running time of the algorithm. By Theorem 1, an (n, p, q)-lopsided universal set can be constructed in time $O(\binom{p+q}{p} \cdot 2^{o(p+q)} n \log n)$. For each $U \in \mathcal{U}$ and $v \in U$, $\mathrm{val}_U(v)$ can be computed in polynomial time. Since $p := k$ and $q := \min\{n - k, kd\}$, the total running time of the algorithm is $\binom{p+q}{p} \cdot 2^{o(p+q)} n^{O(1)} = \binom{k+kd}{k} \cdot 2^{o(kd)} n^{O(1)} = (e + ed)^k 2^{o(kd)} n^{O(1)}$. Note that $\binom{n}{r} \leq \left(\frac{en}{r}\right)^r$ holds for any r and n such that $1 \leq r \leq n$ [18]. $\square$

The same argument can be used for MIN α-FCGP with $\alpha \in [0, 1/3]$. In fact, we can show the following lemma as with Lemma 1.

Lemma 3. *If $0 \leq \alpha \leq 1/3$, it holds that $\sum_{a \in A} \mathrm{val}_U(a) \geq \varphi_G(A)$ for any $U \subseteq V$ and $A \subseteq U$.*

As with Theorem 2, we obtain the following theorem by Lemmas 2 and 3.

Theorem 3. *There exists an $(e + ed)^k 2^{o(kd)} n^{O(1)}$-time algorithm for MIN α-FCGP with $\alpha \in [0, 1/3]$.*

4   FPT Algorithms for Max (Min) Connected α-FCGP

In this section, we address the connected version of MAX (MIN) α-FCGP, called MAX (MIN) CONNECTED α-FCGP. The problem additionally requires the connectivity of a solution of MAX (MIN) α-FCGP.

4.1   FPT Algorithms Parameterized by $k + \Delta$

We give an FPT algorithm parameterized by the solution size k plus the maximum degree Δ of an input graph.

For a connected induced subgraph with k vertices, we can check the value in polynomial time. Since every connected induced subgraph with k vertices can be enumerated in time $(e(\Delta - 1))^{k-1} n^{O(1)}$ [15], we obtain the following theorem.

Theorem 4. *For every fixed $\alpha \in [0, 1]$, MAX (MIN) CONNECTED α-FCGP can be solved in time $(e(\Delta - 1))^{k-1} n^{O(1)}$.*

4.2 FPT Algorithms Parameterized by $k + d$

In this section, we design a fixed-parameter tractable algorithm for the MAX (MIN) CONNECTED α-FCGP with respect to the solution size k and the degeneracy d. This can be shown by using as a subroutine the $2^{O(k)}n^{O(1)}$-time algorithm for the VERTEX-WEIGHTED MAX (MIN) CONNECTED k-SUBGRAPH problem given by Betzler [1].

Theorem 5. *There exists a* $(1 + d)^k 2^{o(kd)+O(k)}n^{O(1)}$*-time algorithm for* MAX CONNECTED α-FCGP *with* $\alpha \in [1/3, 1]$ *and* MIN CONNECTED α-FCGP *with* $\alpha \in [0, 1/3]$.

Proof. We only address the maximization version since the algorithm for the minimization version can be obtained with minor modifications.

Let S^* be an optimal solution of MAX CONNECTED α-FCGP. Then for an (n, p, q)-lopsided universal set $\mathcal{U}$ where $p := k$ and $q := \min\{n - k, kd\}$, there exists $\tilde{U} \in \mathcal{U}$ that satisfies $S^* \subseteq \tilde{U}$ and $(\bigcup_{s \in S^*} \mathrm{PN}_\lambda(s) \setminus S^*) \cap \tilde{U} = \emptyset$.

As with the arguments of Lemmas 1 and 2, we can see that $\sum_{a \in A} \mathrm{val}_U(a) \leq \varphi_G(A)$ for any $A \subseteq V$ and $\varphi_G(S^*) = \sum_{v \in S^*} \mathrm{val}_{\tilde{U}}(v)$ holds for an optimal solution S^*.

For each $U \in \mathcal{U}$, we find a connected vertex set $S \subseteq U$ of size k that maximizes $\sum_{v \in S} \mathrm{val}_U(v)$. Then for $\tilde{U} \in \mathcal{U}$, we have:

$$\varphi_G(S) = \sum_{v \in S} \mathrm{val}_{\tilde{U}}(v) \geq \sum_{v \in S^*} \mathrm{val}_{\tilde{U}}(v) = \varphi_G(S^*).$$

Since S is a connected vertex set of size k, $\varphi_G(S) = \varphi_G(S^*)$ holds.

By Theorem 1, an (n, p, q)-lopsided universal set can be constructed in time $O(\binom{p+q}{p} \cdot 2^{o(p+q)} n \log n)$. For each $U \in \mathcal{U}$, S can be computed in time $2^{O(k)}n^{O(1)}$ by solving VERTEX-WEIGHTED MAX (MIN) CONNECTED k-SUBGRAPH on G such that the weight of a vertex is defined by $\mathrm{val}_U(v)$ [1]. Since $p := k$ and $q := \min\{n - k, kd\}$, the total running time of the algorithm is $2^{O(k)}n^{O(1)} \cdot \binom{p+q}{p} \cdot 2^{o(p+q)} n^{O(1)} = 2^{o(kd)+O(k)}(1 + d)^k n^{O(1)}$.   □

4.3 Subexponential-Time FPT Algorithm on Apex-Minor-Free Graphs

In this section, we design a randomized subexponential-time FPT algorithm for MAX (MIN) CONNECTED α-FCGP on apex-minor-free graphs for every fixed α. Note that the same problem in a restricted setting $\alpha = 0$ is studied in Fomin et al. [8], which mentions that it admits a randomized subexponential-time FPT algorithm parameterized by k on apex-minor-free graphs, though they do not provide a concrete proof. Our result is obtained independently from theirs (see also [20][2]). In the following, we give a formal proof of the randomized subexponential-time FPT algorithm. To show this, we use the following theorem presented by Fomin et al. [9].

---

[2] The result was presented in an unreviewed domestic symposium, and [20] is the preprint for the symposium.

Theorem 6 ([9]). *Given an n-vertex apex-minor-free graph and a positive integer k, we can sample a vertex subset $A \subseteq V$ satisfying the following property:*

- *The treewidth of the subgraph $G[A]$ induced by A is $O(\sqrt{k} \log k)$.*
- *For any vertex subset $X \subseteq V$ of size k or less inducing a connected graph of
  G, the probability that $X \subseteq A$ is at least $(2^{O(\sqrt{k} \log^2 k)})^{-1}$.*

Now, we introduce VERTEX-WEIGHTED MAX (MIN) CONNECTED α-
FCGP[3], which is a generalization of the MAX (MIN) CONNECTED α-FCGP.
Given a vertex- and edge-weighted graph $G = (V, E)$ and a positive integer k,
the problem is to find a connected vertex subset $S \subseteq V$ of size k that maximizes
(minimizes) $\psi_G(S) := (1 - \alpha) \cdot w(S) + \alpha \cdot w(S, V(G) \setminus S) + \sum_{s \in S} w_s$ where w_v
is the weight of a vertex v.

Then we show the following lemma.

Lemma 4. *Let $G = (V, E)$ be an edge-weighted graph and $A \subseteq V$ be a set
of vertices in G. For each vertex in A, define the vertex weights as $w_v =
\sum_{\{u,v\} \in E \wedge u \notin A} \alpha w_{uv}$. Then for any $S \subseteq A$, $\psi_{G[A]}(S) = \varphi_G(S)$ holds.*

Proof. For G, A, S, and w_v, we have:

$$\psi_{G[A]}(S) = (1 - \alpha) \cdot w(S) + \alpha \cdot w(S, A \setminus S) + \sum_{v \in S} w_v$$

$$= (1 - \alpha) \cdot w(S) + \alpha \cdot w(S, A \setminus S) + \sum_{v \in S} \sum_{\{u,v\} \in E \wedge u \notin A} \alpha w_{uv}$$

$$= (1 - \alpha) \cdot w(S) + \alpha \cdot w(S, V \setminus S) = \varphi_G(S).$$

□

For VERTEX-WEIGHTED MAX (MIN) CONNECTED α-FCGP, we can design
a fixed-parameter algorithm parameterized by treewidth $\mathtt{tw}$. This can be shown
by standard dynamic programming over a nice tree decomposition.

Lemma 5. *Given a tree decomposition of width $\mathtt{tw}$, there exists a $\mathtt{tw}^{O(\mathtt{tw})} n^{O(1)}$-
time algorithm for VERTEX-WEIGHTED MAX (MIN) CONNECTED α-FCGP.*

From Theorem 6 and Lemma 5, we can design a randomized subexponential
time fixed-parameter algorithm parameterized by the solution size k for MAX
(MIN) CONNECTED α-FCGP on apex-minor-free graphs.

Theorem 7. *For every fixed $\alpha \in [0,1]$, there exists a $2^{O(\sqrt{k} \log^2 k)} n^{O(1)}$-time
randomized algorithm for MAX (MIN) CONNECTED α-FCGP on apex-minor-
free graphs.*

Proof. Let S^* be an optimal solution of MAX (MIN) CONNECTED α-FCGP.
By Theorem 6, we can obtain a vertex subset A such that $S^* \subseteq A$ and the

---
[3] VERTEX-WEIGHTED MAX (MIN) CONNECTED α-FCGP is also called ANNOTATED
MAX (MIN) CONNECTED α-FCGP [14].

treewidth of $G[A]$ is at most $O(\sqrt{k}\log k)$ in $2^{O(\sqrt{k}\log^2 k)}n^{O(1)}$ with high probability. A tree decomposition of width at most $O(\sqrt{k}\log k)$ can be computed in time $2^{O(\sqrt{k}\log k)}n$ [2].

By Lemma 4, it is sufficient to solve VERTEX-WEIGHTED CONNECTED MAX (MIN) α-FCGP in $G[A]$ with vertex weight $w_v = \sum_{\{u,v\}\in E \wedge u \notin A} \alpha w_{uv}$. From Lemma 5, we can find an optimal solution S^* of VERTEX-WEIGHTED CONNECTED MAX (MIN) α-FCGP in $G[A]$ in time $(\sqrt{k}\log k)^{O(\sqrt{k}\log k)}n^{O(1)} = 2^{O(\sqrt{k}\log^2 k)}n^{O(1)}$. Since $A \subseteq V$ contains an optimal solution with probability at least $(2^{O(\sqrt{k}\log^2 k)})^{-1}$, by iterating $2^{O(\sqrt{k}\log^2 k)}n^{O(1)}$ times, an optimal solution can be found with high probability. Thus, the total running time is $2^{O(\sqrt{k}\log^2 k)}n^{O(1)}$. $\qquad\square$

4.4 FPT-AS for Connected α-FCGP

In this section, we show that MAX CONNECTED α-FCGP with $\alpha \in [1/3,1]$ and MAX CONNECTED α-FCGP with $\alpha \in [1/3,1]$ admit *FPT approximation schemes (FPT-AS)* when parameterized by k. To show this, we give a simple observation.

Observation 1 *For any vertex subset $S \subseteq V$, it holds that:*

$$\varphi(S) = \alpha \sum_{v\in S} \deg_w(v) + (1-3\alpha)w(S).$$

Then we give an FPT-AS for MAX CONNECTED α-FCGP.

Theorem 8. MAX CONNECTED α-FCGP *with $\alpha \in [1/3,1]$ on graphs with constant edge-weights admit an FPT-AS when parameterized by k.*

Proof. Let $\mathcal{T}_k$ be the set of trees with k vertices in G. Then we define $T^* := \arg\max_{T\in\mathcal{T}_k} \sum_{v\in V(T)} \deg_w(v)$. We denote by OPT and ALG the optimal value and the value of a solution the algorithm outputs, respectively.

In the algorithm, we first find a tree T^*. This can be found in time $2^{O(k)}n^{O(1)}$ by applying Betzler's algorithm for VERTEX-WEIGHTED MAX (MIN) CONNECTED k-SUBGRAPH [1] to the vertex-weighted graph G' obtained from G by defining the vertex weight of v as the weighted degree $\deg_w(v)$ of v. Then if T^* satisfies the following inequality, we output it as a solution of MAX CONNECTED α-FCGP.

$$\sum_{v\in V(T^*)} \deg_w(v) \geq \frac{(3-1/\alpha)\binom{k}{2}w_{\max}}{\epsilon}.$$

Otherwise, we have $\Delta \leq \sum_{v\in V(T^*)} \deg_w(v) < \frac{(3-1/\alpha)w_{\max}\binom{k}{2}}{\epsilon}$. Then we apply the algorithm in Theorem 4 that runs in time $\Delta^{O(k)}n^{O(1)}$.

The total running time is dominated by the second case. Thus, it runs in time

$$2^{O(k)}n^{O(1)} + \Delta^{O(k)}n^{O(1)} = \left(\frac{(3-1/\alpha)w_{\max}\binom{k}{2}}{\epsilon}\right)^{O(k)}n^{O(1)} = \left(\frac{(3-1/\alpha)w_{\max}k}{\epsilon}\right)^{O(k)}$$
$$n^{O(1)}.$$

Finally, we analyze the approximation ratio of the algorithm. Since it outputs an exact solution in the second case, the approximation ratio depends on the first case. Let S^* be an optimal solution with $\varphi(S^*) = \text{OPT}$. Using Observation 1 and $1 - 3\alpha \leq 0$, we have:

$$\frac{\text{ALG}}{\text{OPT}} = \frac{\alpha \sum_{v \in V(T^*)} \deg_w(v) + (1 - 3\alpha)w(V(T^*))}{\alpha \sum_{v \in S^*} \deg_w(v) + (1 - 3\alpha)w(V(S^*))}$$

$$\geq \frac{\alpha \sum_{v \in V(T^*)} \deg_w(v) + (1 - 3\alpha)\binom{k}{2}w_{\max}}{\alpha \sum_{v \in S^*} \deg_w(v)}$$

$$\geq 1 - (3 - \frac{1}{\alpha}) \cdot \frac{\binom{k}{2}w_{\max}}{\sum_{v \in V(T^*)} \deg_w(v)} \geq 1 - \epsilon.$$

Note that $(1 - 3\alpha) \leq 0$ because $\alpha \in [1/3, 1]$. Thus, the algorithm achieves the approximation ratio of $1 - \epsilon$. ☐

An FPT-AS for MIN CONNECTED α-FCGP can be obtained in a similar way, but additional observations are requisite.

Theorem 9. MIN CONNECTED α-FCGP *with* $\alpha \in (0, 1/3]$ *on graphs with constant edge-weights admit an FPT-AS when parameterized by* k.

Proof. Let $T^* := \arg\min_{T \in \mathcal{T}_k} \sum_{v \in T} \deg_w(v)$ and $S^* \subseteq V$ be an optimal solution in G. Similarly to the maximization version, we first find T^* in time $2^{O(k)}$ and determine whether $\sum_{v \in V(T^*)} \deg_w(v) \geq \frac{(1/\alpha - 3)\binom{k}{2}w_{\max}}{\epsilon}$. If it is true, we output T^*.

Otherwise, $\sum_{v \in V(T^*)} \deg_w(v) < \frac{(1/\alpha - 3)\binom{k}{2}w_{\max}}{\epsilon}$ holds. Let $\Delta^* = \max_{v \in S^*} \deg(v)$. Then, we have:

$$\text{OPT} = (1 - \alpha)w(S^*) + \alpha w(S^*, V \setminus S^*)$$
$$\geq \alpha w(S^*, V \setminus S^*) \geq \alpha(\Delta^* - k + 1)w_{\min}.$$

On the other hand, from Observation 1 and $1 - 3\alpha \leq 0$,

$$\text{OPT} \leq \varphi(V(T^*)) \leq \alpha \sum_{v \in V(T^*)} \deg_w(v) + (1 - 3\alpha)w(V(T^*))$$

$$< \alpha \frac{(1/\alpha - 3)\binom{k}{2}w_{\max}}{\epsilon} + (1 - 3\alpha)\binom{k}{2}w_{\max}$$

$$= (1 - 3\alpha)\left(1 + \frac{1}{\epsilon}\right)\binom{k}{2}w_{\max}$$

Therefore, we have the following inequality.

$$\alpha(\Delta^* - k + 1)w_{\min} < (1 - 3\alpha)\left(1 + \frac{1}{\epsilon}\right)\binom{k}{2}w_{\max}$$

$$\Delta^* < (1/\alpha - 3)\left(1 + \frac{1}{\epsilon}\right)\binom{k}{2}\frac{w_{\max}}{w_{\min}} + k - 1.$$

This means that any optimal solution contains no vertex of degree more than $\gamma = (1/\alpha - 3)\left(1 + \frac{1}{\epsilon}\right)\binom{k}{2}\frac{w_{\max}}{w_{\min}} + k - 1$. Let D be the set of vertices of degree more than γ. Moreover, let H be a vertex- and edge-weighted graph defined by $G[V \setminus D]$ with the vertex weight $w_v = \sum_{u \in N(v) \cap D} \alpha w_{uv}$ for each $v \in V \setminus D$. Note that $\Delta(H) \leq \gamma$. Since $V(H) = V \setminus D$ contains an optimal solution S^* of instance (G, k) of MIN CONNECTED α-FCGP, by Lemma 4, $\psi_H(S^*) = \varphi_G(S^*)$ holds. Since the algorithm in Theorem 4 can be applied to VERTEX-WEIGHTED MIN CONNECTED α-FCGP, we can obtain an optimal solution S^* in time $\Delta(H)^{O(k)} n^{O(1)} = \gamma^{O(k)} n^{O(1)}$.

In the following, we evaluate the running time of the algorithm. We first find T^* in time $2^{O(k)} n^{O(1)}$. Since the first case immediately outputs a solution, the running time is dominated by the second case. The graph H can be computed in polynomial time, and thus, the total running time is as follows:

$$2^{O(k)} n^{O(1)} + \gamma^{O(k)} n^{O(1)} = \left((1/\alpha - 3)\left(1 + \frac{1}{\epsilon}\right)\binom{k}{2}\frac{w_{\max}}{w_{\min}} + k - 1\right)^{O(k)} n^{O(1)}$$

$$= \left(\frac{k\,(1/\alpha - 3)\,w_{\max}}{\epsilon w_{\min}}\right)^{O(k)} n^{O(1)}.$$

Finally, we analyze the approximation ratio of the algorithm. Since the algorithm outputs an exact solution in the second case, it is sufficient to show the first case. By using Observation 1 and $1 - 3\alpha \geq 0$, we have:

$$\frac{\text{ALG}}{\text{OPT}} = \frac{\alpha \sum_{v \in V(T^*)} \deg_w(v) + (1 - 3\alpha) w(V(T^*))}{\alpha \sum_{v \in S^*} \deg_w(v) + (1 - 3\alpha) w(V(S^*))}$$

$$\leq 1 + \left(\frac{1}{\alpha} - 3\right) \cdot \frac{\binom{k}{2} w_{\max}}{\sum_{v \in V(T^*)} \deg_w(v)} \leq 1 + \epsilon.$$

This completes the proof. $\qquad\qquad\qquad\qquad\qquad\qquad\qquad\qquad\qquad\qquad\qquad\quad\square$

References

1. Betzler, N.: Steiner tree problems in the analysis of biological networks. Master's thesis, Universität Tübingen (2006)
2. Bodlaender, H.L., Drange, P., Dregi, M.S., Fomin, F.V., Lokshtanov, D., Pilipczuk, M.: A c^k n 5-approximation algorithm for treewidth. SIAM J. Comput. **45**(2), 317–378 (2016)
3. Bonnet, E., Escoffier, B., Paschos, V.T., Tourniaire, E.: Multi-parameter analysis for local graph partitioning problems: using greediness for parameterization. Algorithmica **71**(3), 566–580 (2015)
4. Cai, L.: Parameterized complexity of cardinality constrained optimization problems. Comput. J. **51**(1), 102–121 (2008)
5. Cygan, M., et al.: Parameterized Algorithms. Springer, Cham (2015)
6. Duarte, G.L., et al.: Computing the largest bond and the maximum connected cut of a graph. Algorithmica **83**(5), 1421–1458 (2021)

7. Escoffier, B., Gourvès, L., Monnot, J.: Complexity and approximation results for the connected vertex cover problem in graphs and hypergraphs. J. Discrete Algorithms **8**(1), 36–49 (2010)
8. Fomin, F.V., Golovach, P.A., Inamdar, T., Koana, T.: FPT approximation and subexponential algorithms for covering few or many edges. In: Leroux, J., Lombardy, S., Peleg, D. (eds.) 48th International Symposium on Mathematical Foundations of Computer Science, MFCS 2023, August 28 to September 1, 2023, Bordeaux, France, volume 272 of LIPIcs, pp. 1–8. Schloss Dagstuhl - Leibniz-Zentrum für Informatik, (2023)
9. Fomin, F.V., Lokshtanov, D., Marx, D., Pilipczuk, M., Pilipczuk, M., Saurabh, S.: Subexponential parameterized algorithms for planar and apex-minor-free graphs via low treewidth pattern covering. SIAM J. Comput. **51**(6), 1866–1930 (2022)
10. Fomin, F.V., Lokshtanov, D., Panolan, F., Saurabh, S.: Efficient computation of representative families with applications in parameterized and exact algorithms. J. ACM **63**(4), 1–60 (2016)
11. Garey, M.R., Johnson, D.S.: The rectilinear steiner tree problem is NP complete. SIAM J. Appl. Math. **32**, 826–834 (1977)
12. Hajiaghayi, M.T., Kortsarz, G., MacDavid, R., Purohit, M., Sarpatwar, K.: Approximation algorithms for connected maximum cut and related problems. Theor. Comput. Sci. **814**, 74–85 (2020)
13. Keil, J.M., Brecht, T.B.: The complexity of clustering in planar graphs. J. Comb. Math. Comb. Comput. **9**, 155–159 (1991)
14. Koana, T., Komusiewicz, C., Nichterlein, A., Sommer, F.: Covering many (or few) edges with k vertices in sparse graphs. In: 39th International Symposium on Theoretical Aspects of Computer Science, STACS 2022, March 15-18, 2022, Marseille, France (Virtual Conference), volume 219 of LIPIcs, pp. 1– 18. Schloss Dagstuhl - Leibniz-Zentrum für Informatik (2022)
15. Komusiewicz, C., Sorge, M.: An algorithmic framework for fixed-cardinality optimization in sparse graphs applied to dense subgraph problems. Discret. Appl. Math. **193**, 145–161 (2015)
16. Lick, D.R., White, A.T.: k-degenerate graphs. Can. J. Math. **22**(5), 1082–1096 (1970)
17. Matula, D.W., Beck, L.L.: Smallest-last ordering and clustering and graph coloring algorithms. J. ACM **30**(3), 417–427 (1983)
18. Panolan, F., Yaghoubizade, H.: Partial vertex cover on graphs of bounded degeneracy. In: Kulikov, A.S., Raskhodnikova, S. (eds.) Computer Science - Theory and Applications. Lecture Notes in Computer Science, vol. 13296, pp. 289–301. Springer, Cham (2022). https://doi.org/10.1007/978-3-031-09574-0_18
19. Shachnai, H., Zehavi, M.: Parameterized algorithms for graph partitioning problems. Theory Comput. Syst. **61**(3), 721–738 (2017)
20. Yamada, S., Hanaka, T.: A subexponential-time algorithm for connected graph partitioning problems [Translated from Japanese]. In: Summer LA Symposium 2023, July 3 – July 5, Hokkaido, Japan, Proceedings, pp. 14S-1 – 14S-3 (2023). (in Japanese)

Approximation Algorithms

Sequencing Stochastic Jobs with a Single Sample

Puck te Rietmole[1] and Marc Uetz[2]([✉])

[1] Department of Mathematics and Computer Science, TU Eindhoven, Eindhoven, The Netherlands

[2] Mathematics of Operations Research, University of Twente, Enschede, The Netherlands
m.uetz@utwente.nl

Abstract. This paper revisits the single machine scheduling problem to minimize total weighted completion times. The twist is that job sizes are stochastic from unknown distributions, and the scheduler has access to only a *single* sample from the distributions. For this restricted information regime, we analyze the simplest and probably only reasonable scheduling algorithm, namely to schedule by ordering the jobs by weight over sampled processing times. In general, this algorithm can be tricked by adversarial input distributions, performing in expectation arbitrarily worse even in comparison to choosing a random schedule. The paper suggests notions to capture the idea that this algorithm, on reasonable inputs, should exhibit a provably good expected performance.

Keywords: Stochastic scheduling · Approximation · Sampling

1 Introduction, Motivation, and Model

A classical result in the literature on scheduling algorithms is due to Smith [15], showing how to minimize the total weighted completion time $\sum_j w_j C_j$ of n independent, non-preemptive jobs with weights w_j and processing times p_j on a single machine. They should be scheduled in the so-called WSPT order, that is, jobs must be ordered by non-increasing ratios w_j/p_j. The same is still true when the jobs' processing times are not known in advance, but instead governed by independent random variables P_j. Indeed, when minimizing the expected total weighted completion times, $\mathbb{E}[\sum_j w_j C_j] = \sum_j w_j \mathbb{E}[C_j]$, the same exchange argument shows that scheduling the jobs in order of non-increasing ratios $w_j/\mathbb{E}[P_j]$ is optimal in expectation [12]. Likewise, also for more general scheduling models, specifically on more than a single machine, some of the algorithmic framework that has been developed for computing approximately optimal solutions to deterministic scheduling problems, e.g. [1,4], can be generalized to the more general setting with stochastic processing times, e.g. [2,3,5,7,9,14]. However basically all of the previously cited approximation

© The Author(s), under exclusive license to Springer Nature Switzerland AG 2024
A. Basu et al. (Eds.): ISCO 2024, LNCS 14594, pp. 235–247, 2024.
https://doi.org/10.1007/978-3-031-60924-4_18

algorithms for stochastic scheduling problems, and also Rothkopf's result [12] need to assume that the expected processing times $\mathbb{E}[P_j]$ are known exactly.

This paper asks the simple question what can be done if this is *not* the case. We ask how much of the optimality of Smith's result [15] can be recovered if the processing times p_j that the scheduler uses to determine the schedule, is in fact a *sample from an unknown distribution* P_j. In that setting, one should naturally minimize the expected total weighted completion times of jobs $\mathbb{E}[\sum_j w_j C_j]$. We can equivalently see the problem as a *stochastic* single machine scheduling problem, but with the assumption that the scheduler has access to only a *single sample* of the processing time distributions.

This question relates to some recent work on prophet inequalities, where strong results can be recovered with access to a single sample only [10,13]. It can also be interpreted from the perspective of learning augmented algorithms, see [6], but then with a minimalist assumption about the learning of stochastic job sizes, namely through one sample p'_j from P_j only, for all jobs j.

It is maybe no surprise that, without any further assumptions, one can define malicious input distributions that yield "wrong" samples with high probability, rendering an algorithm that follows these samples to have arbitrarily bad performance. Such an example is given below in Sect. 2. Our main result is the identification of three different sufficient conditions on the input distributions that rule out these undesirable effects. This is formalized using a notion of *relative optimality gap*, a scaled variation on the usual notion of multiplicative approximation algorithms.

2   Preliminaries: Single Machine Scheduling by Samples

We consider n jobs with independent processing time distributions P_j and weights w_j, $j = 1, \ldots, n$. The jobs have to be scheduled non-preemptively on a single machine with the goal to minimize $\sum_j w_j C_j$, where C_j denotes the completion time of job j. Let $p := (p_1, \ldots, p_n)$ denote possible realizations of $P := (P_1 \ldots P_n)$. We denote by $I = (w, P)$ an instance. If processing time realizations $p \sim P$ are known, the only optimal solutions are the sequences in so-called WSPT order, ratios w_j/p_j non-increasing [15].

If the processing times are stochastic, the solution is a stochastic scheduling policy that is non-anticipatory in the sense of [8], meaning that it cannot use information about actual realizations $p_j \sim P_j$ before scheduling a job j. If Π denotes such a scheduling policy, let $C_j^\Pi(p)$ be job j's completion time under policy Π for realization p, then the expected cost of policy Π on instance $I = (w, P)$ is

$$\text{cost}_I(\Pi) := \mathbb{E}_{p \sim P}\left[\sum_j w_j C_j^\Pi(p)\right].$$

Also let WSPT(p) be the minimal cost for realization p, achieved by a WSPT order. Our goal is to minimize the *expected regret of* Π, which is minimizing

$$\mathbb{E}_{p \sim P}\left[\sum_j w_j C_j^\Pi(p) - \text{WSPT}(p)\right] = \text{cost}_I(\Pi) - \mathbb{E}_{p \sim P}\left[\text{WSPT}(p)\right].$$

In other words, we effectively seek a policy Π minimizing $\text{cost}_I(\Pi)$. Since the processing times are independent across jobs, it is clear (for the single machine setting) that it suffices to consider so-called static list scheduling policies [11], meaning that jobs are scheduled in the order of a fixed priority list which is determined ex ante, using the given information about processing time distributions P_j. We will therefore simply refer to a scheduling policy as (scheduling) algorithm. If the distributions P_j are known, WSEPT, that is scheduling in order of non-increasing ratios $w_j/\mathbb{E}[P_j]$ is the policy that minimizes the expectation of the total weighted completion times [12], hence also the expected regret.

For convenience let us write $\mathbb{E}[\ ]$ instead of $\mathbb{E}_{p\sim P}[\ ]$, if no ambiguity arises. This paper addresses the information regime where the random variables P_j are unknown, hence also $\mathbb{E}[P_j]$ is unknown. The only information available to the scheduler is one sample $p'_j \sim P_j$, for all jobs $j = 1,\ldots,n$. In this regime, the only reasonable algorithm that exploits the given information is -arguably- to take the sample p'_j as a proxy for $\mathbb{E}[P_j]$.

Definition 1 (Algorithm SAM). *Schedule jobs in non-increasing order weight over sampled processing time w_j/p'_j.*

Our goal is to analyze this algorithm's expected cost, resp. expected regret. The following example gives a malicious input instance showing that, in general, the expected cost as well as expected regret achieved by SAM can be unbounded.

Example 1. Consider instance I with $n - 1$ jobs with deterministic processing time $\varepsilon > 0$ and job n with processing time

$$P_n = \begin{cases} 0 & \text{with probability } 1 - \frac{1}{M}, \\ M^2 & \text{with probability } \frac{1}{M}, \end{cases}$$

where M is large. Let $w_j = 1$ for all jobs.   △

For the following discussion let n be fixed and consider $\varepsilon \to 0$. As $\mathbb{E}[P_n] = M$, the algorithm that minimizes $\mathbb{E}[\sum_j C_j]$ is to schedule all ε-jobs first, with expected cost M and expected regret equal to $(1 - 1/M)(n - 1)\varepsilon \to 0$. As to SAM, observe that with probability $1 - 1/M$ the sample for job n will be $p'_n = 0$, which yields SAM to schedule job n before all ε-jobs. In this case, the expected cost of SAM is nM, while WSPT(p) has expected cost M. With probability only $1/M$, the sample $p'_n = M^2$ and SAM schedules job n last, and just like WSPT(p) achieves expected cost M. Therefore, SAM's expected cost equals

$$\text{cost}_I(\text{SAM}) = \mathbb{E}_{p'\sim P}\mathbb{E}_{p\sim P}\left[\sum_j w_j C_j^{\text{SAM}(p')}(p)\right]$$

$$= \left(1 - \frac{1}{M}\right)nM + \frac{1}{M}M = nM - (n-1) = \Theta(nM),$$

and the expected regret over the offline optimum WSPT(p) is of the same order of magnitude, as one readily verifies that the expected regret of SAM w.r.t. WSPT(p) equals $\left(1 - \frac{1}{M}\right)[nM - M] = (n-1)(M-1) = \Theta(nM)$.

Example 1 also shows more, however. If we make the assumption that the adversary should be required to be non-anticipatory as e.g. in [9], then the omniscient adversary WSPT(p) is ruled out.

However for Example 1, also the weaker, non-anticipatory adversary WSEPT achieves an expected cost that is equal to M, so that the expected regret relative to WSEPT is still $\Theta(nM)$. That being said, the least to ask for would be an algorithm that compares favourably to an adversary that does *not* make use of the processing time samples. In other words, an algorithm that is at least as good as RND, the adversary that picks *any schedule uniformly at random*.

Definition 2 (Algorithm RND). *Schedule jobs in any of the $n!$ sequences uniformly at random.*

The bad news is that Example 1 is built so that algorithm SAM has expected regret of the same order of magnitude even in comparison to RND, because comparing SAM with RND yields an expected regret equal to $\left(1 - \frac{1}{M}\right)\left[nM - \frac{n+1}{2}M\right] + \frac{1}{M}\left[M - \frac{n+1}{2}M\right] = \frac{1}{2}(n-1)(M-2) = \Theta(nM)$. So the fundamental question is, which is the right adversary under the given information regime, and what are additional assumptions on the input distributions that allow to bound the expected regret of the "natural" algorithm SAM.

3   Relative Optimality Gap and Better Than Random Schedules

Here we define a notion for approximation algorithms in terms of expected regret and with respect to the optimal non-anticipatory scheduling policy that has access to the input distributions P_j, that is, WSEPT. For a given problem instance I, denote by L_I the lowest possible expected cost achieved by the optimal WSEPT schedule, and by H_I the highest possible expected cost, achieved by the reverse sequence. For simplicity, let us exclude for the rest of the paper instances I where $L_I = H_I$, in which case trivially all algorithms have the same expected cost. We define:

Definition 3 (rog). *The relative optimality gap of an algorithm Π on a problem instance $I \in \mathcal{I}$ is defined as*

$$\mathrm{rog}_I(\Pi) := \frac{\mathrm{cost}_I(\Pi) - L_I}{H_I - L_I}.$$

For a set of problem instances $\mathcal{I}$, define $\mathrm{rog}_{\mathcal{I}}(\Pi) := \sup_{I \in \mathcal{I}} \mathrm{rog}_I(\Pi)$.

Note that if $\mathrm{rog}_{\mathcal{I}}(\Pi) \leq \alpha$ with $\alpha \leq 1$, it implies that the expected regret of Π is at most an α-fraction of the maximum expected regret. In that sense, the rog captures the idea of approximation algorithms in terms of regret. In terms of expected cost, that translates as follows: $\mathrm{rog}_{\mathcal{I}}(\Pi) \leq \alpha$, for some $\alpha \leq 1$, means that there is no instance I where $\Pi's$ expected cost exceeds that of an optimal solution by more than $\alpha(H_I - L_I)$. It corresponds to a β-approximation algorithm, with $\beta = \beta(I) = 1 + \alpha(H_I/L_I - 1)$. Note that this implies that rog is more meaningful in comparison to the notion of a β-approximation algorithm w.r.t. costs whenever $H_I - L_I$ is small relative to L_I. (Of course this argument works vice versa, too.) One advantage of working with the rog is that it yields a meaningful statement for every single instance, which is not the case e.g. when we have a 2-approximation algorithm for an instance with $H_I \leq 2L_I$.

As long as the set of input instances $\mathcal{I}$ does include instances such as Example 1, we have that (again, considering $\varepsilon \to 0$)

$$\mathrm{rog}_{\mathcal{I}}(\mathtt{SAM}) \geq \frac{nM - (n-1) - M}{nM - M} \to 1 \ (\text{for } M \to \infty \text{ and any } n \geq 2).$$

Not surprisingly, the randomized algorithm RND has an rog equal to $1/2$, and it turns out that an algorithm Π with $\mathrm{rog}_{\mathcal{I}}(\Pi) \leq 1/2$ exactly means that algorithm Π is better than random. To formalize this, define a partial order on the set of algorithms as follows.

Definition 4. *If $\mathcal{I}$ is a set of problem instances, and Π and Π' are two scheduling algorithms, write $\Pi \leq_{\mathcal{I}} \Pi'$ if $\mathrm{cost}_I(\Pi) \leq \mathrm{cost}_I(\Pi')$ for every problem instance $I \in \mathcal{I}$.*

Note that Example 1 shows that not even $\mathtt{SAM} \leq_{\mathcal{I}} \mathtt{RND}$ holds, if $\mathcal{I}$ contains all problem instances. The following lemma defines the benchmark that we aim for in our subsequent analysis of algorithm SAM.

Lemma 1. *For any instance I we have $\mathrm{rog}_I(\mathtt{RND}) = 1/2$. Moreover, for scheduling algorithm Π and any set of input instances $\mathcal{I}$, we have $\Pi \leq_{\mathcal{I}} \mathtt{RND} \Leftrightarrow \mathrm{rog}_{\mathcal{I}}(\Pi) \leq \mathrm{rog}_{\mathcal{I}}(\mathtt{RND})$.*

Proof. For given instance $I = (w, P)$ assume w.l.o.g. that $w_1/\mathbb{E}[P_1] \geq \cdots \geq w_n/\mathbb{E}[P_n]$. The first claim follows because $L_I = \sum_{j=1}^{n}\sum_{k=j}^{n} w_k\mathbb{E}[P_j]$, and any two jobs j, k have probability $1/2$ of being ordered either way by RND, so by canceling the terms $w_j\mathbb{E}[p_j]$

$$\mathrm{cost}_I(\mathtt{RND}) - L_I = \sum_{j=1}^{n-1}\sum_{k=j+1}^{n} \frac{1}{2}w_j\mathbb{E}(P_k) - \frac{1}{2}w_k\mathbb{E}(P_j) = \frac{1}{2}(H_I - L_I).$$

The second claim then follows directly by using the definition of rog, since $\mathrm{rog}_I(\Pi) \leq \frac{1}{2}$ precisely means that $\mathrm{cost}_I(\Pi) \leq \frac{1}{2}(L_I + H_I) = \mathrm{cost}_I(\mathtt{RND})$. □

Note that the second statement in Lemma 1 is not true when comparing two arbitrary scheduling algorithms Π and Π', as $\mathrm{rog}_{\mathcal{I}}(\Pi) \leq \mathrm{rog}_{\mathcal{I}}(\Pi')$ does not imply that $\Pi \leq_{\mathcal{I}} \Pi'$.

For what follows it is convenient to realize that the relative optimality gap may be expressed in terms of the extra cost for scheduling pairs of jobs in the incorrect order. To that end, for a given instance $I = (w, P)$, we denote this extra cost by

$$\Delta_{jk} := w_j\mathbb{E}[P_k] - w_k\mathbb{E}[P_j].$$

Then Δ_{jk} is the cost for scheduling k before j, instead of j before k. Write $\mathbb{P}[\Pi : k \to j]$ for the probability that Π schedules job k before j.

Lemma 2. *Let Π be a scheduling algorithm for instance $I = (w, P)$, and assume w.l.o.g. that $w_1/\mathbb{E}[P_1] \geq \cdots \geq w_n/\mathbb{E}[P_n]$, then*

$$\mathrm{rog}_I(\Pi) = \sum_{j,k:j<k} \mathbb{P}[\Pi : k \to j] \frac{\Delta_{jk}}{\sum_{h,\ell:h<\ell} \Delta_{h\ell}}. \tag{1}$$

Proof. The proof follows from observing that $H_I - L_I = \sum_{h,\ell:h<\ell} \Delta_{h\ell}$, and since, again by canceling the terms $w_j\mathbb{E}[P_j]$,

$$\mathrm{cost}_I(\Pi) - L_I = \sum_{j=1}^{n-1} \sum_{k=j+1}^{n} (\mathbb{P}[\Pi : j \to k] - 1) w_k \mathbb{E}[P_j] + \mathbb{P}[\Pi : k \to j] w_j \mathbb{E}[P_k]$$

$$= \sum_{j=1}^{n-1} \sum_{k=j+1}^{n} \mathbb{P}[\Pi : k \to j] \Delta_{jk}.$$

$\square$

Now we can derive a simple and intuitive bound on the rog by bounding the probability of scheduling jobs in the incorrect order.

Theorem 1. *If Π is an algorithm for instance I, if there exists some $\kappa \leq 1$, so that for all pairs of jobs $\mathbb{P}[\Pi : j \to k] \geq \kappa$ whenever $w_j/\mathbb{E}[p_j] > w_k/\mathbb{E}[p_k]$, then $\mathrm{rog}_I(\Pi) \leq 1 - \kappa$.*

Proof. The proof follows from Lemma 2 and $\mathbb{P}[\Pi : k \to j] \leq (1 - \kappa)$, and because $\Delta_{jk} = 0$ whenever $w_j/\mathbb{E}[p_j] = w_k/\mathbb{E}[p_k]$. $\square$

Note that the inverse statement does not hold, namely, rog $\leq \alpha$ does not necessarily imply that all pairs of jobs are scheduled in the correct order with probability at least $1 - \alpha$.

4   On Uniform Randomization as Adversary

Our goal is to bound the rog from above. It turns out, however, that without strong assumptions on the input distributions (e.g. Sect. 5.4), showing that an algorithm Π has $\mathrm{rog}(\Pi) \leq 1/2$, which by Lemma 1 is equivalent to showing that Π performs at least as good as RND, is essentially the best we can hope for: Considering an instance with two jobs with equal weights and "almost identical" input distributions, the samples will suggest the wrong order in essentially 50% of the cases, and hence the expected performance of any non-anticipatory algorithm must be close to that of random. This effect can be leveraged to any other instance, by appending to a given instance two additional such jobs, but with weights an order of magnitude larger than that of any other job.

That said, when showing $\mathrm{rog}(\Pi) \leq 1/2$, the benchmark that we effectively consider is the random adversary RND. In general, one might also consider other adversaries, for example those that also take into account the job weights. However, Lemma 1 reconfirms that, in a relative sense, the expected cost of RND is independent of the problem instance. Comparing favourably against this instance independent benchmark shows that we have successfully leveraged the fact that we have access to samples. Other, reasonable adversaries, e.g. the algorithm that schedules by largest weight first, or a randomized variation, depend non-trivially on the problem instance, so that showing that an algorithm Π outperforms such an adversary would generally not allow to distinguish between Π performing well, or the adversary performing poorly.

5   Well Behaved Input Distributions

Here we derive our main results, namely three classes of rather natural assumptions on input distributions that allow to make use of Theorem 1 to obtain performance guarantees for sampling based algorithm SAM. We will argue in Sect. 6 that these results are tight in a mild sense. An assumption that we make henceforth is that the distributions for P_j all have a density f_j. The same ideas and results also work for discrete distributions, however.

If p'_j, p'_k are the sampled processing times, then SAM schedules j before k if $p'_j \leq \frac{w_j}{w_k} p'_k$. By independence of processing times we have the following.

Observation 1. *Consider scheduling algorithm* SAM, *then the probability for scheduling two jobs j, k in order $j \to k$ is*

$$\mathbb{P}[\,\mathtt{SAM} : j \to k\,] = \int_0^\infty f_k(y) \int_0^{\frac{w_j}{w_k} y} f_j(x)dxdy. \tag{2}$$

5.1   Symmetric Processing Time Distributions

Intuitively speaking, if all distributions P_j are symmetric around their means $\mathbb{E}[\,P_j\,]$, then for each sampled pair p'_j, p'_k that gives rise to an incorrect ordering of j and k, by symmetry, there exist samples p''_j, p''_k that appear with equal probability and give rise to the correct ordering. The proof of the following theorem confirms that this intuition is essentially correct. Let us first recall what symmetry of non-negative random variables means.

Definition 5. *For a random variable P on $\mathbb{R}$ with density f and positive and finite expected value E, P is symmetric if $f(E - x) = f(E + x)$ for all x.*

Note that in our context $P \geq 0$, which implies that $f(x) = 0$ for all $x \leq 0$ and all $x \geq 2E$.

Theorem 2. *Consider instances $\mathcal{I}$ so that all processing time distributions P_j are symmetric for all jobs $j = 1, \ldots, n$, then $\mathrm{rog}_{\mathcal{I}}(\mathtt{SAM}) \leq 1/2$.*

Proof. Pick an instance I with symmetric processing time distributions, and recall by Theorem 1 that it suffices to prove that for any pair of jobs j, k with $w_j/\mathbb{E}[\,P_j\,] > w_k/\mathbb{E}[\,p_k\,]$, we have $\mathbb{P}[\,\mathtt{SAM} : j \to k\,] \geq \frac{1}{2}$. For convenience write E_j and E_k for $\mathbb{E}[\,P_j\,]$ and $\mathbb{E}[\,P_k\,]$. Making use of (2), we apply linear variable substitutions. Substituting x by $x' = w_k(x - E_j)$, y by $y' = w_j(y - E_k)$, and noting that $dx' = w_k\, dx$ and $dy' = w_j\, dy$, we get

$$\mathbb{P}[\,\mathsf{SAM}: j \to k\,] = \frac{1}{w_j w_k} \int_{-w_j E_k}^{\infty} f_k\left(E_k + \frac{y'}{w_j}\right) \int_{-w_k E_j}^{y' + w_j E_k - w_k E_j} f_j\left(E_j + \frac{x'}{w_k}\right) dx'\, dy'.$$

From symmetry of P_k it follows that $f_k(E_k + \frac{y'}{w_j}) = 0$ for $y' > w_j E_k$. Thus, we can replace the upper boundary of the outer integral by $w_j E_k$. Renaming x' back to x and y' to y, gives

$$\mathbb{P}[\,\mathsf{SAM}: j \to k\,] = \frac{1}{w_j w_k} \int_{-w_j E_k}^{w_j E_k} f_k\left(E_k + \frac{y}{w_j}\right) \int_{-w_k E_j}^{y + w_j E_k - w_k E_j} f_j\left(E_j + \frac{x}{w_k}\right) dx\, dy.$$

And symmetrically also

$$\mathbb{P}[\,\mathsf{SAM}: k \to j\,] = \frac{1}{w_j w_k} \int_{-w_k E_j}^{w_k E_j} f_j\left(E_j + \frac{x}{w_k}\right) \int_{-w_j E_k}^{x + w_k E_j - w_j E_k} f_k\left(E_k + \frac{y}{w_j}\right) dy\, dx.$$

The final two expressions can be interpreted as surface integrals in $\mathbb{R}^2$. The integration domains are isosceles right-angled triangles. These triangles are shown in Fig. 1. We denote the triangle corresponding to $\mathbb{P}[\,\mathsf{SAM}: j \to k\,]$ with T_{jk}, and analogously for T_{kj}. Due to the linear transformations, the symmetry of P_j and P_k translates into the fact that the integrals over these triangles remain the same

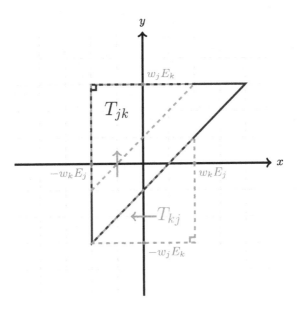

Fig. 1. Two triangles representing the integration domains for the probabilities of SAM scheduling in each pairwise order. The diagonal strip where the triangles do not overlap is the difference in the integration domains, proving $\mathbb{P}[\,\mathsf{SAM}: j \to k\,] \geq \mathbb{P}[\,\mathsf{SAM}: k \to j\,]$.

after reflection through either of the coordinate axes. Mirroring the triangle T_{kj} twice, once in the x-axis, and once in the y-axis, aligns its right-angle corner with the right-angle corner of the triangle T_{jk}. This is illustrated in Fig. 1. This geometric operation can also be done formally by changing the order of integration in the term for $\mathbb{P}[\,\mathsf{SAM} : k \to j\,]$, using Fubini's theorem. It results in

$$\mathbb{P}[\,\mathsf{SAM} : k \to j\,] = \frac{1}{w_j w_k} \int_{w_j E_k - 2w_k E_j}^{w_j E_k} f_k(E_k + \frac{y}{w_j}) \int_{-w_k E_j}^{y - w_j E_k + w_k E_j} f_j(E_j + \frac{x}{w_k}) dx dy.$$

Geometrically, the assumption that $w_j E_k > w_k E_j$ yields that triangle T_{kj} has shorter side-lengths than triangle T_{jk}. This means that the integration domains of the two otherwise identical integrals differ by the diagonal strip in Fig. 1, and since the integrand is non-negative, $\mathbb{P}[\,\mathsf{SAM} : j \to k\,] \geq \mathbb{P}[\,\mathsf{SAM} : k \to j\,]$. Therefore, as required, $\mathbb{P}[\,\mathsf{SAM} : j \to k\,] \geq \frac{1}{2}$. □

5.2   Identical Processing Time Distributions up to Scaling

Here we show that if all distributions P_j are identical up to scaling, then algorithm SAM performs better than random.

Definition 6. *Let $g(\cdot)$ be any probability density function over $\mathbb{R}_{\geq 0}$ with expectation 1. Define for $\lambda_j > 0$, $f_j(x) := \lambda_j g(\lambda_j x)$. Call an instance of the scheduling problem* shape-uniform *if the processing time distributions P_j all have densities $f_j(\cdot)$, for the same $g(\cdot)$ and $\lambda_j > 0$, $j = 1, \dots, n$.*

Note that $\mathbb{E}[\,P_j\,] = 1/\lambda_j$, $j = 1, \dots, n$, for any shape-uniform instance. Also note that exponentially distributed processing times where $f_j(x) = \lambda_j e^{-\lambda_j x}$ are contained as a special case of shape-uniformity via $g(x) = e^{-x}$.

Theorem 3. *Consider instances $\mathcal{I}$ so that all processing time distributions P_j are shape-uniform with the same underlying probability density function $g(\cdot)$, then $\mathrm{rog}_{\mathcal{I}}(\mathsf{SAM}) \leq 1/2$.*

Proof. Consider any two jobs j, k and assume w.l.o.g. that $w_j / \mathbb{E}[\,P_j\,] \geq w_k / \mathbb{E}[\,P_k\,]$, which is equivalent to $w_j \lambda_j \geq w_k \lambda_k$. Again we make use of Theorem 1 and Lemma 1, and show that $\mathbb{P}[\,\mathsf{SAM} : j \to k\,] \geq \mathbb{P}[\,\mathsf{SAM} : k \to j\,]$. Making use of (2), and substituting x by $\lambda_j x$ and y by $\lambda_k y$, we get

$$\mathbb{P}[\,\mathsf{SAM} : j \to k\,] = \int_0^{\infty} g(y) \int_0^{\frac{w_j \lambda_j}{w_k \lambda_k} y} g(x) dx dy.$$

Symmetrically, we get the same term for $\mathbb{P}[\,\mathsf{SAM} : k \to j\,]$, and subtracting gives

$$\mathbb{P}[\,\mathsf{SAM} : j \to k\,] - \mathbb{P}[\,\mathsf{SAM} : k \to j\,] = \int_0^{\infty} g(y) \int_{\frac{w_k \lambda_k}{w_j \lambda_j} y}^{\frac{w_j \lambda_j}{w_k \lambda_k} y} g(x) dx\, dy.$$

Now since $w_j \lambda_j \geq w_k \lambda_k$, and since the integrand is non-negative everywhere, this last term must be non-negative, which yields the claim. □

5.3   Translations of Identical Processing Time Distributions

Here we show that SAM performs better than random if all processing time distributions are identical up to having different expectations. This result is restricted to the special case of uniform weights $w_j = 1$, $j = 1, \ldots, n$, however.

Definition 7. *Let $g(\cdot)$ be the density of some random variable with expectation $E > 0$, so that $g(x) = 0$ for all $x < 0$. Call the processing times translation-identical for $g(\cdot)$, if processing time P_j has density $f_j(x) = g(x - E_j + E)$, for $E_j \geq E$.*

Note that by definition, $\mathbb{E}[P_j] = E_j$ for $j = 1, \ldots, n$.

Theorem 4. *Consider instances $\mathcal{I}$ with uniform weights $w_j = 1$ and translation-identical processing time distributions P_j, for some $g(\cdot)$, then $\mathrm{rog}_{\mathcal{I}}(\mathrm{SAM}) \leq 1/2$.*

Proof. Again we use Theorem 1 and Lemma 1, and show that $\mathbb{P}[\,\mathrm{SAM}:j \to k\,] \geq \mathbb{P}[\,\mathrm{SAM}:k \to j\,]$ for any two jobs j, k with $\mathbb{E}[P_j] \leq \mathbb{E}[P_k]$. For the sake of the proof assume w.l.o.g. that $f_j(x) = g(x)$ and $f_k(x) = g(x - a)$, $a = E_k - E \geq 0$, are the density functions for P_j and P_k, so that $\mathbb{E}[P_k] = a + \mathbb{E}[P_j] = a + E$. By the same arguments as in the proof of Theorem 3, and using the substitution $y - a$ for y, we easily see that

$$\mathbb{P}[\,\mathrm{SAM}:j \to k\,] - \mathbb{P}[\,\mathrm{SAM}:k \to j\,] = \int_0^\infty g(y) \int_{y-a}^{y+a} g(x)\,dx\,dy.$$

As the integrand is non-negative, and as $a \geq 0$, the term is non-negative.   □

The following example shows, maybe surprisingly, that the above result does not generalize to the setting with arbitrary weights. (The example uses finite discrete distributions for simplicity, but could be adapted accordingly).

Example 2. Consider an instance I with $n = 2$ jobs, weights $w_1 = 1$ and $w_2 = 2$, and processing time distributions

$$P_1 = \begin{cases} 1, & \text{with probability } 1 - \frac{1}{M}, \\ M^2 & \text{with probability } \frac{1}{M}, \end{cases}$$

and P_2 identically distributed as P_1, but independently and translated by $1 + \varepsilon$ so that $\mathbb{E}[P_2] = \mathbb{E}[P_1] + 1 + \varepsilon$.   △

Here, again assume that M is large and take limits for $\varepsilon \to 0$. Observe that Algorithm SAM schedules job 1 first whenever the sampled processing time $p_1' = 1$, independent of the sample for P_2. This happens with probability $(1 - \frac{1}{M})$. Scheduling job 1 first has maximal expected cost which equals $5M^2$. The optimal solution is to schedule job 2 first, with expected cost $4M^2$. One readily verifies that

$$\mathrm{rog}_I(\mathrm{SAM}) = \frac{(1 - \frac{1}{M})5M^2 + \frac{1}{M}4M^2 - 4M^2}{5M^2 - 4M^2} = \frac{M^2 - M}{M^2} \to 1 \text{ (for } M \to \infty\text{).}$$

5.4   Exponential Distributions and α-Separated Priorities

So far we showed that using the processing time samples is at least as good as random, arguably the least one would hope for. In light of Sect. 4, this cannot be substantially improved without making additional assumptions on the input. Here we show how to derive such a qualitatively better result. We consider the single machine scheduling problem with arbitrary weights w_j and exponentially distributed processing times, so P_j has a distribution with density $f_j(x) = \lambda_j e^{-\lambda_j x}$, $\lambda_j > 0$. Define $\pi_j := w_j \lambda_j = w_j / \mathbb{E}[P_j]$ for all $j = 1, \ldots, n$, the priority of job j. We know from Sect. 5.2 that $\mathrm{rog}_\mathcal{I}(\mathtt{SAM}) \leq 1/2$ if $\mathcal{I}$ are the instances with exponentially distributed processing times. Under an additional assumption on the priorities this can be improved. Using (2) and the definition of the exponential distribution, elementary calculus yields the following.

Lemma 3. *If processing times are exponentially distributed, then for any pair of jobs j, k, $\mathbb{P}[\,\mathtt{SAM} : j \rightarrow k\,] = \pi_j / (\pi_j + \pi_k)$.*

Then call the priorities $\pi_1, \ldots, \pi_n$ of an instance with weights w_j and exponentially distributed processing times α-*separated*, whenever the following is true for all pairs of jobs j, k: Either $\pi_j = \pi_k$, or π_j and π_k are at least a factor α apart, i.e., $\max\{\pi_j/\pi_k, \pi_k/\pi_j\} \geq \alpha$. In other words, we have groups of jobs with identical priorities in each group, and the priorities across groups are at least a factor α apart. The intuition is that either $\pi_j = \pi_k$, in which case the order of these two jobs should not matter, or π_j and π_k are far apart, in which case algorithm $\mathtt{SAM}$ should have a high(er) probability for scheduling these two jobs in the correct order, hence leading to a better rog bound. This intuition is correct.

Theorem 5. *Consider instances $\mathcal{I} = (w, P)$ with exponentially distributed processing times and α-separated priorities π_j, $\alpha \geq 1$, then $\mathrm{rog}_\mathcal{I}(\mathtt{SAM}) \leq 1/(1+\alpha)$.*

Proof. To use Theorem 1 for an instance I, it suffices to consider pairs of jobs j, k with $\pi_j \neq \pi_k$, because otherwise the contribution of this pair of jobs j, k to $\mathrm{rog}_I(\mathtt{SAM})$ is indeed zero in (1). So take any pair of jobs j, k with $\pi_j > \pi_k$, meaning that $w_j / \mathbb{E}[P_j] > w_k / \mathbb{E}[P_k]$ and the order $j \rightarrow k$ is optimal, then by Lemma 3, $\mathbb{P}[\,\mathtt{SAM} : j \rightarrow k\,] = \frac{\pi_j}{\pi_j + \pi_k} = \frac{1}{1 + \pi_k/\pi_j} \geq \frac{1}{1 + 1/\alpha} = \frac{\alpha}{\alpha+1}$, where the inequality is true because $\pi_j \geq \alpha \pi_k$. The claim now follows by Theorem 1.   □

6   Conclusions

The results of this paper are a first attempt to study scheduling models under uncertainty when the information regime for unknown processing times is very limited. This is in contrast to the common assumption that there are sufficient samples available to rely on the law of large numbers. We believe such a lack of information is a challenge that occurs in many practical settings, albeit perhaps not precisely in the way modelled in this paper. This paper shows that even in the extreme setting of a single sample, performance guarantees can still be derived. Moreover, the three classes of processing time distributions for which we derive performance guarantees are quite realistic from a practical viewpoint. We hope the paper can serve as a starting point to study also other, low information settings.

We finally note that our results for the model at hand can even be called tight in some mild sense. First note that Example 1 can be tweaked from discrete to continuous distributions. The counterexample does its job because there are processing time distributions with (recall Example 1 had unit weights $w_j = 1$)

(i) two different shapes, one of them asymmetric,
(ii) two different means.

Moreover, from Example 2 we conclude that as soon as we have non-uniform weights and different means, not even the *same* distributions around different means allow for positive results, as long as these distributions are asymmetric and are not shape-uniform. One may wonder about instances with identical expected processing times, yet different distributions. But this is either not interesting because all schedules yield the same expected cost (in case of unit weights), or it allows to replicate an example exactly analogous to Example 1 by exchanging the roles of weights and expected processing times, which yields the same lower bound for instances with arbitrary weights and identical expected processing times.

There are some open ends to take this work further. That includes the analysis for adversaries other than uniformly at random, the identification of other classes of distributions that would allow for positive results, and a less restrictive information regime with two or more samples, which would allow for algorithms that use a proxy for both expected processing time and variance. However observe that Example 1 is robust even against taking a polynomial number of samples.

Acknowledgements. This work was done while the first author was MSc student at Utrecht University. The first author thanks Rob Bisseling for advice and support, and Jan Fortuin Sr. for inspiration. The second author thanks Wouter Fokkema and Ruben Hoeksma for helpful discussions that initiated this work, and José Verschae for advice.

References

1. Chekuri, C., Motwani, R., Natarajan, B., Stein, C.: Approximation techniques for average completion time scheduling. SIAM J. Comput. **31**, 146–166 (2001)
2. Gupta, V., Moseley, B., Uetz, M., Xie, Q.: Greed works-online algorithms for unrelated machine stochastic scheduling. Math. Oper. Res. **45**(2), 497–516 (2020)
3. Gupta, V., Moseley, B., Uetz, M., Xie, Q.: Corrigendum: greed works-online algorithms for unrelated machine stochastic scheduling. Math. Oper. Res. **46**(3), 1230–1234 (2021)
4. Hall, L.A., Schulz, A.S., Shmoys, D.B., Wein, J.: Scheduling to minimize average completion time: off-line and on-line approximation algorithms. Math. Oper. Res. **22**, 513–544 (1997)

5. Jäger, S.: An improved greedy algorithm for stochastic online scheduling on unrelated machines. Discret. Optim. **47**, 100753 (2023)
6. Lindermayr, A., Megow, N.: Algorithms with predictions. https://algorithms-with-predictions.github.io/. Accessed 02 Aug 2023
7. Megow, N., Uetz, M., Vredeveld, T.: Models and algorithms for stochastic online scheduling. Math. Oper. Res. **31**(3), 513–525 (2006)
8. Möhring, R.H., Radermacher, F.J., Weiss, G.: Stochastic scheduling problems I: general strategies. ZOR - Zeitschrift für Oper. Res. **28**, 193–260 (1984)
9. Möhring, R.H., Schulz, A.S., Uetz, M.: Approximation in stochastic scheduling: the power of LP-based priority policies. J. ACM **46**, 924–942 (1999)
10. Pashkovich, K., Sayutina, A.: Single sample prophet inequality for uniform matroids of rank 2 (2023). https://arxiv.org/abs/2306.17716
11. Pinedo, M.: Scheduling: Theory, Algorithms, and Systems, 5th edn. Springer, Heidelberg (2016)
12. Rothkopf, M.H.: Scheduling with random service times. Manage. Sci. **12**, 703–713 (1966)
13. Rubinstein, A., Wang, J., Weinberg, S.: Optimal single-choice prophet inequalities from samples. In: Vidick, T. (ed.) 11th Innovations in Theoretical Computer Science Conference (ITCS 2020). Leibniz International Proceedings in Informatics (LIPIcs), vol. 151, pp. 60:1–60:10. Schloss Dagstuhl – Leibniz-Zentrum für Informatik, Dagstuhl (2020)
14. Skutella, M., Uetz, M.: Stochastic machine scheduling with precedence constraints. SIAM J. Comput. **34**, 788–802 (2005)
15. Smith, W.E.: Various optimizers for single-stage production. Naval Res. Logist. Q. **3**, 59–66 (1956)

The Thief Orienteering Problem
on Series-Parallel Graphs

Andrew Bloch-Hansen[(✉)] and Roberto Solis-Oba

Western University, London, ON, Canada
`ablochha@uwo.ca, solis@csd.uwo.ca`

Abstract. In the thief orienteering problem an agent called a *thief* carries a knapsack of capacity W and has a time limit T to collect a set of items of total weight at most W and maximum profit along a simple path in a weighted graph $G = (V, E)$ from a start vertex s to an end vertex t. There is a set I of items each with weight w_i and profit p_i that are distributed among $V \setminus \{s, t\}$. The time needed by the thief to travel an edge depends on the length of the edge and the weight of the items in the knapsack at the moment when the edge is traversed.

There is a polynomial-time approximation scheme for the thief orienteering problem on directed acyclic graphs that produces solutions that use time at most $T(1 + \epsilon)$ for any constant $\epsilon > 0$. We give a polynomial-time algorithm for transforming instances of the problem on series-parallel graphs into equivalent instances of the thief orienteering problem on directed acyclic graphs; therefore, yielding a polynomial-time approximation scheme for the thief orienteering problem on this graph class that produces solutions that use time at most $T(1 + \epsilon)$ for any constant $\epsilon > 0$.

Keywords: Thief Orienteering Problem · Knapsack Problem · Approximation Algorithm · Approximation Scheme · Series-Parallel

1 Introduction

Let $G = (V, E)$ be a weighted graph with n vertices, where two vertices $s, t \in V$ are designated the *start* and *end* vertices, respectively. Let there be a set I of items, where each item $i_j \in I$ has a non-negative integer weight w_j and profit p_j. Each vertex $u \in V \setminus \{s, t\}$ stores a subset $S_u \subseteq I$ of items such that $S_u \cap S_v = \emptyset$ for all $u \neq v$ and $\bigcup_{u \in V \setminus \{s,t\}} S_u = I$. Additionally, every edge $e = (u, v) \in E$ has a length $d_{u,v} \in \mathbb{Q}^+$.

In the *thief orienteering problem* (ThOP) the goal is for an agent called a *thief* to travel a simple path in G between s and t within a given time $T \in \mathbb{Q}^+$ while collecting items in a knapsack with capacity $W \in \mathbb{Z}^+$ taken from the vertices along the path of total weight at most W and maximum total profit.

Andrew Bloch-Hansen and Roberto Solis-Oba were partially supported by the Natural Sciences and Engineering Research Council of Canada, grants 6636-548083-2020 and RGPIN-2020-06423, respectively.

The time needed to travel between two adjacent vertices u, v depends on the length of the edge connecting them and on the weight of the items in the knapsack when the edge is traveled; specifically, the travel time between adjacent vertices u and v is $d_{u,v}/\mathcal{V}$ where $\mathcal{V} = \mathcal{V}_{\max} - w(\mathcal{V}_{\max} - \mathcal{V}_{\min})/W$, w is the current weight of the items in the knapsack, and $\mathcal{V}_{\min}$ and $\mathcal{V}_{\max}$ are the minimum and maximum velocities of the thief.

ThOP is a generalization of the knapsack and longest path problems that has not been extensively studied, but related travelling problems such as the travelling thief problem [3] and some variants of orienteering [11] are well-studied and have applications in areas such as route planning [9] and circuit design [3].

ThOP was first formulated in 2018 by Santos and Chagas [12] and several heuristics have since been designed for it [4,5,8]. In 2023, Bloch-Hansen et al. [2] proved that there exists no approximation algorithm for the thief orienteering problem with constant approximation ratio unless $P = NP$, and they presented a polynomial-time approximation scheme (PTAS) for ThOP when the input graph G is directed and acyclic (DAG) that produces solutions that use time at most $T(1 + \epsilon)$ for any constant $\epsilon > 0$.

In this paper we consider ThOP on series-parallel graphs. This graph class has applications in areas such as scheduling [13], VLSI [6], and electrical circuits [7], and our motivation for studying ThOP on this graph class was production optimization, an important problem in the manufacturing industry. The manufacturing of a product might require several stages involving different resources, equipment, and personnel. The goal of production optimization is to design and schedule the stages needed to manufacture a product at minimum cost.

A directed graph can be used to model the different manufacturing possibilities for a product, with vertices representing manufacturing stages and edges denoting the order in which the stages need to be performed [14]. Every vertex has attributes like cost of the manufacturing stage, resources needed, completion time, and personnel required. Edges indicate time needed to move the (partial) product from one stage to another. The best plan for manufacturing a product then corresponds to a path in the directed graph from the vertex corresponding to the start of the manufacturing process to the vertex corresponding to the completion of the product. This best path must satisfy several constraints like production time, resource cost, and number of personnel needed to fabricate the product, and so this problem can be modelled with the thief orienteering problem. Due to the sequential nature of many manufacturing processes, the different production plans corresponding to the various ways to manufacture a process can be conveniently modelled using series-parallel graphs [1].

Another application for ThOP includes system reliability, which is fundamental in system design. A series system consists of a sequence S of subsystems, and the failure of any subsystem causes the failure of the entire system. To prevent this, redundant subsystems are added in parallel to ensure a minimum level of reliability. These systems are called series-parallel systems [10]. To determine the optimal number of redundant subsystems, a series-parallel graph can be used to model the problem. Each vertex in the graph corresponds to a different number of possible copies of each subsystem. Vertices have costs indicating the cost of the redundant subsystems and they also have profits indicating the reliability achieved with the corresponding redundant systems. An edge from a vertex u to an adjacent vertex v in the sequence S is given length equal to the number of subsystems represented by u. The goal is to find a minimum cost path from a vertex s representing the beginning of the sequence S to a vertex corresponding to the end of the sequence, that achieves a failure rate no larger than a maximum bound W, and that uses at most a given number T of additional subsystems. This problem can also be modelled with the thief orienteering problem on series-parallel graphs.

Our strategy for dealing with series-parallel graphs is to first transform them into DAGs and then to use the PTAS of [2]. The transformation into DAGs is not easy as we need to preserve all simple paths from s to t while avoiding the formation of cycles. To achieve this, we create copies of the vertices and edges of the input graph and carefully select the directions of the edges so every simple undirected path in the input graph has a corresponding, equivalent directed path in the DAGs produced by our algorithm.

The main challenge in achieving polynomial running time when transforming this class of graphs into DAGs is to preserve all the different paths from s to t without adding a very large number of vertices and edges or creating any cycles. We show how to overcome these challenges by exploiting the special structure of series-parallel graphs.

In the rest of the paper we present our algorithm to transform instances of ThOP on series-parallel graphs into equivalent instances of ThOP on DAGs.

2   Thief Orienteering on Series-Parallel Graphs

A series-parallel graph $G = (V, E, t_1, t_2)$ has two terminal vertices t_1, t_2 and is defined inductively:

- $G = (\{t_1, t_2\}, (t_1, t_2), t_1, t_2)$ is series-parallel.
- if $G_1 = (V_1, E_1, t_1, t_2)$ and $G_2 = (V_2, E_2, t_1', t_2')$ are series-parallel then the *series composition* $G = (V_1 \cup V_2, E_1 \cup E_2, t_1, t_2')$ is series-parallel if $t_2 = t_1'$.
- if $G_1 = (V_1, E_1, t_1, t_2)$ and $G_2 = (V_2, E_2, t_1', t_2')$ are series-parallel then the *parallel composition* $G = (V_1 \cup V_2, E_1 \cup E_2, t_1, t_2)$ is series-parallel if $t_1 = t_1'$ and $t_2 = t_2'$. A graph G created from a parallel composition is a *parallel graph*.

We can take advantage of the properties of a series-parallel graph G in order to transform it into the desired DAG $\mathcal{G}$ using a polynomial number of additional vertices and edges such that the DAG has the same set of simple paths between s and t as G. Note that the start and end vertices s and t might not be the two terminals in a series-parallel graph.

2.1   Removing Vertices with Degree 1

Let G be an undirected series-parallel graph. If there is only one simple path from s to t or if all vertices in G have degree less than 3, then directing the edges away from s and towards t produces the desired DAG. Therefore, in the sequel we assume that at least one vertex in G has degree 3 or higher and that there is more than one simple path in G from s to t.

Consider a vertex u of degree 1, where $u \neq s$ and $u \neq t$. A simple path from s to t cannot route through u, so we delete these vertices u, and any subsequent vertices $v \neq s$, $v \neq t$ with degree 1 that are exposed from this process.

If s has degree 1 we direct away from s all edges along a simple path to the closest vertex v of degree at least 3; we rename v to s as clearly a path from s to t must route through v. Similarly, if t has degree 1 let v be the closest vertex to t with degree at least 3; direct all edges away from v along the simple path to t and rename v to t.

2.2   Cut Vertices

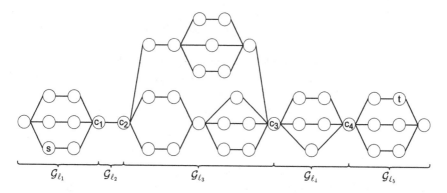

Fig. 1. Identifying the subgraphs $\mathcal{G}_{\ell_x}$ in a series-parallel graph.

A *cut vertex* is a vertex whose removal from a connected graph G disconnects it into at least 2 non-empty connected components. Note that for series-parallel graphs, a cut vertex will always be a terminal of some series-parallel subgraph of G whose removal splits G into exactly 2 connected components.

Consider a shortest path p between s and t. Let c_1, c_2, ..., c_{k-1} be the cut vertices, other than s and t, along p in increasing order of distance to s (see Fig. 1). Let $c_0 = s$ and $c_k = t$. If the removal of c_0 splits G into 2 connected components, we delete the component that does not contain t. Similarly, if the removal of c_k splits G into 2 connected components, we delete the component that does not contain s.

For each c_x from $x = 1, 2, ..., k$, we define the graph $\mathcal{G}_{\ell_x}$ that includes the cut vertices c_{x-1} and c_x and all vertices and edges in all simple paths between c_{x-1} and c_x. We transform each graph $\mathcal{G}_{\ell_x}$ into a DAG as described below. We then combine these DAGs to produce the DAG $\mathcal{G}$ for the input graph G.

2.3   Transforming $\mathcal{G}_{\ell_x}$ into a DAG

Consider one of the graphs $\mathcal{G}_{\ell_x}$. If $\mathcal{G}_{\ell_x}$ contains no parallel subgraphs, or if $\mathcal{G}_{\ell_x}$ contains neither s nor t, then transforming $\mathcal{G}_{\ell_x}$ into the required DAG that contains paths between c_{x-1} and c_x equivalent to those paths in G is achieved by simply directing all edges in $\mathcal{G}_{\ell_x}$ away from c_{x-1} and towards c_x. So, in the sequel we assume that $\mathcal{G}_{\ell_x}$ contains at least one parallel graph with s and/or t.

A parallel subgraph G_i of $\mathcal{G}_{\ell_x}$ with terminals a_i and b_i is a *maximal* parallel subgraph of $\mathcal{G}_{\ell_x}$ if G_i is the parallel subgraph of $\mathcal{G}_{\ell_x}$ with the largest number of vertices and edges that has terminals a_i and b_i. For example, in Fig. 2 (left) the maximal parallel subgraph G_i with terminals a_i and b_i includes 12 vertices.

Let G_1, G_2, ..., G_p be the maximal parallel subgraphs of $\mathcal{G}_{\ell_x}$ indexed such that if G_j is a subgraph of G_i then $1 \leq j < i \leq p$ (note that G_j and G_i can have at most one terminal in common). To transform $\mathcal{G}_{\ell_x}$ into a DAG, we process the maximal parallel subgraphs G_1, ..., G_p in $\mathcal{G}_{\ell_x}$ in increasing order of index. If $\mathcal{G}_{\ell_x}$ contains both s and t, then any path σ from s to t in G must stay within $\mathcal{G}_{\ell_x}$ and so the rest of G can be deleted.

In the sequel we refer to multiple different parallel subgraphs at the same time, so we describe a consistent notation. The index i corresponds to the maximal parallel subgraph G_i of $\mathcal{G}_{\ell_x}$ that is currently being processed, the index j corresponds to a maximal parallel subgraph G_j contained within G_i, and the index k corresponds to a maximal parallel subgraph G_k that contains G_i. All the maximal parallel subgraphs with indices $j < i$ have already been processed and we have yet to process the maximal parallel subgraphs with indices $k > i$.

To simplify the description of our algorithm we first modify $\mathcal{G}_{\ell_x}$ as follows. If vertex a is a common terminal of two or more maximal parallel subgraphs G_{a_1}, G_{a_2}, ..., G_{a_r}, then we modify $\mathcal{G}_{\ell_x}$ by replacing a with r vertices a_1, a_2, ..., a_r. Vertex a_g, for all $g = \{1, 2, ..., r\}$, is adjacent to all neighbours u of a in G_{a_g} that are not in any other maximal subgraph G_{a_h}, $h \neq g$; the length of edge (a_g, u) is the same as the length of (a, u). Each one of these vertices a_g has the same set of items as a and vertices a_h and a_{h+1} are connected with an edge (a_h, a_{h+1}) of length 0 for all $h = 1, 2, ..., r - 1$. Note that this transformation creates some paths in which the same item appears multiple times. We will fix this later by merging all vertices a_h back into a single vertex a. We say that terminal vertices a_1, a_2, ..., a_r are *entangled*.

Important Note. We always assume that for each maximal parallel subgraph G_i of $\mathcal{G}_{\ell_x}$ the terminal a_i is on the left and terminal b_i is on the right, as shown in the figures.

Below we explain how to transform $\mathcal{G}_{\ell_x}$ into a DAG using three steps.

Step 1: Edges Close to s and t

Consider the maximal parallel subgraph G_i of $\mathcal{G}_{\ell_x}$ with terminals a_i and b_i. If G_i is the lowest indexed maximal parallel subgraph of $\mathcal{G}_{\ell_x}$ containing s, then for each simple path p from s to a_i that does not route through either t or b_i, the edges of p are directed away from s and towards a_i. Similarly, for each simple path p from s to b_i that does not route through either t or a_i, the edges of p are directed away from s and towards b_i (see Fig. 2).

If G_i is the lowest indexed maximal parallel subgraph of $\mathcal{G}_{\ell_x}$ containing t, then direct to t the edges in each simple path p from a_i to t or from b_i to t that does not route through either s, a_i, or b_i (see Fig. 2).

Iif G_i is the lowest indexed maximal parallel subgraph of $\mathcal{G}_{\ell_x}$ containing s and t, then for each simple path p from s to t that does not route through either a_i or b_i, the edges in p are directed away from s and towards t (see Fig. 2).

Step 2: Transforming a Parallel Subgraph G_i of $\mathcal{G}_{\ell_x}$ into Several DAGs

(a) Let $\hat{G}_i = (\hat{V}_i, \hat{E}_i, a_i, b_i)$ be the subgraph of G_i that includes the terminals a_i and b_i of G_i, all vertices of G_i that do not belong to any maximal parallel subgraph G_j of G_i, and all undirected edges between these vertices. For example, in the left side of Fig. 2 $\hat{G}_i$ does not include the simple paths from s to the terminals or from the terminals to t, as these edges were directed during Step 1.

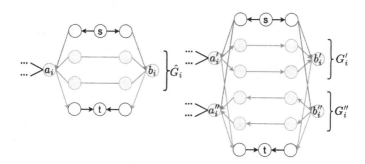

Fig. 2. (Left) The maximal parallel subgraph G_i is shown before being processed by Step 2; $\hat{G}_i$ consists of a single connected parallel subgraph (shown in blue). (Right) After G_i has been processed by Step 2 the simple paths between a_i' and b_i' are directed from a_i' to b_i' and the simple paths between b_i'' and a_i'' are directed from b_i'' to a_i''. Any edges incident on the original terminals a_i and b_i before processing G_i are now incident on the corresponding terminals of G_i' and G_i''. (Color figure online)

The subgraph $\hat{G}_i$ consists of one or more components. Multiple components are present when G_i contains multiple maximal parallel subgraphs; since each maximal parallel subgraph G_j of G_i must have been processed before G_i, $\hat{G}_i$ does not include the directed edges of the DAGs created for the maximal subgraphs G_j. Let the connected components of $\hat{G}_i$ be $\hat{G}_{i_1}, \hat{G}_{i_2}, ..., \hat{G}_{i_q}$ (see Fig. 3).

We create two copies of each $\hat{G}_{i_w}$: G_{i_w}' and G_{i_w}''. Each of the edges in these copies has the same length as the corresponding edge in $\hat{G}_{i_w}$, and the two copies u' and u'' of a vertex u in $\hat{G}_{i_w}$, store the same items as u.

For the remainder of the description of the algorithm, we use the following notation. A vertex v' marked with the prime symbol $(')$ represents the copy of a vertex v that belongs to the first copy G_ℓ' of a subgraph $\hat{G}_\ell$ and a vertex v'' marked with the double prime symbol $('')$ represents the copy of vertex v that belongs to the second copy G_ℓ'' of $\hat{G}_\ell$.

For each undirected edge (u, v) where u is in $\hat{G}_{i_w}$ and v is in G_i but not $\hat{G}_{i_w}$ (see the purple edges in Fig. 3), delete (u, v) from G_i. If v was not a duplicate

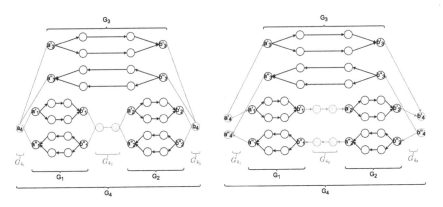

Fig. 3. (Left) The maximal parallel subgraph $G_i = G_4$ is shown before being processed by Step 2 and since the maximal parallel subgraphs G_1, G_2, and G_3 are each contained within G_4 (so each is transformed into DAGs before G_4 is processed) then $\hat{G}_i$ consists of multiple components. The vertices and edges of $\hat{G}_{4_1}$, $\hat{G}_{4_2}$, and $\hat{G}_{4_3}$ are shown in blue. (Right) After G_i has been processed by Step 2, the components of $\hat{G}_i$ have been duplicated and their incident edges have been directed. (Color figure online)

vertex (this occurs when s or t is a terminal of a maximal parallel subgraph G_j contained within G_i) then add the undirected edge (u', v) to G'_{i_w} and add the undirected edge (u'', v) to G''_{i_w}. If v was a duplicate vertex, then after removing both edges (u, v') and (u, v'') add the undirected edge (u', v') to G'_{i_w} and add the undirected edge (u'', v'') to G''_{i_w}.

(b) For each vertex v contained in G_i but not in $\hat{G}_i$, that is adjacent to a_i or b_i, we proceed as follows: (i) for each directed edge (a_i, v) of G_i (remember that some edges of G_i were assigned a direction in Step 1) delete (a_i, v) and add directed edges (a'_i, v) to G'_{i_1} and (a''_i, v) to G''_{i_1}, and for each directed edge (v, a_i) of G_i delete (v, a_i) and add directed edges (v, a'_i) to G'_{i_1} and (v, a''_i) to G''_{i_1}; (ii) for each directed edge (b_i, v) of G_i delete (b_i, v) and add directed edges (b'_i, v) to G'_{i_q} and (b''_i, v) to G''_{i_q}, and for each directed edge (v, b_i) of G_i delete (v, b_i) and add directed edges (v, b'_i) to G'_{i_q} and (v, b''_i) to G''_{i_q}. For example, in Fig. 2 (left) the red and green directed edges are incident on a_i and b_i but in Fig. 2 (right) the red and green directed edges (which have been duplicated) are incident on a'_i, a''_i, b'_i, and b''_i.

Now the subgraphs G'_{i_1}, G'_{i_2}, ..., G'_{i_q} are all connected and hence we refer to them simply as G'_i, and all the subgraphs G''_{i_1}, G''_{i_2}, ..., G''_{i_q} are connected and we refer to them as G''_i.

(c) We transform each G'_i and G''_i into DAGs by directing every undirected edge in G'_i from left to right and every undirected edge in G''_i from right to left (see Fig. 2).

The above steps (a)–(c) allow paths to traverse from the left terminal of G'_i to its right terminal and from the right terminal of G''_i to its left terminal.

After the edges of G_i' and G_i'' have been directed, any undirected edges (u, v) where u is a terminal a_i or b_i of G_i and v is not in G_i need to be added to the directed graphs G_i' and G_i''. These edges will allow paths to traverse to and from the newly created G_i' and G_i'' and any adjacent maximal parallel subgraphs and/or any maximal parallel subgraph G_k containing G_i. Note that G_k might contain within it several maximal parallel subgraphs that were transformed into DAGs, so sometimes the edges that we will add to G_i' and G_i'' to connect them to G_k are incident on vertices that have already been processed and sometimes they are incident on vertices that have not been processed yet.

Observation 1. *Each terminal a_i and b_i of a maximal parallel subgraph G_i of $\mathcal{G}_{\ell_x}$ has at most one neighbor v that is contained in $\mathcal{G}_{\ell_x}$ but not contained in G_i.*

Observation 1 follows from the definitions of series and parallel compositions and from the creation of entangled vertices for shared terminals.

(d) We add undirected edges to G_i' and G_i'' as follows:

- For each undirected edge (a_i, v') where v' is not in G_i, delete (a_i, v') and add undirected edge (a_i', v') to G_i'; for each undirected edge (a_i, v'') where v'' is not in G_i, delete (a_i, v'') and add undirected edge (a_i'', v'') to G_i''. Note that the vertices v' and v'' mentioned above correspond to copies of a vertex that has already been processed.
- For each unprocessed vertex v adjacent to a_i that is not in G_i, add undirected edges (a_i', v) and (a_i'', v) to G_i' and G_i'', respectively.
- The same two steps above are repeated for each vertex v adjacent to b_i that is not in G_i, and the new edges are incident to b_i' and b_i'' in G_i' and G_i''. Each component $\hat{G}_{i_w}$ is then deleted, so that the terminals a_i and b_i and all undirected edges of $\hat{G}$ are removed. Note that the duplicate copies of each of these vertices and edges still exist in G_i' and G_i''.

Step 3: Edges Connecting Nested Parallel Graphs Containing s or t
Recall that since G_i is a maximal parallel subgraph, it was created from the parallel composition of a group of series-parallel subgraphs $G_{i_1}, G_{i_2}, ..., G_{i_m}$. Let the *parallel components* $pc_{i1}, pc_{i2}, ..., pc_{im}$ of G_i be these series-parallel subgraphs, excluding the terminals of G_i (see Fig. 4).

If a parallel component pc_{ih} either contains no maximal parallel subgraphs or it does not contain s or t, then the edges in the DAGs for pc_{ih} have already been correctly directed in Steps 1 and 2. Therefore, we only consider parallel components pc_{ih} of G_i that contain at least one maximal parallel subgraph and s and/or t. Each maximal parallel subgraph G_j in a parallel component pc_{ih} of G_i was previously processed, and so the two terminals a_j and b_j of G_j were transformed into four terminals a_j', a_j'', b_j', and b_j''.

Intuitively, the algorithm that we describe below modifies the edges of the DAGs corresponding to a parallel component pc_{ih} of G_i containing s or t such that (i) when a maximal parallel subgraph G_j of pc_{ih} contains s but not t, paths in the DAGs corresponding to G_i are directed outwards away from G_j;

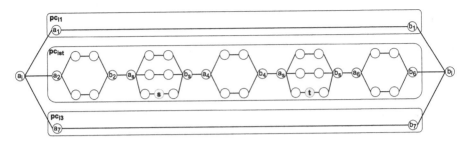

Fig. 4. This maximal parallel subgraph G_i of G contains three parallel components pc_{i1} and pc_{ist}, and pc_{i3}; for simplicity, pc_{i1} and pc_{i3} do not contain many vertices. Note that the parallel component pc_{ist} contains both s and t.

(ii) when a maximal parallel subgraph G_j of pc_{ih} contains t but not s, paths in the DAGs corresponding to G_i are directed inwards towards G_j; and (iii) when a maximal parallel subgraph G_j of pc_{ih} contains both s and t, paths in the DAGs corresponding to G_i can travel away from s through G_j and possibly through several maximal parallel subgraphs containing G_j in increasing order of index, then traverse exactly one parallel component of one of these maximal parallel subgraphs, and finally return to t.

If a parallel component pc_{ih} of G_i contains s or t, we modify the DAGs corresponding to the parallel component pc_{ih} as follows:

(a) Let G_s and G_t be the smallest indexed maximal parallel subgraphs that contain s and t, respectively. If pc_{ih} contains at least one of s and t, but no maximal parallel subgraph G_j of pc_{ih} contains both s and t, then:

- If a simple path σ_1 exists between a_i and a terminal of G_s that does not traverse through G_t or b_i, we add a directed edge (u_a'', a_i') (see the red edges on the left side of Fig. 5), where u_a is a vertex of pc_{ih} adjacent to a_i (see Observation 1). Then, we delete G_j' for each maximal parallel subgraph G_j of pc_{ih} reachable by σ_1 such that $G_j \neq G_s$ and G_s is not contained within G_j. These steps allow paths to traverse from s to the left terminal of G_i and then onto the right terminal of G_i without creating cycles.
  Similarly, if a simple path σ_2 exists between b_i and G_s that does not traverse through G_t or a_i, we add a directed edge (u_b', b_i''), where u_b is a vertex of pc_{ih} adjacent to b_i. Then, we delete G_j'' for each maximal parallel subgraph G_j of pc_{ih} reachable by σ_2 such that $G_j \neq G_s$ and G_s is not contained within G_j. These steps allow paths to traverse from s to the right terminal of G_i and then onto the left terminal of G_i without creating cycles.
- If a simple path σ_1 exists between a_i and a terminal of G_t that does not traverse through G_s or b_i, we add a directed edge (a_i'', u_a'), where u_a is a vertex of pc_{ih} adjacent to a_i (see Observation 1). Then, we delete G_j'' for each maximal parallel subgraph G_j of pc_{ih} reachable by σ_1 such that $G_j \neq G_t$ and G_t is not contained within G_j.
  If a simple path σ_2 exists between b_i and G_t that does not traverse through G_s or a_i, we add a directed edge (b_i', u_b'') (see Fig. 5), where u_b is a vertex of

pc_{ih} adjacent to b_i, and we delete G'_j for each maximal parallel subgraph G_j of pc_{ih} reachable by σ_2 such that $G_j \neq G_t$ and G_t is not contained within G_j. These two changes allow paths to traverse both terminals of G_i before reaching t.

– We delete G''_j for each maximal parallel subgraph G_j contained within pc_{ih} such that $G_j \neq G_s$, $G_j \neq G_t$, G_s and G_t are not contained within G_j, and G_j is reachable from a simple path between G_s and G_t that does not traverse through either a_i or b_i. This step simplifies the DAGs.

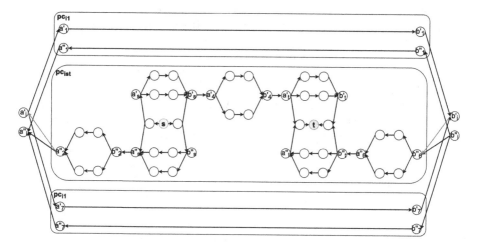

Fig. 5. The parallel component pc_{ist} from Fig. 4 containing both s and t, but not in the same maximal parallel subgraph G_j. The edges highlighted in red have been added. (Color figure online)

(b) If any parallel component pc_{ih} of G_i contains both s and t within the same maximal parallel subgraph G_j, then we need to change a few edges incident on the terminals of G_i (see Fig. 6):

– For each directed edge (u'', a''_i), where u'' is in G''_i but not in pc_{ih}, delete the edge and add the directed edge (a''_i, u'), where u' is the other duplicate of u''. This allows paths to traverse from s towards the left terminal a''_i of G''_i to then go towards the right through a different parallel component pc'_{im}.

– For each directed edge (b''_i, u''), where u'' is in G''_i but not in pc_{ih}, delete the edge and add the directed edge (u', b''_i), where u' is the other duplicate of u''. These last two changes allow paths from s to a''_i to then traverse through another parallel component pc'_{im} to reach b''_i and then return to t. Note that a path σ' from s to t traveling through b''_i must have already traveled through a''_i, so the edges (u'', a''_i) and (b''_i, u'') cannot be in σ'.

– For each directed edge (u', b'_i), where u' is in G'_i but not in pc_{ih}, delete the edge and add the directed edge (b'_i, u''). This allows paths that traverse from

s towards the right terminal b'_i of G'_i to then go towards the left through a different parallel component pc''_{im}.

- For each directed edge (a'_i, u'), where u' is in G'_i but not in pc_{ih}, delete the edge and add the directed edge (u'', a'_i). These changes allow paths from s to b'_i to then traverse left to reach a'_i and then return to t. Note that a path σ' from s to t traveling through a'_i must have already traveled through b'_i, so the edges (u', b'_i) and (a'_i, u') cannot be in σ'.

- Finally, if G_i is the lowest indexed maximal parallel subgraph that contains G_s, $G_s = G_t$, and neither s nor t is a terminal of G_s, then we delete the edges (u', a'_s), (a'_s, u''), (b'_s, u'), and (u'', b''_s), where u' is in G'_i and u'' is in G''_i but neither are in G_s. Then, add the directed edges (a'_s, u''), (u', a'_s), (u'', b'_s), and (b''_s, u'); these edges allow paths from s that traverse both terminals of G'_i or G''_i to reach t (see Fig. 6).

These changes are done because a path σ from s to t in G entering G_s through b_s must have traveled through a_s and must then travel to t, and so the edges (a''_s, u'') and (u'', b''_s) cannot be in the DAGs as they create cycles. Similarly, (u', a'_s) and (b'_s, u') cannot be in the DAGs.

To see this, note that if the edges (a''_s, u'') and (u'', b''_s) existed then a path σ' that exits G''_s from a''_s and traverses through some parallel component pc''_{ih} to a''_s and then traverses to the right to b''_s would then continue towards b''_i, which then can reach a''_s, creating a cycle. A similar argument shows that (u', a'_s) and (b'_s, u') cannot be in the DAGs.

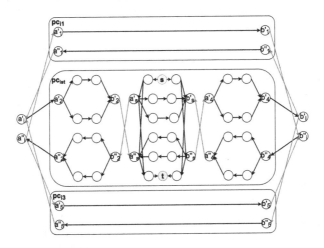

Fig. 6. A parallel component pc_{ist} of G_i contains DAGs for each maximal parallel subgraph G_j contained within G_i. Since pc_{ist} contains both s and t within a maximal parallel subgraph G_j of G_i, the edges highlighted in red have been added. (Color figure online)

After a maximal parallel subgraph G_i has been transformed into the DAGs G'_i and G''_i, the remaining undirected edges from G_i and any isolated vertices are deleted. Additionally, any entangled vertices for which all its corresponding entangled vertices are contained within G_i are merged into a single vertex.

Note that the two DAGs G_i' and G_i'' might each contain a vertex corresponding to every vertex of G_i. Additionally, G_i' and G_i'' might each contain a directed edge corresponding to every undirected edge of G_i. Therefore, G_i' and G_i'' have at most twice as many vertices and edges as G_i.

2.4 Algorithm Analysis

Definition 1. *An undirected path σ from s to t in G and a directed path σ' from s to t in $\mathcal{G}$ are equivalent if both have the same number of vertices, each vertex in σ and the corresponding vertex in σ' store the same set of items, and every edge in σ and its corresponding edge in σ' have the same length.*

We say that a directed path σ *enters* a maximal parallel subgraph $G_i = (V_i, E_i, a_i, b_i)$ when σ includes an edge where one endpoint is either a_i or b_i, the other endpoint u is in G_i, and the edge is directed towards u. Similarly, we say that a directed path σ *exits* G_i when it includes an edge where one endpoint is either a_i or b_i, the other endpoint u is not in G_i, and the edge is directed to u. Finally, we say that a path σ *traverses* through G_i if σ both enters and exits G_i. We extend the notation of entering, exiting, and traversing maximal parallel subgraphs to parallel components as well.

Recall that we assume that in G_i' and G_i'' the terminals a_i' and a_i'' are always drawn on the left and the terminals b_i' and b_i'' are drawn on the right. In the analysis, we use pc_{ih} to refer to the hth parallel component in G_i, and we use pc_{ih}' and pc_{ih}'' to refer to the corresponding directed subgraphs created by the algorithm.

Lemma 1. *Our algorithm transforms an instance of ThOP on an undirected series-parallel graph G into an instance of ThOP on a directed graph $\mathcal{G}$ with no cycles, so all paths from s to t in $\mathcal{G}$ are simple.*

Proof. To prove that there are no cycles in $\mathcal{G}$, we note that since all directed edges in $\mathcal{G}$ correspond to undirected edges in G, a cycle in $\mathcal{G}$ would be either (1) two anti-parallel directed edges in $\mathcal{G}$ corresponding to an undirected edge (u, v) in G or (2) a directed cycle in $\mathcal{G}$ corresponding to a cycle in G.

(1) For each undirected edge (u, v) in G there are no corresponding anti-parallel directed edges in $\mathcal{G}$: When our algorithm creates two copies (u', v') and (u'', v'') of an undirected edge (u, v) of G, these edges are incident on different copies of u and v, thus they do not form a cycle.

(2) For each simple cycle C in G, we show that the corresponding directed edges in $\mathcal{G}$ do not form a cycle. Consider a simple cycle C in G. Observe that C must be fully contained within a parallel subgraph. Let G_i be the lowest indexed maximal parallel subgraph that contains C. Note that C must include both terminals a_i and b_i of G_i and so C contains a path that traverses a parallel component pc_{ih} of G_i from a_i to b_i and a path that traverses a different parallel component pc_{im} of G_i from b_i to a_i.

Recall that the algorithms in Sect. 2.3 change some of the edges of the parallel subgraphs G_i' and G_i'' corresponding to G_i if G_i contains s and/or t. To show that the directed edges in $\mathcal{G}$ corresponding to C do not form a cycle, we must consider five cases depending on whether the parallel subgraph G_i contains s, t, or neither:

- If G_i does not contain s or t, then either (i) the vertices and edges in G_i were not duplicated but instead were directed away from one terminal and towards the other (directed towards the cut vertex closer to t) and this does not form cycles, or (ii) the algorithm in Sect. 2.3 directs the edges in G_i' and G_i'' corresponding to G_i; in this case, there is a path σ' in $\mathcal{G}$ from a_i' to b_i' but since our algorithm does not add any incident edges to b_i' directed towards vertices inside G_i' or G_i'' then σ' cannot be extended to traverse back to a_i' to create a cycle. Similarly, a path σ' in $\mathcal{G}$ from b_i'' to a_i'' cannot be extended to create a cycle.
- If G_i contains s but not t, then by statement (ii) in the above argument there is a path σ' in $\mathcal{G}$ from a_i' to b_i', but there are no incident edges to b_i' directed towards vertices inside G_i' or G_i''. Similarly, a path σ' in $\mathcal{G}$ from b_i'' to a_i'' cannot be extended to create a cycle.
- If G_i contains t but not s, then by statement (ii) in the above argument there is a path σ' in $\mathcal{G}$ from a_i' to b_i' but the algorithm in Sect. 2.3 (a) ensures that the only incident edges to b_i' directed towards G_i' or G_i'' lead only to t, and so σ' cannot be extended to create a cycle. Similarly, a path σ' in $\mathcal{G}$ from b_i'' to a_i'' cannot be extended to create a cycle.
- If G_i contains both s and t but no maximal parallel subgraph G_j contained within G_i contains both s and t, then a combination of the above arguments shows that there is path σ' in $\mathcal{G}$ from a_i' to b_i' (or from b_i'' to a_i'') but σ' cannot be extended to create a cycle.
- If G_i contains both s and t in the same maximal parallel subgraph G_j contained within G_i, then there exists no path σ' in $\mathcal{G}$ from a_i' to b_i', as all paths from a_i' are directed towards t. To see this, note that that the algorithms in Sect. 2.3 (b) include only one outgoing directed edge in $\mathcal{G}$ incident to a_i' and it leads to a_s'', whose only outgoing directed edge leads to t (see Fig. 6).
  However, there is a path σ' from a_i'' to b_i'' because the algorithms in Sect. 2.3(b) add edges from a_i'' towards each of the other parallel components of G_i that lead to b_i'' (see Fig. 6). If there is also a path σ'' in $\mathcal{G}$ from b_i'' that enters G_i'' then σ'' leads to t, but since in this case our algorithm does not add any incident edges to t directed towards a_i'' then σ'' cannot be extended to create a cycle.
  Similarly, a path σ' in $\mathcal{G}$ from b_i' to a_i' corresponding to a subpath in C from b_i to a_i cannot be extended to create a cycle.   □

Lemma 2. *Our algorithm transforms an instance of ThOP on an undirected series-parallel graph G into an instance of ThOP on a directed graph $\mathcal{G}$ such that for every directed path σ' from s to t in $\mathcal{G}$, there is an equivalent simple path σ from s to t in G.*

Proof. Given a path σ' from s to t in $\mathcal{G}$ we include in σ the undirected edges corresponding to the directed edges in σ' and so we only need to show that σ' does not traverse through two vertices storing the same set of items.

For a maximal parallel subgraph G_i of G, one of the duplicate vertices v' is in G_i' and the other duplicate v'' is in G_i''. If G_i does not contain s or t, then no path from s to t in $\mathcal{G}$ visits both v' and v'' because our algorithm does not add edges to $\mathcal{G}$ for a path σ' to exist that traverses from G_i' to G_i'' or from G_i'' to G_i'.

Recall that for parallel components pc_{ih} in a maximal parallel subgraph G_i that contain s or t, but not both within the same maximal parallel subgraph G_j, the algorithm in Sect. 2.3 (a) deletes one of G_j' or G_j'' for each maximal parallel subgraph G_j in G_i, except G_s and G_t; hence the only vertices for which their two duplicate copies exist in G_i are located in the DAGs for G_s and G_t. For G_s' and G_s'', a path σ' from s to t must either traverse and exit G_s' or traverse and exit G_s''; in either case, σ' cannot traverse through the two duplicate vertices of a vertex of G_s. Similarly, a path σ' from s to t must either traverse G_t' and then go to t or traverse G_t'' and then go to t; in either case, σ' cannot traverse through the two duplicate vertices of the same vertex of G_t.

For a parallel component pc_{ih} in G_i that contain s and t within the same maximal parallel subgraph G_j, a path σ' from s to t must traverse one of the parallel components pc_{ih}' or pc_{ih}'' created for parallel component pc_{ih}, but σ' cannot traverse both. To see this, observe that if σ' traverses pc_{ih}'' toward the left terminal a_i'' of G_i'', then by the algorithm in Sect. 2.3 (b) σ' cannot reach pc_{ih}', but it can traverse to the right to a different parallel component pc_{im}' and onto b_i'', from where it still cannot reach pc_{ih}'. Note that when σ' traverses from b_i'' back into pc_{ih}'', σ' continues toward t and this does not create a cycle. Similarly, σ' can traverse pc_{ih}' to the right terminal b_i' from where it cannot reach pc_{ih}'' but it can traverse to the left to a different parallel component pc_{im}'' and onto a_i', from where it must continue toward t. Therefore, σ' is unable to traverse multiple duplicate vertices that correspond to the same vertex in G.   □

Lemma 3. *Our algorithm transforms an instance of ThOP on an undirected series-parallel graph G into an instance of ThOP on a directed graph $\mathcal{G}$ such that for every simple path σ from s to t in G, there is an equivalent directed path σ' from s to t in $\mathcal{G}$.*

Due to space limitations, we have omitted the proof of Lemma 3[1]. However, it can be shown through induction that for a simple undirected path σ from s to t in G that goes through a list $s = t_0, t_1, t_2, ..., t_d, t_{d+1} = t$ of terminals of maximal parallel subgraphs (and hence, includes subpaths $<t_0, t_1>$, $<t_1, t_2>$, ..., $<t_d, t_{d+1}>$), for the subpath of σ from s to any terminal t_i there is an equivalent directed path in $\mathcal{G}$ from s to a vertex corresponding to t_i that can be extended to include a directed subpath equivalent to the next segment $<t_i, t_{i+1}>$.

Theorem 1. *There is a PTAS for the thief orienteering problem when the graph G is an undirected series-parallel graph that produces solutions that use time at most $T(1 + \epsilon)$ for any constant $\epsilon > 0$.*

---

[1] The full paper can be found at www.csd.uwo.ca/~ablochha/SeriesParallel.pdf.

Proof. Our algorithm transforms an undirected series-parallel graph G into a DAG $\mathcal{G}$ by creating at most one additional vertex for each vertex in G and creating two directed edges for each undirected edge in G, and so our algorithm runs in polynomial time. By Theorem 2 in [2], Algorithm 3 in [2] is a PTAS for the thief orienteering problem when the input graph G is a DAG that produces solutions that use time at most $T(1 + \epsilon)$ for any constant $\epsilon > 0$. Therefore, by Lemmas 1, 2, and 3 the combination of our algorithm and Algorithm 3 in [2] is a PTAS for the thief orienteering problem when the input graph G is an undirected series-parallel graph that produces solutions that use time at most $T(1 + \epsilon)$ for any constant $\epsilon > 0$. □

References

1. Abdelkader, R., Abdelkader, Z., Mustapha, R., Yamani, M.: Optimal allocation of reliability in series parallel production system. In: Search Algorithms for Engineering Optimization, pp. 241–258. InTechOpen, Croatia (2013)
2. Bloch-Hansen, A., Page, D., Solis-Oba, R.: A polynomial-time approximation scheme for thief orienteering on directed acyclic graphs. In: Hsieh, S.Y., Hung, L.J., Lee, C.W. (eds.) IWOCA 2023. LNCS, vol. 13889, pp. 87–98. Springer, Cham (2023). https://doi.org/10.1007/978-3-031-34347-6_8
3. Bonyadi, M., Michalewicz, Z., Barone, L.: The travelling thief problem: the first step in the transition from theoretical problems to realistic problems. In: IEEE Congress on Evolutionary Computation (CEC), pp. 1037–1044. IEEE (2013)
4. Chagas, J., Wagner, M.: Ants can orienteer a thief in their robbery. Oper. Res. Lett. **48**(6), 708–714 (2020)
5. Chagas, J., Wagner, M.: Efficiently solving the thief orienteering problem with a max-min ant colony optimization approach. Optim. Lett. **16**(8), 2313–2331 (2022)
6. Chung, F., Leighton, F., Rosenberg, A.: Embedding graphs in books: a layout problem with applications to VLSI design. SIAM J. Algebraic Discrete Methods **1**(8), 33–58 (1987)
7. Duffin, R.: Topology of series-parallel networks. J. Math. Anal. Appl. **2**(10), 303–318 (1965)
8. Faêda, L., Santos, A.: A genetic algorithm for the thief orienteering problem. In: 2020 IEEE Congress on Evolutionary Computation (CEC), pp. 1–8. IEEE (2020)
9. Freeman, N., Keskin, B., Çapar, İ: Attractive orienteering problem with proximity and timing interactions. Eur. J. Oper. Res. **266**(1), 354–370 (2018)
10. Gago, J., Hartillo, I., Puerto, J., Ucha, J.: Exact cost minimization of a series-parallel reliable system with multiple component choices using an algebraic method. Comput. Oper. Res. **40**(11), 2752–2759 (2013)
11. Gunawan, A., Lau, H., Vansteenwegen, P.: Orienteering problem: a survey of recent variants, solution approaches and applications. Eur. J. Oper. Res. **255**(2), 315–332 (2016)
12. Santos, A., Chagas, J.: The thief orienteering problem: formulation and heuristic approaches. In: IEEE Congress on Evolutionary Computation, pp. 1–9 (2018)
13. Valdes, J: Parsing flowcharts and series-parallel graphs. Ph.D. dissertation, Standford University (1978)
14. Wang, J., Wu, X., Fan, X.: A two-stage ant colony optimization approach based on a directed graph for process planning. Int. J. Adv. Manuf. Technol. **80**, 839–850 (2015)

Approximation Algorithm for Job Scheduling with Reconfigurable Resources

Pierre Bergé[1]([✉])[iD], Mari Chaikovskaia[2][iD], Jean-Philippe Gayon[1][iD], and Alain Quilliot[1][iD]

[1] Université Clermont-Auvergne, CNRS, Mines de Saint-Etienne, Clermont-Auvergne-INP, LIMOS, 63000 Clermont-Ferrand, France
`pierre.berge@uca.fr`
[2] IMT Atlantique, LS2N, UMR CNRS 6004, 44307 Nantes, France

Abstract. We consider a scheduling problem with reconfigurable resources. Several types of jobs have to be processed by a set of identical resources (e.g. robots, workers, processors) over a discrete time horizon. In each time period, teams of resources must be formed to process jobs. During a given time period, a team handles one type of job and the number of jobs that can be processed depends on the team size. A resource which is used to perform some job type in a given period may be employed for another job type in the next period. The objective is to determine the minimum number of resources needed to meet a given demand for each job type. We provide a polynomial-time 4/3-approximation algorithm for this strongly NP-hard problem.

Keywords: Reconfiguration · Scheduling · Approximation

1 Introduction

In many industrial contexts, automatized production must adapt itself to a fast evolving demand of a large variety of customized products. One achieves such a flexibility requirement thanks to *reconfiguration*. Once an operation has been performed, related either to the production of some good or to its transportation, one may redesign the infrastructure that supported this operation by adding, removing or replacing some atomic components. Those components may be hardware (robots, instruments), software or human resources. They behave as renewable resources [2,3,9]. Depending on the way one assigns those resources to a given operation, one may not only achieve this operation but also speed it, increase its throughput or lessen its cost, as in the **multi-modal Resource Constrained Project Scheduling Problem** [2].

This work was financed by the French government IDEX-ISITE initiative 16-IDEX-0001 (CAP 20–25) and by the Auvergne-Rhone-Alpes region under the program "Pack Ambition Recherche". It was also supported by the International Research Center "Innovation Transportation and Production Systems" of the I-SITE CAP 20–25 and by Institut Carnot M.I.N.E.S.

A. Basu et al. (Eds.): ISCO 2024, LNCS 14594, pp. 263–276, 2024.
https://doi.org/10.1007/978-3-031-60924-4_20

We consider the following scheduling problem with reconfigurable resources. Several types of jobs (e.g. production operations, transportation tasks) have to be processed by a set of identical resources (e.g. workers, robots, processors) over a discrete time horizon in order to achieve a certain demand. Assigning a number p of resources to some job of type k gives a certain production c_{pk}: all production values, called *capacities*, are given as inputs. In each time period, teams of resources must be formed to process jobs. A resource which is used to perform some type of job k at period t may be employed for another type of job $k' \neq k$ in the next period $t+1$. The objective is to determine the minimum number of resources needed to obtain a certain production for each type of job. This problem is called **Multi_Bot** and is strongly NP-hard [5].

It was first introduced in a warehouse logistics context [6] in collaboration with the MecaBotiX company [12] which designs reconfigurable mobile robots. In this application, resources are mobile robots and jobs consist in moving loads of various types, such as pallets or boxes. Those robots are capable of aggregating into a cluster to form poly-robots that can adapt to the type of product. Other applications can be found in the automotive industry [1].

From a theoretical point of view, **Multi_Bot** is strongly related to some classical scheduling problems of the literature [4,7,10,11,13]. A very well-known one is **Identical-machines scheduling** (**IMS** or $P||C_{\max}$), where the objective is to pack items into a set of identical boxes while minimizing the size of the most filled box [8]. Its high multiplicity variant, denoted by $P|HM(n)|C_{\max}$, encodes as binary inputs the number of items with the same size. $P|HM(n)|C_{\max}$ with polynomially bounded item sizes is a special case of **Multi_Bot**.

The main result of this paper is the presentation of a polynomial-time $\frac{4}{3}$-approximation algorithm for **Multi_Bot**. On one hand, we provide, for industrial applications such as MecaBotiX robots, a fast and efficient heuristic with the strict guarantee that it will not fail on pathological instances more than 33% over the optimum. On the other hand, we extend the theoretical knowledge on approximability of scheduling problems, showing that a generalization of $P|HM(n)|C_{\max}$ admits a constant approximation ratio.

2   Problem Description and Notations

We consider a set of identical resources (for instance robots, workers, etc.) which cooperate in order to process different types of jobs. A *p-resource* is a *configuration* which makes p resources cooperate on the same job. For example, in robotics, it models the fact that p elementary robots can assemble together in order to transport heavy loads. A maximum of P resources may cooperate. The set of configurations is $\mathcal{P} = \{1,\dots,P\}$. There are K job types and let $\mathcal{K} = \{1,\dots,K\}$. For each type $k \in \mathcal{K}$, its *demand* is denoted by d_k: it is the number of jobs of type k which have to be performed throughout the process. In robotics, it models the fact that there exists K types of loads (pallets, boxes...) and that at least d_k loads of type k have to be transported.

All tasks must be executed within a discrete time horizon $\mathcal{T} = \{1,\dots,T\}$. At the beginning of each period $t \in \mathcal{T}$, the resources may be reconfigured in order

to treat different jobs. During a given period t, a p-resource can deal with only a single type $k \in \mathcal{K}$, and its production is given by the capacity c_{pk}. For each type $k \in \mathcal{K}$, there is at least one value p such that c_{pk} is non-zero, *i.e.* at least one p-resource is able to process a job of type k. Our purpose is to minimize the number of resources H used into the whole process. If H_t denotes the number of active resources during period t, then $H = \max_{t \in \mathcal{T}} H_t$.

We denote by x_{pkt} the decision variable representing the number of jobs of type k performed in configuration p at period t. Values K, P, T can be seen as combinatorial inputs while demands and capacities are numerical values.

*Problem 1 (**Multi_Bot**).*

Input: $\mathcal{K} = \{1, \ldots, K\}, \mathcal{P} = \{1, \ldots, P\}$
$\mathcal{T} = \{1, \ldots, T\}$
Capacities $(c_{pk})_{\substack{p \in \mathcal{P} \\ k \in \mathcal{K}}}$ *, Demands* $(d_k)_{k \in \mathcal{K}}$

Objective: $\text{minimize } H = \max_{t \in \mathcal{T}} H_t$

subject to: $\displaystyle\sum_{t \in \mathcal{T}} \sum_{p \in \mathcal{P}} c_{pk} \cdot x_{pkt} \geq d_k$ $\hspace{2cm} \forall k \in \mathcal{K}$

$\displaystyle H_t = \sum_{k \in \mathcal{K}} \sum_{p \in \mathcal{P}} p \cdot x_{pkt}$

$x_{pkt}, H \in \mathbb{N}$ $\hspace{2cm} \forall k \in \mathcal{K}, p \in \mathcal{P}, t \in \mathcal{T}$

Figure 1 illustrates an optimal solution (*i.e.* which minimizes H) for the following instance of **Multi_Bot**. There are two types of jobs ($K = 2$) and three periods ($T = 3$). The maximum size of a performed job is $P = 5$. The demands are $d_1 = 13$ and $d_2 = 10$. The non-zero capacities for jobs of type 1 are $c_{11} = 1$ and $c_{21} = 4$. The demand $d_1 = 13$ is achieved as the schedule contains three 2-resources (total production 12) and one 1-resource (production 1) handling jobs of type $k = 1$. The capacities for jobs of type 2 are linear in p: $c_{p2} = p$. One can check the demand for $k = 2$ is also satisfied. Furthermore, it is not possible to find a solution which achieves $H = 5$.

A *schedule* is a solution of the **Multi_Bot** problem.

$$\mathbf{x} = (x_{pkt})_{p \in \mathcal{P}, k \in \mathcal{K}, t \in \mathcal{T}}$$

In the remainder, we abuse notation when the context is clear: the same vector could be denoted implicitly by $(x_{pkt})_{p,k,t}$ to gain some space. Observe that the size PKT of a schedule $\mathbf{x}$ is polynomial in the input size. However, values c_{pk} and d_k might be exponential in the input size, but also x_{pkt}, H_t and H. Given an instance of **Multi_Bot**,

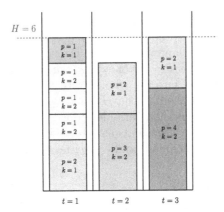

Fig. 1. An example of schedule $\mathbf{x} = (x_{pkt})_{p,k,t}$ with $H = \max_t H_t = 6$

– let H^* denote the optimum value for the objective function H,
– let $\mathbf{x}^* = (x^*_{pkt})_{p,k,t}$ be a schedule reaching this optimum.

In comparison with a schedule $\mathbf{x}$ which is a vector with KPT values, a *packing* $\boldsymbol{\pi} = (\pi_{pk})_{p,k}$ represents the schedule of jobs of type k by p-resources, independently from any time consideration. For example, it can be used to describe the production process during one specific period. This is a vector with KP values. Given some schedule $\mathbf{x} = (x_{pkt})_{p,k,t}$, its *associated packing* is simply given by all values $\pi_{pk} = \sum_{t\in\mathcal{T}} x_{pkt}$, for all $p \in \mathcal{P}, k \in \mathcal{K}$. In particular, the packing associated with the optimal schedule $\mathbf{x}^*$ is denoted by $\boldsymbol{\pi}^* = \left(\sum_{t\in\mathcal{T}} x^*_{pkt}\right)_{p,k}$.

Definition 1. (Volume). *The* volume *of a packing is the total number of resources involved in it. Formally, for $\boldsymbol{\pi} = (\pi_{pk})_{p,k}$,*

$$\mathrm{vol}(\boldsymbol{\pi}) = \sum_{k=1}^{K}\sum_{p=1}^{P} p \cdot \pi_{pk}.$$

For the sake of simplicity, we also use this notion for schedules. The volume of a schedule $\mathbf{x}$ is the volume of its associated packing, i.e. $\sum_{p,k,t} p \cdot x_{pkt}$.

Definition 2. (Maximum). *The* maximum *of a packing $\boldsymbol{\pi}$ is the maximum $p \in \mathcal{P}$ such that some $\pi_{pk} > 0$:*

$$\max(\boldsymbol{\pi}) = \max\{p \in \mathcal{P} : \pi_{pk} > 0 \text{ for some } k\}$$

Definition 3. (Scale). *Given some integer λ, let us call the* big configurations *the p-resources with $p > \frac{2\lambda}{3}$ and the* medium configurations *with $\frac{\lambda}{3} < p \le \frac{2\lambda}{3}$. Naturally, the* small configurations *refer to $p \le \frac{\lambda}{3}$. We define the λ-scale as the number of jobs performed with big configurations in packing $\boldsymbol{\pi}$ plus half the number of jobs performed with medium configurations in $\boldsymbol{\pi}$. It is a half-integer:*

$$\lambda\text{-scale}(\boldsymbol{\pi}) = \sum_{k\in\mathcal{K}}\sum_{p>\frac{2\lambda}{3}} \pi_{pk} + \frac{1}{2}\sum_{k\in\mathcal{K}}\sum_{\frac{\lambda}{3}<p\le\frac{2\lambda}{3}} \pi_{pk} \tag{1}$$

Given a solution $\mathbf{x}$ of **Multi_Bot** using H resources, its associated packing π has a limited H-scale. Indeed, a period of $\mathbf{x}$ contains, in addition with small configurations, at most a big one or two medium ones. Hence, H-scale$(\pi) \leq T$.

As an example, for the packing π associated to the schedule proposed in Fig. 1, we have vol$(\pi) = 17$, max$(\pi) = 4$ and 5-scale$(\pi) = 1 + 4 \cdot \frac{1}{2} = 3$.

3   Identical-Machines Scheduling

We recall the definition of a well-known problem in operations research: **IMS** [8]. To draw parallels between **IMS** with **Multi_Bot**, we use the following syntax. Formally, we are given a set of items $\mathcal{I} = \{1, \ldots, n\}$ and a set of boxes[1] $\mathcal{B} = \{1, \ldots, m\}$. An item i has a certain size s_i. The objective is to pack all items into the boxes such that the maximum size packed into a box is minimized. More precisely, we aim at minimizing the size of the most filled box.

Observe that the minimization function of **Multi_Bot** and **IMS** are similar: in both problems, we aim at decreasing a certain "volume" of the jobs/items put into the most filled box/period. Given an **IMS** instance $\mathcal{J}$, we define:

- its *volume* vol$(\mathcal{J})$ as the total size of its items, *i.e.* vol$(\mathcal{J}) = \sum_{i \in \mathcal{I}} s_i$,
- its *maximum* as the maximum item size: max$(\mathcal{J}) = \max_{i \in \mathcal{I}} s_i$.

A heuristic of **IMS** will be used as a sub-routine for our approximation algorithm dedicated to **Multi_Bot**. It is called LONGEST-PROCESSING-TIME-FIRST (LPT-FIRST) [8]. Its description is relatively simple: first sort the items by decreasing order of their sizes (the largest size comes first): they will be into the boxes in this order. Second, the filling satisfies the following rule: we always select the least filled box to pack the current item. Figure 2 shows the packing obtained with LPT-FIRST for item sizes $(5, 3, 3, 3, 2, 2, 1)$ with three boxes. Graham showed that LPT-FIRST is a $\frac{4}{3}$-approximation algorithm [8]. Moreover, LPT-FIRST is a fast algorithm: it runs only in $O(n(\log m + \log n))$.

In the literature, **IMS** admits other approximation algorithms with a smaller ratio [7,10]. In the remainder, we focus only on LPT-FIRST since this algorithm allows us to identify an approximation algorithm of **Multi_Bot** with our framework.

4   Presentation of the Algorithm

Our idea to design approximation algorithms for the general **Multi_Bot** follows. We call this general framework BOT-APPROX:

- **Step 1.** Compute a polynomial-sized collection Π of packings with structural properties (see Theorem 1 for details),

---

[1] In the classic **IMS** syntax, "boxes" are typically called machines and the "items" are jobs with a certain processing time. We modify this syntax to be consistent with the **Multi_Bot** framework.

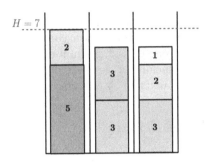

Fig. 2. Output of LPT-FIRST with item sizes $(5, 3, 3, 3, 2, 2, 1)$ and $m = 3$.

- **Step 2.** For each $\pi \in \Pi$, create an **IMS** instance $I(\pi)$ with T boxes and items which are directly obtained from the packing π,
- **Step 3.** Find an approximate solution for all **IMS** instances $I(\pi)$ with LPT-FIRST. Return the one which corresponds to the best schedule.

Step 1 is very fuzzy for now, and we will introduce in Sect. 6 our method for computing this collection of packings. Our objective is to produce a collection Π containing at least one packing $\pi \in \Pi$ which is a good candidate for achieving a satisfying schedule $\mathbf{x}$ when we put its p-resources into the periods. The properties of this collection Π are presented in Theorem 1 (proof in Sect. 6).

Theorem 1. *Given some **Multi_Bot** instance, we can produce in polynomial time $O(KP^3T^3 + KP^7)$ a collection Π of at most $3P$ packings such that:*

- *each packing in Π satisfy all demands d_k, i.e. $\forall k$, $\sum_{p \in \mathcal{P}} c_{pk} \pi_{pk} \geq d_k$,*
- *it contains at least one packing π which satisfies $\mathrm{vol}(\pi) \leq \mathrm{vol}(\pi^*)$,*
- *if $H^* < 3P$, it contains at least one packing π' which satisfies $\mathrm{vol}(\pi') \leq \mathrm{vol}(\pi^*)$, $\max(\pi') \leq H^*$, and H^*-scale$(\pi') \leq T$.*

In the remainder of this algorithm, we apply Steps 2 and 3 for any packing in Π. Eventually, as the size of the collection is at most $3P$, we will obtain a set of at most $3P$ schedules. We simply keep the one which gives the minimum H.

Step 2 is summarized here and we omit some details as it is the most natural part of our algorithm. For some packing $\pi \in \Pi$, a natural idea is, for each value π_{pk}, $p \in \mathcal{P}$, $k \in \mathcal{K}$, to create π_{pk} items of size p, and then to solve **IMS** with T boxes. Said differently, we can transform a packing into an **IMS** instance with exactly $\sum_k \pi_{pk}$ items of size p for each $p \in \mathcal{P}$. In this way, a solution of this **IMS** instance $I(\pi)$ with T boxes correspond to a schedule for the initial **Multi_Bot** instance made up of the jobs performed by π.

Unfortunately, this transformation I might not be achieved in polynomial time as values π_{pk}, which depend on demands d_k, can be exponential in the input size of **Multi_Bot**. To handle this, certain items, called *block items*, will represent several p-resources instead of a single one. We ensure ourselves to obtain a polynomial-sized instance $I(\pi)$ and, at the same time, that each block item size does not exceed one third of $\mathrm{vol}(\pi)/T$. Concretely, the block items are filled greedily until their size exceeds $\frac{\mathrm{vol}(\pi)}{3T}$.

Lemma 1 *(proof omitted). Let π be some packing. The **IMS** instance $I(\pi)$:*

1. *contains $O(PKT)$ items,*
2. *satisfies $\mathrm{vol}(I(\pi)) = \mathrm{vol}(\pi)$ and $\max(I(\pi)) = \max(\pi)$,*
3. *does not contain block items of size greater than $\frac{H^*}{3}$ if $\mathrm{vol}(\pi) \leq \mathrm{vol}(\pi^*)$.*

Eventually, Step 3 consists in applying LPT-FIRST on each **IMS** instance $I(\pi)$ - created at Step 2 - with T boxes. The solutions obtained thus correspond to schedules, as the T boxes of **IMS** represent the periods of **Multi_Bot** and each of them contains a set of performed jobs (represented by the items), characterized by a configuration $p \in \mathcal{P}$ and some type of job $k \in \mathcal{K}$. The schedule with the minimum H offers a $\frac{4}{3}$-approximation for **Multi_Bot** (see Theorems 2 and 3, Sect. 5).

From now on, our objective is to prove Theorem 1 (Sect. 6), but also Theorems 2 and 3 (Sect. 5) which are the keystones for the main result of the paper: BOT-APPROX is a $\frac{4}{3}$-approximation algorithm (Theorem 7, page 13).

5 Approximation Analysis

The objective of this section is to show that, given the **IMS** instances $I(\pi)$ we created with Steps 1 and 2, LPT-FIRST algorithm provides a $\frac{4}{3}$-approximation for **Multi_Bot** on at least one of these instances (Step 3). Indeed, as each $I(\pi)$ contains T boxes, which can be seen as the periods of **Multi_Bot**, any of its solution can be directly converted into a schedule $(x_{pkt})_{p,k,t}$. The creation of collection Π will be treated later in Sect. 6 as it is the most technical part.

Given some packing π with a smaller volume than π^* and which satisfy all demands, LPT-FIRST achieves a $\frac{4}{3}$-approximation under the condition: $H^* \geq 3P$.

Theorem 2 *Consider some **Multi_Bot** instance and a packing π which satisfies all demands. Moreover, $\mathrm{vol}(\pi) \leq \mathrm{vol}(\pi^*)$. If $H^* \geq 3P$, then the schedule $\mathbf{x}$ returned by solving $I(\pi)$ with LPT-FIRST satisfies $H \leq \frac{4}{3}H^*$.*

Proof Let N be the number of items in $I(\pi)$. We know that $N = O(PKT)$ from Lemma 1. We proceed by induction on the number z of items packed by LPT-FIRST. We prove that, for any $0 \leq z \leq N$, the number of resources present in each period (or box) after packing the z most large items is at most $\frac{4}{3}H^*$.

The base case $z = 0$ is trivial. Now, assume we already packed the z first items and we want to pack item $z + 1$. According to Lemma 1, $\mathrm{vol}(I(\pi)) = \mathrm{vol}(\pi) \leq \mathrm{vol}(\pi^*)$, so the least filled period contains less than H^* resources, otherwise the total volume overpasses $\mathrm{vol}(\pi^*)$. The size of item $z + 1$ does not exceed $\frac{H^*}{3}$ (Lemma 1). With the induction hypothesis, this observation shows us that all periods use at most $\frac{4}{3}H^*$ resources after packing $z + 1$ items. $\square$

Now we fix the other side of the tradeoff, *i.e.* when $H^* < 3P$.

Theorem 3 *Consider some* **Multi_Bot** *instance with* $\lambda = H^*$ *and a packing* π *which satisfy all demands. Moreover,* $\text{vol}(\pi) \leq \text{vol}(\pi^*)$, $\max(\pi) \leq \lambda$ *and* $\lambda\text{-scale}(\pi) \leq T$. *If* $H^* < 3P$, *the schedule* $\mathbf{x}$ *returned by solving* $I(\pi)$ *with* LPT-FIRST *verifies* $H \leq \frac{4}{3}H^*$.

Proof Observe that items of size larger than $\frac{H^*}{3}$ are not block items (Lemma 1): they represent a single p-resource. LPT-FIRST packs first the items of sizes $p > \frac{2H^*}{3}$. As $\lambda = H^*$ and $\lambda\text{-scale}(\pi) \leq T$, then there are at most T big items. Moreover, all big items ($p > \frac{2H^*}{3}$) size does not exceed H^*, as $\max(I(\pi)) \leq \lambda = H^*$.

Second, LPT-FIRST packs the medium items of sizes $p > \frac{H^*}{3}$ which are not big. The empty boxes are filled and, when none of them is empty anymore, the remaining medium items are packed upon another medium item, as their boxes are less filled than the ones with big items. If there is a period with at least three medium items, then all other periods containing medium items are made up of two of them at least, otherwise we contradict the principle of LPT-FIRST. However, having such a period with three medium items contradicts the scale criterion as we would have $\lambda\text{-scale}(\pi) > T$. Consequently, there cannot be more than two medium items per period, and the number of resources present in the periods filled with medium items is at most $\frac{4H^*}{3}$ (Fig. 3).

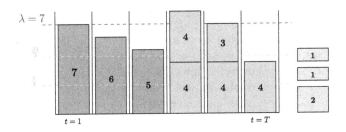

Fig. 3. The result of LPT-FIRST after filling periods with big and medium items first. Items of sizes 1 and 2 ($p \leq \frac{\lambda}{3}$) have still to be treated.

The end of the proof consists in applying again the argument used in the proof of Theorem 2. Now, all periods contain at most $\frac{4H^*}{3}$ resources and the least filled period will contain at most H^* resources in order to satisfy the volume condition. The remaining items have size at most $\frac{H^*}{3}$. □

6   Generation of Packings with Small Volume

This section is dedicated to the proof of Theorem 1. The conclusion of this proof is given in Sect. 6.3.

We focus on the specific formulation of **Multi_Bot** with a single period $T = 1$. We denote it by **Multi_Bot_1P**. Its mathematical formulation can be obtained by simply replacing $T = 1$ into the one given in Problem 1. A solution is thus a packing $(\pi_{pk})_{p,k}$ as the parameter t does not intervene anymore here.

To deal with the requirements of Theorem 1, we define a problem which is more general than **Multi_Bot_1P**. We call it **Multi_Bot_1P**$[\lambda, \tau]$: it adds extra constraints on the shape of the expected packing and, hence, makes two extra parameters $\lambda, \tau \in \mathbb{N}$ intervene, see Problem 2.

Observe that if $\lambda \to +\infty$, the problem "tends to" **Multi_Bot_1P** which always admit a solution. In the remainder, we abuse notation and denote it by $\lambda = \infty$. Our idea consists in solving **Multi_Bot_1P**$[\lambda, \tau]$ for $\tau = T$ and all values $\lambda \in \{1, 2, 3, \ldots, 3P - 1, \infty\}$: if we find a solution, we put it into Π. Concretely, let $\boldsymbol{\pi}(\lambda)$ denote an optimum packing for **Multi_Bot_1P**$[\lambda, T]$ if it exists.

$$\Pi = \{\boldsymbol{\pi}(\lambda) : 1 \leq \lambda \leq 3P - 1 \text{ or } \lambda = \infty\} \qquad (2)$$

*Problem 2 (**Multi_Bot_1P**$[\lambda, \tau]$).*

    Input: $\mathcal{K}, \mathcal{P}, (c_{pk})_{\substack{p \in \mathcal{P} \\ k \in \mathcal{K}}}, (d_k)_{k \in \mathcal{K}}, 1 \leq \lambda \leq 3P$

  Objective: *find a packing* $\boldsymbol{\pi} = (\pi_{pk})_{p,k}$ *which*

                *minimizes H **or** state that no solution exists*

  subject to: $\displaystyle\sum_{p \in \mathcal{P}} c_{pk} \cdot \pi_{pk} \geq d_k$ $\forall k \in \mathcal{K}$

$$\sum_{k \in \mathcal{K}} \sum_{p \in \mathcal{P}} p \cdot \pi_{pk} = H \leq \lambda\tau$$

$$\pi_{pk} = 0 \qquad\qquad\qquad\qquad\qquad \forall k \in \mathcal{K}, p > \lambda$$

$$\sum_{k \in \mathcal{K}} \sum_{p > \frac{2\lambda}{3}} \pi_{pk} + \frac{1}{2} \sum_{k \in \mathcal{K}} \sum_{\frac{\lambda}{3} < p \leq \frac{2\lambda}{3}} \pi_{pk} \leq \tau$$

$$\pi_{pk}, H \in \mathbb{N}, \qquad\qquad\qquad\qquad \forall k \in \mathcal{K}, p \in \mathcal{P}$$

The extra constraints introduced in **Multi_Bot_1P**$[\lambda, \tau]$ say that the volume of the packing solution cannot exceed $\lambda\tau$, its maximum λ, and its λ-scale τ. Collection Π will contain at least 1 packing because $\boldsymbol{\pi}(\infty)$ necessarily exists, and at most $3P$ packings. Problem 2, Observe that putting the optimum solutions for all these problems in Π allows us to meet the requirements of Theorem 1. The following Theorem 4 is a direct consequence of the definition of **Multi_Bot_1P**$[\lambda, \tau]$ (Problem 2).

Theorem 4. *(proof omitted) Packings $\boldsymbol{\pi}(\lambda)$ satisfy the following properties:*

1. for any λ, $\boldsymbol{\pi}(\lambda)$, if it exists, satisfies all demands d_k,
2. $\mathrm{vol}(\boldsymbol{\pi}(\infty)) \leq \mathrm{vol}(\boldsymbol{\pi}^*)$,
3. if $H^* < 3P$, then $\boldsymbol{\pi}(H^*)$ exists and we have: $\mathrm{vol}(\boldsymbol{\pi}(H^*)) \leq \mathrm{vol}(\boldsymbol{\pi}^*)$,
   $\max(\boldsymbol{\pi}(H^*)) \leq H^*$, H^*-scale$(\boldsymbol{\pi}(H^*)) \leq T$.

6.1   Dynamic Programming for the Single-Period Problems

We begin with the generation of packings $\pi(\lambda)$ for $1 \leq \lambda \leq 3P - 1$. We will deal with the case $\lambda = \infty$ in Sect. 6.2. We fix some positive integer λ with $\lambda \leq 3P - 1$. Our objective is to solve **Multi_Bot_1P**$[\lambda, T]$ and, hence, to obtain either some packing $\pi(\lambda)$ or a negative answer. We present a dynamic programming (DP) procedure to achieve this task.

Structure of DP Memory. From now on, variable τ represents a positive half-integer upper-bounded by T and W a positive integer representing the authorized volume, which is at most λT.

We construct a four-dimensional vector `Table`, whose elements are `Table`$[p, k, W, \tau]$ with three integers $0 \leq p \leq \lambda$, $1 \leq k \leq K$, $0 \leq W \leq \lambda T$ and a half-integer $0 \leq \tau \leq T$. Hence, the total size of the vector is $\lambda^2 T^2 K = O(P^2 T^2 K)$. The role of this vector is to help us producing intermediary packings to obtain the solution of **Multi_Bot_1P**$[\lambda, T]$. Intuitively, index k provides us with the types $1, \ldots, k$ of jobs we have to consider. Index p is the maximum configuration which can be used to perform jobs of type k. Eventually, W is a volume limit for the packing and τ is the λ-scale limit.

More formally, we denote by $\pi[p, k, W, \tau]$ a packing which:

- produces only jobs of type $1, 2, \ldots, k$: $\pi_{pk'} = 0$ when $k' > k$,
- uses only configurations $1, 2, \ldots, p$ to process the jobs of type k: $\pi_{p'k} = 0$ when $p' > p$,
- satisfies the demands until d_{k-1}: $\sum_{p'=1}^{\lambda} c_{p'k'} \cdot \pi_{p'k'} \geq d_{k'}$ for all $1 \leq k' < k$,
- has a volume at most W: $\sum_{p'=1}^{\lambda} \sum_{k'=1}^{k} p' \cdot \pi_{p'k'} \leq W$,
- has a λ-scale at most τ: $\lambda\text{-scale}(\pi) \leq \tau$,
- maximizes the number of jobs of type k processed, *i.e.* $\sum_{p'=1}^{\lambda} c_{p'k} \cdot \pi_{p'k}$.

Packing $\pi[p, k, W, \tau]$ does not necessarily exist since reaching the demands might be avoided by the volume and scale conditions. Observe that, by definition, **Multi_Bot_1P**$[\lambda, T]$ admits a solution if and only if $\pi[\lambda, K, \lambda T, T]$ exists and satisfies the demands for jobs of type K. Also, for the specific value $p = 0$, the packing $\pi[0, k, W, \tau]$ cannot use any configuration for jobs of type k, so in fact it does not process jobs of type k at all. We can fix: $\pi[0, k, W, \tau] = \pi[\lambda, k-1, W, \tau]$. The particular case $p = 0$ and $k = 1$ gives an empty packing $\pi[0, 1, W, \tau]$.

The objective of vector `Table` is to contain either the production of jobs of type k processed by $\pi[p, k, W, \tau]$ if it exists, or $-\infty$. More formally,

$$
\texttt{Table}[p, k, W, \tau] = \begin{cases} \sum_{p'=1}^{p} c_{p'k} \pi_{p'k} & \text{if } \pi[p, k, W, \tau] = (\pi_{p'k'})_{p',k'} \text{ exists} \\ -\infty & \text{otherwise} \end{cases} \tag{3}
$$

In addition, we also compute two vectors `Bool` and `Pack` with the same dimension sizes as `Table`. Vector `Bool` simply indicates, with a boolean, whether

$\pi[p, k, W, \tau]$ satisfies the demand d_k for jobs of type k (Bool$[p, k, W, \tau]$ = True) or not (= False). Finally, vector Pack provides us with the number of p-resources performing jobs of type k in $\pi[p, k, W, \tau]$. This allows us in the remainder to retrieve recursively the content of packing $\pi[p, k, W, \tau]$.

Before stating our recursive formula to achieve the DP algorithm, we define an extra function $S_{p,\lambda}$ for any $1 \le p \le \lambda$. Given some half-integer "budget" τ and an integer j, it returns the updated λ-scale budget which remains after adding a number j of p-resources to some packing. Formally, $S_{p,\lambda}(\tau, j) = \tau$ if $p \le \frac{\lambda}{3}$, $\tau - \frac{j}{2}$ if $\frac{\lambda}{3} < p \le \frac{2\lambda}{3}$, and $\tau - j$ if $p > \frac{2\lambda}{3}$.

We construct a **Recursive Formula.** The base case is $p = k = 1$. Packing $\pi[1, 1, W, \tau]$ has no demand to satisfy, therefore it necessarily exists. It corresponds to taking the maximum number of 1-resources which satisfy both the volume and scale conditions. Moreover, Bool$[1, 1, W, \tau]$ = True if and only if Table$[1, 1, W, \tau] \ge d_1$.

$$\forall W, \forall \tau, \quad \text{Table}[1, 1, W, \tau] = \max_{\substack{0 \le j \le W \\ S_{1,\lambda}(\tau,j) \ge 0}} j \cdot c_{11} \qquad (4)$$

For the general statement, we distinguish two cases. First assume that $p = 0$ and $k > 1$. We have $\pi[0, k, W, \tau] = \pi[\lambda, k - 1, W, \tau]$. Hence, Table$[0, k, W, \tau] = 0$ if Table$[\lambda, k - 1, W, \tau] \ge d_{k-1}$, else $-\infty$. As $d_k > 0$, Bool$[0, k, W, \tau]$ = False. Second, assume that $p \ge 1$, $i.e.$ jobs of type k can be processed by p'-resources, with $1 \le p' \le p$. We write:

$$\text{Table}[p, k, W, \tau] = \max_{\substack{0 \le j \le \lfloor \frac{W}{p} \rfloor \\ S_{p,\lambda}(\tau,j) \ge 0 \\ \text{Bool}[\lambda, k-1, W-jp, S_{p,\lambda}(\tau,j)]}} \text{Table}[p - 1, k, W - jp, S_{p,\lambda}(\tau, j)] + j c_{pk} \quad (5)$$

$$\text{Bool}[p, k, W, \tau] = \text{True} \Leftrightarrow \text{Table}[p, k, W, \tau] \ge d_k \qquad (6)$$

Index j represents the number of p-resources performing jobs of type k we might add to our packing. Naturally, this addition should overpass neither the volume constraint $(j \le \frac{W}{p})$, nor the scale one $(S_{p,\lambda}(\tau, j) \ge 0)$. Furthermore, Bool$[\lambda, k - 1, W - jp, S_{p,\lambda}(\tau, j)]$ must be True as it guarantees that demands $d_1, \ldots, d_{k-1}$ can be satisfied before the add-on of j p-resources. This index j selected in Eq. (5) to maximize Table$[p, k, W, \tau]$ is put in Pack$[p, k, W, \tau]$. It may happen that no index j (even $j = 0$) satisfies these three conditions. In this case, we fix Table$[p, k, W, \tau]$ = Pack$[p, k, W, \tau]$ = $-\infty$.

Observe that Table$[p, k, W, \tau]$ = $-\infty$ if and only if Bool$[\lambda, k - 1, W, \tau]$ is False. This makes sense since it means that with volume W and scale τ we are considering, the demands for jobs of type $1, 2, \ldots, k - 1$ could not be achieved.

If Table$[p, k, W, \tau]$ = $-\infty$ or $p = 0$, then Pack$[p, k, W, \tau]$ = $-\infty$ and $\pi[p, k, W, \tau]$ does not exist. Otherwise, packing $\pi[p, k, W, \tau]$ can be recovered recursively:

- if $p = k = 1$ and Pack$[1, 1, W, \tau] = j$, then $\pi[1, 1, W, \tau]$ is (π_{11}) with $\pi_{11} = j$.
- if $p = 0$ and $k > 1$, then $\pi[0, k, W, \tau] = \pi[\lambda, k - 1, W, \tau]$.
- if Pack$[p, k, W, \tau]$ = j, compute recursively packing $\pi[p - 1, k, W - jp, S_{p,\lambda}(\tau, j)]$ and add $\pi_{pk} = j$ to it (Fig. 4).

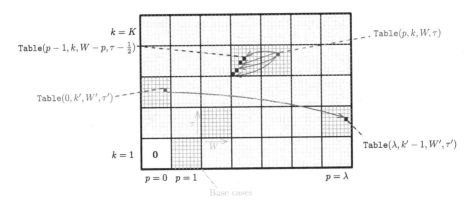

Fig. 4. 2D-projection of the 4D-vector `Table` illustrating the recursive calls

This DP algorithm, that we call BOT-PROGDYN, solves **Multi_Bot_1P** $[\lambda, T]$: first compute tables `Table`, `Bool` and `Pack`, second answer NO if `Table`$[\lambda, K, \lambda T, T] = -\infty$, else retrieve packing $\pi[\lambda, K, \lambda T, T]$ recursively. If $\pi[\lambda, K, \lambda T, T]$ satisfies demand d_K, then return it as $\pi(\lambda)$, else answer NO.

Theorem 5. BOT-PROGDYN *solves* **Multi_Bot_1P**$[\lambda, T]$ *in time* $O(\lambda^3 T^3 K)$.

6.2   Optimum Packing with Unlimited Scale

We focus in this subsection on problem **Multi_Bot_1P**$[\infty, T]$. We aim at producing the packing $\pi(\infty)$ which will be denoted π to simplify notations.

Definition 4 (Optimal configuration for $k \in \mathcal{K}$). *For any* $k \in \mathcal{K}$, *let* $p_0(k)$ *be the* optimal configuration *for jobs of type* k, *i.e. the configuration* $p \in \mathcal{P}$ *which is the most efficient in terms of resources. In brief,* $p_0(k) = \text{argmax}_{p \in \mathcal{P}} \frac{c_{pk}}{p}$.

For any optimum solution $\mathbf{x}^*$ of **Multi_Bot**, the total number of non-optimal configurations used to proceed jobs is polynomially bounded.

Lemma 2. *There is an optimum solution* $\mathbf{x}^*$ *for which, for any* $k \in \mathcal{K}$, $t \in \mathcal{T}$ *and* $p \neq p_0(k)$, $x^*_{pkt} \leq p_0(k) \leq P$.

Idea of the Proof. Any set of p-resources of cardinality $p_0(k)$ can be replaced more efficiently by a set of $p_0(k)$-resources of cardinality p.   □

Our reasoning is based on Lemma 2. Due to page limit, we do not describe all formal details but only the idea beyond this construction of π. We guess, for each $k \in \mathcal{K}$, the number W_k of resources required for the production of jobs of type k in π with configurations $p \neq p_0(k)$. We have $W_k = \sum_{p \neq p_0(k)} p \cdot \pi_{pk} \leq P \frac{P(P-1)}{2}$ (Lemma 2). We determine the most productive packing for W_k (observe that the selection of W_k and $W_{k'}$ are independent) using BOT-PROGDYN on the instance where optimal configurations are withdrawn (their capacitites are put to zero). Finally, for each guess $0 \le W_k \le P\frac{P(P-1)}{2}$, we complete the obtained packing with the necessary number of $p_0(k)$-resources to reach the demand d_k. In this way, for each k, we determine the guess W_k which allows us to minimize the number of resources used, independently from other $k' \ne k$ and eventually obtain an optimum solution for **Multi_Bot_1P**$[\infty, T]$.

Theorem 6 *(proof omitted). Packing* $\pi(\infty)$ *can be computed in* $O(KP^7)$.

6.3   Consequences for Multi_Bot

Combining the results obtained in both Sects. 6.1 and 6.2, we are now ready for the computation of the whole collection Π of packings

Proof of Theorem 1. According to Theorems 5 and 6, we build the collection described in Eq. (2) with the time announced.

By definition, any packing $\pi(\lambda)$, if it exists, satisfies all demands (Problem 2). Packing $\pi(\infty)$, which necessarily exists, satisfies all demands while minimizing the volume. As π^* is a (not necessarily optimal) solution of **Multi_Bot_1P**, then its volume is at least $\mathrm{vol}(\pi(\infty))$. Finally, if $H^* < 3P$, then packing $\pi(H^*)$ exists since π^* is a solution of **Multi_Bot_1P**$[H^*, T]$: it satisfies all demands, its volume is at most TH^*, its maximum at most H^* and its scale at most T. $\square$

Based on Theorems 1, 2 and 3, we present the main result of this paper.

Theorem 7. BOT-APPROX *is a* $\frac{4}{3}$*-approximation algorithm for* **Multi_Bot**.

Proof. If $H^* \ge 3P$, then we know that the collection Π computed at Step 1 contains a packing π with a smaller volume than π^* (Theorem 1). According to Theorem 2, the schedule **x** produced with LPT-FIRST by considering this initial packing π offers the guarantee that $H \le \frac{4}{3}H^*$.

If $H^* < 3P$, then, again from Theorem 1, collection Π contains a packing π' with a smaller volume than π^* such that $\max(\pi') \le H^*$ and H^*-scale$(\pi') \le T$. By Theorem 3, we also obtain a schedule **x**$'$ which gives $H' \le \frac{4}{3}H^*$.

As BOT-APPROX returns the schedule minimizing H among all initial packings $\pi \in \Pi$, we are sure to use at most $\frac{4}{3}H^*$ resources. $\square$

7   Perspectives

This article highlights a constant approximation ratio of $\frac{4}{3}$ for **Multi_Bot**. First, a natural question is whether this value can be decreased. An idea could consist in replacing LPT-FIRST by algorithms which achieve a smaller approximation ratio for **IMS**, such as MULTIFIT ([7], ratio $\frac{13}{11}$). Nevertheless, such a modification implies to develop technically the generation of packings as it needs now to satisfy more restrictive conditions. We believe that such an outcome could be obtained, however, it is necessary to catch up the approximation analysis for MULTIFIT, which is much more involved than the one of LPT-FIRST. Moreover, it will certainly imply an increase of the running time. In our opinion, from the theoretical point of view, handling this question directly for all expected approximation ratio amounts to identifying a PTAS for **Multi_Bot**. We conjecture that a PTAS exists for **Multi_Bot** because the size of the items/bots is part of the combinatorial input of the problem.

More generally, an interesting direction of research is to define properly all satellite problems which could stand around **Multi_Bot**, and, in particular, all problems which lie between **IMS** and **Multi_Bot**. Indeed, perhaps these problems could be natural models for certain industrial applications and they certainly also admit this $\frac{4}{3}$-approximation factor. Finally, the parameterized complexity of **Multi_Bot** is challenging as it consists in assessing the complexity of the problem when some combinatorial input is supposed to be small. As an example, we know that **Multi_Bot** is polynomial-time solvable when $T = 1$, but what about the case $T = O(1)$?

References

1. Battaïa, O., et al.: Workforce minimization for a mixed-model assembly line in the automotive industry. Int. J. Prod. Econ. **170**, 489–500 (2015)
2. Beşikci, U., Bilge, U., Ulusoy, G.: Multi-mode resource constrained multi-project scheduling and resource portfolio problem. Eur. J. Oper. Res. **240**(1), 22–31 (2015)
3. Boysen, N., Schulze, P., Scholl, A.: Assembly line balancing: what happened in the last fifteen years? Eur. J. Oper. Res. **301**(3), 797–814 (2022)
4. Brinkop, H., Jansen, K.: High multiplicity scheduling on uniform machines in FPT-time. CoRR abs/2203.01741 (2022)
5. Chaikovskaia, M.: Optimization of a fleet of reconfigurable robots for logistics warehouses. Ph.D. thesis, Université Clermont Auvergne, France (2023)
6. Chaikovskaia, M., Gayon, J.P., Marjollet, M.: Sizing of a fleet of cooperative and reconfigurable robots for the transport of heterogeneous loads. In: Proceedings of IEEE CASE, pp. 2253–2258 (2022)
7. Coffman, E.G., Jr., Garey, M.R., Johnson, D.S.: An application of bin-packing to multiprocessor scheduling. SIAM J. Comput. **7**(1), 1–17 (1978)
8. Graham, R.L.: Bounds on multiprocessing timing anomalies. SIAM J. Appl. Math. **17**(2), 416–429 (1969). https://doi.org/10.1137/0117039
9. Hartmann, S., Briskorn, D.: An updated survey of variants and extensions of the resource-constrained project scheduling problem. Eur. J. Oper. Res. **297**(1), 1–14 (2022)
10. Hochbaum, D.S., Shmoys, D.B.: Using dual approximation algorithms for scheduling problems: theoretical and practical results. In: FOCS, pp. 79–89 (1985)
11. McCormick, S.T., Smallwood, S.R., Spieksma, F.: A polynomial algorithm for multiprocessor scheduling with two job lengths. Math. Oper. Res. **26**(1), 31–49 (2001)
12. MecaBotiX (2023). https://www.mecabotix.com/
13. Mnich, M., Wiese, A.: Scheduling and fixed-parameter tractability. Math. Program. **154**(1–2), 533–562 (2015)

Network Design on Undirected Series-Parallel Graphs

Ishan Bansal$^{(\boxtimes)}$ ⬤, Ryan Mao ⬤, and Avhan Misra ⬤

Cornell University, Ithaca, NY, USA
{ib332,rwm275,am2934}@cornell.edu

Abstract. We study the single pair capacitated network design problem and the budget constrained max flow problem on undirected series-parallel graphs. These problems were well studied on directed series-parallel graphs, but little is known in the context of undirected graphs. The major difference between the cases is that the source and sink of the problem instance do not necessarily coincide with the terminals of the underlying series-parallel graph in the undirected case, thus creating certain complications. We provide pseudopolynomial time algorithms to solve both of the problems and provide an FPTAS for the budget constrained max flow problem. We also provide some extensions, arguing important cases when the problems are polynomial-time solvable, and describing a series-parallel gadget that captures an edge upgrade version of the problems.

Keywords: Network Design · Series-Parallel Graphs · Pseudopolynomial Algorithm · Approximation Scheme

1 Introduction

Network design problems have played a central role in the field of combinatorial optimization and naturally arise in various applications, including transportation, security, communications, and supply-chain logistics. In this paper, we study two problems in the area of network design, the *Budget-Constrained Max Flow Problem* (BCMFP) and the single pair *Capacitated Network Design Problem* (CapNDP). In both these problems, the input is a multi-graph with costs and capacities on edges, and a source-sink pair of vertices. The algorithmic goal in the BCMFP is to maximize the connectivity between the source and sink while the total cost of purchased edges fits within a given budget. CapNDP is the minimization version of BCMFP where the goal is to find a cheapest subgraph that meets a demanded level of connectivity between the source and sink.

We study the version of these problems where the costs on edges are all-or-nothing referred to in the literature as "fixed-charge". Hence, one has to pay for the entire edge, irrespective of the amount of positive flow being sent across it. These problems are notoriously difficult, and we limit our focus to a class of input graphs that emerge as specific types of electrical networks, known

© The Author(s), under exclusive license to Springer Nature Switzerland AG 2024
A. Basu et al. (Eds.): ISCO 2024, LNCS 14594, pp. 277–288, 2024.
https://doi.org/10.1007/978-3-031-60924-4_21

as series-parallel graphs, which possess favorable algorithmic properties. Both the problems BCMFP and CapNDP have been well studied on directed series-parallel graphs [6, 7] with fully polynomial time approximation schemes (FPTAS) available. However, the situation is more complex in the case of undirected series-parallel graphs as noted in [1]. We present a pseudopolynomial time algorithm for both BCMFP and CapNDP on undirected series parallel graphs and present an FPTAS for BCMFP. We supplement these results with some extensions of these problems, and show that our pseudo-polynomial time algorithm can be implemented in polynomial time if the input capacities are suitably structured. Note that both of the problems are NP-HARD even on simple two vertex graphs, since they capture the knapsack and the min-knapsack problems.

1.1   Related Work

The budget constrained max flow problem was studied by Schwarz and Krumke [7]. They considered different cost variants of the problem including the most difficult all-or-nothing variant. Here, they provided a pseudopolynomial time algorithm when the input graph is a directed series-parallel graph and used common scaling techniques to further provide an FPTAS. They also show that the problem is NP-HARD on series parallel graphs and strongly NP-HARD even on bipartite graphs. Krumke et al. [6] obtained similar results for the single pair capacitated network design problem on directed series parallel graphs.

The main structural tool used in these results was developed by Valdes et al. [8], where they provide a linear time algorithm to produce a decomposition of a series-parallel graphs. Furthermore, by preprocessing the input graph, one can always assume that the terminals of the series parallel graph coincide with the source and sink of the problem instance [7]. Carr et al. [1] observed that the case is slightly more complex for undirected series-parallel graphs as now the terminals of the graph and the source-sink pair of the problem instance need not coincide. The following simple example shows this difference concretely (see Fig. 1). In the undirected case, the graph is not series-parallel with respect to the terminal pair s and t but s and t could be a source-sink pair of the network design problem. Carr et al. [1] provided a pseudopolynomial time algorithm for the CapNDP problem on outerplanar graphs (a sub-family of series-parallel graphs), but incorrectly claimed that this implies an FPTAS for CapNDP on outerplanar graphs. We outline this difficulty in Sect. 4.

Carr et al. [1] also provided an approximation algorithm for the general CapNDP problem by strengthening the standard cut-based LP using the so called knapsack cover constraints. However, no $o(m)$-approximation algorithm is known for the general CapNDP problem where m is the number of edges in the input graph. Chakraborty et al. [2] showed that CapNDP cannot be approximated to within $\Omega(\log \log m)$ unless $NP \subseteq DTIME(m^{\log \log \log m})$.

In recent years, a generalization of BCMFP called budget constrained minimum cost flows has received attention [3–5]. Here, the algorithmic goal is to compute a classical minimum cost-flow while also adhering to a flow-usage budget.

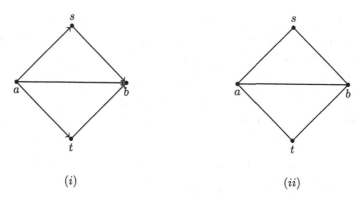

Fig. 1. (i) A directed series-parallel graph with terminals a and b. (ii) An undirected series-parallel graph with terminals a and b.

1.2 Our Results

We provide a pseudopolynomial time algorithm for BCMFP on undirected series parallel graphs in Sect. 3 by designing a dynamic program that makes use of the series-parallel decomposition of a series-parallel graph. In the case of directed series-parallel graphs, [7] considered the minimum cost to flow a value of f between the terminals of a series-parallel graph, and computed this minimum cost using the decomposition of the series-parallel graph. However as also noted in [1], the terminals of an undirected series parallel graph need not coincide with the source and the sink of the problem instance, and so the method used in [7] does not work. Instead, we make use of circulations, and consider the minimum cost to flow a suitably defined circulation in the series-parallel graph. We show how all arising cases can still be handled using a more complicated dynamic program. Since the CapNDP problem is just the minimization version of the BCMFP, this also implies that CapNDP on series parallel graphs can be solved in pseudopolynomial time.

Theorem 1. *There exists pseudopolynomial time algorithms to solve both the BCMFP and the single pair CapNDP on undirected series-parallel graphs.*

In Sect. 4, we leverage the above algorithm to provide an FPTAS for the BCMFP on series-parallel graphs. This is done using common scaling techniques that lowers the capacities on edges so that their values are polynomially bounded. The pseudopolynomial time algorithm then runs in polynomial time as long as all capacities are still integral. But in maintaining integrality, we must lose an ϵ factor in optimality. However, in contrast to the results in [6], a similar FPTAS cannot be obtained for the CapNDP. This is rather intriguing and pinpoints a difficulty in extending earlier results on directed series-parallel graphs to undirected series-parallel graphs.

Theorem 2. *There exists a fully polynomial time approximation scheme for the BCMFP on series-parallel graphs*

In Sect. 5, we consider two extensions of our results. First, we provide instances when the pseudopolynomial algorithm developed in Sect. 3 can be implemented in polynomial time. For instance, if all the capacities on edges are the same, then we can scale down the capacities so that they are all unit (and hence polynomially bounded) without loosing any factor in optimality. This cannot be done even if there are just two distinct values of capacities. Nonetheless, we show that our algorithm can still be implemented in polynomial time in such and more general cases.

Theorem 3. *Suppose the capacities on edges lie in a bounded lattice defined by basis values $d_1, \ldots, d_k$ i.e. $\{\sum_{i=1}^{k} \alpha_i d_i : |\alpha_i| \leq K \text{ and integral}\}$ where k is a constant and K is polynomially bounded, then both the problems BCMFP and CapNDP can be solved in polynomial time.*

Second, we consider a variant of the BCMFP and CapNDP where edges have various levels of upgrades with different cost and capacity values, but only one upgrade can be chosen and purchased. Using parallel edges for each upgrade does not solve this problem since multiple upgrades could then be simultaneously applied. Instead, we provide a simple gadget that is series-parallel and that captures this extension. Hence, all our earlier results hold in this setting as well.

2 Preliminaries

Now some basic definitions and problem statements.

Definition 1 (Graph). *A graph $G = (V, E)$ is a set of vertices V and possibly parallel edges E with non-negative costs $\{c_e\}$ and non-negative capacities $\{u_e\}$ on the edges. Additionally, there are two designated vertices s and t called the source and the sink.*
The cost of a subset of the edges $F \subseteq E$ denoted $c(F)$ is defined as $\sum_{e \in F} c_e$. We denote the number of edges $|E|$ using m and the number of vertices $|V|$ using n.

Definition 2 (Flow). *An s-t flow in a graph G is an assignment of a direction and a non-negative flow value $f_e \leq u_e$ on each edge such that for every vertex v other than s and t, the sum of the flow values on edges directed into v is equal to the sum of the flow values on edges directed out of v. The value of the s-t flow is equal to the sum of the flow values on edges directed out of s minus the sum of the flow values on edges directed into s.*

Definition 3 (Circulation). *Given a graph G and a set of values $\{r_v\}$ called residues on the vertices, an r-circulation is an assignment of a direction and a non-negative flow value $f_e \leq u_e$ on each edge such that for every vertex v, the sum of the flow values on edges directed into v minus the sum of the flow values on edges directed out of v is equal to r_v.*

Definition 4 (BCMFP). *An instance of the budget constrained max flow problem (BCMFP) takes as input a graph $G = (V, E)$ and a non-negative budget B. The goal is to find a set of edges F such that $c(F) \leq B$ that maximizes the max s-t flow in the graph (V, F).*

Definition 5 (CapNDP). *An instance of the capacitated network design problem (CapNDP) takes as input a graph $G = (V, E)$ and a non-negative demand D. The goal is to find a cheapest set of edges F, such that the max s-t flow in the graph (V, F) is at least the demand D.*

Definition 6 (Series-Parallel Graphs). *A series-parallel graph is defined recursively as follows. The graph G with vertex set $\{a, b\}$ and edge set $\{(a, b)\}$ is a series-parallel graph with terminals a and b. If $G_1 = (V_1, E_1)$ and $G_2 = (V_2, E_2)$ are series-parallel with terminals a_1, b_1 and a_2, b_2 respectively, then*

- *The graph obtained by identifying b_1 and a_2 is series-parallel with terminals a_1 and b_2. This graph is the series composition of G_1 and G_2.*
- *The graph obtained by identifying a_1 with a_2 and also b_1 with b_2 is series-parallel with terminals $a_1(= a_2)$ and $b_1(= b_2)$. This graph is the parallel composition of G_1 and G_2.*

There are well known algorithms to determine if a graph is series-parallel, and if so, provide a parse tree (decomposition tree) specifying how the graph is obtained using the above two composition rules [8].

Definition 7 (FPTAS). *A fully polynomial time approximation scheme (FPTAS) for a maximization problem is an algorithm which takes as input a problem instance and a parameter $\epsilon > 0$ and returns a solution which is within a $1/1 + \epsilon$ factor of the optimal solution, and runs in time polynomial in the size of the instance and $1/\epsilon$.*

3   Pseudopolynomial Algorithm

In this section we present a pseudopolynomial time algorithm to solve BCMFP and CapNDP on undirected series-parallel graphs, thus proving Theorem 1.

Proof (Proof of Theorem 1). Let f_{max} be the maximum possible s-t flow in the series-parallel graph G. We will compute suitably defined circulations in subgraphs of G that arise in its series-parallel decomposition. Let $G' = (V', E')$ be a subgraph of G which is in the series-parallel decomposition tree of G. Let a and b be its terminals. We associate with G' a 2, 3 or 4-tuple $R_{G'} = (r_a, r_s, r_t, r_b)$ where the term r_s is included only if G' contains s and $s \neq a, b$, and similarly for the the term r_t. The states of our dynamic program (DP) will be $(G', R_{G'})$ for all possible tuples $R_{G'}$ such that $r \in R_{G'} \Rightarrow |r| \leq f_{max}$ and $\sum_{r \in R_{G'}} r = 0$. Note that the tuple $R_{G'}$ also defines the residues of a circulation in G' by setting $r_v = 0$ for all $v \in G \setminus \{a, s, t, b\}$. We restrict our attention to circulations where the absolute values of the residues are bounded by f_{max} since we can assume that in the optimal solution, the total flow into or out of any vertex is bounded by f_{max}.

For each such state, we find the minimum cost subset of edges $F' \subseteq E'$ such that the subgraph (V', F') admits an $R_{G'}$-circulation. When it is clear from context, we abuse notation to denote both the set of edges F' and its cost $c(F')$ as $DP(G', R_{G'})$. We now use the series-parallel decomposition tree of G to find $DP(G', R_{G'})$. Clearly if G' was a leaf of the decomposition tree, then we could find F' easily (as every leaf of the decomposition tree is a two vertex graph with one edge). Indeed if the graph has two vertices a and b and a single edge $e = \{a, b\}$ with cost c_e and capacity u_e. Then, if $|r_a| \leq u_e$, $F' = \{e\}$ and $c(F') = c_e$. Else we set $F' = \emptyset$ and $c(F') = \infty$. We consider the following cases.

– **Case 1** $R_{G'}$ is a 2-tuple i.e. $R_{G'} = (r_a, r_b)$
  • **Case 1.1** G' is formed using a series composition of G_1 and G_2.

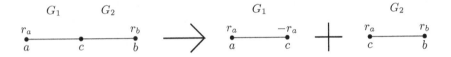

$$DP(G', R_{G'}) = DP(G_1, (r_a, -r_a)) + DP(G_2, (r_a, r_b))$$

  • **Case 1.2** G' is formed using a parallel composition of G_1 and G_2.

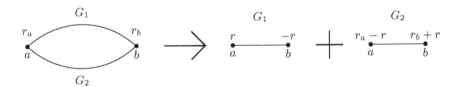

$$DP(G', R_{G'}) = \min_{-f_{max} \leq r \leq f_{max}} DP(G_1, (r, -r)) + DP(G_2, (r_a - r, r_b + r))$$

– **Case 2** $R_{G'}$ is a 3-tuple i.e. $R_{G'} = (r_a, r_s, r_b)$
  • **Case 2.1** G' is formed using a series composition of G_1 and G_2 splitting at s.

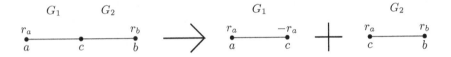

$$DP(G', R_{G'}) = DP(G_1, (r_a, -r_a)) + DP(G_2, (r_a + r_s, r_b))$$

- **Case 2.2** G' is formed using a series composition of G_1 and G_2 not splitting at s.

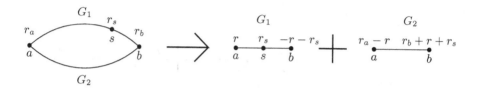

$$DP(G', R_{G'}) = DP(G_1, (r_a, r_s, -r_a - r_s)) + DP(G_2, (r_a + r_s, r_b))$$

- **Case 2.3** G' is formed using a parallel composition of G_1 and G_2.

$$DP(G', R_{G'}) = \min_{-f_{max} \le r \le f_{max}} DP(G_1, (r, r_s, -r - r_s)) + DP(G_2, (r_a - r, r_b + r + r_s))$$

- **Case 3** $R_{G'}$ is a 4-tuple i.e. $R_{G'} = (r_a, r_s, r_t, r_b)$
  - **Case 3.1** G' is formed using a series composition of G_1 and G_2 splitting at s (or t).

$$DP(G', R_{G'}) = DP(G_1, (r_a, r_s, -r_a - r_s)) + DP(G_2, (r_t + r_a + r_s, r_b))$$

  - **Case 3.2** G' is formed using a series composition of G_1 and G_2 splitting s and t into different subgraphs.

$$DP(G', R_{G'}) = DP(G_1, (r_a, r_s, -r_a - r_s)) + DP(G_2, (r_a + r_s, r_t, r_b))$$

- **Case 3.3** G' is formed using a series composition of G_1 and G_2 not splitting s and t into different subgraphs.

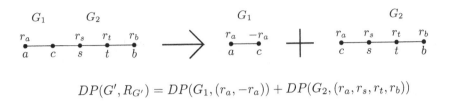

$$DP(G', R_{G'}) = DP(G_1, (r_a, -r_a)) + DP(G_2, (r_a, r_s, r_t, r_b))$$

- **Case 3.4** G' is formed using a parallel composition of G_1 and G_2 with s and t in the same parallel component.

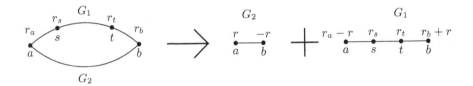

$$DP(G', R_{G'}) = \min_{-f_{max} \le r \le f_{max}} DP(G_1, (r, -r)) + DP(G_2, (r_a - r, r_s, r_t, r_b + r))$$

- **Case 3.5** G' is formed using a parallel composition of G_1 and G_2 with s and t in different parallel components.

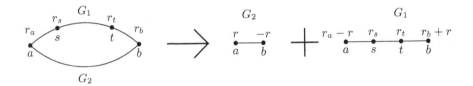

$$DP(G, R_{G'}) = \min_{-f_{max} \le r \le f_{max}} DP(G_1, (r, r_s, -r - r_s)) + DP(G_2, (r_a - r, r_t, r_b + r_s + r))$$

We have exhausted all cases barring any symmetric ones and have shown how the DP table can be built. The total number of states in the DP are $O(mf_{max}^3)$ since the number of subgraphs G' in the series-parallel decomposition tree is $O(m)$ [8] and the number of tuples $R_{G'}$ for a fixed subgraph G' is bounded by $O(f_{max}^3)$ (by the condition that $\sum_{r \in R_{G'}} r = 0$). The above cases show that each state of the DP table can be calculated in $O(f_{max})$ time. Hence, the total running time of computing the DP table is $O(mf_{max}^4)$. Finally, the optimal solution for the BCMFP will correspond to the largest value r such that the cost of $DP(G, (0, r, -r, 0))$ is bounded by the budget B. The optimal solution for the CapNDP will correspond to the value of $DP(G, (0, D, -D, 0))$. This concludes the proof. □

4   Fully Polynomial Time Approximation Scheme

In this section we prove Theorem 2 and provide an FPTAS for the BCMFP on series-parallel graphs by leveraging the pseudopolynomial time algorithm presented in the previous section.

Proof (Proof of Theorem 2). Consider the following question,

    Given a budget C and a flow value R, is there a subset of edges $E' \subseteq E$ of cost at most C such that there is a flow of value R between s and t in (V, E')?

The DP presented in the previous section will also be able to provide the subset of edges E' if the answer to the above question is YES. Furthermore, the DP will run in time $O(mR^4)$ since we are only interested now in tuples $R_{G'}$ whose entries are bounded in absolute value by R.

Let OPT be the max flow possible within the budget B and let $E^* \subseteq E$ be an optimal subset of the edges. Let $\epsilon > 0$ be the desired level of accuracy and let $\epsilon' = \min(1, \epsilon/3)$. For any $M \in [1, F]$, define new capacities on the edges of G as:

$$u_e^M := \left\lfloor \frac{mu_e}{M\epsilon'} \right\rfloor$$

We now run the DP to answer the question above with capacities on edges given by u_e^M and $R = \lceil m/\epsilon' \rceil$ and the budget B remaining the same. As observed earlier, the running time of the DP will be $O(m^5/\epsilon'^4)$. Additionally, if $M \leq OPT/(1+\epsilon')$, then the answer to the above question will be YES. This is because the set of edges E^* itself achieves a flow of value at least R. Indeed, let $S \subseteq V$ such that $s \in S$ and $t \notin S$. Define $\delta(S)$ to be the set of edges in E with one endpoint in S and the other endpoint in $V \setminus S$. Define the capacity across a cut with respect to a set of edges F to be $u_F(S) = \sum_{e \in \delta(S) \cap F} u_e$ and the scaled capacity across this cut to be $u_F^M(S) = \sum_{e \in \delta(S) \cap F} u_e^M$. This scaled capacity across any cut S that disconnects s and t with respect to the optimal solution E^* is given by,

$$
\begin{aligned}
u_{E^*}^M(S) &= \sum_{e \in \delta(S) \cap E^*} \left\lfloor \frac{mu_e}{M\epsilon'} \right\rfloor \\
&\geq \frac{m}{M\epsilon'} u_{E^*}(S) - m \\
&\geq \frac{m}{M\epsilon'} OPT - m \\
&\geq \frac{m}{\epsilon'} \qquad\qquad \text{(if } M \leq OPT/(1+\epsilon') \text{)}
\end{aligned}
$$

But since $u_{E^*}^M(S)$ is integral, we obtain $u_{E^*}^M(S) \geq \lceil m/\epsilon' \rceil = R$. Since the cut was arbitrary, we see that (V, E^*) admits a flow of value at least R.

We can run a binary search in the search space $1, 1 + \epsilon', (1 + \epsilon')^2, \dots$ up to F to find the largest M' such that we get YES as the answer to the question above when capacities are $u_e^{M'}$. From the arguments above, we know $M' \geq OPT/(1 + \epsilon')^2$. Also let E' be the corresponding optimal edge set found by running the DP (on the instance where $R = \lceil m/\epsilon' \rceil$ and capacities are $u_e^{M'}$). For each cut S separating the source and the sink in (V, E'), we have,

$$u_{E'}(S) = \sum_{e \in \delta(S) \cap E'} u_e$$

$$\geq \frac{M'\epsilon'}{m} u_{E'}^{M'}(S) \geq \frac{M'\epsilon'}{m} \frac{m}{\epsilon'} \geq \frac{OPT}{(1+\epsilon')^2}$$

$$\geq \frac{OPT}{(1+\epsilon)} \qquad\qquad \text{since } \epsilon' = \min(1, \epsilon/3)$$

The running time of the entire algorithm is polynomial in m, $1/\epsilon$ and $\log(f_{max})$. since we are running the DP at most $\log(f_{max})$ times and each run takes time $O(m^5/\epsilon^4)$. This concludes the proof. □

Remark: The above method does not provide an FPTAS for the CapNDP on undirected series parallel graphs (in contrast to the results in [6] for directed graphs). This is because we do not have the luxury of scaling the capacities in the CapNDP, as doing so would affect the feasible region of the problem instance. The objective function values can be scaled down and hence, only costs can be scaled in the CapNDP. In [6], the authors provided a different pseudopolynomial time algorithm with running time polynomial in m and the maximum cost on an edge c_{max}. They could then scale down the costs to provide an FPTAS for CapNDP on directed series parallel graphs. It is currently unknown if the same can be done for undirected series-parallel graphs. In [1], the authors provided a pseudopolynomial time algorithm for the CapNDP on outerplanar graphs and claimed that this can be used to obtain an FPTAS. However, this claim is erroneous since their pseudopolynomial time algorithm also had running time polynomial in m and f_{max}. It is thus an interesting open question as to whether an FPTAS exists for CapNDP on undirected series parallel graphs or whether it is APX-HARD. Such questions are interesting from a complexity theory perspective as well [9].

5   Further Extensions

5.1   Capacities from a Lattice

We argue that the pseudopolynomial time algorithm presented in Sect. 3 can be implemented in polynomial time under some assumptions on the capacities on edges. For instance, if the number of distinct values of capacities on edges is small, then intuitively it feels like the problem should be easier. We substantiate this by proving Theorem 3. Let the capacities on the edges lie on a lattice defined by some non-negative numbers acting as a basis $d_1, \ldots, d_k$, i.e. every capacity u_e is expressible as $u_e = \alpha_1 d_1 + \cdots + \alpha_k d_k$ where $|\alpha_i| \leq K$ for all i and α_i are integers. We assume that the number of basis elements k is a constant, and the value of K is polynomially bounded. We show that the algorithm provided in Sect. 3 can be implemented in polynomial time and hence, both the BCMFP and CapNDP are polynomial time solvable.

Proof (Proof of Theorem 3). The DP has states $(G', R_{G'})$ where G' is a subgraph in the series-parallel decomposition tree of the input graph G, and $R_{G'}$ defines the residues in a circulation on G'. These residues are computed for all integer values from $-F$ to F where F is the max s-t flow in the graph G. We argue that not every integer value from $-F$ to F need to be considered. Indeed for any demanded flow R, let E^* be an optimal solution to the CapNDP and consider the maximum s-t flow in the graph $H = (V, E^*)$. Consider now a flow decomposition of this maximum flow. Each s-t path in this flow decomposition saturates an edge of the graph H. Hence, the flow value on any edge in the graph H lies in the set

$$\{\sum_{i=1}^k \alpha_i d_i \; : \; |\alpha_i| \le mK \text{ and integral}\}$$

This implies that in any subgraph G' arising in the series-parallel decomposition tree of G, the residues on any node can be expressed as $\sum_{i=1}^k \alpha_i d_i$ where each $|\alpha_i|$ is bounded by $m^2 K$. The number of such residues is at most $O((m^2 K)^k)$. Thus instead of considering $O(F^3)$ many values for $R_{G'}$, we only need to consider polynomially many. This completes the proof. □

5.2 Edge Upgrades

We consider a variant of the BCMFP and CapNDP problems where each edge has some upgrade options wherein a cost can be paid to increase the capacity of the edge by a certain amount. However, at most one upgrade can be applied. Note that adding parallel edges does not solve this variant since then we cannot control how many upgrades are applied. We provide a gadget reduction from this variant to the original BCMFP and CapNDP. The gadget we provide is a series-parallel graph and hence, all our earlier results in this paper extend to this new variant of BCMFP and CapNDP on undirected series parallel graphs.

Formally, for every edge e in G, there are various choices of costs and capacities provided as input $(c_e^1, u_e^1), \ldots, (c_e^k, u_e^k)$, and at most one of these choices can be purchased by a solution. We depict below the gadget construction to capture this variant when there are 2 or three choices, and this easily generalizes to more choices. We assume that $u^i \ge u^j$ if $i \ge j$ (Figs. 2 and 3).

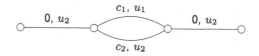

Fig. 2. An edge upgrade gadget with two choices

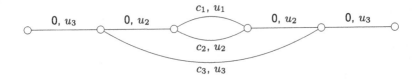

Fig. 3. An edge upgrade gadget with three choices

Note that, for k upgrade choices of an edge, the above reduction creates $k + 2(k - 1)$ edges and $2k$ nodes, thus increasing the size of the input graph linearly. Furthermore, we are replacing an edge of the graph with a series-parallel graph. Hence, if the original graph was series-parallel, then after applying our reduction the graph will remain series-parallel.

References

1. Carr, R.D., Fleischer, L.K., Leung, V.J., Phillips, C.A.: Strengthening integrality gaps for capacitated network design and covering problems. Technical report, Sandia National Lab. (SNL-NM), Albuquerque, NM, USA; Sandia ... (1999)
2. Chakrabarty, D., Chekuri, C., Khanna, S., Korula, N.: Approximability of capacitated network design. Algorithmica **72**, 493–514 (2015)
3. Holzhauser, M., Krumke, S.O., Thielen, C.: Budget-constrained minimum cost flows. J. Comb. Optim. **31**(4), 1720–1745 (2016)
4. Holzhauser, M., Krumke, S.O., Thielen, C.: A network simplex method for the budget-constrained minimum cost flow problem. Eur. J. Oper. Res. **259**(3), 864–872 (2017)
5. Holzhauser, M., Krumke, S.O., Thielen, C.: On the complexity and approximability of budget-constrained minimum cost flows. Inf. Process. Lett. **126**, 24–29 (2017)
6. Krumke, S.O., Noltemeier, H., Schwarz, S., Wirth, H.C., Ravi, R.: Flow improvement and network flows with fixed costs. In: Kall, P., Lüthi, H.J. (eds.) Operations Research Proceedings 1998, vol. 1998, pp. 158–167. Springer, Heidelberg (1999). https://doi.org/10.1007/978-3-642-58409-1_15
7. Schwarz, S., Krumke, S.O.: On budget-constrained flow improvement. Inf. Process. Lett. **66**(6), 291–297 (1998)
8. Valdes, J., Tarjan, R.E., Lawler, E.L.: The recognition of series parallel digraphs. In: Proceedings of the Eleventh Annual ACM Symposium on Theory of Computing, pp. 1–12 (1979)
9. Woeginger, G.J.: When does a dynamic programming formulation guarantee the existence of a fully polynomial time approximation scheme (FPTAS)? INFORMS J. Comput. **12**(1), 57–74 (2000)

Online Graph Coloring with Predictions

Antonios Antoniadis, Hajo Broersma, and Yang Meng[(✉)]

Faculty of Electrical Engineering Mathematics and Computer Science,
University of Twente, P. O. Box 217, 7500 AE Enschede, The Netherlands
{a.antoniadis,h.j.broersma,y.meng}@utwente.nl

Abstract. We introduce learning augmented algorithms to the online
graph coloring problem. Although the simple greedy algorithm FIRSTFIT
is known to perform poorly in the worst case, we are able to establish a
relationship between the structure of any input graph G that is revealed
online and the number of colors that FIRSTFIT uses for G. Based on this
relationship, we propose an online coloring algorithm FIRSTFITPREDIC-
TIONS that extends FIRSTFIT while making use of machine learned predictions. We show that FIRSTFITPREDICTIONS is both *consistent* and
smooth. Moreover, we develop a novel framework for combining online
algorithms at runtime specifically for the online graph coloring problem.
Finally, we show how this framework can be used to robustify FIRST-
FITPREDICTIONS by combining it with any classical online coloring algorithm (that disregards the predictions).

Keywords: learning augmented algorithms · online algorithms ·
online graph coloring · first fit

1 Introduction

Before we will properly define the concepts we use throughout this paper in
Sect. 1.1, let us start with a less formal introduction to the subject, together
with some background and motivation.

Graph coloring is a central topic within graph theory that finds its origin
in the notorious Four Color Problem, dating back to 1852. Since then graph
coloring has developed into a mature research field with numerous application
areas, ranging from scheduling [20], and memory allocation [2] to robotics [7].
In the online version of the problem, the vertices of a (usually unknown) graph
arrive online one by one, together with the adjacencies to the already present
vertices. Upon the arrival of each vertex v, a color has to be irrevocably assigned
to v. The goal is to obtain a proper coloring, that is a coloring in which no
two adjacent vertices have the same color. The challenge is to design a coloring
strategy which keeps the total number of assigned colors as small as possible.

In this paper, we introduce the online graph coloring problem to learning
augmented algorithms. We assume that alongside the arrival of each vertex,
the algorithm also obtains a prediction $P(v)$ (of unknown quality) on the color

that should be assigned to v. These predictions may be used by an algorithm to obtain a coloring which uses fewer colors, if the predictions turn out to be relatively accurate. At the same time, the algorithm should maintain a worst-case guarantee to safeguard against the case, in which the predictions are inaccurate.

Graph coloring is notoriously hard, already in the *offline setting*, where the whole input graph is known in advance. Let $\chi(G)$ denote the *chromatic number* of a graph G, that is the minimum number of distinct colors needed to obtain a proper coloring of G. A straightforward observation is that $\chi(G)$ is greater than or equal to $\omega(G)$, which is defined as the cardinality of a maximum clique in G. However, there even exist triangle-free graphs (so with no cliques of cardinality 3) with an arbitrarily large chromatic number. In case $\chi(H) = \omega(H)$ for every induced subgraph H of a graph G, then G is called a *perfect graph*. Perfect graphs are known to be $\chi(G)$-colorable in polynomial-time via semidefinite programming [8]. An interesting special case of perfect graphs is the class of bipartite graphs. These graphs admit a proper coloring using only 2 colors, and in the offline setting such a 2-coloring can be computed in linear time, for example via breadth first search. However, in general it is an NP-complete problem to decide whether a given graph admits a proper coloring using k colors, for any fixed $k \geq 3$ [15]. With respect to approximation algorithms, there is a polynomial-time algorithm using at most $O(n(\log \log n)^2/(\log n)^3) \cdot \chi(G)$ [11] colors for a graph on n vertices, and it is NP-hard to approximate the chromatic number within a factor $n^{1-\epsilon}$ for all $\epsilon > 0$ [24].

The problem becomes even more challenging in the *online setting* where, as mentioned, vertices arrive online one by one and an algorithm has to irrevocably assign a color to each vertex upon its arrival – while only having knowledge of the subgraph revealed so far. Any online algorithm may require at least $(2n/(\log n)^2)\chi(G)$ colors in the worst case [12], where n is the number of vertices of the input graph – and this is true even for bipartite graphs (so with chromatic number 2). Restricting the input graphs even further, for instance to P_6-free bipartite graphs (containing no path on six vertices as an induced subgraph), no online algorithm can guarantee a coloring with constantly many colors [10]. In such cases, bounding the number of colors used from above by a function of the chromatic number is not feasible. To this end, a line of research has focused on developing so-called *online-competitive algorithms*. Applied to any graph G, such online algorithms are guaranteed to produce a proper coloring, the number of colors of which is bounded from above by a function of the number of colors used by the *best possible online algorithm* for G. As an example, in the previously mentioned setting of P_6-free bipartite graphs, there exists an online algorithm that uses at most twice as many colors as the best possible online algorithm [5,21]. A good source for more information regarding online graph coloring is the following book chapter due to Kierstead [16]. We will come back to this in Sect. 3.3, where we review and utilize several known results, including more recent work.

Probably the conceptually simplest online algorithm for graph coloring is the greedy algorithm, which is known as FIRSTFIT. Suppose we have a total order over the set of available colors. Then upon arrival of each vertex, FIRSTFIT assigns to it the smallest color according to that order, among the ones that maintain a proper coloring. This algorithm has been extensively analyzed in the literature, also for particular graph classes [10, 13]. Although FIRSTFIT performs well for many practically relevant inputs, for example for interval graphs and for the complement of bipartite graphs, it can be very sensitive to the order in which the vertices of G are revealed. For instance, in a popular example where G is a complete balanced bipartite graph $K_{n,n}$ minus a perfect matching, there is a specific permutation on the arrival of the vertices of G for which FIRSTFIT requires n colors (whereas $\chi(G) = 2$) [14].

That FIRSTFIT performs well in some practical scenarios, can be attributed to the fact that real-world graph coloring instances rarely resemble worst-case inputs. It is often the case that either the structure of the input graph or the permutation in which the vertices are revealed can be exploited by a heuristic or a machine-learning approach in order to yield reasonably good colorings, despite the inherent worst-case difficulty of the problem. However, and not too surprisingly, such approaches tend to come without a worst-case guarantee. In the (hopefully rare) cases where the input diverges substantially from the expected structure, the resulting coloring could be arbitrarily poor.

In this paper, we design an algorithm that incorporates predictions of unknown quality obtained by such a machine-learned approach. It produces a relatively good coloring in case the predictions turn out to be accurate, while at the same time providing a worst-case guarantee comparable to the best classical online algorithm that does not make use of the predictions. Our work falls within the context of *learning augmented algorithms*.

Learning augmented algorithms is a relatively new and very active field. The main goal is to develop algorithms combining the respective advantages of machine-learning approaches and classical worst case algorithm analysis. A plethora of online problems have already been investigated through the learning augmented algorithm lens, including for example, caching [19], facility location [6], ski-rental [23], or various scheduling problems [17, 23] to name just a few. To the best of our knowledge, learning augmented algorithms have not been studied for graph coloring problems to date. For a more extensive discussion on learning augmented algorithms, we refer the interested reader to a recent survey [22].

A common approach to developing learning augmented algorithms is to design an algorithm that attempts to follow the predictions, in some sense. At the same time, this algorithm should be robustified by appropriately combining it with a classical algorithm that disregards the predictions. At a high level, this is our approach for graph coloring as well. However, the nature of the problem poses several novel challenges. First of all, already assigned colors may significantly restrict the choice of colors for the next and future assignments of the algorithm. This is in contrast to settings where an algorithm can, at some cost, move to any arbitrary configuration, for instance, in problems with an underlying metric. The fact that it is not possible for an algorithm to move to any possible configuration also rules out a robustification approach by combining algorithms in an experts-like setting. See [1] for more information. Secondly, existing online algorithms for online graph coloring do not possess a particular monotonicity property that tends to be a crucial ingredient in robustifying algorithms for other problems. In particular, it is possible that running an algorithm only on the graph induced by a suffix of the input permutation requires significantly more colors than running the same algorithm on the graph of the complete input permutation. This further complicates the robustification, since one can not simply use a classical algorithm as a fall back option upon recognizing that the predictions are of insufficient quality.

1.1 Preliminaries

In this section, we formally define the problem setting and its associated prediction model. We start with the concepts related to (offline) graph coloring.

Definition 1 (Graph coloring [4]). *A k-vertex coloring, or simply a k-coloring, of a graph G is a mapping $\phi : V(G) \to S$, where S is a set of k colors. A k-coloring is proper if no two adjacent vertices are assigned the same color. A graph is k-colorable if it admits a proper k-coloring. The minimum k for which a graph G is k-colorable is called its chromatic number, denoted by $\chi(G)$. An optimal coloring of G is a proper $\chi(G)$-coloring.*

Online graph coloring describes the setting, in which the vertices of G arrive one by one in an online fashion. Upon arrival, the vertices have to be irrevocably and properly colored.

Definition 2 (Online graph coloring [21]). *An online graph (G, π) is a graph G together with a permutation $\pi = v_1, v_2, \ldots, v_n$ of $V(G)$. An online coloring algorithm takes an online graph (G, π) as input and produces a proper coloring of $V(G)$, where the color of a vertex v_i is chosen from a universe $\mathcal{U}$ of available colors and the choice depends only on the subgraph of G induced by $\{v_1, v_2, \ldots, v_i\}$ and the colors assigned to $v_1, v_2, \ldots, v_{i-1}$, for $1 \le i \le n$.*

Throughout the paper, and unless otherwise specified, algorithm refers to a *deterministic* algorithm. We next define the notions of *competitiveness, online competitiveness* and *competitive ratio*, which are used to evaluate the performance of algorithms for online graph coloring.

Definition 3 (Competitiveness [10], online competitiveness [9] and competitive ratio [18]). *Let $AOL(G)$ be the set of all online coloring algorithms for a graph G and let $\Pi(G)$ be the set of all permutations of $V(G)$. For an algorithm $A \in AOL(G)$ and a permutation $\pi \in \Pi(G)$, the number of colors used by A when $V(G)$ gets revealed according to π is denoted by $\chi_A(G, \pi)$. The A-chromatic number of G is the largest number of colors used by the online algorithm A for the graph G, denoted by $\chi_A(G)$. That is,*

$$\chi_A(G) = \max_{\pi \in \Pi(G)} \chi_A(G, \pi).$$

For a graph G, the online chromatic number $\chi_{OL}(G)$ is the minimum number of colors used for G, over all algorithms of $AOL(G)$. That is,

$$\chi_{OL}(G) = \min_{A \in AOL(G)} \chi_A(G).$$

Let $\mathcal{G}$ be a family of graphs and $AOL(\mathcal{G})$ be the set of online algorithms for $\mathcal{G}$. For some $A \in AOL(\mathcal{G})$, if there exists a function such that $\chi_A(G) \le f(\chi(G))$, (resp. $\chi_A(G) \le f(\chi_{OL}(G))$) holds for every $G \in \mathcal{G}$, then A is competitive (resp. online competitive) on $\mathcal{G}$. Furthermore, the competitive ratio of an algorithm $A \in AOL(\mathcal{G})$ over a class of graphs $\mathcal{G}$ is the maximum of $\frac{\chi_A(G)}{\chi(G)}$ for all $G \in \mathcal{G}$.

We note that the notion of competitiveness used here follows the literature on online graph coloring and contrasts the definition commonly used for other online problems, where an algorithm is said to be competitive if it attains a constant competitive ratio. We complete this section by presenting the basic definitions associated with the setting, involving predictions on the colors.

Predictions and Prediction Error. Here we assume that alongside the disclosure of each vertex v, the algorithm also obtains a prediction $P : V(G) \to \mathcal{U}$ on the color of v, where $\mathcal{U}$ is the set of available colors. These predictions are aimed at obtaining a reasonable coloring. They may stem from a machine-learning approach based on past inputs or training data, or from a simple heuristic known to perform well in practice. The quality of the obtained predictions is measured by means of a *prediction error*. This *prediction error* is defined naturally to be the (smallest) number of vertices that obtained wrong predictions.

Definition 4 (Prediction error). *Given an online graph* (G, π), *let* $\mathcal{O}(G)$ *be the set of all optimal colorings of* G, *where* $O \in \mathcal{O}$ *assigns color* $O(v) \in \mathcal{U}$ *to vertex* v. *Then the* prediction error *for online graph* (G, π) *is given by*

$$\eta(G) = \min_{O \in \mathcal{O}(G)} \Sigma_{v \in V(G)} |\{P(v)\} \setminus \{O(v)\}|.$$

In the following, we drop the dependence on G when the underlying online graph is clear from the context. We use the notation (G, π, P) to refer to an *online graph with predictions*, where G is the underlying graph, π the permutation in which V is revealed, and P the set of associated predictions.

Following the literature, we say that an algorithm is α-*consistent* if it attains a competitive ratio of α in the case that the predictions are perfect ($\eta = 0$), and *robust* if it independently of the prediction error obtains a competitive ratio within a constant factor of that of the best classical online algorithm. Furthermore, we say that an algorithm is *smooth* if its competitive ratio degrades at a rate that is at most linear in the prediction error. Note that the notion of robustness also extends to the case where the best known classical online algorithm A

is "only" online competitive. In this scenario, any algorithm that is guaranteed to use at most a constant number of colors more than what A uses is *robust*.

We note that one can trivially obtain an optimal coloring when the predictions are perfect (in other words when $\eta = 0$) by just coloring each vertex v with color $P(v)$ upon arrival. This already is a 1-consistent algorithm. However, when the obtained predictions are only slightly off, this algorithm may not even be a valid algorithm for online graph coloring. Indeed, consider the case where only one vertex v receives a wrong prediction $P(v)$ but is adjacent to a vertex u with $P(u) = P(v)$.

1.2 Our Contribution

Our first contribution lies in establishing a relationship between the structure of an online graph (G, π) and the amount of colors used by FIRSTFIT for G. We emphasize that this result is independent of predictions and might be of broader interest. More specifically, in Sect. 2 we show that if FIRSTFIT uses x colors for G, then there exists a set $V' \subseteq V$ of vertices of size $|V'| = x + q$, for some $0 \le q \le x - 2$, such that V' can be partitioned into $q + 1$ non-trivial subsets, each of which is a *clique* in G (that is, each subset consists of at least two vertices which are all pairwise adjacent in G). Our result is even constructive, i.e., we present such an algorithm for finding V' and a partitioning.

Theorem 1. *Let (G, π) be an online graph for which FIRSTFIT uses x colors. Then, there exists a set $V' \subseteq V$ of size $|V'| = x + q$ with $0 \le q \le x - 2$, such that V' can be partitioned into $q + 1$ non-trivial subsets of vertices, each of which is a clique.*

Our second contribution is to develop a 1-consistent and smooth algorithm for online graph coloring with predictions, called FIRSTFITPREDICTIONS in Subsect. 3.1. Consider the setting where the algorithm, upon the reveal of a vertex v also obtains a prediction on the color that v should be colored with in an optimal coloring. We give an algorithm that employs FIRSTFIT with a distinct color palette for each subgraph induced by the set of vertices that obtained the same prediction. By carefully utilizing the aforementioned structural result, we are able to associate the number of colors used by the algorithm with the number of wrong predictions obtained. More specifically, we are able to show that the number of colors used by the algorithm differs from that of an optimal coloring by at most the number of wrong predictions (implying that if the predictions are perfect, the algorithm actually recovers an optimal coloring, even though the quality of the predictions is not a priori known to the algorithm).

Theorem 2. *Assume that FIRSTFITPREDICTIONS uses $x(G)$ colors for some online graph with predictions (G, π, P) whose chromatic number is unknown to the algorithm, then*

$$x(G) \le \eta(G) + \chi(G).$$

Our third contribution is a novel framework for combining different online graph coloring algorithms in Subsect. 3.2. Earlier frameworks developed for other online problems do not seem to carry over to the online graph coloring problem. Our framework allows us to robustify our algorithm by combining it with a classical algorithm that disregards the predictions. We show that the number of colors used by the combination of the two algorithms is within a factor of 2 from that of the best performing of the two on this input instance. This implies that this combination is a 2-consistent, smooth and robust algorithm.

Although in this paper we only use the framework to combine two algorithms, we prove the result for combining any number t of online algorithms (at a loss of t in the competitive ratio). Given an online graph (G, π) and an online coloring algorithm A, let $A(G)$ denote the number of colors A uses for G.

Theorem 3. *Let $A_1, A_2, \ldots, A_t$ be online graph coloring algorithms that may or may not make use of the predictions. Then, there exists a meta-algorithm A that combines $A_1, A_2, \ldots, A_t$, such that for any online graph with predictions (G, π, P) it holds that*

$$A(G) \leq t \cdot \min_{1 \leq i \leq t} A_i(G).$$

The generality of our result allows us to obtain learning augmented algorithms for online graph coloring for different graph classes in Subsect. 3.3.

2   A Structural Theorem About FIRSTFIT

This section is devoted to proving Theorem 1, which establishes a relationship between the number of colors used by FIRSTFIT for an online graph (G, π) and a partition of a subset of V into cliques. As mentioned, our proof is constructive and implies an efficient algorithm for finding such a partition. In the proof we assume that FIRSTFIT uses $x \geq 2$ colors which are ordered as $c_0 < c_1 < \ldots < c_{x-1}$. We use $N(u)$ to denote the neighbors of a vertex u, i.e., the set of vertices that are adjacent to u.

Proof of Theorem 1. Let $t_{x-1} \in V$ be a vertex for which FIRSTFIT uses color c_{x-1}. By the definition of FIRSTFIT vertex t_{x-1} must be adjacent to vertices $t_0, t_1, \ldots, t_{x-2}$, such that t_i is colored with color $c(t_i) = c_i$ for all $i = 0, \ldots, x-2$. Let $S = \bigcup_{i=0}^{x-1} \{t_i\}$ and let $N^-(u) = \{w \in N(u) \cap S \mid c(w) < c(u)\}$ be its *neighborhood of smaller-colored vertices within S*, $\forall u \in S$.

Note that the set of vertices $V_0' = \{t_i \in S \mid N^-(t_i) = \{t_0, t_1, \ldots, t_{i-1}\}\}$ is a clique. Also note that $|V_0'| \geq 2$, since $\{t_0, t_{x-1}\} \subseteq V_0'$. If $V_0' = S$, then the theorem directly follows for $q = 0$. So, assume for the remainder of this proof that $V_0' \neq S$. Let $S' = S \setminus V_0' = \{u_1, u_2, \ldots, u_\ell\}$, in which the vertices of S' are ordered by increasing color. We describe an algorithm for partitioning S' into q subsets, with the property that each of these subsets of S' together with one distinct vertex from $V \setminus S$ is a clique of size at least two. Note that this implies the theorem since $1 \leq |S'| \leq x - 2$ and thus $1 \leq q \leq x - 2$.

For every $h \in \{1, ..., \ell\}$ let $\alpha(u_h) \in S'$ be a vertex of maximal color with $c(\alpha(u_h)) < c(u_h)$ such that $\alpha(u_h)$ is not adjacent to u_h, thus $\alpha(u_h) \notin N^-(u_h)$. And let $\beta(u_h) \in V \setminus S$ be a vertex with $c(\beta(u_h)) = c(\alpha(u_h))$ that is adjacent to u_h. Note that such vertices must exist: if $\alpha(u_h)$ did not exist, then u_h would be contained in V_0'; and if $\beta(u_h)$ did not exist, FIRSTFIT would have assigned a smaller color, namely $c(\alpha(u_h))$, to u_h.

The algorithm proceeds iteratively over the vertices of S' in order of increasing color. For each vertex $u_h \in S'$, if $\beta(u_h) \notin V_j'$ for all j with $1 \leq j < h$, then a new clique $V_h' = \{u_h, \beta(u_h)\}$ is created. Else, there exists a j with $1 \leq j < h$ such that $\beta(u_h) \in V_j'$. In this case, set $V_j' := V_j' \cup \{u_h\}$, in other words, add u_h to V_j'. We will show that this creates a larger clique. But first note that by the definition of the algorithm each different $\beta(u_h)$ is added to exactly one such set V_j', and each such set V_j' contains exactly one specific $\beta(u_h)$. Thus, the output is indeed a partition of $S' \bigcup \cup_h \{\beta(u_h)\}$.

It remains to argue that upon termination, each set V_j' is a clique. We will prove the stronger statement that throughout the execution of the algorithm each set V_j' is a clique, and consists of $\beta(u_j)$ and a subset of vertices of S' with a color strictly larger than $c(\beta(u_j)) = c(\alpha(u_j))$. This invariant clearly holds upon creation of such a set V_j', since it is created as a clique $\{u_j, \beta(u_j)\}$ and $c(u_j) > c(\beta(u_j))$. Assume that the invariant holds up to some iteration $h-1$. Now consider iteration h during which u_h gets added to V_j'. By the definition of the algorithm $\beta(u_h) \in V_j'$, and thus $\beta(u_h) = \beta(u_j)$. And by the definition of $\beta(u_h)$, it is adjacent to u_h. It remains to show that u_h is adjacent to all the vertices in $V_j' \setminus \{\beta(u_j)\}$, and that $c(u_h) > c(\beta(u_j))$. The latter directly follows from our ordering and the fact that $c(u_h) > c(u_j) > c(\beta(u_j))$. For the former, recall that $\alpha(u_h)$ is defined as the vertex of S' of maximal color that is not adjacent to u_h. In other words, $N^-(u_h)$ contains a vertex of each color strictly between $c(\alpha(u_h))$ and $c(u_h)$, and therefore each vertex of S' with such a color is adjacent to u_h. By the induction hypothesis V_j' only contains such vertices (except for vertex $\beta(u_j) = \beta(u_h)$ whose adjacency to u_h has already been argued).

3   Algorithmic Results

In this section, we focus on deriving and analysing learning augmented algorithms for online graph coloring. We introduce a consistent and smooth algorithm in Subsect. 3.1, and show how it can be robustified in Subsect. 3.2. Finally, in Subsect. 3.3 we argue how it can be used to obtain learning augmented algorithms for online coloring of specific graph classes.

Due to space constraints most of the remaining proofs are deferred to the full version of the paper.

3.1   FIRSTFITPREDICTIONS (FFP)

Throughout this section we assume that the predicted colors are chosen from the set $\{c_0, c_1, c_2, \ldots\}$. Given an online graph with predictions (G, π, P), upon revealing of a new vertex v with prediction $P(v) = c_i$, the algorithm FIRSTFIT-PREDICTIONS (FFP for short) employs FIRSTFIT with a distinct color palette associated with c_i. We use $C(i) = \{c_i^0 = c_i, c_i^1, c_i^2, \ldots\}$ to denote the *color palette* associated with color c_i, implying a natural ordering according to the superscripts. Keeping the colors of each such palette distinct enables us to associate the total number of colors used by FFP to the total prediction error.

> FIRSTFITPREDICTIONS: When a new vertex v is revealed with prediction $P(v) = c_i$, assign to v the smallest-superscript eligible color $c \in C(i)$.

FFP implies a partition of the vertices of G (and the subgraphs of G induced by the vertices that have been revealed so far) based on their color palettes.

Definition 5. *We say that a vertex v belongs to color palette $C(i)$, if it was assigned a color $c \in C(i)$ by FFP (or equivalently, it received the prediction c_i). We use $G_i = (V_i, E_i)$ to denote the subgraph of G induced by the set of vertices of color palette $C(i)$.*

Note that an alternative, equivalent description of FFP is that it colors each induced subgraph G_i of G using FIRSTFIT with color palette $C(i)$. Also note that it is without loss of generality to assume that the color palettes are distinct: every time a specific color is predicted for the first time, one can "rename" it to a new, unused color (if required) and define the corresponding color palette accordingly. Finally, note that FFP does not require any information on the chromatic number $\chi(G)$ of the graph G. We can now relate the number of colors used by FFP in each color palette to the number of prediction errors within that color palette.

Lemma 1. *Fix an optimal coloring $O \in \mathcal{O}(G)$, let $\eta_i(G)$ be the number of vertices v of color palette $C(i)$ for which $O(v) \neq P(v)$, and let $x_i(G)$ be the number of distinct colors used by FFP for vertices of $C(i)$. Then*

$$x_i(G) \leq \eta_i(G) + 1.$$

Lemma 1 plays a central role in the proof of Theorem 2. Theorem 2 shows that FFP never uses more than $\eta(G) + \chi(G)$ colors for an online graph G with predictions. The next result shows that there exist graphs for which this amount of colors may indeed be used.

Lemma 2. *For every integer $k \geq 2$, there exists an online graph with predictions (G, π, P) and $\chi(G) = k$ for which FFP uses $x(G) = \eta(G) + k$ colors.*

Theorem 2 and Lemma 2 directly imply the following result.

Theorem 4. *The competitive ratio of FFP is $1 + \frac{\eta(G)}{\chi(G)}$.*

Theorem 4 directly implies the 1-consistency and smoothness of algorithm FFP: if $\eta(G) = 0$, then FFP produces a $\chi(G)$-coloring and is therefore optimal; furthermore the number of assigned colors by FFP grows linearly with the prediction error. Nevertheless, FFP is not a robust algorithm. Indeed, for example, suppose we are given a bipartite online graph of order n with predictions (G, π, P) such that $\eta(G) = \Theta(n)$. Then FFP would use $\Theta(n)$ colors. But there exist classical online algorithms (without predictions) [5,16,18] that can color any bipartite graph with $O(\log n)$ colors. In the next section, we present how FFP can be robustified by elegantly combining it with a classical algorithm.

3.2 ROBUSTFIRSTFITPREDICTIONS(RFFP)

A common approach for robustifying a consistent algorithm is to appropriately combine it with a classical algorithm that disregards the predictions. A first such attempt for robustifying FFP would be to switch to some classical online coloring algorithm A, once the number of colors used by FFP becomes larger than some predetermined threshold T. Recall that $\chi_A(G)$ denotes the number of colors that algorithm A uses for an online graph with predictions (G, π, P), where $\pi = v_1, v_2, v_3, \ldots, v_n$. Furthermore, assume that FFP would for the first time use $T + 1$ colors upon arrival of some vertex v_i. Then, by switching to a classical online coloring algorithm A (using the same set of colors) for the restriction of π to the *suffix-subgraph* G' induced by $\{v_i, v_{i+1}, \ldots, v_n\}$, the total number of colors used by the combined algorithm would be at most $T + \chi_A(G')$. Similarly to the deterministic combination result for problems with an underlying metric [1], this would already give a robust algorithm, if we can assume that A is weakly monotone in the following sense.

Definition 6. *Let $A(G, \pi)$ be the number of colors A uses for (G, π), where $\pi = v_1, v_2, \ldots, v_n$. Let $\pi(i)$ be the suffix $v_i, v_{i+1}, \ldots, v_n$ of π, and let $(G(i), \pi(i))$ be the online subgraph of (G, π) induced by the vertices in $\pi(i)$. We say that A is weakly monotone (resp. monotone) if for any i, $A(G(i), \pi(i)) \leq c \cdot A(G, \pi)$ for some constant c (resp. for $c = 1$).*

Unfortunately, and perhaps somewhat surprisingly, we are not aware of any weakly monotone online graph coloring algorithm with a non-trivial guarantee on the number of used colors. (In the full version we present instances showing that both FIRSTFIT and BICOLORMAX (see [5]) are not weakly monotone, even on specific classes of bipartite graphs which are known to admit online competitive coloring algorithms.)

We are able to circumvent this issue related to the non-monotonicity by reserving a distinct color palette for each algorithm during the combination. This, however, has the side-effect that after switching to an algorithm A in some round r, it is possible that upon arrival of a vertex v the algorithm A itself does not increase its number of used colors (by using a color that was already employed before round r), but the combined algorithm does. This rules out an expert-setting approach for combining the algorithms (see [1,3] for more information), but we are still able to bound the total number of colors used by the combined algorithm. More specifically, we are able to combine FFP with $t - 1$ classical algorithms and obtain an algorithm ROBUSTFIRSTFITPREDICTIONS (RFFP *for short*) which uses a number of colors bounded from above by the expression in Theorem 3.

Proof of Theorem 3. For $1 \leq i \leq n$, let $G(i)$ denote the (online) graph induced by $\{v_1, v_2, \ldots, v_i\}$. For any algorithm B, let $c(B, i)$ be the color that B assigns to vertex v_i.

In the following, we restrict each of the algorithms $A_1, A_2, \ldots, A_t$ to use its own distinct color palette, where we assume a total ordering of the colors within each palette. Meta-algorithm A is defined as the algorithm that upon arrival of vertex v_i (and the accompanying prediction $P(v_i)$) colors it with color $c(\text{ALG}_i, i)$, where $\text{ALG}_i \in \{A_1, A_2, \ldots, A_t\}$ is an algorithm realizing $\min_{1 \leq j \leq t} A_j(G(i))$.

Note that since each A_i produces a proper coloring and uses a distinct color palette, the resulting coloring after applying A is proper as well. It remains to argue about the number of colors it would use for G.

Let ALG be ALG_n, and for any algorithm A_i, let $d(i)$ be the maximal index such that $A_i(G(d(i))) \leq \text{ALG}(G(d(i)))$. Note that by the definition of A, it will not use any color from A_i's color palette on vertices $v_{i+1}, v_{i+2}, \ldots, v_n$. Thus, A uses at most $A_i(G(d(i)))$ colors from A_i's color palette. Overall, this gives

$$A(G) \leq \sum_{i=1}^{t} A_i(G(d(i))).$$

By the definition of $d(i)$, the above is at most

$$\sum_{i=1}^{t} \text{ALG}(G(d(i))).$$

Since an online algorithm cannot alter any assigned color, $\text{ALG}(G(j)) \leq \text{ALG}(G(j + 1))$ for all $j \in \{1, 2, \ldots, n - 1\}$. This implies $\text{ALG}(G(d(i))) \leq \text{ALG}(G)$, which concludes the proof.

In the full version, we show that the result is tight for this meta-algorithm A.

In the previous result, we have been combining deterministic algorithms. However, Theorem 3 easily extends to randomized algorithms as well, assuming that one can simulate the execution of all algorithms simultaneously. The following directly follows from Theorem 3, Jensen's inequality and the concavity of the minimum function.

Corollary 1. *Let $A_1, A_2, \ldots, A_t$ be randomized algorithms for online graph coloring that may or may not make use of the predictions, and assume that one can simulate the execution of these algorithms simultaneously. Then, there exists a* *(randomized) meta-algorithm A that combines $A_1, A_2, \ldots, A_t$ and for any online graph (G, π)*

$$\mathbb{E}(A(G)) \leq t \cdot \min_{1 \leq i \leq t} \mathbb{E}(A_i(G)).$$

Theorem 3 implies that we can combine FFP with a c-competitive classical algorithm (perhaps on a specific class of input graphs) and obtain a $2\min\{1 + \frac{\eta(G)}{\chi(G)}, c\}$-competitive algorithm. Assume that the (learning augmented) algorithm has knowledge that the input graph belongs to a specific class of graphs such that (i) all graphs of this class have chromatic number k and (ii) there exists a classical online algorithm that is online competitive on this class, with function $f(\cdot)$. In that specific case, a slight adaptation in the proof of Theorem 3 even gives us a 1-consistent algorithm that is at the same time robust.

Corollary 2. *For some fixed k, assume that the algorithm FFP is aware that the input graph belongs to a class $\mathcal{C}$ of graphs such that $\chi(G) = k$ for all $G \in \mathcal{C}$. Moreover, assume that a classical online algorithm A_1 is known for all graphs of class $\mathcal{C}$. Then, there exists a meta-algorithm A' that combines FFP with A_1, and for any online graph $(G, \pi, P) \in \mathcal{C}$,*

$$A'(G) \leq 3\min\{\text{FFP}(G), A_1(G)\} \text{ if } \eta(G) > 0, \text{ and}$$
$$A'(G) = k \text{ otherwise.}$$

3.3 Results for Specific Classes of Graphs

Given that the input graph belongs to a specific graph class (and this is known to the algorithm a priori), we can obtain more refined results. In this section, we review some interesting cases for which classical (deterministic) online algorithms are known.

Theorem 5. *There exist (different) algorithms for online coloring bipartite graphs with predictions obtaining a competitive ratio of 1 if $\eta(G) = 0$, and $3\min\{\frac{\eta(G)}{2} + 1, X\}$ otherwise, where X is $\Theta(\log n)$ for general bipartite graphs, $\chi_{OL}(G) - \frac{1}{2}$ for bipartite P_6-free graphs, $2\chi_{OL}(G) - \frac{1}{2}$ for bipartite P_7-free graphs, $\frac{3(\chi_{OL}(G)+1)^2}{2}$ for bipartite P_8-free graphs, and $3(\chi_{OL}(G)+1)^2$ for bipartite P_9-free graphs.*

Theorem 6. *There exist (different) algorithms for online coloring chordal graphs, d-inductive graphs, graphs of treewidth d and disk graphs with predictions obtaining a competitive ratio of 1 if $\eta(G) = 0$, and $2\min\{\frac{\eta(G)}{\chi(G)} + 1, X\}$ otherwise, where $X = O(\log n)$ is the competitive ratio of the respective classical online algorithm.*

Theorem 7. *There exist (different) algorithms for online coloring I_s-free graphs and $K_{1,t}$-free graphs for $t \geq 3$ with predictions obtaining a competitive ratio of 1 if $\eta(G) = 0$, and $2\min\{\frac{\eta(G)}{\chi(G)} + 1, X\}$ otherwise, where X is $t - 1$ for $K_{1,t}$-free graphs, and $\frac{s}{2}$ for I_s-free graphs.*

4   Discussion

It would be interesting to investigate whether a learning augmented algorithm with a consistency better than 2 is possible, when the chromatic number of the input graph is not known to the algorithm. Furthermore, all presented algorithms are smooth and the number of used colors grows linearly with the prediction error. It is an open question whether there exist any learning augmented algorithms which achieves the same consistency, but with a better dependence on the prediction error.

References

1. Antoniadis, A., Coester, C., Eliás, M., Polak, A., Simon, B.: Online metric algorithms with untrusted predictions. In: ICML 2020, vol. 119, pp. 345–355 (2020)
2. Barenboim, L., Drucker, R., Zatulovsky, O., Levi, E.: Memory allocation for neural networks using graph coloring. In: ICDCN 2022, pp. 232–233 (2022)
3. Blum, A., Burch, C.: On-line learning and the metrical task system problem. Mach. Learn. **39**(1), 35–58 (2000)
4. Bondy, J.A., Murty, U.S.R.: Graph Theory with Applications. Macmillan Education, UK (1976)
5. Broersma, H., Capponi, A., Paulusma, D.: A new algorithm for on-line coloring bipartite graphs. SIAM J. Discret. Math. **22**(1), 72–91 (2008)
6. Cohen, I.R., Panigrahi, D.: A general framework for learning-augmented online allocation. CoRR, abs/2305.18861 (2023)
7. Demange, M., Ekim, T., de Werra, D.: A tutorial on the use of graph coloring for some problems in robotics. Eur. J. Oper. Res. **192**(1), 41–55 (2009)
8. Grötschel, M., Lovász, L., Schrijver, A.: Polynomial algorithms for perfect graphs. In: Topics on Perfect Graphs. North-Holland Mathematics Studies, vol. 88, pp. 325–356 (1984)
9. Gyárfás, A., Király, Z., Lehel, J.: Online competitive coloring algorithms. Technical report, TR-9703-1 (1997)
10. Gyárfás, A., Lehel, J.: On-line and first fit colorings of graphs. J. Graph Theory **12**(2), 217–227 (1988)
11. Halldórsson, M.M.: A still better performance guarantee for approximate graph coloring. Inf. Process. Lett. **45**(1), 19–23 (1993)
12. Halldórsson, M., Szegedy, M.: Lower bounds for on-line graph coloring. Theor. Comput. Sci. **130**(1), 163–174 (1994)
13. Irani, S.: Coloring inductive graphs on-line. Algorithmica **11**(1), 53–72 (1994)
14. Johnson, D.S.: Worst case behavior of graph coloring algorithms. In: Proceedings of SEICCGTC, pp. 513–527. Utilitas Mathematica (1974)
15. Karp, R.M.: Reducibility among combinatorial problems. In: Miller, R.E., Thatcher, J.W., Bohlinger, J.D. (eds.) Complexity of Computer Computations. The IBM Research Symposia Series, pp. 85–103. Springer, Cham (1972). https://doi.org/10.1007/978-1-4684-2001-2_9
16. Kierstead, H.A.: Coloring graphs on-line. In: Fiat, A., Woeginger, G.J. (eds.) Online Algorithms. LNCS, vol. 1442, pp. 281–305. Springer, Heidelberg (1996). https://doi.org/10.1007/BFb0029574
17. Lattanzi, S., Lavastida, T., Moseley, B., Vassilvitskii, S.: Online scheduling via learned weights. In: Proceedings of SODA 2020, pp. 1859–1877 (2020)

18. Lovász, L., Saks, M.E., Trotter, W.T.: An on-line graph coloring algorithm with sublinear performance ratio. Discret. Math. **75**(1–3), 319–325 (1989)
19. Lykouris, T., Vassilvitskii, S.: Competitive caching with machine learned advice. In: ICML 2018, vol. 80, pp. 3302–3311 (2018)
20. Marx, D.: Graph coloring problems and their application in scheduling. Periodica Polytech. Electr. Eng. **48**, 11–16 (2004)
21. Micek, P., Wiechert, V.: An on-line competitive algorithm for coloring bipartite graphs without long induced paths. Algorithmica **77**(4), 1060–1070 (2017)
22. Mitzenmacher, M., Vassilvitskii, S.: Algorithms with predictions. Commun. ACM **65**(7), 33–35 (2022)
23. Purohit, M., Svitkina, Z., Kumar, R.: Improving online algorithms via ML predictions. In: NeurIPS 2018, pp. 9684–9693 (2018)
24. Zuckerman, D.: Linear degree extractors and the inapproximability of max clique and chromatic number. Theory Comput. **3**(1), 103–128 (2007)

Integer Programming for Machine Learning

Neuron Pairs in Binarized Neural Networks Robustness Verification via Integer Linear Programming

Dymitr Lubczyk[1] and José Neto[2(✉)]

[1] University of Amsterdam, Science Park 904, 1098 XH Amsterdam, The Netherlands
`dymitr.lubczyk@student.uva.nl`
[2] Samovar, Télécom SudParis, Institut Polytechnique de Paris,
19 place Marguerite Perey, 91120 Palaiseau, France
`jose.neto@telecom-sudparis.eu`

Abstract. In the context of classification, robustness verification of a neural network is the problem which consists in determining if small changes of inputs lead to a change of their assigned classes. We investigate such a problem on binarized neural networks via an integer linear programming perspective. We namely present a constraint generation framework based on disjunctive programming and complete descriptions of polytopes related to outputs of neuron pairs. We also introduce an alternative relying on specific families of facet defining inequalities. Preliminary experiments assess the performance of the latter approach against recent single neuron convexification results.

Keywords: Disjunctive programming · Cutting-plane · Robustness verification

1 Introduction

Nowadays Deep Neural Networks (DNNs) turn out to be successful in diverse domains such as computer vision, natural language processing, machine translation, etc. (see, e.g., [8,15]). One of the reasons for this is their great expressiveness [10], which comes however at the price of being hard to reason about [21]. The latter motivated research directed towards the assessment and improvement of the robustness of DNNs for their use in the context of critical AI systems. Another important drawback of many DNNs lies in the fact that they resort to important amounts of computational and energy resources. Through the use of binarized weights and a simple activation function, binarized neural networks (BNNs) appear as an interesting option in the context of resource-constrained systems. However, already challenging optimization problems in the context of ReLU DNNs (such as verification or training) may seem even harder for BNNs due to their inherent discrete features. In this paper we focus on robustness verification of BNNs. Our main contributions are

A. Basu et al. (Eds.): ISCO 2024, LNCS 14594, pp. 305–317, 2024.
https://doi.org/10.1007/978-3-031-60924-4_23

- the presentation of a constraint generation algorithm based on disjunctive programming and complete descriptions established for polytopes related to outputs of neuron pairs,
- the design and evaluation of a constraint generation algorithm relying on specific families of facet defining inequalities to solve a robustness verification problem for BNNs.

The paper is organized as follows. In Sect. 2 we introduce a robustness verification problem for BNNs and point out works related to ours. Polyhedral results are reported in Sect. 3. Preliminary computational experiments are presented in Sect. 4 before we conclude in Sect. 5. Due to length restrictions proofs are omitted from this extended abstract.

2   Robustness Verification Problem in BNNs

2.1   Description of BNNs

A BNN is a special type of DNN having for activation function:

$$\text{sign}(x) = \begin{cases} 1 & \text{if } x \geq 0 \\ -1 & \text{otherwise} \end{cases} \qquad \text{for all } x \in \mathbb{R}.$$

Given a vector $x \in \mathbb{R}^n$, $\sigma(x) \in \mathbb{B}^n$, with $\mathbb{B} = \{-1; 1\}$, denotes the vector such that $(\sigma(x))_i = \text{sign}(x_i)$, for all $i \in [n]$, where $[n] = \{1, 2, \ldots, n\}$.

A BNN has some number $s + 1$ of ordered layers, the last one being called the *output layer* and the others *hidden layers*. Each layer k, with $k \in [s+1]$, has some number n_k of neurons and an associated weight matrix $W^k \in \mathbb{B}^{n_k} \times \mathbb{B}^{n_{k-1}}$ which is computed during a training phase. Given an input vector $y^0 \in \mathbb{B}^{n_0}$, an output vector $x^{s+1} \in \mathbb{Z}^{n_{s+1}}$ is computed using the recursion

$$x^k = W^k y^{k-1}, \forall k \in [s+1]$$
$$y^k = \sigma(x^k), \forall k \in [s].$$

In the context of classification, each entry of the output vector x^{s+1} is associated to one class and the input y^0 is assigned to the class corresponding to the largest entry of x^{s+1}. (In what follows, we may assume ties are broken arbitrarily.) Given a BNN $\mathcal{B}$, the output vector obtained with the input y^0 will be denoted by $\mathcal{B}(y^0)$.

2.2   Robustness Verification Problem

Let $\mathcal{B}$ represent a BNN as described above, and let $z \in \mathbb{B}^{n_0}$ be an input vector whose class is ℓ and satisfying $(\mathcal{B}(z))_\ell > (\mathcal{B}(z))_j$ for all $j \in [n_{s+1}] \setminus \{\ell\}$. Given a target class $t \in [n_{s+1}] \setminus \{\ell\}$ and a neighborhood $\Omega(z)$ of z in $\mathbb{R}^{n_0}$, the BNN $\mathcal{B}$ will be said *locally robust at z w.r.t. the target t* if, for all $y^0 \in \Omega(z)$, the ℓth

entry of $\mathcal{B}\left(y^0\right)$ is larger than the tth entry. So, checking local robustness of $\mathcal{B}$ at z w.r.t. target t reduces to solving the following problem

$$\max_{x^k,y^k} x_t^{s+1} - x_\ell^{s+1} \tag{1a}$$

$$\text{s.t.} \quad x^k = W^k y^{k-1}, \forall k \in [s+1] \tag{1b}$$

$$y^k = \sigma(x^k), \forall k \in [s] \tag{1c}$$

$$y^0 \in \Omega(z) \tag{1d}$$

$$x^k \in \mathbb{R}^{n_k}, k \in [s+1] \tag{1e}$$

$$y^k \in \mathbb{B}^{n_k}, k \in \{0\} \cup [s] \tag{1f}$$

$\mathcal{B}$ is locally robust at z w.r.t. target t if and only if the optimal objective of (1) is negative. Note that even though the chosen neighborhood $\Omega(z)$ may be convex, problem (1) is not, due to constraints (1c).

2.3 Related Work

The importance of robustness verification problems of DNNs stimulated much research efforts leading to the development of solution approaches relying on diverse search strategies, optimization methods and satisfiability modulo theories (SMT), see e.g. [2,4,16,17] and references therein. Among the vast literature in the field, the share dedicated to BNNs seems rather limited.

A BNN can be represented as a Boolean formula, a feature allowing the use of SAT solvers to verify robustness [5,12,19,20]. An SMT based approach extending the Reluplex method [13] is proposed by Amir et al. [1] to support sign activation functions (in addition to ReLU or max-pooling). Khalil et al. [14] design a heuristic to identify adversarial examples for BNNs that is based on the solution of several integer linear programs associated with the layers. This procedure aims at identifying adversarial examples, namely in the context of adversarial training to strengthen robustness of BNNs. However, it does not solve the verification problem exactly and it is rather proposed as an alternative to the solution of an exact formulation of the problem as a mixed integer linear program (MILP) introduced in the same reference but which does not scale to handle large BNNs. Efforts have been dedicated to strengthen relaxations based on such MILP formulations so as to improve performance further by leveraging BNN specificities. Building upon similar techniques to the ones used by Anderson et al. [2] who introduced strong MILP formulations for robustness verification of ReLU based networks, Han and Gómez [9] derive an ideal formulation of a polytope related to the output of a single neuron in BNNs (details are given in Sect. 3.1). Their work is concurrent with the one (we became aware of later) by Lyu and Huchette [18], the latter investigating also some extensions, such as handling zero weights (in addition to weights in $\mathbb{B}$). Han and Gómez's work [9] is the closest to and originally motivated ours: overall objective is to investigate MILP techniques and polyhedral structures to take into account correlations between the outputs of pairs of neurons in the same layer of a BNN.

3   Disjunctive Programming Based View to Neuron Pairs in BNNs

In this section, after recalling Han and Gómez results about single neuron convexification [9], we present a disjunctive programming based framework to deal with neuron pairs and also present specific families of facet defining inequalities for a polytope related to neuron pairs. Due to length restrictions, we focus on the case when the number of inputs and the number of neurons of each hidden layer in the BNN are even, i.e. n_k is even, for all $k \in \{0\} \cup [s]$. Given a set $X \subseteq \mathbb{R}^n$, $\mathrm{conv}(X)$ stands for its convex hull.

3.1   Single Neuron Convexification

Let n denote a positive integer. We define the set

$$S_1 = \left\{ (\boldsymbol{y}, t) \in \mathbb{B}^n \times \mathbb{B} \colon t = \mathrm{sign}\left(\sum_{i=1}^n y_i \right) \right\}$$

corresponding to all the possible input/output pairs of the sign function with n binarized inputs and all the weights of value one on the inputs. Note that assuming all the weights have value one instead of possibly different values in $\mathbb{B}$ is with no loss of generality (resorting to simple substitutions). Han and Gómez determined a complete description of $\mathrm{conv}(S_1)$

Theorem 1. [9] *If n is even, then the convex hull of S_1 is given by*

$$\frac{n}{2}(t - 1) \le \sum_{i=1}^n \min\{y_i, t\} \tag{2a}$$

$$\sum_{i=1}^n \max\{y_i, t\} \le -2 + (n + 2)\frac{t + 1}{2} \tag{2b}$$

$$(\boldsymbol{y}, t) \in [-1, 1]^{n+1} \tag{2c}$$

Although the number of *linear* inequalities corresponding to (2) is exponential in n, the separation problem can be solved efficiently (in linear time). These inequalities were used in [9] to strengthen a linear relaxation of (1). However, they do not account for potential correlations between the outputs of different neurons. We introduce hereafter a framework aiming at the generation of inequalities taking into account such correlations to strengthen relaxations of (1).

3.2   Disjunctive Programming Based Approach for Neuron Pairs

Consider the following set defined similar to S_1 but for neuron pairs.

$$S_2 = \left\{ (\boldsymbol{y}, t_1, t_2) \in \mathbb{B}^n \times \mathbb{B} \times \mathbb{B} \colon \begin{array}{l} t_1 = \mathrm{sign}\left(\sum_{i=1}^n y_i\right) \\ t_2 = \mathrm{sign}\left(\left(\sum_{i=1}^k y_i\right) - \left(\sum_{i=k+1}^n y_i\right)\right) \end{array} \right\} \tag{3}$$

where $k \in \{0\} \cup [n]$. Note that any set of the form

$$\left\{ (\boldsymbol{y}, t_1, t_2) \in \mathbb{B}^n \times \mathbb{B} \times \mathbb{B} : \begin{array}{l} t_1 = \mathrm{sign}\left(\sum_{i=1}^{n} w_{1,i} y_i\right) \\ t_2 = \mathrm{sign}\left(\sum_{i=1}^{n} w_{2,i} y_i\right) \end{array} \right\}$$

with weights $w_{q,j} \in \mathbb{B}$ for all $q \in \{1,2\}$ and $j \in [n]$, can be represented as (3) (resorting to substitutions and a reordering of the variable indices).

Computational experiments on small instances illustrate the fact that the size of an ideal description of S_2 may be significantly much larger (in terms of the number of inequalities) than S_1, at least when restricted to the original space of variables (see Table 1).

Table 1. Number of facet-defining inequalities of conv (S_2) for $n \in \{4, 6, 8, 10\}$ and $k \in \{0\} \cup [n]$ obtained with PORTA [6]. The number of facets of conv (S_1) for $n = 4, 6, 8, 10$ is $26, 78, 274, 1046$, respectively.

n \\ k	0	1	2	3	4	5	6	7	8	9	10
10	1299	2191	10793	11455	25485	27276	21769	19107	3300	1092	1046
8	345	555	1799	2097	2923	2650	857	292	274		
6	99	149	311	337	223	87	78				
4	33	39	35	21	26						

This may suggest that families of inequalities that are valid for conv (S_2) could contribute to strengthen further relaxations of (1) already including (2). So far we could not derive a complete description of conv (S_2) and we alternatively present hereafter a disjunctive programming based approach to solve the separation problem w.r.t. conv (S_2) in polynomial time. (Given a polyhedron $P \subseteq \mathbb{R}^n$ and a point $\boldsymbol{x} \in \mathbb{R}^n$, the corresponding separation problem is to determine whether $\boldsymbol{x} \in P$, and, if not, to find an inequality that is valid for P and violated by $\boldsymbol{x}$.) The proposed method is based on complete descriptions of polytopes derived from disjunctions of S_2 using the following inequalities.

$$\sum_{i=1}^{n} y_i \geq 0 \qquad (t_1+)$$

$$\sum_{i=1}^{n} y_i \leq -2 \qquad (t_1-)$$

$$\left(\sum_{i=1}^{k} y_i\right) - \left(\sum_{i=k+1}^{n} y_i\right) \geq 0 \qquad (t_2+)$$

$$\left(\sum_{i=1}^{k} y_i\right) - \left(\sum_{i=k+1}^{n} y_i\right) \leq -2 \qquad (t_2-)$$

Inequalities (t_1+) and (t_1-) (resp. (t_2+) and (t_2-)) lead to a disjunction of S_2 w.r.t. the sign of t_1 (resp. t_2). Note that, due to the assumed even parity of n, the right-hand side can be decreased from -1 to -2 in (t_1-) and (t_2-). The four inequalities above lead us to consider the sets

$$Z_{\bullet \blacktriangleright} = \left\{ (y, t_1, t_2) \in \mathbb{B}^n \times \mathbb{B} \times \mathbb{B} : \begin{array}{l} y \text{ satisfies inequalities } (t_1 \bullet) \text{ and } (t_2 \blacktriangleright) \\ t_1 = \operatorname{sign}\left(\sum_{i=1}^{n} y_i\right) \\ t_2 = \operatorname{sign}\left(\left(\sum_{i=1}^{k} y_i\right) - \left(\sum_{i=k+1}^{n} y_i\right)\right) \end{array} \right\}$$

with $\bullet, \blacktriangleright \in \{+, -\}$. An important property which was used in [9] to get an ideal description of conv (S_1) was the fact that the matrix corresponding to the system of constraints composed of $-1 \le t \le 1$, $-1 \le y \le 1$ and (t_1+) or (t_1-) is totally unimodular. This, however, no longer holds if either inequality (t_2+) or (t_2-) is added to such a system. Anyhow, we can show that a simple ideal description can still be determined for all of the above sets $Z_{\bullet \blacktriangleright}$.

Proposition 1. *Assuming n is even the following holds.*

$$\operatorname{conv}(Z_{++}) = \left\{ (y, 1, 1) : \begin{array}{l} (t_1+), (t_2+), y \in [-1, 1]^n \\ \sum_{i=1}^{k} y_i \ge 1 \text{ if } k \text{ is odd} \end{array} \right\}$$

$$\operatorname{conv}(Z_{+-}) = \left\{ (y, 1, -1) : \begin{array}{l} (t_1+), (t_2-), y \in [-1, 1]^n \\ \sum_{i=k+1}^{n} y_i \ge 2 \text{ if } k \text{ is even} \end{array} \right\}$$

$$\operatorname{conv}(Z_{-+}) = \left\{ (y, -1, 1) : \begin{array}{l} (t_1-), (t_2+), y \in [-1, 1]^n \\ \sum_{i=k+1}^{n} y_i \le -2 \text{ if } k \text{ is even} \end{array} \right\}$$

$$\operatorname{conv}(Z_{--}) = \left\{ (y, -1, -1) : \begin{array}{l} (t_1-), (t_2-), y \in [-1, 1]^n \\ \sum_{i=1}^{k} y_i \le -3 \text{ if } k \text{ is odd and } k \ge 3 \end{array} \right\}$$

Proof. We just report here the proof for conv (Z_{++}) since it is similar for the other convex hulls. Let T denote the set

$$T = \left\{ (y, 1, 1) : \begin{array}{l} (t_1+), (t_2+), y \in [-1, 1]^n \\ \sum_{i=1}^{k} y_i \ge 1 \text{ if } k \text{ is odd} \end{array} \right\}.$$

Firstly, we show $Z_{++} \subseteq T$ which implies then conv $(Z_{++}) \subseteq T$ because T is convex. Let $(\overline{y}, 1, 1) \in Z_{++}$. By definition, $\overline{y}$ satisfies (t_1+) and (t_2+), implying $\sum_{i=1}^{k} \overline{y}_i \ge 0$. If k is odd, then, due to $\overline{y} \in \mathbb{B}^n$, we have $\sum_{i=1}^{k} \overline{y}_i \ge 1$, and thus $(\overline{y}, 1, 1) \in T$.

We now prove $T \subseteq \operatorname{conv}(Z_{++})$. Obviously, $T \cap \mathbb{B}^{n+2} \subseteq Z_{++}$. It is then sufficient to prove that all the extreme points of T belong to $\mathbb{B}^{n+2}$. Let $(\hat{y}, 1, 1)$ denote an extreme point of T. Then $\hat{y}$ must verify with equality n linearly independent inequalities from the system S composed of (t_1+), (t_2+), $-1 \le y \le 1$, and, if k is odd: $\sum_{i=1}^{k} y_i \ge 1$. We can distinguish the following cases.

- Case 1: k is even, or k is odd with $\sum_{i=1}^{k} \hat{y}_i > 1$.
  - Subcase 1.1: $\mathbf{1}^\top \hat{y} = 0$. Then $\hat{y}$ is also an extreme point of the polyhedron $Q_{11} \subseteq \mathbb{R}^n$ defined by the system: $\mathbf{1}^\top y = 0, \sum_{i=k+1}^{n} y_i \leq 0, -1 \leq y \leq 1$, which is totally unimodular. Thus $\hat{y}$ is integral, has at least $n-2$ entries in $\mathbb{B}$ and at most two zero entries. Since n is even and $\mathbf{1}^\top \hat{y} = 0$, the number of zero entries cannot be odd. If $\hat{y}$ has two zero entries: $\hat{y}_p = \hat{y}_q = 0$ with $(p,q) \in [n]^2$, $p \neq q$, then necessarily $\sum_{i=k+1}^{n} \hat{y}_i = 0$ and (using our assumptions on k with $\mathbf{1}^\top \hat{y} = 0$), k must be even and either $\{p,q\} \subseteq [k]$ or $\{p,q\} \subseteq \{k+1,\ldots,n\}$. Now, let $\hat{y}^1 = \hat{y} + \epsilon (e_p - e_q)$ and $\hat{y}^2 = \hat{y} - \epsilon (e_p - e_q)$, where e_i stands for the ith unit vector and $\epsilon \in]0, \frac{1}{2}[$. We can check that $\hat{y}^1$ and $\hat{y}^2$ belong to Q_{11} and $\hat{y} = \frac{1}{2}(\hat{y}^1 + \hat{y}^2)$, i.e. a contradiction with $\hat{y}$ being an extreme point of Q_{11}. Consequently, $\hat{y}$ has no zero entries and $\hat{y} \in \mathbb{B}^n$.
  - Subcase 1.2: $\mathbf{1}^\top \hat{y} > 0$. Then $\hat{y}$ is an extreme point of the polyhedron defined by the system: $(t_2+), -1 \leq y \leq 1$, which is totally unimodular. Thus, $\hat{y}$ is integral with at least $n-1$ entries in $\mathbb{B}$ and at most one entry with value zero. If $\hat{y}$ has one entry equal to zero, then (t_2+) must be verified with equality, which is not possible because n is assumed to be even. So, $\hat{y} \in \mathbb{B}^n$.
- Case 2: k is odd and $\sum_{i=1}^{k} \hat{y}_i = 1$.
  - Subcase 2.1: $\mathbf{1}^\top \hat{y} = 0$. Then $\hat{y}$ is also an extreme point of the polyhedron Q_{21} defined by the system: $\sum_{i=1}^{k} y_i = 1, \sum_{i=k+1}^{n} y_i = -1, -1 \leq y \leq 1$. Since the matrix defining this system is totally unimodular we can deduce that $\hat{y}$ is integral with at least $n-2$ entries in $\mathbb{B}$ and at most 2 entries with value zero. Since $\mathbf{1}^\top \hat{y} = 0$ and n is even, the number of zero entries must be even. If $\hat{y}$ has two zero entries $\hat{y}_p = \hat{y}_q = 0$, with $(p,q) \in [n]^2$, $p \neq q$, then the equations in the definition of Q_{21} imply that either $\{p,q\} \subseteq [k]$ or $\{p,q\} \subseteq \{k+1,\ldots,n\}$. But either case leads to a contradiction, similar to Subcase 1.1 above. Thus, $\hat{y} \in \mathbb{B}^n$.
  - Subcase 2.2: $\mathbf{1}^\top \hat{y} > 0$. Then, $\hat{y}$ is an extreme point of the polyhedron Q_{22} defined by the system: $\sum_{i=1}^{k} y_i = 1, \sum_{i=k+1}^{n} y_i \leq 1, -1 \leq y \leq 1$, which is totally unimodular. We can deduce that $\hat{y}$ is integral with at least $n-2$ entries in $\mathbb{B}$ and at most 2 entries with value zero. Using the odd parity of k, we can deduce that $\hat{y}$ cannot have exactly one entry with value zero. If $\hat{y}$ has two zero entries $\hat{y}_p = \hat{y}_q = 0$, with $(p,q) \in [n]^2$, $p \neq q$, then the equation in the definition of Q_{22} implies that either $\{p,q\} \subseteq [k]$ or $\{p,q\} \subseteq \{k+1,\ldots,n\}$. The rest of the proof is similar to Subcase 1.1 above.

$\square$

Note that $\text{conv}(S_2) = \text{conv}\left(\cup_{\bullet, \blacktriangleright \in \{+,-\}} \text{conv}(Z_{\bullet \blacktriangleright})\right) = \text{conv}\left(\cup_{\bullet, \blacktriangleright \in \{+,-\}} Z_{\bullet \blacktriangleright}\right)$. Thus, the derivation of an extended formulation of $\text{conv}(S_2)$ from Proposition 1 with disjunctive programming techniques is straightforward. The latter can be used to design a separation procedure w.r.t. $\text{conv}(S_2)$.

3.3 Separation Procedure w.r.t. conv (S_2)

We describe hereafter a generic procedure to solve the separation problem w.r.t. the convex hull of a finite number of nonempty polytopes and that we used in our experiments w.r.t. conv (S_2).

Let $(\mathcal{P}_i)_{i=1}^{\ell}$ denote a finite family of polytopes in $\mathbb{R}^n$ with

$$\mathcal{P}_i = \{x \in \mathbb{R}^n : A_i x \leq b_i\} \text{ for all } i \in [\ell],$$

with $A_i \in \mathbb{R}^{m_i \times n}$, $b_i \in \mathbb{R}^{m_i}$, $m_i \in \mathbb{N}$. Let $\mathcal{P}$ stand for the convex hull of the union of the polytopes $(\mathcal{P}_i)_{i=1}^{\ell}$: $\mathcal{P} = \text{conv}\left(\cup_{i=1}^{\ell}\mathcal{P}_i\right)$. The problem of determining whether some given point $\hat{x} \in \mathbb{R}^n$ belongs to $\mathcal{P}$ reduces to solving the linear program

$$(\text{SEP}) \begin{cases} \min_{\lambda, y, \epsilon^+, \epsilon^-} - \sum_{i=1}^{n} \epsilon_i^+ + \epsilon_i^- \\ \hat{x} = \sum_{i=1}^{\ell} y_i + \epsilon^+ - \epsilon^- \\ A_i y_i \leq \lambda_i b_i, i = 1, \ldots, \ell \\ \sum_{i=1}^{\ell} \lambda_i = 1 \\ \lambda \in \mathbb{R}_+^{\ell}, \epsilon^+, \epsilon^- \in \mathbb{R}_+^n, \\ y_i \in \mathbb{R}^{m_i}, i = 1, \ldots, \ell. \end{cases}$$

One can check that $\hat{x} \in \mathcal{P}$ holds if and only if the optimal objective value of (SEP) is zero. Consider then the dual problem:

$$(\text{DSEP}) \begin{cases} \max_{c, \alpha, \gamma} c^\top \hat{x} + \gamma \\ c = A_i^\top \alpha_i, \forall i \in [\ell] \\ \alpha_i^\top b_i + \gamma \leq 0, \forall i \in [\ell] \\ c \in [-1; 1]^n, \gamma \in \mathbb{R}, \alpha_i \in \mathbb{R}_+^{m_i}. \end{cases}$$

Note that for any feasible solution (c, α, γ) of (DSEP) the inequality $c^\top x \leq -\gamma$ is valid for $\mathcal{P}_i, \forall i \in [\ell]$. Let $Z_{\text{DSEP}}^{\star}$ denote the optimal objective of (DSEP). Using strong duality in linear programming, the separation problem w.r.t. $\mathcal{P}$ reduces to solving (DSEP): either $Z_{\text{DSEP}}^{\star} = 0$ and in that case $\hat{x} \in \mathcal{P}$. Otherwise $Z_{\text{DSEP}}^{\star} > 0$ and a violated inequality is given by an optimal solution $(\bar{c}, \bar{\alpha}, \bar{\gamma})$ of (DSEP): $\bar{c}^\top \hat{x} > -\bar{\gamma}$. Taking for $\mathcal{P}_i$ the polytopes conv $(Z_{\bullet \blacktriangleright})$ with $\bullet, \blacktriangleright \in \{+, -\}$, the approach described above leads to the next result.

Proposition 2. *The separation w.r.t. conv (S_2) can be solved in polynomial time.*

3.4 Facet Defining Inequalities

Designing a cutting-plane algorithm based on the separation procedure described above (Sect. 3.3) to generate constraints can lead to poor performance in terms of computational time, as we could observe in preliminary experiments. This led us to consider the alternative of proceeding to the separation over specific families of inequalities. We introduce hereafter four families of inequalities stemming from studies based on the framework described in Sect. 3.2–3.3 and that we used in our experiments. We also provide sufficient conditions for these inequalities to be facet defining and study the corresponding separation problem. In what follows, we assume $n \geq 4$, n even.

Proposition 3. *The following inequalities, together with the specified conditions on the set I are valid for* conv (S_2).

$$\left(|I| - \left\lceil \frac{k-1}{2} \right\rceil\right)(t_1 + t_2) - \sum_{i \in I} y_i \leq |I|, I \subseteq [k], |I| \geq \left\lceil \frac{k-1}{2} \right\rceil \tag{5}$$

$$\left(\left\lceil \frac{k-3}{2} \right\rceil - |I|\right)(t_1 + t_2) + \sum_{i \in I} y_i \leq |I|, I \subseteq [k], |I| \geq \left\lceil \frac{k-3}{2} \right\rceil \tag{6}$$

$$\left(|I| - \left\lceil \frac{n-k-2}{2} \right\rceil\right)(t_1 - t_2) - \sum_{i \in I} y_i \leq |I|, I \subseteq [n]\backslash[k], |I| \geq \left\lceil \frac{n-k-2}{2} \right\rceil \tag{7}$$

$$\left(\left\lceil \frac{n-k-2}{2} \right\rceil - |I|\right)(t_1 - t_2) + \sum_{i \in I} y_i \leq |I|, I \subseteq [n]\backslash[k], |I| \geq \left\lceil \frac{n-k-2}{2} \right\rceil \tag{8}$$

Proposition 4. *The following properties hold.*

(i) *Assume $k < \frac{n}{2}$, and let $I \subseteq [k]$ such that $|I| > \lceil \frac{k-1}{2} \rceil$. Then (5) is facet defining for* conv (S_2).

(ii) *Assume $3 \leq k < \frac{n}{2}$, and let $I \subseteq [k]$ such that $|I| > \lceil \frac{k-3}{2} \rceil$. Then (6) is facet defining for* conv (S_2).

(iii) *Assume $\frac{n}{2}+1 < k \leq n-2$, and let $I \subseteq [n]\backslash[k]$ such that $|I| > \lceil \frac{n-k-2}{2} \rceil$. Also assume $k < n - 2$ if $|I| > 1$. Then (7)–(8) are facet defining for* conv (S_2).

Proposition 5. *The separation problem w.r.t. (5)–(8) can be solved in polynomial time.*

4   Computational Experiments

In this section we provide preliminary computational results to assess the performance of constraint generation procedures relying on results from [9] and the families of inequalities (5)–(8) to verify robustness of BNNs.

4.1   Evaluated Methods and Setup

We consider solving (1) with a constraint generation algorithm, starting with the relaxation:

$$\max_{x^{s+1}, y^k} x_t^{s+1} - x_\ell^{s+1} \tag{9a}$$

$$\text{s.t. } x^{s+1} = W^{s+1} y^s \tag{9b}$$

$$\|y^0 - z\|_1 \leq \epsilon \tag{9c}$$

$$x^{s+1} \in \mathbb{R}^{n_{s+1}} \tag{9d}$$

$$y^k \in \mathbb{B}^{n_k}, k \in \{0\} \cup [s] \tag{9e}$$

for some fixed value $\epsilon > 0$. Starting with (9), two options to generate constraints at each iteration are considered:

- `single`: for each inequality of type (2a) and (2b), we check if one is violated, and if so, one with largest violation is generated (for each type).
- `approx`: in addition to the procedure `single`, for each pair of neurons and for each of the four types of inequalities (5)–(8), we check if one is violated, and if so, one with largest violation is generated (for each type).

Due to length restrictions we only report results for three configurations of BNNs: 64×2, 128×2 and 256×1, where the first number denotes the number of neurons per hidden layer (common to all hidden layers) and the second number is the number of layers. Each BNN has 784 inputs and 10 outputs (each one corresponding to a digit). The BNNs have been trained on the MNIST dataset as described in [7], using the methodology from [11]. The training process of BNNs was conducted on the DAS-5 cluster [3]. All networks have been trained to an accuracy rate of approximately 75%. 25 images from the MNIST dataset are used when performing robustness verification, and the target class is always selected so that it differs from the predicted class. The reported results are always averaged over this set of instances. The reported results were obtained using a computer with an Apple M1 processor and 8 GB of RAM. Gurobi 10 (with default options) is used to solve the optimization problems.

4.2   Computational Results

We first evaluate the efficiency of the contraint generation methods to determine (with certainty) the robustness status of a BNN, while restricting the number of iterations to 20 and considering different values for the parameter ϵ defining the neighborhood in (9): $\epsilon \in \{11, 12, \ldots, 20\}$. By one *iteration* we mean the application of the separation procedures for each neuron (w.r.t. (2a)–(2b)), and also for each pair of neurons in the case of `approx` (w.r.t. (5)–(8)).

Figure 1 (resp. 2) displays the verification accuracy, i.e. the proportion of images for which the robustness status could be settled depending on ϵ (resp. the objective value after 20 iterations).

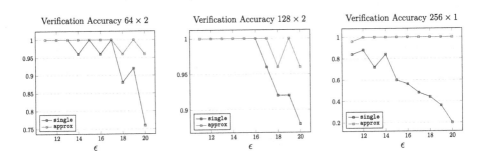

Fig. 1. Verification accuracy

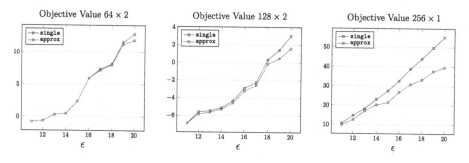

Fig. 2. Objective value after 20 iterations (starting with (9))

The method `approx` clearly improves the verification abilities of the solver for the considered configurations of BNNs. Its overall verification accuracy - averaged over all epsilons and networks - equals 99.3%, whereas the single method verifies 86.1% with especially poor performance for 256×1 BNNs. It is important to notice that when starting with the formulation (9) and keeping the integrality constraints the used integer programming solver may add many cuts (such as Gomory cuts). In order to better assess the potential improvement of `approx` over `single`(i.e. independently of cuts added by the solver), in what follows we report results obtained by relaxing the integrality constraints of 9.

Another observation from experiments we carried out is that `approx` may be much more time consuming than `single`. This led us to investigate an alternative constraint generation strategy denoted by `approx-q` with $q \in \{1, 5\}$. It differs from `approx` by the fact that it generates at most q inequalities per neuron and per iteration. We report in Figures and the evolution of the objective value of the continuous relaxation depending on the number of iterations and time respectively, within the limit of 20 iterations and for $\epsilon = 5$. The ratio of the number of cuts added compared with `single` is in the following ranges: $[1.3, 1.57]$ for `approx-1`, $[2.45, 2.5]$ for `approx-5` and $[5.6, 8.15]$ for `approx`. The objective value obtained wth `approx-5` within 20 iterations tends to be close to `approx` but

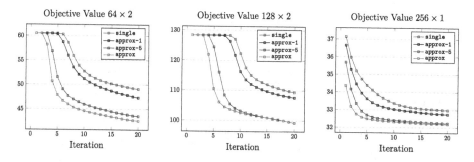

Fig. 3. Evolution of objective value of continuous relaxation (starting with (9)) w.r.t. the number of iterations

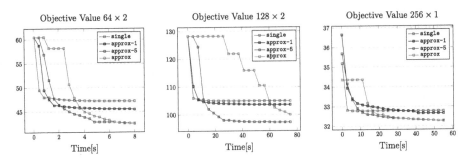

Fig. 4. Evolution of objective value of continuous relaxation (starting with (9)) w.r.t. computation time

with fewer cuts added. On the other hand, it seems that `approx-1` delivers only a slight improvement over `single`, and this is even more stressed for deeper networks (i.e. instances with two hidden layers for the results reported here). `single` or `approx-1` appear to converge much faster than the other methods but they are not able to reach the bounds of the same quality (Figs. 3 and 4).

5   Conclusion

In this paper we addressed a robustness verification problem for BNNs via a constraint generation algorithm. We namely introduced a constraint generation framework relying on disjunctive programming and complete descriptions established for polytopes defining a special disjunction related to the outputs of neuron pairs. Considering the limitations of the latter approach due to high computation times we proposed an alternative constraint generation algorithm relying on specific families of facet defining inequalities. Our preliminary computational results illustrate improvements in terms of verification accuracy over recent convexification results for a single neuron. Ongoing research is directed towards alternative constraint generation strategies and further polyhedral studies related to the outputs of two or more neurons.

References

1. Amir, G., Wu, H., Barrett, C., Katz, G.: An SMT-based approach for verifying binarized neural networks. In: Groote, J.F., Larsen, K.G. (eds.) TACAS 2021. LNCS, vol. 12652, pp. 203–222. Springer, Cham (2021). https://doi.org/10.1007/978-3-030-72013-1_11
2. Anderson, R., Huchette, J., Ma, W., Tjandraatmadja, C., Vielma, J.P.: Strong mixed-integer programming formulations for trained neural networks. Math. Program. **183**, 3–39 (2020)
3. Bal, H., et al.: A medium-scale distributed system for computer science research: infrastructure for the long term. Computer **49**(5), 54–63 (2016)

4. Bunel, R., Turkaslan, I., Torr, P.H.S., Kohli, P., Mudigonda, P.K.: A unified view of piecewise linear neural network verification. In: Neural Information Processing Systems (2017)

5. Cheng, C., Nührenberg, G., Ruess, H.: Verification of binarized neural networks. CoRR abs/1710.03107 (2017). http://arxiv.org/abs/1710.03107

6. Christof, T., Löbel, A.: Porta - polyhedron representation transformation algorithm. https://porta.zib.de/

7. Deng, L.: The MNIST database of handwritten digit images for machine learning research. IEEE Signal Process. Mag. **29**(6), 141–142 (2012)

8. Goldberg, Y.: A primer on neural network models for natural language processing. J. Artif. Intell. Res. **57**, 345–420 (2016)

9. Han, S., Gómez, A.: Single-neuron convexifications for binarized neural networks. University of Southern California (2021). https://optimization-online.org/?p=17148

10. Hornik, K., Stinchcombe, M., White, H.: Multilayer feedforward networks are universal approximators. Neural Netw. **2**(5), 359–366 (1989)

11. Hubara, I., Courbariaux, M., Soudry, D., El-Yaniv, R., Bengio, Y.: Binarized neural networks. In: Lee, D., Sugiyama, M., Luxburg, U., Guyon, I., Garnett, R. (eds.) Advances in Neural Information Processing Systems, vol. 29. Curran Associates, Inc. (2016)

12. Jia, K., Rinard, M.C.: Efficient exact verification of binarized neural networks. CoRR abs/2005.03597 (2020). https://arxiv.org/abs/2005.03597

13. Katz, G., Barrett, C., Dill, D.L., Julian, K., Kochenderfer, M.J.: Reluplex: a calculus for reasoning about deep neural networks. Formal Methods Syst. Des. **60**, 87–116 (2022). https://doi.org/10.1007/s10703-021-00363-7

14. Khalil, E.B., Gupta, A., Dilkina, B.: Combinatorial attacks on binarized neural networks. In: International Conference on Learning Representations (ICLR) (2019)

15. Krizhevsky, A., Sutskever, I., Hinton, G.E.: ImageNet classification with deep convolutional neural networks. In: Pereira, F., Burges, C., Bottou, L., Weinberger, K. (eds.) Advances in Neural Information Processing Systems, vol. 25. Curran Associates, Inc. (2012)

16. Lin, W., et al.: Robustness verification of classification deep neural networks via linear programming. In: 2019 IEEE/CVF Conference on Computer Vision and Pattern Recognition (CVPR), pp. 11410–11419 (2019)

17. Liu, C., Arnon, T., Lazarus, C., Strong, C., Barrett, C., Kochenderfer, M.J.: Algorithms for Verifying Deep Neural Networks (2021)

18. Lyu, B., Huchette, J.: Verifying binarized neural networks: convex relaxations, mixed-integer programming, and consistency. https://bochuanbob.github.io/BNN_MIP.pdf

19. Narodytska, N., Kasiviswanathan, S., Ryzhyk, L., Sagiv, M., Walsh, T.: Verifying properties of binarized deep neural networks. In: AAAI Conference on Artificial Intelligence (AAAI) (2018)

20. Narodytska, N., Zhang, H., Gupta, A., Walsh, T.: In search for a sat-friendly binarized neural network architecture. In: International Conference on Learning Representations (ICLR) (2020)

21. Szegedy, C., et al.: Intriguing properties of neural networks. In: International Conference on Learning Representations (ICLR) (2014)

Optimal Counterfactual Explanations for k-Nearest Neighbors Using Mathematical Optimization and Constraint Programming

Claudio Contardo[1], Ricardo Fukasawa[2(⊠)], Louis-Martin Rousseau[3], and Thibaut Vidal[3]

[1] Department of Mechanical, Industrial and Aerospace Engineering, Concordia University, Montreal, Canada
[2] Department of Combinatorics and Optimization, University of Waterloo, Waterloo, Canada
rfukasawa@uwaterloo.ca
[3] Mathematics and Industrial Engineering, Polytechnique Montreal, Montreal, Canada

Abstract. Within the topic of explainable AI, counterfactual explanations to classifiers have received significant recent attention. We study counterfactual explanations that try to explain why a data point received an undesirable classification by providing the closest data point that would have received a desirable one. Within the context of one the simplest and most popular classification models—k-nearest neighbors (k-NN)—the solution to such optimal counterfactual explanation is still very challenging computationally. In this work, we present techniques that significantly improve the computational times to find such optimal counterfactual explanations for k-NN.

1 Introduction

k-Nearest Neighbors (k-NN) stands as one of the most popular and simplest machine learning (ML) classification models. In k-NN, we are given a set of n *data points* or *observations* (given as points in $\mathbb{R}^d$). Each such observation has an associated label, taken from a set $\mathcal{L}$. Finally, we are also given an integer $k \geq 1$. A new unseen data point $x \in \mathbb{R}^d$ is classified by looking at which label appears more frequently among the labels of its k closest observations.

In this paper, we study the problem of providing counterfactual explanations for k-NN. Counterfactual explanations in ML play a key role in the interpretability of the models. They provide answers to the following fundamental question: "What is the smallest change that should be applied to a sample point x to shift its label from an undesirable classification to a desirable one?". Counterfactual explanations have been proposed for a variety of ML models, such as tree ensembles [7] and linear classifiers [11], among others. Recent surveys on counterfactual explanations can be found in [2,5].

A. Basu et al. (Eds.): ISCO 2024, LNCS 14594, pp. 318–331, 2024.
https://doi.org/10.1007/978-3-031-60924-4_24

Despite its algorithmic simplicity, the only counterfactual explanatory model for k-NN has been proposed in [4] via a mixed-integer program (MIP), but as mentioned in that paper, "Explanations of nearest-neighbor predictors can be obtained in short times for small sample sizes, but do not scale as well to large sample sizes". We are interested in applying techniques from mathematical optimization and constraint programming to develop more efficient counterfactual models and algorithms for k-NN. The contributions of our article can therefore be summarized as follows:

1. We introduce a filtering mechanism aimed at reducing the dimension of these models without compromising their correctness.
2. We introduce two relaxation bounds and an incremental solution approach, which exploits them, to accelerate the resolution of the models.
3. We assess the relative performances of the proposed models and methods.

The remainder of this article is organized as follows. Section 2 presents the basic problem definition and formulation. Section 3 focuses on the first dimensionality reduction technique called *filtering*. Section 4 presents the other dimensionality reduction techniques (*partition* and *sampling*). Computational results are presented in Sect. 5 and final discussions in Sect. 6.

2 Problem Definition

We start by formally defining the k-nearest neighbors (k-NN) problem. In k-NN, we are given a set $\{(x^i, y(x^i)) : i = 1, \ldots, n\}$ of labeled data, where $\mathcal{O} = \{x^i : i = 1, \ldots, n\} \subseteq \mathbb{R}^d$ is the set of *data points* or *observations*, and $y(x^i) \in \mathcal{L}$, where $\mathcal{L} = \{0, 1\}$ is the set of possible labels. We assume, WLOG, that 0 is an undesirable label and 1 is a desirable one. We are also given a dissimilarity measure $dist(x, w) \geq 0$ that establishes the dissimilarity between two points $x, w \in \mathbb{R}^d$. Throughout this work, we assume $dist(x, y)$ to be $||x - y||_1$. Two observations with a dissimilarity close to zero are therefore interpreted as being similar. Finally, we are also given an integer $k \geq 1$.

The way the k-NN classifier works for a new, unseen data point $x \notin \mathcal{O}$ is as follows. Let $N := \{1, \ldots, n\}$ and $N(k, x)$ be the index set of the k closest points in $\mathcal{O}$ to x according to the dissimilarity measure $dist$, i.e., $|N(k, x)| = k$ and $dist(x, x^i) \leq dist(x, x^j), \forall i \in N(k, x), j \in N \setminus N(k, x)$. Let $N_v(k, x) = \{i \in N(k, x) : y(i) = v\}$ for $v \in \mathcal{L}$ (here and throughout this work we will abuse notation and use $y(i)$ to represent $y(x^i)$). The k-NN classifier returns the majority class in $N(k, x)$, i.e., $c(x) = \arg\max\{|N_v(k, x)| : v \in \mathcal{L}\}$. We will assume that ties are not possible, for instance by assuming that k is odd. The k-NN may be ill-defined if two or more points have the same dissimilarity, in which case there may be multiple sets $N(k, x)$. We will assume an *optimistic* definition: In case of ties, we will assume that $N(k, x)$ corresponds to the subset with the minimum number of observations labeled as 0 (the choice is arbitrary if there are multiple such subsets).

We now proceed to define the counterfactual k-NN problem (cnt-k-NN). In addition to the input of k-NN, the cnt-k-NN problem receives as input a point

x^0 such that $c(x^0) = 0$. In the cnt-k-NN we wish to find the point $x \in \mathbb{R}^d$ such that $c(x) = 1$ and minimizing $dist(x, x^0)$. We define some notation (that will be used later) to make the dependence on $x^1, \ldots, x^n$ explicit. Let $N_v := \{i \in N : y(i) = v\}$ for $v \in \mathcal{L}$. The following formulation is given in [4]:

$$
\text{cnt}(N_0, N_1) = \begin{cases}
\min & \delta_0 & \text{(1a)} \\
\text{s.t.} & \delta_0 = dist(x, x^0) & \text{(1b)} \\
& \delta_i = dist(x, x^i), \forall i \in N_0 \cup N_1 & \text{(1c)} \\
& \lambda_i = 1 \Rightarrow \delta_i \leq \Delta, \forall i \in N_0 \cup N_1 & \text{(1d)} \\
& \lambda_i = 0 \Rightarrow \delta_i \geq \Delta, \forall i \in N_0 \cup N_1 & \text{(1e)} \\
& \sum_{i \in N_0 \cup N_1} \lambda_i = k & \text{(1f)} \\
& \sum_{i \in N_0} \lambda_i \leq \lfloor k/2 \rfloor & \text{(1g)} \\
& \lambda \in \{0,1\}^{N_0 \cup N_1}; x \in \mathbb{R}^d; \delta \in \mathbb{R}_+^{\{0\} \cup N_0 \cup N_1}; \Delta \in \mathbb{R}_+ & \text{(1h)}
\end{cases}
$$

We denote by $X(N_0, N_1)$ the set of x for which there exist δ, Δ, λ such that $(x, \delta, \Delta, \lambda)$ is feasible for (1).

Note that while formulation (1) is not a MIP directly, it can be input into a MIP solver by replacing constraint (1d) with $\delta_i \leq \Delta + M(1 - \lambda_i)$ and constraint (1e) with $\delta_i \geq \Delta - M\lambda_i$. Moreover, since Gurobi [6] accepts constraints like $y = |x|$ and reformulates them automatically in a MIP, the norm constraints can also be passed to it.

Note that, in [4], constraint (1e) was written as $\delta_i \geq \Delta - M\lambda_i + \epsilon$ for some $\epsilon > 0$. However, such formulation will lead to an infeasible model for instance when $n > k$ and $x^1 = \ldots = x^n$, with $y(1) = \ldots = y(n) = 1$, even though any x is feasible in that case. We also note that x is feasible for (1) if and only if $c(x) = 1$ according to an optimistic k-NN classifier.

In addition, (1) can be solved using a constraint programming (CP) solver directly. Note, however, that CP solvers assume that all variables are required to be integer, so we scale the variables x to assume that it only assumes integer variables. Such scaling and integrality may cause some possible solutions to (1) to be lost.

The main challenge in solving problem (1) is that n may be very large, becoming too big to solve using either MIP or CP. In addition, for MIP, constraints (1d) and (1e) are "Big-M" type constraints, which are not a very good structure to have in MIP.

In what follows, we present techniques that were applied to help reducing the size of the problem.

3   Filtering

The first idea for reducing the size of (1) relies on a basic observation that, if one of the points x^i is "too far" from x^0, then it will not be one of the k nearest neighbors of a point that is "close" to x^0.

Formally, we assume that we have already found a feasible solution $\bar{x} \in X(N_0, N_1)$. Any solution we are interested in must be closer to x^0 than $\bar{x}$ and so, the following lemma allows us to eliminate points x^i from (1) which are too far.

Lemma 1. *Suppose that $\bar{x} \in X(N_0, N_1)$ and let $\bar{d} := ||\bar{x} - x^0||$. Assume that we can compute $\bar{\mu} \geq 0$ such that, for every x such that $||x - x^0|| \leq \bar{d}$, there exist k points $x^{i_1}, \ldots, x^{i_k}$ such that $||x - x^{i_l}|| \leq \bar{\mu}$, for all $l = 1, \ldots, k$.*

Then if x^i is such that $||x^i - x^0|| > \bar{d} + \bar{\mu}$, x^i cannot be one of the k nearest neighbors of x, for any x such that $||x - x^0|| \leq \bar{d}$.

Proof. The Lemma follows from a simple application of triangle inequality, since
$$||x^0 - x^i|| \leq ||x - x^0|| + ||x^i - x|| \Rightarrow ||x^i - x|| \geq ||x^i - x^0|| - ||x - x^0|| > \bar{d} + \bar{\mu} - \bar{d} = \bar{\mu}$$
Therefore, $x^{i_1}, \ldots, x^{i_k}$ are closer to x than x^i ☐

We call this operation of removing points x^i based on Lemma 1 *filtering*.

We propose two basic filtering methods to compute $\bar{\mu}$. *Filter 1* is based on finding the k nearest neighbors of $\bar{x}$. Suppose that the k nearest neighbors of $\bar{x}$ are all within a distance $\varepsilon \geq 0$ of it. Then, for any point such that $||x - x^0|| \leq \bar{d}$, we get that $||x - x^{i_l}|| \leq ||x - x^0|| + ||x^{i_l} - \bar{x}|| + ||\bar{x} - x^0|| \leq \bar{d} + \bar{d} + \varepsilon$. So we may apply Lemma 1 with $\bar{\mu} = 2\bar{d} + \varepsilon$.

Filter 2 is based on spending some more computational effort to find $\bar{\mu}$, to derive a second bound that can also be used to further reduce the size of (1). What we will do is that, for each i, we solve the following problem

$$\mu_i = \max ||x - x^i||$$
$$\text{s.t.} \quad ||x - x_0|| \leq \bar{d} \tag{2}$$

Now, we can pick $\bar{\mu}$ to be the k-th smallest μ_i value, and then can apply Lemma 1 with that value for $\bar{\mu}$. Note that (2) is still a MIP, but it is a relatively simple one that a modern MIP solver like Gurobi can be expected to solve in a not so large computational time.

4 Partition and Sampling Relaxations

While the filtering procedures in Sect. 3 may help in reducing the number n of points, more often than not, n will still be relatively large. In this section, we propose two relaxations of $\text{cnt}(N_0, N_1)$ to further reduce the size of problem (1).

4.1 Sampling-Based Relaxation for Dealing with N_0

The first observation is that, if we remove some of the points in N_0, we get a relaxation of $\text{cnt}(N_0, N_1)$. This is formalized in the following lemma.

Theorem 1. *For any $U_0 \subseteq N_0$ such that $|U_0 + N_1| \geq k$, $X(U_0, N_1) \supseteq X(N_0, N_1)$, or in other words, $\text{cnt}(U_0, N_1)$ is a relaxation of $\text{cnt}(N_0, N_1)$.*

Proof. Let us consider the case when $N_0 \setminus U_0 = \{l\}$.

Suppose that $x \in X(N_0, N_1)$ and let δ, Δ, λ be the values for which $(x, \delta, \Delta, \lambda)$ is feasible for cnt(N_0, N_1).

If $\lambda_l = 0$, then $x \in X(U_0, N_1)$ follows by picking the corresponding components of $(x, \delta, \Delta, \lambda)$ that exist in cnt(U_0, N_1).

If $\lambda_l = 1$. Let x^t be the $(k+1)$-th nearest neighbor of x in cnt(N_0, N_1). We can then set $\lambda' = \lambda - e_l + e_t$, and note that λ' satisfies constraints (1f) and (1g), since $\sum_{i \in N_0 \cup N_1} \lambda_i = \sum_{i \in N_0 \cup N_1} \lambda_i'$ and $\sum_{i \in N_0} \lambda_i' \leq \sum_{i \in N_0} \lambda_i$.

We can, therefore, obtain that x is feasible for cnt(U_0, N_1).

For the more general case, when $|N_0 \setminus U_0| \geq 2$, let $l \in N_0 \setminus U_0$. We have shown that $X(U_0, N_1) \supseteq X(U_0 \cup \{l\}, N_1)$ and, by induction on $|N_0 \setminus U_0|$, $X(U_0 \cup \{l\}, N_1) \supseteq X(N_0, N_1)$, so the result follows. □

Theorem 1 allows us to remove a huge number of the points in N_0 and only be left with a very small number of samples from the set N_0. For this reason, we call this relaxation the *sampling* relaxation. Unfortunately, we cannot do the same for points in N_1, which can still leave us with a problem of relatively large size to solve.

4.2 Partition-Based Relaxation for Dealing with N_1

To cope with a large number of points in N_1, we propose a partitioning-based approach. The main idea is that we will partition N_1 into different sets and then each set will be treated in a unified way.

To formalize the approach, we first start by noting that (1) can be modified so that (1f) becomes $\sum_{i \in N_0 \cup N_1} \lambda_i \geq k$. We call this new constraint (1f)' and the resulting problem (1)'.

Lemma 2. *The optimal value of* (1) *and* (1)' *are equal.*

Proof. It is easy to see that (1)' is a relaxation of (1).

Now let $(x', \delta', \Delta', \lambda')$ be optimal for (1)'

Let $I := \{i \in 1, \ldots, n : \lambda_i' = 1\} = \{i_1, \ldots, i_l\}$ for some $l \geq k$. Assume WLOG that $dist(x^{i_j}, x^0) \leq dist(x^{i_{j+1}}, x^0)$ for all $j = 1, \ldots, l-1$.

Note that $dist(x^{i_k}, x^0) \leq dist(x^{i_j}, x^0) \leq \Delta'$ for all $j \geq k$. Then, we can set $\bar{\Delta} = dist(x^{i_k}, x^0)$, $\bar{\lambda}_{i_j} = 0$ for all $j > k$ and $\bar{\lambda}_{i_j} = 1$, for all $j \leq k$.

Then $(x', \delta', \bar{\Delta}, \bar{\lambda})$ is a solution to (1) of the same cost. □

Now let us consider $\mathcal{S} = \{S^1, \ldots, S^p\}$ a partition of $N_0 \cup N_1$, such that, for all $t = 1, \ldots, p$, we have:

(P.1) $S^t \neq \emptyset$;
(P.2) Either $S^t \subseteq N_0$ (in which case $y(S^t) = 0$) or $S^t \subseteq N_1$ (in which case $y(S^t) = 1$);
(P.3) If $y(S^t) = 0$ then $|S^t| = 1$.

We can now formulate the partition-based relaxation of $\text{cnt}(N_0, N_1)$ as:

$$
\text{rel-cnt}(\mathcal{S}, N_0, N_1) =
\begin{cases}
\min & \delta_0 & \text{(3a)} \\
\text{s.t.} & \delta_0 = dist(x, x^0) & \text{(3b)} \\
& \delta_i = dist(x, x^i), \forall i \in N_0 \cup N_1 & \text{(3c)} \\
& \mu_t = \min_{i \in S^t} \delta_i, \forall t = 1, \ldots, p & \text{(3d)} \\
& \lambda_t = 1 \Rightarrow \mu_t \leq \Delta, \forall t = 1, \ldots, p & \text{(3e)} \\
& \lambda_t = 0 \Rightarrow \mu_t \geq \Delta, \forall t = 1, \ldots, p & \text{(3f)} \\
& \sum_{t=1}^{p} |S^t| \lambda_t \geq k & \text{(3g)} \\
& \sum_{t=1,\ldots,p:y(S^t)=0} \lambda_t \leq \lfloor k/2 \rfloor & \text{(3h)} \\
& \lambda \in \{0,1\}^p; x \in \mathbb{R}^d; \delta \in \mathbb{R}_+^{\{0\} \cup N_0 \cup N_1}; \Delta \in \mathbb{R}_+ & \text{(3i)}
\end{cases}
$$

Let $X'(\mathcal{S}, N_0, N_1)$ be the set of x such that there exist $(\delta, \Delta, \mu, \lambda)$ such that $(x, \delta, \Delta, \mu, \lambda)$ is feasible for $\text{rel-cnt}(\mathcal{S}, N_0, N_1)$.

Formulation $\text{rel-cnt}(\mathcal{S}, N_0, N_1)$ considers that either we pick all of the set S^t or none of it to be part of our nearest neighbor set. Moreover, we consider the distance to the set to be the smallest distance to any point in the set.

Before proving that this also leads to a relaxation, a few points to note are as follows. Even though the number of δ and x variables has not changed, the number of binary variables can be significantly smaller than the number of binary variables in $\text{cnt}(N_0, N_1)$. Therefore, it is reasonable to expect that solving $\text{rel-cnt}(\mathcal{S}, N_0, N_1)$ may become significantly cheaper to solve computationally.

In addition, (3) can be input directly into a CP solver. However, while constraint (3d) and (3f) can be easily linearized, constraint (3e) cannot, so writing a MIP to solve (3) is not straightforward and will likely require the addition of extra binary variables, which will defeat the purpose of considering the partition in the first place. The following theorem shows that such partition leads to a relaxation of the problem.

Theorem 2. *Suppose* $\mathcal{S} = (S^1, \ldots, S^p)$ *is a partition of* $N_0 \cup N_1$ *satisfying* **(P.1)**, **(P.2)** *and* **(P.3)**. *Also, suppose* $p \geq 2$ *and* $y(S^{p-1}) = y(S^p) = 1$.
Let $\mathcal{S}' = (S^1, \ldots, S^{p-1} \cup S^p)$. *Then* $\mathcal{S}'$ *also satisfies* **(P.1)**, **(P.2)** *and* **(P.3)**. *Moreover* $X'(\mathcal{S}, N_0, N_1) \subseteq X'(\mathcal{S}', N_0, N_1)$, *that is,* $\text{rel-cnt}(\mathcal{S}', N_0, N_1)$ *is a relaxation of* $\text{rel-cnt}(\mathcal{S}, N_0, N_1)$.

Proof. Property **(P.1)** for $\mathcal{S}'$ follows directly. Properties **(P.2)** and **(P.3)** for $\mathcal{S}'$ follow since the only new set in $\mathcal{S}'$ is $S^{p-1} \cup S^p$, which is obtained by the union of two sets that have $y(S^{p-1}) = y(S^p) = 1$.

Let $(\bar{x}, \bar{\delta}, \bar{\Delta}, \bar{\mu}, \bar{\lambda})$ be feasible for $\text{rel-cnt}(\mathcal{S}, N_0, N_1)$.

Set $\hat{\lambda}_t = \bar{\lambda}_t$, $\hat{\mu}_t = \bar{\mu}_t$, for all $t = 1, \ldots, p-2$. And set $\hat{\lambda}_{p-1} = \max\{\bar{\lambda}_{p-1}, \bar{\lambda}_p\}$, $\hat{\mu}_p = \min\{\bar{\mu}_{p-1}, \bar{\mu}_p\}$.

It is clear that (3h) is satisfied, since $\hat{\lambda}_t = \bar{\lambda}_t$ for all $t : y(S^t) = 0$.

Now let's look at (3g). Notice that

$$\sum_{t=1}^{p-2} |S^t|\hat{\lambda}_t + (|S^{p-1}| + |S^p|)\hat{\lambda}_{p-1} \geq \sum_{t=1}^{p} |S^t|\bar{\lambda}_t \geq k$$

Now constraints (3f) and (3e) are immediately satisfied if $t < p - 1$, since those sets did not change.

If $\hat{\lambda}_{p-1} = 1$, then assume WLOG that $\bar{\lambda}_p = 1$. Therefore, $\bar{\Delta} \geq \bar{\mu}_p \geq \hat{\mu}_p$, so (3e) is satisfied.

If $\hat{\lambda}_{p-1} = 0$, then $\bar{\lambda}_{p-1} = \bar{\lambda}_p = 0$. Therefore, $\bar{\Delta} \leq \bar{\mu}_{p-1}$ and $\bar{\Delta} \leq \bar{\mu}_p$, so $\bar{\Delta} \leq \hat{\mu}_{p-1}$. So constraint (3f) is also satisfied.

Thus, $(\bar{x}, \bar{\delta}, \bar{\Delta}, \hat{\mu}, \hat{\lambda})$ is feasible for rel-cnt$(\mathcal{S}', N_0, N_1)$, i.e. $\bar{x} \in X'(\mathcal{S}', N_0, N_1)$

□

Convex Hull Relaxation for MIPs. As was mentioned in Sect. 4, formulating problem (3) as a MIP will require the addition of extra binary variables, which is not desirable. Therefore, we propose a further relaxation that can be modeled as a MIP, based on relaxing constraint (3e) to consider the distance to the convex hull of points in S^t, instead of the minimum distance to the points in S^t.

This can be achieved by deleting constraint (3e) and replacing it with the following constraints (and adding the corresponding decision variables):

$$\sum_{j=1}^{d} |y_j^t - x_j| \leq \Delta + M(1 - \lambda_t), \forall t = 1, \ldots, p \tag{4}$$

$$y^t = \sum_{i \in S^t} \theta_i^t x^i, \forall t = 1, \ldots, p \tag{5}$$

$$\sum_{i \in S^t} \theta_i^t = 1, \forall t = 1, \ldots, p \tag{6}$$

$$y^t \in \mathbb{R}^d, \theta_i^t \in [0, 1], \forall t = 1, \ldots, p, i \in S^t \tag{7}$$

Since the distance to the convex hull of S^t is always at most the minimum distance to the points in S^t (as $S^t \subseteq conv(S^t)$), the fact that we remain with a relaxation follows trivially. The ensuing relaxation can now be modeled as a MIP, with the expense of adding extra continuous variables and constraints only, which is preferable to adding extra binary variables.

4.3   An Iterative Algorithm

We can put Theorems 1 and 2 together and pick some $B \subseteq N_0$ and a partition $\mathcal{S} = (S^1, \ldots, S^p)$ of $B \cup N_1$ satisfying **(P.1)**, **(P.2)** and **(P.3)**, and it follows that rel-cnt$(\mathcal{S}, B, N_1)$ is a relaxation of cnt(N_0, N_1). If the solution $\bar{x}$ to rel-cnt$(\mathcal{S}, B, N_1)$ is in $X(N_0, N_1)$, then $\bar{x}$ is optimal for cnt(N_0, N_1). We now proceed to try to find how to modify $\mathcal{S}$ and/or B if that is not the case. We will use the notation $S(i)$ to denote the element of the partition $\mathcal{S}$ that contains i. Recall that we assume that the choice of the k nearest neighbors is *optimistic*, that is, it is the choice of $N(k, x)$ that minimizes the number of observations labeled as 0.

The following theorem determines how to proceed if $\bar{x} \notin X(N_0, N_1)$.

Theorem 3. *Let $B \subseteq N_0$ and $\mathcal{S} = (S^1, \ldots, S^p)$ be a partition of $B \cup N_1$ satisfying (P.1), (P.2) and (P.3). Let $(\bar{x}, \bar{\delta}, \bar{\Delta}, \bar{\mu}, \bar{\lambda})$ be optimal for rel-cnt$(\mathcal{S}, B, N_1)$, such that $\bar{x} \notin X(N_0, N_1)$. Then there exists $i \in N(k, \bar{x})$ such that either $i \in N_0 \backslash B$ or $|S(i)| > 1$.*

Proof. Suppose that for all $i \in N(k, \bar{x})$ either $i \in B$ or $i \in N_1$ and $|S(i)| = 1$.

The first thing to note is that, for all $i \in N(k, \bar{x})$, $S(i)$ exists. Moreover, property (P.3) implies that $i \in B \Rightarrow |S(i)| = 1$, so for all $i \in N(k, \bar{x})$, $|S(i)| = 1$, therefore $S(i) = \{i\}$.

To simplify notation (and without loss of generality), we assume that the points in $\mathcal{O}$ are ordered by nondecreasing distance to $\bar{x}$, that is, such that $\bar{\delta}_i \leq \bar{\delta}_{i+1}$ for all $i = 1, \ldots, n-1$. Moreover, we assume (also WLOG) that $\{1, \ldots, k\} = N(k, \bar{x})$ and that $S^i = S(i) = \{i\}$ for all $i = 1, \ldots, k$.

Since $\bar{x} \notin X(N_0, N_1)$, we must have $|N_0(k, \bar{x})| \geq \lfloor \frac{k}{2} \rfloor + 1$.

From (3h), (P.2) and $B \subseteq N_0$ we have that:

$$\sum_{t=1,\ldots,k:t\in B} \bar{\lambda}_t \leq \sum_{t=1,\ldots,p:y(S^t)=0} \bar{\lambda}_t \leq \lfloor k/2 \rfloor$$

Now $\{1, \ldots, k\} \cap B$ is precisely $N_0(k, \bar{x}) = N(k, \bar{x}) \cap B$. This implies that there exists $l \in N_0(k, \bar{x})$ such that $\bar{\lambda}_l = 0$.

Claim: There exists $t > k$ with $y(S^t) = 1$ such that $\bar{\lambda}_t = 1$

Proof. Constraints (3g), (3h) and (P.3) imply

$$k \leq \sum_{t=1,\ldots,p:y(S^t)=1} |S^t|\bar{\lambda}_t + \sum_{t=1,\ldots,p:y(S^t)=0} \bar{\lambda}_t \leq \sum_{t=1,\ldots,p:y(S^t)=1} |S^t|\bar{\lambda}_t + \left\lfloor \frac{k}{2} \right\rfloor$$

which implies

$$\sum_{t\in N_1(k,\bar{x})} \bar{\lambda}_t + \sum_{t=k+1,\ldots,p:y(S^t)=1} |S^t|\bar{\lambda}_t = \sum_{t=1,\ldots,p:y(S^t)=1} |S^t|\bar{\lambda}_t \geq \left\lfloor \frac{k}{2} \right\rfloor + 1.$$

Now since $|N_0(k, \bar{x})| \geq \lfloor \frac{k}{2} \rfloor + 1$, we have $|N_1(k, \bar{x})| \leq \lfloor \frac{k}{2} \rfloor$ and the result follows.
□

Let $\Delta' = \min_{i=k+1,\ldots,n:y(i)=1} \bar{\delta}_i$ be the smallest distance from $\bar{x}$ to any of the points x^i for $i \in N_1 \setminus N(k, \bar{x})$ (from the previous claim, at least one such point exists).

Claim: $\Delta' = \bar{\Delta} = \bar{\delta}_l$

Proof. From (3d) and (3f) it follows that $\bar{\mu}_l = \bar{\delta}_l \geq \bar{\Delta}$. Let $t > k$ be such that $y(S^t) = 1$ and $\bar{\lambda}_t = 1$. Thus it follows from (3e) and the fact that S^t can only contain indices in $\{k+1, \ldots, n\}$ that $\bar{\Delta} \geq \bar{\mu}_t = \min_{i\in S^t} \bar{\delta}_i \geq \Delta' \geq \bar{\delta}_l$.
□

Now let l' be the index in $k+1,\ldots,n$ with $y(l') = 1$ and $\Delta' = \bar{\delta}_{l'}$. But then $N(k,\bar{x}) \setminus \{l\} \cup \{l'\}$ is still a set of k nearest neighbors of $\bar{x}$ but it has fewer observations labeled as 0, which contradicts our optimistic choice of $N(k,\bar{x})$ $\square$

These results lead to Algorithm 1 which is an iterative algorithm for solving $\mathrm{cnt}(N_0, N_1)$. We note that the following corollary is also immediate from Theorems 1 and 2:

Algorithm 1: Algorithm to solve (3)

1 Set $\mathcal{S} = \{N_1\}$
2 $B \leftarrow \emptyset$
3 $LB \leftarrow 0$
4 $UB \leftarrow \infty$
5 **while** $UB - LB > \epsilon$ **do**
6 $\quad$ Let $\bar{x}, \bar{\lambda}$ be optimal for rel-cnt$(\mathcal{S}, B, N_1)$, with value $\bar{z}$
7 $\quad$ $LB \leftarrow \max\{LB, \bar{z}\}$
8 $\quad$ **if** $\bar{x}$ is feasible for $\mathrm{cnt}(N_0, N_1)$ **then**
9 $\quad\quad$ $UB \leftarrow \bar{z}$
10 $\quad$ **else**
11 $\quad\quad$ Let $\{i_1, \ldots, i_k\} = N(k, \bar{x})$
12 $\quad\quad$ **for** $l = 1, \ldots, k$ **do**
13 $\quad\quad\quad$ **if** $y(i_l) = 1$ and $|S(i_l)| > 1$ **then**
14 $\quad\quad\quad\quad$ $\mathcal{S} \leftarrow \mathcal{S} \setminus \{S(i_l)\} \cup \{\{i_l\}, \{S(i_l) \setminus i_l\}\}$
15 $\quad\quad\quad$ **if** $y(i_l) = 0$ and $i_l \notin B$ **then**
16 $\quad\quad\quad\quad$ $B \leftarrow B \cup \{i_l\}$
17 $\quad\quad\quad\quad$ $\mathcal{S} \leftarrow \mathcal{S} \cup \{\{i_l\}\}$

Corollary 1. *Let $\bar{z}_q$ be the bound obtained when line 6 of Algorithm 1 is executed for the q-th time. Then $\bar{z}_{q+1} \geq \bar{z}_q$.*

We end this section by commenting that the partition-based relaxation and its refinement in Algorithm 1 shares similarities with dynamic discretization [13] approaches, where discretization points are clustered and clusters are subsequently refined. In addition, partition-based approaches to stochastic programs also share a similar philosophy [10,12].

5   Computational Experiments

To test the several different approaches, we performed computational experiments in several datasets from the literature. We present the results in this section. All implementations were made in Python 3.8, using Gurobi 8.1.1 [6] as a MIP solver and CP-SAT from OR-Tools [9] as a constraint programming solver. All experiments were run in single-thread mode in a machine with Intel Xeon Gold 6142 CPU 2.60 GHz processors, and 264 GB of RAM.

5.1   Instances

We took the 9 base instances that were used in the paper [7] consisting on a variety of data sets from different sources and applications. Each of these base instances may have several thousands of data points, so we generated our instances based on picking (at random) a subset of n points for n being the minimum between the number of data points and a parameter $TCAP \in \{500, 1000, \ldots, 4000\}$. The selection of the points was made via a direct call to the `train_test_split` function in the `sk-learn` package [8]. In addition, we tested each instance by picking different values of $k \in \{5, 15, 25, 35\}$.

We also note that each instance contains specification of possible actionability constraints that were already implemented in [4]. These restrict some features to being fixed (e.g. gender at birth cannot be changed), or restricted to being increasing only (e.g. age), and other types of restrictions. These are implemented as explicit bound constraints on the x variables and are taken as part of the input as well.

5.2   Evaluation

To evaluate each possible approach, we use performance profiles [3], which are distribution functions of a performance metric (time). A performance profile is computed as follows. Given a set of methods M, and a set of instances I, we compute, for each $m \in M$, $i \in I$, the desired metric t_{im}. We then compute the best performance for each instance $\tau_i := \min_{m \in M} t_{im}$, and then for each $m \in M$, $i \in I$, we compute the ratio $r_{im} = \frac{t_{im}}{\tau_i}$. Such ratio represents how many times worse than the best a method performed in an instance. With this, the performance profile consists of a set of lines (one for each $m \in M$) where a point (x, y) in a line represent that in y percent of the instances, the ratio r_{im} was at most x. It can be seen as a cumulative graph of the r_{im} values.

These types of graphs help identify benchmark trends by normalizing "easy" and "hard" instances in similar ways. They give immediate ways to identify better approaches that are more informative than averages. An easy way to read such graphs is to see which lines are on top. *Upper and to the left* means better. We considered a time limit of 3600 s for all approaches and, for the performance profiles, we considered $t_{im} = \infty$ if $t_{im} \geq 3600$, with $\frac{\infty}{\infty} := \infty$.

5.3   Effect of Filtering

We test our filters by taking as an initial feasible solution $\bar{x} = x^i$ for each data point in the input for which $y(x^i) = 1$. Three different filtering settings were tested: No filter (`f0`), Filter type 1 (`f1`) and Filter type 2 (`f2`).

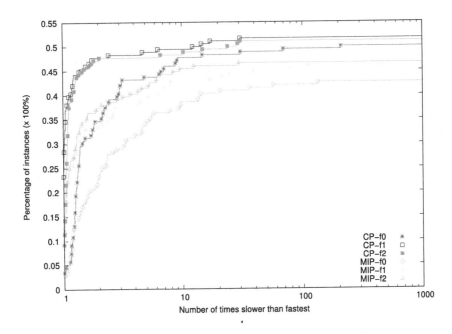

Fig. 1. Results of different filters for baseline approaches

The methods tested are: CP and MIP which just solve problem (1) directly using CP/MIP solvers; CP-part and MIP-part which solve problem (3) using Algorithm 1. Note that MIP-part uses also the convex hull relaxation from Sect. 4.2. Approaches MIP-part and CP-part were used by initializing B to have the data points x^i with $y(x^i) = 0$ among the $2k$ nearest neighbors of x^0, and initializing S to having one set for each of $2k$ nearest neighbors of x^0, plus one set containing all remaining elements in N_1.

Figures 1 and 2 show the results for both the baseline approaches (CP and MIP), as well as for the approaches based on Algorithm 1. It is clear from these figures that the idea of filtering helps significantly, with F1 being the best setting for CP, and F2 the best for MIP, MIP-part, CP-part.

It is also clear that CP-based approaches significantly outperform MIP-based ones. This may be due to the fact that LP-relaxations for the MIP-based formulations are bad (lower bounds at the root are often 0.0, which is the trivial lower bound for minimizing distance), typically giving little information about what the integer solution should look like.

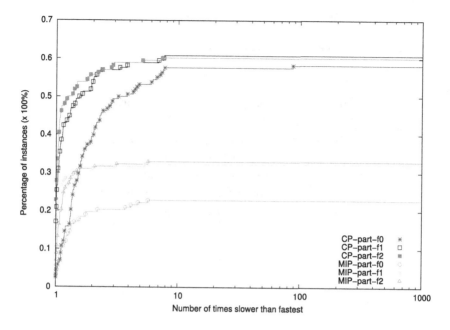

Fig. 2. Results of different filters for partition and sampling relaxations

5.4 Overall Results

In this section, we compare the best filter settings of each approach against each other. In addition, we leave the setting MIP-f0 in the experiments, since this is the baseline MIP formulation that was proposed in [4], so it represents the best in the literature so far.

An additional approach was tested, with the rationale that using a single partition S^t for all elements in N_1 may not be too good, since it may treat data points that are very far from each other in a uniform manner. With that in mind, we also tested the -km variation, which initializes S with clusters computed using the k-means algorithm implemented in [8].

The average (shifted geometric average with a shift of 1 s) solution time decreased from 780 s in the baseline MIP-f0 to 451 s in CP-part-f2 (for the purposes of average computation, a solution time of 3600 s were considered for all runs which did not terminate within the time limit). We use shifted geometric mean since it reduces influence of outliers that are "too big/small". See [1] for more on the topic. Table 1 shows average computing times for the different methods.

Figure 3 shows that both Algorithm 1 and filtering made a significant difference in improving solution times for the problem (we only plot in that figure the 4 MIP approaches and 4 CP approaches with lowest average time, plus the baseline MIP-f0). In addition, for MIP-based approaches, the best option is to either use the more aggressive filtering, or turn filtering off altogether, and give

Table 1. Average computing times (in seconds) for different approaches

Method	T(s)	Method	T(s)
MIP-f0	780.0	CP-f0	601.35
MIP-f1	643.5	CP-f1	472.83
MIP-f2	590.91	CP-f2	495.59
MIP-part-f0	1692.6	CP-part-f0	607.3
MIP-part-f1	1052.7	CP-part-f1	464.88
MIP-part-f2	1037.6	CP-part-f2	450.66
MIP-part-km-f0	1057.4	CP-part-km-f0	641.4
MIP-part-km-f1	696.92	CP-part-km-f1	499.6
MIP-part-km-f2	710.88	CP-part-km-f2	513.57

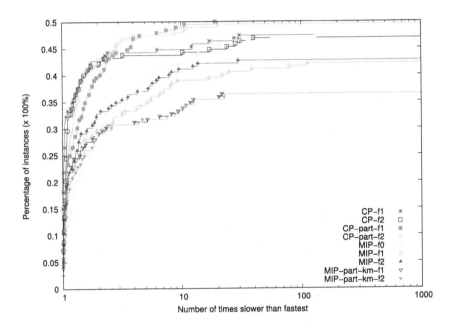

Fig. 3. Overall comparison of approaches

the resulting problem to a MIP solver directly. Once more, we suspect that while the initial LP relaxation is bad, the further relaxation that needed to be made in Sect. 4.2 weakens it even more, while making the solution to each subproblem not fast enough.

6   Conclusion

In this work, we proposed several techniques to speed up the computation of counterfactual explanations for k-NN. The overall combined effect of these techniques was an average reduction of 42% in running time.

While performing more refined techniques to reduce the dimension may lead to additional time savings, other possible research directions are designing better dual bounds for MIP formulations. In addition, a more integrated approach, where solution to previous MIPs/CPs can be utilized to speed up computation of similar, but more refined MIPs/CPs is a promising area of improvement.

References

1. Achterberg, T.: Constraint integer programming. Ph.D. thesis, TU Berlin (2007)
2. Artelt, A., Hammer, B.: On the computation of counterfactual explanations – a survey (2019). https://arxiv.org/abs/1911.07749
3. Dolan, E.D., Moré, J.J.: Benchmarking optimization software with performance profiles. Math. Program. **91**, 201–213 (2002)
4. Forel, A., Parmentier, A., Vidal, T.: Explainable data-driven optimization: from context to decision and back again. In: 40th International Conference on Machine Learning (ICML 2023) (2023). https://publications.polymtl.ca/56889/
5. Guidotti, R.: Counterfactual explanations and how to find them: literature review and benchmarking. Data Min. Knowl. Discov. 1–55 (2022)
6. Gurobi Optimization, LLC: Gurobi Optimizer Reference Manual (2023). https://www.gurobi.com
7. Parmentier, A., Vidal, T.: Optimal counterfactual explanations in tree ensembles. In: International Conference on Machine Learning, pp. 8422–8431. PMLR (2021)
8. Pedregosa, F., et al.: Scikit-learn: machine learning in python. J. Mach. Learn. Res. **12**(Oct), 2825–2830 (2011)
9. Perron, L., Didier, F.: CP-SAT. https://developers.google.com/optimization/cp/cp_solver/
10. Roland, M., Forel, A., Vidal, T.: Adaptive partitioning for chance-constrained problems with finite support (2023). https://arxiv.org/abs/2312.13180
11. Russell, C.: Efficient search for diverse coherent explanations. In: Proceedings of the Conference on Fairness, Accountability, and Transparency, pp. 20–28 (2019)
12. Song, Y., Luedtke, J.: An adaptive partition-based approach for solving two-stage stochastic programs with fixed recourse. SIAM J. Optim. **25**(3), 1344–1367 (2015)
13. Vu, D.M., Hewitt, M., Boland, N., Savelsbergh, M.: Dynamic discretization discovery for solving the time-dependent traveling salesman problem with time windows. Transp. Sci. **54**(3), 703–720 (2020)

Applications

Surrogate Constraints for Synchronized Energy Production/Consumption

Fatiha Bendali, Alejandro Olivas Gonzales, Alain Quilliot$^{(\boxtimes)}$,
and Helene Toussaint

LIMOS CNRS 6158, Labex IMOBS3, Clermont-Ferrand, France
{bendali,alain.quilliot,helene.toussaint}@isima.fr

Abstract. We deal here with job scheduling under encapsulated renewable and non-renewable resource constraints, while relying on a case study related to energy production. Such a context requires synchronizing resource production and consumption in order to both minimize production costs and efficiently achieve the jobs. It most often involves distinct players with their own agenda and non-shared information. We adopt the point of view of the job scheduler and short-cut part of the production level with the help of surrogate estimators.

Keywords: Scheduling · Complexity · Energy Management

1 Introduction

Multi-level decision models (see [3]) usually involve several players, independent from each other or tied together by some hierarchical or collaborative links, who share the decision with respect to a given system. They aim at providing best scenarii in case of a common authority or at in finding a compromise else.

Handling such a model is usually complex. Standard approaches rely on decomposition schemes, hierarchical (Benders,...) or horizontal (Lagrangean relaxation) (see [3]). In both cases a major difficulty remains, related to the way one may retrieve information from the different levels in order to make them interact. Moreover in practice, it may be utopian, for both technical and economical reasons, to assume that all players will agree on a common agenda. *Game Theory* provides us with a useful tool for the distribution of the costs among them. But it does not help a given player in making its own decision under incomplete information. So, when the focus is on a specific player, a trend is to bypass some levels of the global model and replace them by surrogate estimators. Present contribution gets along with this trend. For the sake of clarity, we start from an applicative context involving the joint management of a local photovoltaic (PV) plant and the consumption of resulting energy by a consortium of users (industrial players, services providers, ...) according to their own purposes. This context derives from the activities of IMOBS3 (*Innovative Mobility*) Labex in Clermont-Ferrand, about intelligent vehicles, and from the

© The Author(s), under exclusive license to Springer Nature Switzerland AG 2024
A. Basu et al. (Eds.): ISCO 2024, LNCS 14594, pp. 335–347, 2024.
https://doi.org/10.1007/978-3-031-60924-4_25

national PGMO program promoted by power company EDF. The fact is that both market deregulation and emergent technologies currently induce the rise of local renewable energy producers-consumers (factories, farms and even individual householders) who make self-consumption become an issue (see [7]).

So we consider here on one side, a production manager who runs a PV-Plant (Photovoltaic Plant) and who not only distributes the energy among end-users but also buys and sells it on the market. On the other side, one or several players schedule specific jobs that need energy. Both meet in order to perform *recharging transactions*: in order to keep the users from wasting time while waiting for recharge, the PV-Plant relies on a set of batteries, so that users only need to switch from a battery to another one. We suppose that this *plug out/in* operation is instantaneous. Limited storage and recharge capacities require carefully synchronizing the time-dependent energy production and its consumption. Many researchers showed interest in the decision problems raised by the management of renewable energy, but they most often focused either on production scheduling (see [2,6]) or on consumption (see [4]). Few dealt with the interaction between both (see [1]). We study it here, while adopting the point of view of the job scheduler. Though in practice uncertainty is part of the challenge we suppose, for the sake of simplicity, that our system behaves in a deterministic way.

Starting from this context, we set in Sect. 2 a bi-level **PVSync** model as an extension of the well-known **RCPSP**: *Resource Project Scheduling Problem*(see [8]). We cast it into the MILP (*Mixed Integer Linear Programming*) framework and study its complexity. Next we analyze in Sect. 3 the *Virtual Battery* relaxation before addressing in Sect. 4 the algorithmic issue while adopting the point of view of the job scheduler: We first short-cut the battery level through *Virtual Battery* constraints and next the whole production level through a parametric allocation mechanism. We devote Sect. 5 to numerical experiments.

2   The PVSync Problem

This section is devoted to our case study.

The *Job Scheduler* Side: We consider a set of *jobs* $\mathbf{J} = \{1,\ldots,J\}$ to be performed exactly once within a set $\mathbf{N} = \{1,\ldots,N\}$ of unit length periods, together with a set $<<$ of precedence pairs (j_1,j_2): $(j_1,j_2) \in$ $<<$ (we also write $j_1 << j_2$) means that j_1 must be finished when j_2 starts. Any job j requires t_j periods and is constrained by a time window $\{Min_j,\ldots,Max_j\}$, consistent with the $<<$ precedence relation: the starting period of j must be no smaller that Min_j and its ending period no larger than Max_j. We do not allow *Preemption*: Once a job starts, it must be run without any interruption. The Job Scheduler wants to minimize the sum $SchCost$ of the completion times of the jobs.

The *PV-Plant* Side: Every job j requires some amount of (electric) energy e_j. In order to implement a *self-consumption* policy, the job scheduler interacts with a *PV-Plant*, that means on a photovoltaic facility provided with a set $\mathbf{K} = \{1,\ldots,K\}$ of identical batteries. This PV-Plant produces its own energy that it distributes among the currently idle batteries or sells on the market. In case this self-produced energy is not enough, the PV-Plant can also buy energy. Then, every time a job j starts, the PV-Plant assigns it a currently idle battery loaded with at least e_j, that it get back when the job is finished. Plugging a battery in/out a job j is instantaneous and may take place only at the junction between two consecutive periods. Running a job requires exactly one battery and a battery cannot be simultaneously used by several jobs, so that no more than K jobs may be running at the same time. A battery is *idle* at a given period if it is not used by any job during this period. Then it may be recharged. Energy stored inside the batteries can be neither transferred to another battery nor sold. We denote by C the *storage* capacity of a battery and by C^R its *recharge* capacity, that means the maximal amount of energy that a battery may receive during 1 period. The initial load of battery $k \in \{1, \ldots, K\}$ is denoted by H_k^{Init}.

At every period i, the PV-Plant decides about energy purchase and sale, about the way it distributes energy among the idle batteries, and about the way it assigns batteries among starting jobs. We denote by R_i the expected production of the PV-Plant at period i, by P_i the energy purchase unit price and by S_i the energy sale unit price. Of course we have: $P_i \geq S_i$. Then the goal of the PV-Plant is to minimize a *self-consumption* cost $PVCost$, equal to the difference between the energy purchase cost and the profit derived from the sales.

The PV-Plant and the Job Scheduler are most often distinct players, provided with their own agenda. We adopt here the point of view of the Job Scheduler and we formulate resulting **PVSync** problem as follows:

PVSync: {Schedule the jobs (*leader* decision) in such a way that:

- Every job j is run without any interruption within its time window.
- For any $(j_1, j_2) \in \; <<$ (such that $j_1 << j_2$), j_2 ends before j_2 starts.
- Hybrid cost $\alpha.SchCost + PVCost$ is minimal, where α is a *scaling* coefficient and $PVCost$ is the optimal value of the following sub-problem:
  {**PV-Plant:** Decide about the purchases, the sales and the assignment of the batteries to the jobs (*follower* decision) in such a way that:
  * Every time a job j is performed, it is provided with a battery $k(j)$ loaded with at least e_j energy.
  * The global energy load of the batteries at the end of period N must be at least equal to $\sum_{k \in K} H_k^{Init}$.
  * The difference $PVCost$ between the energy purchase cost and the profit derived from the sales is minimal. } }

Remark 1: PVSync and the **RCPSP** problem: If the PV-Plant and the Job Scheduler merge, then **PVSync** extends the well-known **RCPSP**: *Resource Constrained Scheduling Problem* (see [8]), with *encapsulated* resources, renewable batteries and non-renewable energy embedded into the batteries.

An Example: Let us suppose that $J = \{A, B, C, D, E\}$, with respective durations $t_j = 2, 1, 2, 3, 1$ and energy requirements $e_j = 5, 5, 4, 9, 4$. Jobs A and B

must be run between periods 3 and 5, job C between periods 5 and 8, job D between periods 2 and 7, and job E between periods 7 and 10. We are provided with 2 batteries k_1 and k_2, with respective initial loads equal to 7 and 6. We have: $C = 12$, $C^R = 3$. The time space is divided into 10 periods and $\alpha = 2$. Production data together with energy amounts respectively bought, sold and distributed to the batteries at the different periods are given by Table 1.

Table 1. Prices, Production, Purchase, Sales, Loadings.

i	1	2	3	4	5	6	7	8	9	10	*	1	2	3	4	5	6	7	8	9	10
P_i	2	3	7	7	3	2	6	7	4	2	Bought	1	0	0	0	1	0	0	0	0	1
S_i	1	2	4	4	1	1	3	3	2	1	Sold	0	1	3	5	0	0	1	4	2	0
R_i	4	4	3	5	2	6	4	4	4	5	To k_1	2	*	*	*	3	3	*	*	3	*
*	*	*	*	*	*	*	*	*	*	*	To k_2	3	3	*	*	*	3	*	2	3	*

- Battery k_2 consecutively runs jobs A and B at periods 3, 4, 5 and comes back to the PV-Plant at the end of period 5. It reloads until period 7, runs job E and comes back to the PV-Plant.
- Battery k_1 runs job D at period 2, 3, 4 and comes back to the PV-Plant at the end of period 4. Then it reloads before running job C at periods 8, 9.

2.1   A MILP Model

Let us now suppose that the PV-Plant and the Job Scheduler follow the rule of some global decider. So we adapt standard **RCPSP** MILP: *Mixed Integer Linear Program* formulation corresponding to the case when the time space is explicitly divided into unit-time periods and rely on:

- A $\{0, 1\}$-valued vector $Z = (Z_{j,i}, i = 1, \ldots, N, j = 1, \ldots, J)$ that tells us at which period i a job j starts.
- A $\{0, 1\}$-valued vector $Y = (Y_{j,k}, k = 1, \ldots, K, j = 1, \ldots, J)$: $Y_{j,k} = 1$ means that battery k is assigned to job j.
- A $\{0, 1\}$-valued vector $X = (X_{j,k,i}, j = 1, \ldots, J, k = 1, \ldots, K, i = 1, \ldots, N)$ that links the jobs, the periods and the batteries: $X_{j,k,i} = 1$ means that battery k is assigned to job j, starting at period i.
- An auxiliary $\{0, 1\}$-valued vector $T = (T_{k,i}, k = 1, \ldots, K, i = 1, \ldots, N)$: $T_{k,i} = 1$ means that battery k is active at period i

We represent the activity of the PV-Plant through 3 non negative vectors:

- $U = (U_i, i = 1, \ldots, N)$: U_i means the energy purchase at period i.
- $V = (V_i, i = 1, \ldots, N)$: V_i means the energy sale at period i.
- $Q = (Q_{k,i}, k = 1, \ldots, K, i = 1, \ldots, N)$: $Q_{k,i}$ means the energy distributed to battery k at period i.

Finally, we link the PV-Plant, the jobs and the batteries through a last variable:

- $W = (W_{k,i}, k = 1, \ldots, K, i = 0, \ldots, N) \geq 0$ with rational values: $W_{k,i}$ represents the energy inside battery k at the end of period i.

Then we get:

PVSync_MILP Model:{Compute Z, Y, T, X, U, V, Q, W that meet:

- Minimize: $\sum_i (P_i.U_i - S_i.V_i) + \alpha.(\sum_{j,i} i.Z_{j,i})$. (E1)
- $\forall i \in \mathbf{N}, j \in \mathbf{J}$: $Z_{j,i} = \sum_k X_{j,k,i}$. (E2)
- $\forall k \in \mathbf{K}, j \in \mathbf{J}$: $Y_{j,k} = \sum_i X_{j,k,i}$. (E3)
- $\forall k \in \mathbf{K}, i1 \in \mathbf{N}$: $T_{k,i_1} = \sum_{j,i=i_1-t_j+1,\ldots,i_1} X_{j,k,i}$. (E4)
- $\forall j \in \mathbf{J}$: $\sum_k Y_{j,k} = 1 = \sum_i Z_{j,i} = 1$. (E5)
- $\forall j \in \mathbf{J}, i \in \mathbf{N} \ s.t \ (i < Min_j) \vee (i > 1 + Max_j - t_j)$: $Z_{j,i} = 0$. (E6)
- $\forall k \in \mathbf{K}, i \in \mathbf{N}$: $Q_{k,i} \leq C^R.(1 - T_{k,i})$. (E7)
- $\forall k \in \mathbf{K}, i \in \mathbf{N}$: $W_{k,i} \leq C$. (E8)
- $\forall k \in \mathbf{K}$: $W_{k,0} = H_k^{Init}$ and $W_{k,N} \geq H_k^{Init}$. (E9)
- $\forall k \in \mathbf{K}, i \in \mathbf{N}, i \geq 1$: $W_{k,i} = W_{k,i-1} + Q_{k,i} - (\sum_j e_j.X_{j,k,i})$. (E10)
- $\forall i \in \mathbf{N}$: $U_i + R_i = V_i + (\sum_k Q_{k,i})$. (E11)
- $\forall j_1, j_2 \in \mathbf{J} \ s.t. \ j_1 << j_2$: $\sum_i i.Z_{j_1,i} + t_{j_1} \leq \sum_i i.Z_{j_2}$. (E12)}

Proposition 1: *Solving **PVSync_MILP** solves the **PVSync** problem.*

Proof: One easily turns any feasible solution of **PVSync** into a feasible solution of **PVSync_MILP** with same cost: (E2) means that if job j starts at period i, then there is exactly one battery assigned to j. (E3) means that if battery k is assigned to job j, then there exists a unique i such that j starts with k at i. (E4) means that if battery k is active at period i_1, then there exists a unique job j starting with k at the beginning of some period i s.t. $i \leq i_1 \leq i+t_j-1$. (E5) means that any job j must be provided with exactly 1 battery and 1 starting period. (E6) is the standard time window constraint. (E7, E8) are standard capacity constraints. (E9) expresses the initial and final load requirements. (E10, E11) distribute energy over the time between purchase, sale, storage and consumption. (E12) expresses the precedence constraints.

Conversely, (E5) assigns 1 battery k and 1 starting period i to every job and variables $X_{j,k,i}$ in (E2, E3, E4) characterize the periods when a battery is active or idle. It comes that (E7, E11) control the energy loaded into any battery at any period, as well as the *Non-Preemption* constraint. We conclude. **End-Proof**.

2.2 Some Complexity Results

Once again we suppose that the PV-Plant and the Job Scheduler are under the rule of a global decider. Then we may distinguish inside the **PV-Sync** problem a scheduling (Z), a battery (Y) and a production level (U, V, Q).

The *Scheduling* Level: Restricting **PVSync** to its scheduling level means short-cutting the PV-plant and only considering the jobs, together with K identical batteries that lose their container status and behave as machines in the standard scheduling sense. Notice that resulting problem does not contain the NP-Complete *Multi-Machine Scheduling* problem (see [5]), since it involves a time space explicitly divided into periods. Let us consider the following setting:

PVSync-Sched:

- **Inputs:**
  * The period set $\mathbf{N} = \{1, \ldots, N\}$ and the battery set $\mathbf{K} = \{1, \ldots, K\}$
  * The job set $\mathbf{J}$: A job j requires t_j periods and $\sum_j t_j \leq K.N$.
- **Outputs:** Schedule the jobs of $\mathbf{J}$ inside the periods $1, \ldots, N$, in such a way that no more than K jobs are performed during a same period i.

Theorem 1: *PVSync-Sched is NP-Complete.*

Proof: Let us consider a **Bin Packing** instance BP defined by a set $\mathbf{I} = \{1, \ldots, I\}$ of *items*, every item i provided with a weight w_i, and by a set $\mathbf{B} = \{1, \ldots, B\}$ of identical *boxes*, all with a same capacity CAP. A solution of BP consists in an assignment $i \to b(i)$ from $\mathbf{I}$ into $\mathbf{B}$ such that: $\forall b \in B: \sum_{i \ s.t. \ b(i)=b} w_i \leq CAP$. **Bin Packing** is strongly polynomial (see [5]): There exists a polynomial function $x, y \to P(x, y)$ such that even if we impose $Max(CAP, \sum_i w_i) \leq P(I, B)$, then **Bin Packing** remains NP-Complete. Then we turn BP into an instance $PVSS$ of **PVSync-Sched** by setting: $K = B$; $N = CAP$; $J = I$; $t_i = w_i$ for every i and conclude. **End-Proof**

The *Battery* Level: Let us now suppose that the jobs of $\mathbf{J}$ have been scheduled and that we must provide them with batteries. We focus on the constraints (E3, E4, E5, E9, E10), while supposing infinite production rates R_i and restricting ourselves to 2 batteries. More precisely, we set:

PVSync-Battery Problem:

- **Inputs:**
  * The period set $\mathbf{N} = \{1, \ldots, N\}$.
  * The job set $\mathbf{J} = \{1, \ldots, J\}$: A job j requires e_j energy and is scheduled inside a single period $i(j) \in \{1, \ldots, N-1\}$. We suppose that $J = 2.(N-1)$ and that no more than 2 jobs simultaneously run.
  * Two identical batteries k_1, k_2, with a same initial load H_0.
- **Outputs:** Assign the batteries to the jobs while meeting the energy requirements.

Theorem 2: *PVSync-Battery is NP-Complete.*

Proof: Let us restrict ourselves to the case when $H_0 = \frac{\sum_j e_j}{2}$. For any i there must exist exactly 2 jobs $j(i)$ and $\bar{j}(i) \geq e_{j(i)}$ scheduled at period i. We see that k_1 and k_2 must be active during all periods $1, \ldots, N-1$ and can only recharge at period N. Thus, solving our problem means partitioning the period set $1, \ldots, N-1$ into 2 subsets N_1 and N_2 such that $\sum_{i \in N_1}(e_{\bar{j}(i)} - e_{j(i)}) = \sum_{i \in N_2}(e_{\bar{j}(i)} - e_{j(i)})$. We get a **2-Partition** instance and conclude. **End-Proof.**

3 The *Virtual Battery* Relaxation

Let us come back to the general **PVSync** MILP model and let us set, for any i:

- $\hat{Q}_i = \sum_k Q_{k,i}$; $\hat{W}_i = \sum_k W_{k,i}$; $\hat{C} = K \cdot C$;
- $I_i = K - \sum_{j,i1 \ s.t. \ i1 \leq i \leq i1 + t_j - 1} Z_{j,i}.$ (E13)

Those aggregated variables must satisfy:

– $\forall i \in \mathbf{N}$:	$0 \leq I_i \leq K$	(E13.1)
– $\forall i \in \mathbf{N}$:	$\hat{Q}_i \leq C^R.I_i$ (*Virtual Battery* Constraint E7.1)	
–	$\hat{W}_0 = \sum_k H_k^{Init}$; $\hat{W}_N \geq \hat{W}_0.$	(E9.1)
– $\forall i \in \mathbf{N}$:	$\hat{W}_i \leq \hat{C}.$	(E8.1)
– $\forall i \in \mathbf{N}$:	$U_i + R_i = V_i + \hat{Q}_i.$	(E11.1)
– $\forall i \in \mathbf{N}$:	$\hat{W}_i = \hat{W}_{i-1} + \hat{Q}_i - (\sum_j e_j.X_{j,k,i}).$	(E10.1)

Then we get the following relaxation **PVSync-Merge** of **PVSync**:

PVSync-Merge: {Compute $Z, I, U, V, \hat{W}, \hat{Q}$ such that (E1) is minimized under the constraints (E6, E12, E13, E13.1, E7.1, E8.1, E9.1, E10.1, E11.1).}

PVSync-Merge aggregates the batteries into a single *virtual* one with storage capacity $\hat{C}$, that distributes energy among the active jobs, while simultaneously receiving energy in a way consistent with the *Idle* vector I (E7.1).

Remark 2: Even if Z may be extended into a feasible solution of **PVSync**, the value of **PVSync-Merge** is usually smaller than the value of **PVSync**.

3.1 Complexity of the *Virtual Battery* Constraint

Let us consider the following satisfiability problem **PVSync-Merge_Weak**:
  PVSync-Merge_Weak

- **Inputs:**
    * A period set $\mathbf{N} = \{1, 2, 3, 4\}$; A unit-period job set $\mathbf{J} = \{1, \ldots, J\}$: A job j requires e_j energy.
    * *Battery* parameters K, $H_0 = \sum_k H_k^{Init}$, $C^R = \frac{H_0}{J}$ and $C = \frac{H_0}{K}$.
- **Outputs:** We suppose that the PV-Plant cannot produce, but buys energy according to infinite purchase costs in periods 1 and 3 and null purchase costs in periods 2, 4. Then we want to schedule under a null cost the jobs of $\mathbf{J}$ inside the periods $\{1, 2, 3, 4\}$, in such a way that (E7.1, E8.1, E9.1, E10.1, E11.1, E13, E13.1)) are satisfied:

Theorem 3: *PVSync-Merge_Weak is NP-Complete.*

Proof: The *Idle Battery* constraint, the storage capacity equal to H_0 and the null cost target imply that all jobs must be performed at periods 1, 3. Thus solving **PVSync-Idle** means partitioning $\mathbf{J}$ into two subsets J_1, J_2 such that $\sum_{j \in J_1} e_j = \sum_{j \in J_2} e_j$. The NP-Completeness (see [5]) of **2-Partition** allows us to conclude. **End-Proof**

342   F. Bendali et al.

3.2 Augmenting PVSync-Merge

It may happen that a schedule vector Z solution of **PVSync-Merge** cannot be extended into a feasible solution of **PVSync**. So we introduce the following constraints (E14, E15):

- (E14) imposes that we may feed the jobs j starting at some period i while relying on the initial loads H_k^{Init} and a full load strategy. For any k, we denote by $B(k)$ the number of batteries $k\prime$ such that $H_{k\prime}^{Init} \geq H_k^{Init}$ and get:
$$\forall i \in \mathbf{N}, k \in \mathbf{K}: \quad B(k) \geq \sum j \ \ s.t. \ \ e_j \geq H_k^{Init} + (i-1).C^R Z_{j,i}. \quad \text{(E14)}$$
- (E15) is a non necessary constraint requiring, at any period i, that some batteries have been idle long enough during the periods preceding i in order to be able to feed the jobs starting at i. For any job j and any number of periods π, we set $\delta(j,\pi) = 1$ if $\pi \cdot C^R \geq e_j$, that means if π periods are enough to load a battery with the energy required by j. Then we get:
$$\forall i1, i2 \in \mathbf{N}, i1 < i2: \quad I_{i1} \geq \sum_j Z_{j,i2} \cdot \delta(j, i2 - i1). \quad \text{(E15)}$$

We denote by **PVSync-Merge_Augment** (no longer a relaxation of **PVSync**) the **PVSync-Merge** problem augmented with constraints (E14, E15).

Remark 3: PVSync-Merge_Augment does not ensure us that a schedule vector Z can be extended into a feasible solution of **PVSync**.

4 Scheduling the Jobs Under Surrogate Constraints

As previously told, our purpose is to design a job scheduling model **PVSync_Surr** that bypasses the production level of **PVSync**, this in order to fit with both collaborative contexts and the high complexity of **PVSync**. Bypassing the production level means that we accept, though we are provided with the basic information about the battery characteristics K, C, C^R production rates $R_i, i = 1, \ldots, N$ and the prices $P_i, S_i, i = 1, \ldots, N$, that we do not know what will be the decisions related to the way energy is going to be bought, sold and distributed among the batteries. It comes that such a model should focus on the search for the schedule vector Z and the auxiliary vector I, that derives from Z the number of *idle* batteries that may be in recharge at a given period. Yet, in order to ease the interaction with its partner (the PV-Plant side), the Job Scheduler needs some kind of anticipation of the behavior the PV-Plant. The Job Scheduler will get it through some vector $E = E_i, i = 1, \ldots, N$ providing it, for any i, with an estimation $E_i = \frac{\hat{Q}_i}{I_i}$ of the mean amount of energy loaded into the idle batteries at period i. Clearly, the Job Scheduler cannot know this value but it knows that, if this value were independent on i, then it would be equal to $E^{Mean} = \frac{\sum_j e_j}{K.N - \sum_j e_j}$. So we design the **PVSync_Surr** model as a parametric **PVSync_Surr**(E) model, meaning that the Job Scheduler tries to guess those values $E_i, i = 1, \ldots, N$ all along its interaction with the PV-Plant.

4.1   Surrogate Constraints

Our problem here is to make sure (as much as possible) that it will be possible to extend any feasible solution Z, I of **PVSync_Surr**(E) into a feasible solution of **PVSync**. In order to do it, we first address a weaker issue and characterize the conditions which make possible to extend any vectors Z, I satisfying (E6, E12, E13, E13.1) into a feasible solution of **PVSync-Merge**. Given such a pair Z, I, let us set:

- $Conso(i_1, i_2, Z) = \sum_{j,i \ s.t. \ i_1 \leq i \leq i_2} e_j \cdot Z_{j,i}$ = the energy consumed by the jobs starting no sooner than i_1 and no later than i_2.
- $ProdMax(i_1, i_2, Z) = C^R \cdot (\sum_{i \ s.t. \ i_1 \leq i \leq i_2} I_i$ = the maximal energy that the virtual battery may receive during periods $i_1, \ldots, i_2$.

Theorem 4. *Given Z, I that meet (E6, E12, E13, E13.1). They may be extended into a feasible solution of the* ***PVSync-Merge*** *problem if and only if:*

- *For any i: $H_0 + ProdMax(1, i, Z) \geq Conso(1, i+1, Z)$.*   (E16)
- *For any i, i_1, s.t. $i \leq i_1$: $\hat{C} + ProdMax(i, i_1, Z) \geq Conso(i, i_1 + 1, Z)$.*   (E17)

Proof. Above conditions (E16, E17) are clearly necessary and the key point is about sufficiency. We proceed by induction on i, and suppose the converse. We suppose that there exists i_1 such that we could schedule all jobs starting no later than i_1, and that at the end of period i_1, we cannot feed the jobs j starting in $i_1 + 1$ with enough energy. We may impose our production strategy to be such that, at any period i, we load the virtual battery with as much energy as possible, taking into account the *Virtual Battery* constraint (E7.1) and the storage capacity $\hat{C}$. Then we see, by moving backward from i_1 to 1, that either we reach some period i such that (E17) is violated or we reach period $i = 1$ in such a way that (E16) is violated. We conclude. **End-Proof**

Moreover, we see that if parameter vector $E = (E_i, i = 1, \ldots, N)$ is given with the meaning: $E_i = \frac{\hat{Q}_i}{I_i}$, then Z and I must be such that:

- $\forall i \in \mathbf{N}$: $E_i \leq C^R$.   (E7.2)
- $\forall i \in \mathbf{N}$: $K.C \geq \sum_k H_k^{Init} + \sum_{i1 < i} E_{i1}.I_{i1} \geq \sum_{j,i1 \ s.t. \ i1 \leq i+1} e_j \cdot Z_{j,i1}$.   (E16.1)
- $\forall i, i_1$, s.t. $i \leq i_1$: $\hat{C} + \sum_{i \ s.t. \ i_1 \leq i \leq i_2} I_i \cdot E_i \geq \sum_{j,i \ s.t. \ i_1 \leq i \leq i_2} e_j \cdot Z_{j,i}$.   (E17.1)

Constraints of the Parametric PVSync_Surr(E) Model: It comes in a natural way that the surrogate constraints (E7.2, E14, E15, E16, E16.1, E17, E17.1) will be added to (E6, E12, E13, E13.1) in order to define the constraints imposed to Z, I inside any **PVSync_Surr**(E) model.

4.2   A Surrogate Estimator

We must now provide **PVSync_Surr**(E) with an objective function. In order to do it, let us first consider some optimal solution $Z, I, U, V, \hat{Q}, \hat{W}$ of **PVSync-Merge** and let us denote, for any i, by E_i the average energy amount $\frac{\hat{Q}_i}{I_i}$ distributed among the idle batteries. Then:

Lemma 1. *For any i, $P_i.U_i - S_i.V_i$ is equal to $\Phi_i(E_i)$, Φ_i being the convex function defined by: $\Phi_i(e) = P_i \cdot (e.I_i - R_i)$ if $I_i \cdot e \geq R_i$ and $\Phi_i(e) = S_i \cdot (e.I_i - R_i)$ else.*

Proof. Function Φ is piecewise linear, with an increasing slope (because of our assumption $\forall i, P_i \geq S_i$). So it is convex. If $I_i \cdot E_i \geq R_i$ then $P_i \geq S_i$ implies that V_i is going to be null. Conversely, if $I_i \cdot E_i \leq R_i$ then $P_i \geq S_i$ implies that U_i is going to be null. We conclude. **End-Proof**

Linearizing the convex function Φ_i yields the following quadratic program:
   PVSync_No_Prod Quadratic Model:
{Compute Z, I, E together with an unsigned rational vector $\Pi = (\Pi_i, i = 1, \ldots, N)$ under the constraints:

- Minimize $\sum_i \Pi_i + \alpha.(\sum_{j,i} i.Z_{j,i})$.   (E1.1)
- $\forall i \in \mathbf{N}$: $\Pi_i \geq P_i \cdot (E_i.I_i - R_i)$.   (E18)
- $\forall i \in \mathbf{N}$: $\Pi_i \geq S_i \cdot (E_i.I_i - R_i)$.   (E18.1)
- (E6, E2, E13, E13.1, E16.1, E16.2) are satisfied.}

Theorem 5. *PVSync_No_Prod is equivalent to PVSync-Merge.*

Proof. We turn any feasible solution Z, I, E of **PVSync_No_Prod** into a feasible solution of **PVSync-Merge** by following the proof of Theorem 4. Constraints (E18, E18.1) mean the minimization of $\sum_i \Phi_i(E_i)$ and so Lemma 1 tells us that minimizing (E1) means minimizing (E1.1). **End-Proof**

So we get our parametric model **PVSync_Surr(E:)** by introducing an additional vector $\Pi = (\Pi_i, i = 1, \ldots, N)$ and setting:

   PVSync_Surr(E) : {Compute Z, I, Π that minimizes $\sum_i \Pi_i + \alpha.(\sum_{j,i} i.Z_{j,i})$ under the constraints (E6, E7.2, E12, E13, E13.1, E14, E15, E16, E16.1, E17, E17.1, E18, E18.1).}

4.3   An Algorithm Management of PVSync

Let us denote by **PVSync_No_Prod(Z, I)** the rational linear program with unknown vector E, Π that we derive from **PVSync_No_Prod** by fixing Z and I. Then previous sections lead us to handle **PVSync** by initializing E and iteratively solving **PVSync_Surr(E)** and next updating E. In order to minimize the risk of failure, we iterate several trials along the following algorithmic scheme:

PVSync_Surr **Procedure:** *Trial* is the number of trials that are allowed.
Initialization: Initialize E while considering that E_i takes a same value
   $E^{Mean} = \frac{\sum_j e_j}{K.N - \sum_j e_j}$ for every period i; *NotStop*; *Counter* = 1;
While: *NotStop* and *Counter* $\leq$ *Trial* do
   1. Solve **PVSync_Surr(E)** and get vectors Z, I; *Counter* = *Counter* + 1;
   2. Solve **PVSync_No_Prod(Z, I)** and get a vector $\bar{E}$;   (I1)
   3. Check the feasibility of Z in the **PVSync** sense;   (I2)

4. If $\bar{E} = E$ or if Z has already been generated then *Stop* else set $E = \bar{E}$; Keep the best feasible schedule Z ever obtained.

PVSync_Surr works according to the centralized paradigm. In order to adapt it to a collaborative context, we should replace (I1, I2) by instructions expressing a negotiation between the job scheduler and the PV-Plant.

5   Numerical Experiments

Purpose: Evaluate the behavior of **PVSync-Merge** and *PVSync_Surr*.

Technical Context: A processor IntelCore i5-6700@3.20 GHz, 16 Gb RAM, C++, Linux and the library CPLEX20.1.

Instances: We choose $N = 40, \ldots, 60$ and $M = 2, \ldots, 6$, split **N** into M macro-periods related to different mean production rates and prices and generate coefficients R_i, P_i, S_i accordingly. We put stress on the instances while ensuring their feasibility: We set $C = 2.\mathrm{Sup}_j\, e_j$, $K = \lambda.J.(\frac{t^{Mean}}{N})$, where t^{Mean} is the average duration of the jobs and λ is some control parameter, and generate C^R in such a way that batteries globally receive at least $\beta.J.C$ energy units, β being a control parameter. We describe 10 instances in Table 2.

Outputs: For any instance:

- We apply CPLEX (Table 2) to **PVSync_MILP** and get (≤ 1 CPU h), a lower bound LBG, an upper bound UBG and a CPU time (in sec.) TG. We do not provide optimality gaps since LBG, UBG may be negative.
- We apply (Table 3) the *PVSync_Surr* procedure with $Trial = 1$. We denote by $WP1$ resulting **PVSync** value. We do the same with $Trial = 8$ and denote by $WP8$ related value. We provide related CPU times $TP8$ as well as the number $Trial$ of trials that have been effectively performed. We show in boldface the best value, among $UBG, WP1$ and $WP8$.
- We solve (Table 3) **PVSync-Merge** and **PVSync-Merge_Augment** and get cost values WM and WMA, as well as boolean indicators IM and IMA: $IM = 1$ ($IMA = 1$) tells us that the solution Z of **PVSync-Merge** (**PVSync-Merge_Augment**) is feasible in the **PVSync** sense. We denote by TM the running time induced by **PVSync-Merge**.

Table 2. Instance Characteristics and Behavior of the **PVSync** MILP Model.

Id	N	J	M	t^{Mean}	K	α	β	λ	*	LBG	UBG	TG
1	40	21	3	4	4	1	2	0,5	*	−239.6	−235.9	3600
2	40	23	4	5	4	0,5	3	1	*	−85.5	−82.4	3600
3	40	20	5	6	5	0,2	4	2	*	−1439.3	−1311.6	3600
4	40	24	3	4	5	1	2	0,5	*	488.4	619.1	3600
5	40	32	4	6	4	0,5	3	1	*	212.9	248.3	3600
6	40	34	5	8	4	0,2	4	2	*	177.4	198.3	3600
7	60	43	4	5	3	1	3	1	*	1738.5	1760.3	3600
8	60	47	6	10	3	0,5	4	2	*	Fail	Fail	Fail
9	60	53	4	5	5	1	3	1	*	571.3	3742.1	3600
10	60	61	6	10	5	0,5	4	2	*	−52.9	155.2	3600

Table 3. The *PVSync_Surr* procedure and the **PVSync-Merge** Relaxation.

Id	UBG	WP1	WP8	TP8 (s)	Trial	WM	WMA	IM	IMA	TM (s)
1	**−235.9**	−216.6	**−235.9**	48.2	8	−250.3	−235.9	1	1	100.0
2	**−82.4**	−75.8	−78.5	71.8	6	−99.7	−88.6	0	1	195.1
3	**−1311.6**	−1165.5	−1268.4	76.8	5	−1311.6	−1311.6	1	1	228.5
4	619.1	Fail	**600.3**	62.6	8	574.0	608.2	0	0	169.1
5	**248.3**	**248.3**	**248.3**	131.8	2	222.3	228.6	1	1	328.6
6	**198.3**	221.1	205.6	83.4	3	172.8	177.3	1	1	237.6
7	**1760.3**	Fail	**1760.3**	179.5	5	1705.0	1705.0	0	0	775.8
8	Fail	1625.5	**1588.3**	300.6	6	1487.5	1525.3	1	1	1322.5
9	3742.1	1987.5	**1925.0**	358.1	8	1810.3	1885.0	0	1	1845.4
10	155.2	155.2	**150.3**	178.6	4	139.0	139.0	1	1	680.4

Comments: The global ILP model is in trouble, even on small instances. Solving **PVSync** while relying on the parametric *PVSync_No_Prod* scheme seems rather efficient, even when $Trial = 1$. It outperforms the MILP model in 4 instances and fails only twice, when $Trial = 1$. As for the **PVSync-Merge** relaxation, its value is close to the optimal *PVSync* value but it also yields a non negligible failure rate, that **PVSync-Merge_Augment** reduces only partially.

Acknowledgement. Present work was funded by French ANR: National Agency for Research, and Labex IMOBS3, as well as by PGMO Program.

References

1. Bendali, F., Mole, K.E., Mailfert, J., Quilliot, A., Toussaint, H.: Synchronizing energy production and vehicle routing. RAIRO-OR **55**(4), 2141–2163 (2021)
2. Chen, Z.L.: Integrated production and outbounds distribution scheduling: review and extensions. Oper. Res. **58**, 130–148 (2010)

3. Dempe, S., Kalashnikov, V., Perez-Valdez, G., Kalashikova, N.: Bi-level Programming Problems Theory. Springer, Cham (2015)
4. Erdelic, T., Caric, T.: A survey on the electric vehicle routing problem: variants and solutions. J. Adv. Transp. (2019)
5. Garey, M.R., Johnson, D.S.: Computers and Intractability: A Guide to the Theory of NP-Completeness. Freeman and Co. Ed. (1979)
6. Irani, S., Pruhs, K.: Algorithmic problems in power management. SIGACT News **36**(2), 63–76 (2003)
7. Luthander, R., Widen, J., Nilsson, D., Palm, J.: Photovoltaic self-consumption in buildings: a review. Appl. Energy **142**, 80–94 (2015)
8. Orji, M.J., Wei, S.: Project scheduling under resource constraints: a recent survey. Int. J. Eng. Res. Technol. **2**(2) (2013)

A Robust Two-Stage Model for the Urban Air Mobility Flight Scheduling Problem

Tom Portoleau[(✉)] and Claudia D'Ambrosio

LIX CNRS, École Polytechnique, Institut Polytechnique de Paris, Paris, France
{portoleau,dambrosio}@lix.polytechnique.fr

Abstract. Thanks to recent technical progress, it is now possible to consider air mobility for people at the scale of a city. In this work, we focus on a robust strategic planning problem in an urban air mobility context. For this purpose, we model this problem as a two-stage robust optimization problem in which we aim at computing a schedule that is the least costly, from the user's point of view, to repair in its worst case scenario. The problem is solved with the adversarial Benders method. The subproblem obtained by this method is then solved by three different heuristics. The first one is based on a local search that relies on a scenario dominance rule that allows to significantly reduce the search space. The different parameters involved in this local search have been selected following an experimental analysis. The two others heuristics are based on a more constrained variant of the subproblem. These heuristics are then compared experimentally.

Keywords: Urbain Air Mobility · Two-Stage Optimization · Robust Optimization · Air Traffic Management

With recent technical advances, it is now possible to move people by air in widely spread urban areas over short distances. Thus, a literature has started to emerge to frame and describe the new related technical issues. In [10], the authors propose a study in which they aim to give an overview of the different research fields on the emerging topic of urban air mobility (UAM). To this end, results from several areas of the UAM research community have been gathered and are presented to provide an overview of the relevant technical issues surrounding its implementation. This overview considers vehicle-related aspects, such as aircraft requirements and a classification of aircraft for intra- and inter-urban passenger transport, and, then, examines potential barriers to their introduction. The study of technical issues includes certification, safety, and policy issues, as well as challenges in the area of traffic management and ground infrastructure requirements. In addition, they also examine the literature on operational concepts, possible market structures, and interaction with existing transport systems. Between all the different areas explored in this paper, we fall into the field of operations research, and, more specifically, the solution of the flight scheduling problem, which concerns two different time horizons. First, we consider the so-called strategic problem, which consists in offline computing of a scheduling for a set of flights on a given shared network and which satisfies a set of technical constraints. Second, we are also interested in the so-called tactical problem (also known as the deconfliction problem) in which we try to recalculate the schedules in response to different kinds of disruptions while trying to modify the initial planning as little as possible. These problems are inspired by the well-established strategic and tactical planning in classic aviation, for which a vast literature has already been proposed [5,8]. However, these problems do not apply straightforwardly to the UAM context, where electric vertical take-off and landing (eVTOL) vehicles will be used. In this framework, several approaches have been proposed. For example, in [11], the authors compare decentralized approaches, in which the routing and the planning of each trip is made on the fly following specific rules and policies, against centralized approaches, in which all the decision concerning the flights are taken at once. In [3], the authors solve the strategic problem by adding a constraint limiting the number of vehicles present in predefined zones, without really solving the deconfliction problem. To solve this problem, we can cite [12] and [6], which, in both cases, use decentralized approaches, as opposed to [9], which proposes a centralized approach with a Mixed Integer Linear Programming (MILP) formulation. In this work, we also consider a centralized approach, to compute a solution to the strategic problem in such a way that the deconfliction is the least costly, using methods from robust optimization. As far as we know, this is the first time that this problem is considered with this approach.

To be more specific, we will consider that the first level of decisions (here-and-now decisions) will be the strategic level and that the second level of decisions (wait-and-see decisions), which may depend on the realization of uncertainties, will be the tactical level. In [1], the authors show that multi-stage problems are NP-hard even with only two decision levels. Thus, we introduce in this paper two heuristics to solve the problem at stake.

1   Problem Modeling

In this section, the problem at stake is formally introduced. First, we present a deterministic version of the problem to introduce the notation, the different constraints, and the constraint programming model. The network setting is given by a directed graph $G = (V, E)$ in which the vertices are waypoints and the arcs lie on the air corridors. The distance between two vertices $x \in V$ and $x' \in V$ is denoted by $d(x, x')$. The flights to schedule are denoted by a set $\mathcal{F}$. For each flight $i \in \mathcal{F}$, its route through the network is denoted by the ordered set $\mathcal{P}_i \subseteq V$. The starting waypoint and the destination waypoint are respectively denoted by s_i and e_i (and correspond to vertiports, where flights can takeoff or land). We aim at computing a schedule for the flights which satisfies the following constraints. A vehicle cannot stop mid-air, so once the trip has started, it cannot be interrupted. For any flight i and any waypoint $x \in \mathcal{P}_i$ such that i goes through the arc $(x, \text{succ}_i(x)) \in E$ where $\text{succ}_i(x)$ is the successor of x in $\mathcal{P}_i$, the authorized speed of a vehicle must belong to a given interval $\underline{v}_{i,x}, \overline{v}_{i,x}$, both taking positive values for any $i \in \mathcal{F}$ and $x \in \mathcal{P}_i$. And finally, we have the separations constraints, that is to say, the constraints such that the distance between any pair of vehicles cannot be lower than a given distance D at any time. It is assumed that the distance between any pair of waypoints is always larger than D. The decision variables of the problem are interval variables $T := \{T_{i,x} : i \in \mathcal{F}, x \in \mathcal{P}_i\}$, with a fixed length such that $T_{i,x}$ denotes the interval of time in which the vehicle of the flight i will be passing through the node x. Each $T_{i,x}$ has its lower end bounded by two values, denoted by $\underline{T}_{i,x}$ and $\overline{T}_{i,x}$, which are the earliest and the latest arrival time of i at waypoint x given the distance between x and its predecessor in $\mathcal{P}_i$ (denoted $\text{pre}_i(x)$) and the speed limitations. Using interval variables instead of simple time variables adds a level of robustness to the problem and allows us to take into account micro speed variations for extra safety. The length of intervals is set to 1 in our model which is enough to achieve that while degrading the objective as least as possible. We now introduce a first set of constraints which will be denoted by $\text{Sched}(T)$:

$$\texttt{LengthOf}(\texttt{T}_{\texttt{i,x}}) = 1 \qquad\qquad \forall i \in \mathcal{F}, \forall x \in \mathcal{P}_i, \quad (1)$$

$$\underline{T}_{i,x} \le \texttt{StartOf}(\texttt{T}_{\texttt{i,x}}) \le \overline{T}_{i,x} \qquad\qquad \forall i \in \mathcal{F}, \forall x \in \mathcal{P}_i, \quad (2)$$

$$\underline{T}_{i,x} = \texttt{StartOf}(\texttt{T}_{\texttt{i,pre}_{\texttt{i}}(\texttt{x})}) + \frac{d(\text{pre}_i(x), x)}{\overline{v}_{i,x}} \qquad\qquad \forall i \in \mathcal{F}, x \in \mathcal{P}_i \setminus \{s_i\}, \quad (3)$$

$$\overline{T}_{i,x} = \texttt{StartOf}(\texttt{T}_{\texttt{i,pre}_{\texttt{i}}(\texttt{x})}) + \frac{d(\text{pre}_i(x), x)}{\underline{v}_{i,x}} \qquad\qquad \forall i \in \mathcal{F}, x \in \mathcal{P}_i \setminus \{s_i\}, \quad (4)$$

$$\bigvee (\texttt{StartOf}(\texttt{T}_{\texttt{i,x}}) - \texttt{EndOf}(\texttt{T}_{\texttt{j,x}}) \ge \tau_{\texttt{i,j,x}},$$

$$\texttt{StartOf}(\texttt{T}_{\texttt{j,x}}) - \texttt{EndOf}(\texttt{T}_{\texttt{i,x}}) \ge \tau_{\texttt{j,i,x}}) \quad \forall (i,j,x) \in \texttt{CONFLICTS}, \quad (5)$$

$$\underline{T}_{i,x}, \overline{T}_{i,x} \in \mathbb{R}_+, \; T_{i,x} \text{ is an interval variable of } \mathbb{R}_+. \qquad\qquad (6)$$

Constraint (1) ensures that the time interval is exactly one unit of time. Given that $\text{pre}_i(x)$ is the waypoint that precedes x in the route $\mathcal{P}_i$, Constraints (2)–(4) guarantee that a vehicle never stops and moves at an authorized speed. Finally, Constraints (5) are the separation constraints. The set $\texttt{CONFLICTS}$ is the set of triplets (i, j, x) such that node x is in the route of both flights i and j, namely, $x \in \mathcal{P}_i$ and $x \in \mathcal{P}_j$. Those are the specific cases in which separation must be checked. The value of parameter $\tau_{i,j,x}$ depends on several parameters, such as the distance, angle and legal speed between waypoints, and the nature of the conflict, which depends on the setting of the conflict (for instance if the vehicles are taking the same arc, or if they must go through the same waypoint from two different precedent waypoints, etc.). As it is not the focus of this work, the details of these constraints will not be discussed here, for more explanation, the reader is referred to [9].

By solving this constraint satisfaction problem, we obtain a feasible flight schedule for our problem. As stated earlier, our objective is to approach this problem from a robust perspective. To be more specific, we now consider a two-stage variation of this problem, with a budgeted uncertainty on the availability of the vehicles to takeoff at the planned time, with the objective of minimizing the sum of the arrival time of each flight and the cost of re-optimization after the uncertainty has been revealed. On the first level of decision (also referred to as the strategic level) the goal is to compute an initial schedule, independently of the realization of the uncertainty. On the second level of decision (the tactical level or the recourse) the goal is to repair (if necessary) the schedule such that all constraints are satisfied while degrading as little as possible the arrival time of the flights.

In our context, we suppose that the uncertainty set is a budgeted scenario set. Indeed, for a given budget Γ, we assume that, for each flight i decided at the strategic level, the sum of their respective delays γ_i is at most Γ.

The problem at stake, the Two-Stage Robust Urban Air Mobility Planning Problem can now be written as follows:

$$\min_{\hat{T}} \max_{\gamma} \min_{T} \sum_{i \in \mathcal{F}} \text{EndOf}(T_{i,e_i})$$

$$\text{s.t.} \quad \text{Sched}(\hat{T}) \qquad \qquad \text{(strategic decision),} \quad (7)$$

$$\text{StartOf}(T_{i,s_i}) \geq \text{StartOf}(\hat{T}_{i,s_i}) + \gamma_i \qquad \qquad \forall i \in \mathcal{F}, \quad (8)$$

$$\sum_{i \in \mathcal{F}} \gamma_i \leq \Gamma \qquad \qquad \forall i \in \mathcal{F}, \quad (9)$$

$$\text{Sched}(T) \qquad \qquad \text{(tactical decision),} \quad (10)$$

$$\hat{T}_{i,x}, T_{i,x} \text{ are interval variables} \qquad \forall i \in \mathcal{F}, \forall x \in \mathcal{P}_i.$$

The sets of Constraints (7) and (10) as introduced earlier in this section ensure that the schedules induced by the first and second stage variables, $\hat{T}$ and T, respectively, are feasible solutions.

Note that we consider only potential delays in the departure time of each flight. The scenarios corresponding to other disruptions considered in [9], like priority flights to be accommodated or an intruder, cannot be easily defined. Thus, they are not considered in this work.

2 Adversarial Benders Decomposition Method

The problem described above is typically hard to solve exactly, and one way to do it is by using the adversarial Benders Method (also called cutting-plane method) introduced in [7]. The idea of this method is to approximate the set of scenarios by a subset of them, that grows at each iteration of the method. Let us consider a generic robust problem $\min_{x \in X} \max_{\omega \in \Omega} z(x, \omega)$, where X is the set of feasible solutions and Ω a set of scenarios. The problem is decomposed into two different problems. First the master problem $\mathcal{M}(U)$ which can be written $\min_{x \in X} \max_{\omega \in U} z(x, \omega)$ where U is a finite subset of Ω. Then, the subproblem (also called adversarial problem, or pessimization problem) $\mathcal{S}(x)$ aims at computing the worst scenario for a given solution x. More formally, $\mathcal{S}(x)$ is $\arg\max_{\omega \in \Omega} z(x, \omega)$.

Using the same notation as the previous section, the original problem is decomposed into a master problem and a subproblem. The master problem $\mathcal{M}(U)$ can be written as follows:

$$\min \bar{z}$$

$$\text{s.t. } \texttt{Sched}(\hat{T}), \tag{11}$$

$$z_l \geq \sum_{i \in \mathcal{F}} \texttt{EndOf}(T_{i,e_i,l}) \qquad \forall l \leq |U|, \tag{12}$$

$$\bar{z} \geq z_l \qquad \forall l \leq |U|, \tag{13}$$

$$\texttt{StartOf}(T_{i,s_i,l}) \geq \texttt{StartOf}(\hat{T}_{i,s_i}) + \gamma_{i,l} \qquad \forall i \in \mathcal{F}, \forall l \leq |U|, \tag{14}$$

$$\texttt{Sched}(T_l) \qquad \forall l \leq |U|, \tag{15}$$

$$\hat{T}_{i,x}, T_{i,x,l} \text{ are interval variables,} \qquad \forall i \in \mathcal{F}, \forall x \in \mathcal{P}_i, \forall l \leq |U|,$$

$$\bar{z}, z_l \in \mathbb{R} \qquad \forall l \leq |U|,$$

where the second-level decision variables T are now indexed by the scenarios contained in the finite set U. In this context, T_l in Constraint (15) denotes, for a given l, all the $T_{i,x,l}$, with $i \in \mathcal{F}$ and $x \in \mathcal{P}_i$.

Then, the subproblem $\mathcal{S}(\hat{T})$ reads:

$$\max_{\gamma} \min_{T} \sum_{i \in \mathcal{F}} \texttt{EndOf}(T_{i,e_i})$$

$$\text{s.t. } \texttt{StartOf}(T_{i,s_i}) \geq \texttt{StartOf}(\hat{T}_{i,s_i}) + \gamma_i \qquad \forall i \in \mathcal{F}, \tag{16}$$

$$\sum_{i \in \mathcal{F}} \gamma_i \leq \Gamma \qquad \forall i \in \mathcal{F}, \tag{17}$$

$$\texttt{Sched}(T), \tag{18}$$

$$T_{i,x} \text{ are interval variables} \qquad \forall i \in \mathcal{F}, \forall x \in \mathcal{P}_i.$$

Once again, this problem is typically hard to solve. Thus, we propose two different heuristics. The first one is a local search based on the fact that, for a fixed γ, the resulting problem is easy to solve for a constraint programming solver (numerical results are shown in the experimentation section). The second heuristic is based on the fact that, under the assumption that at the second decision level the order of vehicles cannot be changed, the subproblem $\mathcal{S}(\hat{T})$ can be dualized, thus allowing the elimination of one level, obtaining a minmax adversarial problem.

2.1 Local Search

In this section, we introduce the heuristic based on local search. Since, for a fixed γ, the scheduling problem is, in practice, easy to solve, we will consider that a solution is defined by a scenario $\gamma = (\gamma_i)_{i \in \mathcal{F}}$.

First, let us notice that it is possible to significantly reduce the size of the search space with the following result:

Proposition 1. *Given a value $\Gamma > 0$, there exists an optimal solution γ^* of $\mathcal{S}(\hat{T})$ such that $\sum_i \gamma_i^* = \Gamma$.*

Proof. Let $\gamma^* = (\gamma_1^*, \gamma_2^*, ..., \gamma_{|\mathcal{F}|}^*)$ be an optimal solution of the problem $\mathcal{S}(\hat{T})$ such that $\sum_i \gamma_i^* < \Gamma$, and let $\delta = \Gamma - \sum_i \gamma_i^*$. Let us consider a new solution $\gamma' = (\gamma_1^* + \frac{\delta}{|\mathcal{F}|}, \gamma_2^* + \frac{\delta}{|\mathcal{F}|}, ..., \gamma_{|\mathcal{F}|}^* + \frac{\delta}{|\mathcal{F}|})$. Then, if T^* is the best schedule under scenario γ^*, then the schedule T' obtained by delaying each flight departure time from T^* by $\frac{\delta}{|\mathcal{F}|}$ is optimal under scenario γ'. Finally, we have $\sum_{i \in \mathcal{F}} \texttt{EndOf}(T'_{i,e_i}) = \sum_{i \in \mathcal{F}} \texttt{EndOf}(T^*_{i,e_i}) + \delta > \sum_{i \in \mathcal{F}} \texttt{EndOf}(T^*_{i,e_i})$, which, since both γ^* and γ' are solutions of $\mathcal{S}(\hat{T})$, contradicts the hypothesis that γ^* is optimal. $\qquad\square$

Using this result, we now define the neighbourhood of a solution γ as

$$\mathcal{N}(\gamma) = \{\gamma' : \|\gamma - \gamma'\|_1 < G, \sum_i \gamma_i = \sum_i \gamma'_i = \Gamma\},$$

where G is a parameter whose value is going to be discussed in the experimental section of this paper. An element of this set is chosen randomly by distributing a random amount of budget from a γ_i equally to a random subset of elements. At each iteration of the search, this neighbourhood is sampled such that each i is chosen at least once. After evaluating all the sampled elements, the best objective improving neighbour will be used as a base solution for the next iteration. Otherwise, i.e., if no better neighbour is found, the search restarts, until the time limit is reached.

2.2 Fixed Order Approximation

In this section, we introduce two methods to solve a more constrained variant of the problem. It is now assumed that the ordering decided at the strategic level is fixed and cannot be changed at the tactical level.

Under this hypothesis, the disjunction (5) which models the ordering decision between two flights on a shared waypoint is no longer needed in the subproblem and can be replaced by a regular inequality, depending on the first level decision. The subproblem can now easily be expressed as the minmax problem:

$$\max_{\gamma} \min_{t} \sum_{i \in \mathcal{F}} t_{i,e_i}$$

$$\text{s.t.} \sum_{i \in \mathcal{F}} \gamma_i \leq \Gamma \qquad\qquad \forall i \in \mathcal{F}, \quad (19)$$

$$t_{i,s_i} \geq \texttt{StartOf}(\hat{T}_{i,s_i}) + \gamma_i \qquad\qquad \forall i \in \mathcal{F}, \quad (20)$$

$$t_{i,x} - t_{i,\text{pre}_i(x)} \leq \frac{d(\text{pre}_i(x), x)}{\overline{v}_{i,x}} \qquad \forall i \in \mathcal{F}, \forall x \in \mathcal{P}_i \setminus \{s_i\}, \quad (21)$$

$$t_{i,x} - t_{i,\text{pre}_i(x)} \geq \frac{d(\text{pre}_i(x), x)}{\underline{v}_{i,x}} \qquad \forall i \in \mathcal{F}, \forall x \in \mathcal{P}_i \setminus \{s_i\}, \quad (22)$$

$$t_{i,x} - t_{j,x} \geq \tau_{i,j,x} \qquad \forall (i,j,x) \in \texttt{ORDER}, \qquad (23)$$
$$t_{i,x} \in \mathbb{R}_+ \qquad \forall i \in \mathcal{F}, \forall x \in \mathcal{P}_i,$$
$$\gamma_i \in \mathbb{R}_+ \qquad \forall i, \forall x \in \mathcal{P}_i,$$

where $t_{i,x}$ denotes the start of the time interval in which the flight i passes through waypoint x, and $\texttt{ORDER}$ is a set of ordered triplets representing conflicts disambiguated conflicts. For instance if a triplet (i,j,x) is in $\texttt{CONFLICT}$, then $\texttt{ORDER}$ must contain either (i,j,x) or (j,i,x) depending on which flight between i and j has been scheduled to pass first through waypoint x. In this formulation, there are no interval variables anymore because the solving methods that will be introduced in this section are based on solving mathematical programs.

The first method consists of solving a quadratic problem, obtained by considering the dual problem of the previous formulation. The resulting quadratic program reads:

$$\max \sum_{(i,j,x)} \tau_{i,j,x} c_{i,j,x} + \sum_{i,x} \left(\overline{p}_{i,x} \frac{d(\text{pre}_i(x),x)}{\overline{v}_{i,x}} + \underline{p}_{i,x} \frac{d(\text{pre}_i(x),x)}{\underline{v}_{i,x}} \right)$$
$$+ \sum_i (\texttt{StartOf}(\hat{\mathtt{T}}_{\mathtt{i,s_i}}) + \gamma_i) g_i$$

$$\text{s.t.} \quad \texttt{First}_{\mathtt{i,x}} = \sum_{\substack{j \in \mathcal{F}, \\ (i,j,x) \in \texttt{ORDER}}} \mathtt{c_{i,j,x}} - \sum_{\substack{j \in \mathcal{F}, \\ (j,i,x) \in \texttt{ORDER}}} \mathtt{c_{i,j,x}} \qquad \forall i \in \mathcal{F}, \forall x \in \mathcal{P}_i, \tag{24}$$

$$\sum_{i \in \mathcal{F}} \gamma_i \leq \Gamma, \tag{25}$$

$$g_i + \texttt{First}_{\mathtt{i,s_i}} - \underline{\mathtt{p}}_{\mathtt{i,succ_i(s_i)}} - \overline{\mathtt{p}}_{\mathtt{i,succ_i(s_i)}} \geq 0 \qquad \forall i \in \mathcal{F}, \tag{26}$$

$$\texttt{First}_{\mathtt{i,e_i}} + \underline{\mathtt{p}}_{\mathtt{i,e_i}} + \overline{\mathtt{p}}_{\mathtt{i,e_i}} \geq -1 \qquad \forall i \in \mathcal{F}, \tag{27}$$

$$\texttt{First}_{\mathtt{i,x}} + \underline{\mathtt{p}}_{\mathtt{i,x}} + \overline{\mathtt{p}}_{\mathtt{i,x}} - \underline{\mathtt{p}}_{\mathtt{i,succ_i(x)}} - \overline{\mathtt{p}}_{\mathtt{i,succ_i(x)}} \geq 0, \forall i \in \mathcal{F}, \forall x \in \mathcal{P}_i \setminus \{s_i, e_i\}, \tag{28}$$

$$g_i, \underline{p}_{i,x}, c_{i,j,x} \in \mathbb{R}_-, \overline{p}_{i,x}, \gamma_i \in \mathbb{R}_+ \qquad \forall i \in \mathcal{F}, \forall x \in \mathcal{P}_i,$$

where $\texttt{First}_{\mathtt{i,x}}$ is a substitution variable to make other constraints easier to read. Basically it is the number of times a flight i is scheduled first in a conflict, minus the number of times it is scheduled second. Then g_i are dual variables of Constraints (20), $\overline{p}_{i,x}$ and $\underline{p}_{i,x}$ are dual variables of Constraints (21) and (22), and $c_{i,j,x}$ are dual variables of Constraints (23).

The second method is based on the following observation.

Proposition 2. *It is possible to compute the increase in the objective function when a flight i is delayed by a time γ_i with the recursive formula*

$$Delay(i, \gamma_i) = \sum_{i', i < i'} \max(\gamma_i - w_{i,i'}, 0) + Delay(i', \max(\gamma_i - w_{i,i'}, 0)),$$

where $w_{i,i'}$ is the time between the two flights arrival at a waypoint. It is unique because paths are shortest paths.

The idea is that when a flight i is scheduled before another i' at a waypoint, the total delay induced by delaying i will delay equally the flight i', unless there is a waiting time $w_{i,i'}$ that can absorb a part of said delay.

From this one can note that the objective is a sum of continuous, increasing, piecewise-linear functions of γ and can now be modeled as a mixed-integer linear program with the Multiple Choice formulation [4]:

$$\max \sum_{i \in \mathcal{F}} \sum_{s=1}^{s_i} (\alpha_{i,s} z_{i,s} + \mathtt{offset}_{i,s} y_{i,s}) \tag{29}$$

$$x_i = \sum_{s=1}^{s_i} z_{i,s} \qquad \forall i \in \mathcal{F}, \tag{30}$$

$$\sum_{i \in \mathcal{F}} x_i \leq \Gamma, \tag{31}$$

$$b_{i,s-1} y_{i,s} \leq z_{i,s} \qquad \forall i \in \mathcal{F}, s \in \{1, \ldots, s_i\}, \tag{32}$$

$$b_{i,s} y_{i,s} \geq z_{i,s} \qquad \forall i \in \mathcal{F}, s \in \{1, \ldots, s_i\}, \tag{33}$$

$$\sum_{s=1}^{s_i} y_{i,s} = 1 \qquad \forall i \in \mathcal{F}, \tag{34}$$

$$x_i \in \mathbb{R}, z_{i,s} \in \mathbb{R}, y_{i,s} \in \{0,1\} \qquad \forall i \in \mathcal{F}, \tag{35}$$

where, for a given flight $i \in \mathcal{F}$, $b_{i,s}$ is the sth breakpoint, $\alpha_{i,s}$ the value of the slope on segment s, $\mathtt{offset}_{i,s}$ the offset of the piece on segment s (for $i \in \mathcal{F}, s = 1, \ldots, s_i$), x_i part of the budget allocated to flight i, $y_{i,s}$ binary variable, is 1 if x is in segment s, 0 otherwise. In this context, the breakpoints are the waiting times $w_{i,i'}$ and the slopes are the number of other flights being delayed by allotting budget to i. The last heuristic consists of solving this program.

3   Experimental Results

3.1   Instances

Since there is little to no work on this problem, we are inspired by the instances introduced in [9] which deals with the tactical subproblem similar to the one in this paper (but does not consider the strategic and adversarial aspects). There are three different types of networks used in the instances, each with a different topology, so that, for the same number of waypoints and flights, the number of conflicts between flights varies significantly. Reduced versions of these networks are shown on Fig. 1. Because of its shape, the grid network will, on average, have less conflict than the metroplex and airport networks, which have choke points where flights are forced to cross. This will influence the computation time of the different methods. Each network is constructed so that it contains

25 nodes and respects the shape presented in Fig. 1. For each network, 100 instances, that is a set of flights $\mathcal{F}$, and their routes $\mathcal{P}_i$, are built by generating 100 flights each, whose starting point and arrival point are randomly chosen. For each instance, the uncertainty budget Γ is set to the values $\{1, 5, 10, 20\}$. Concerning the computation time limits, the master problem is not constrained, and the local search method and the quadratic program solution are limited to 2 min of computation. The constraint programming (CP) problems have been solved with CP Optimizer version 12.9.0 and the mathematical optimization problems have been solved with CPLEX version 12.9.0 running on Linux Ubuntu 16.04.4 with a Intel Xeon E5-2695 v4 2.10GHz cores CPU. The code and the instances are available here[1].

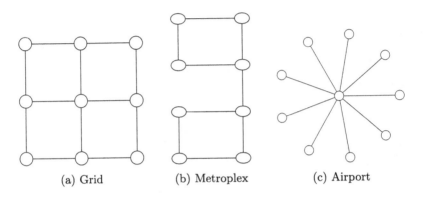

(a) Grid   (b) Metroplex   (c) Airport

Fig. 1. Small examples of the three topologies

3.2   Preliminary Results for the Local Search Heuristic

In this section, we present preliminary results to justify the choices of the different parameters involved in the local search. First of all, it is preferable that, in the neighbourhood we construct, each γ_i has been selected at least once to be the one whose value will be decreased. For this purpose, given that the selection of the γ_i follows a uniform distribution, to make sure that on average each γ_i is drawn at least once we set the size of the sampling set to be $K = (ln(0.1))/(ln(\frac{|\mathcal{F}|-1}{|\mathcal{F}|}))$. Then, in order to determine values that make the heuristic based on local search efficient, this heuristic was used to solve the instances whose parameters are slightly different than the ones introduced in the previous section. Those experiments were run with $\Gamma = 10$. As for the initialization value of γ, denoted by γ^0, two possibilities were considered: either all the budget was allowed to only one random γ_i with a uniform distribution ($\gamma^0 = (0, ..., \Gamma, ..., 0)$), denoted by γ^F, or the budget is evenly split between all the γ_i ($\gamma^0 = (\frac{\Gamma}{|\mathcal{F}|}, \frac{\Gamma}{|\mathcal{F}|}, ..., \frac{\Gamma}{|\mathcal{F}|})$), denoted by γ^S. The values of G were taken in the set $\{1, 2, ..., 9, 10, 20, .., 50 = |\mathcal{F}|\}$. For

---

[1] https://github.com/TomPortoleau/RobustUAM.

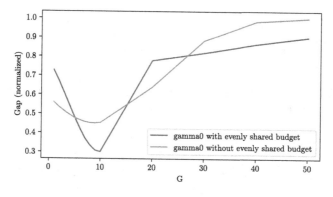

Fig. 2. Performance comparison between initial values of γ^0 for different values of G.

each run, we computed the percentage gap between the objective value of the solution obtained if the scenario was known from the beginning, and the solution returned by the heuristic. All the values were then normalized. The results are shown in Fig. 2. One can observe that the heuristics produce better solution when $\gamma_0 = \gamma^F$ and the value of G is low, whereas the opposite occurs when the value of G tends to $\mathcal{F}$. Overall, one can observe that for both values of γ^0, the heuristics perform best with $G = 10$, with a significant advantage when $\gamma^0 = \gamma^S$. These are the values that will be chosen in the next section. Now we compare, for a given strategic planning and scenario, the computation time needed by a MILP formulation proposed in [9] and our CP formulation to find the best tactical solution. We consider the same instances proposed in the same paper, by varying number of flights. The results are presented in Table 1. We notice that the computation time of the CP model increases less than that of the MILP model when the number of flights increases. This justifies the use of this model in our local search approach.

Table 1. Computation time comparison between MILP and CP formulation

Instance	Model	MILP			CP		
	$|\mathcal{F}|$	10	100	500	10	100	500
Grid	avg time(s)	<0.1	0.19	0.82	<0.1	<0.1	<0.1
Airport	avg time(s)	<0.1	0.35	0.79	<0.1	<0.1	<0.1
Metroplex	avg time(s)	<0.1	0.29	0.86	<0.1	<0.1	<0.1

3.3 Methods Comparison

In this section, we compare the three different heuristics. In order to evaluate the robustness of the solutions, for each instance, 100 scenarios have been randomly drawn from a uniform distribution. The value of the gap is computed the same way as the one described in the previous section. However, it is not normalized here. We also look at the number of iteration of the adversarial Benders method for each heuristic. The results are displayed in Table 2.

Table 2. Comparisons between the Local Search (LS), the Quadratic Program (FOAQ), and the piecewise linear formulation (FOAMC)

Γ		LS		FOAQ		FOAMC	
		10	20	10	20	10	20
Grid	avg	39.2%	68.1%	53.1%	72.3%	53.1%	72.3%
	time (s)	356	676	158	546	193	580
	iter.	6.50	9.12	5.24	5.93	5.24	5.93
Metro.	avg	52.3%	112%	78.4%	183%	78.4%	183%
	time (s)	607	1356	301	811	492	1045
	iter.	8.07	8.42	4.96	5.65	5.24	5.93
Airport	avg	64.5%	119%	289%	316%	289%	TimeOut
	time (s)	634	1486	478	929	1457	TimeOut
	iter.	8.32	11.05	4.60	7.89	5.24	TimeOut

For each topology (Column 1), statistics are reported on the generated instances. In particular, for each of the methods Local Search (LS) and Fixed Order Approximation (FOA), we report the results for values of Γ varying in $\{10, 20\}$. First, we can observe that, for all instances and all methods, the gap increases with the value of Γ. This can be explained by the fact that, for higher uncertainty budgets, the solution becomes more and more pessimistic [2], and not very suitable for scenarios with few deviations from the strategic planning. This suggests that, for practical applications, it is best not to overestimate the uncertainty budget, as this would lead to under-utilization of the network. Then, we can notice that, for all the networks, the calculated gap is larger with the FOA methods than with the LS method. This can be explained by the fact that the FOA methods solve a more constrained version of the problem, and struggles to calculate bad scenarios for the less constrained problem. Furthermore, it appears that the value of Γ has little impact on the number of iterations of the adversarial Benders method, independently of the FOA method. The fact that there is little iteration count of the method is explained by the fact that the sub-problem is solved by a heuristic. Thus, the heuristic may fail to compute a scenario that degrades the solution of the master problem while the optimum is not yet reached. As mentioned in Sect. 4.1, the nature of the network has an impact on the performance of our methods. Indeed, one can note that the time elapsed until the best solution is found and the quality of the said solution is lower with the Grid network than with the Airport and Metroplex networks. This is due to the fact that the number of conflicts is higher with the latter networks, which leads to an increase in the number of constraints in the master problem, increasing its solving time as well. Overall, it appears that the heuristic based on local search produces better solutions, but takes more time to converge than the heuristics based on the approximation of the problem.

4   Conclusion

In this work, we propose a robust approach to solve the urban air mobility flights planning problem and model the problem as a two-stage problem with budgeted uncertainty. We decompose the problem using the adversarial Benders method and present three different heuristics to solve the resulting subproblem. The first one is based on a local search that relies on a scenario dominance rule that allows us to significantly reduce the search space. The different parameters involved in this local search have been selected following an experimental analysis. The next heuristics are based on a more constrained approximation of the subproblem, which is then rewritten as a quadratic program and a mixed-integer linear program. All the proposed methods are then compared experimentally. Although the methods that consider an approximate version of the problem seem to produce good solutions faster, the heuristic based on local search computes overall better solutions in the given time. We believe that this heuristic can be further improved with additional dominance rules on the scenarios.

References

1. Ben-Tal, A., Goryashko, A., Guslitzer, E., Nemirovski, A.: Adjustable robust solutions of uncertain linear programs. Math. Program. **99**(2), 351–376 (2004)
2. Bertsimas, D., Sim, M.: The price of robustness. Oper. Res. **52**(1), 35–53 (2004)
3. Chin, C., Gopalakrishnan, K., Egorov, M., Evans, A., Balakrishnan, H.: Efficiency and fairness in unmanned air traffic flow management. IEEE Trans. Intell. Transp. Syst. (2021)
4. Croxton, K.L., Gendron, B., Magnanti, T.L.: A comparison of mixed-integer programming models for nonconvex piecewise linear cost minimization problems. Manag. Sci. **49**(9), 1268–1273 (2003)
5. Ikli, S., Mancel, C., Mongeau, M., Olive, X., Rachelson, E.: The aircraft runway scheduling problem: a survey. Comput. Oper. Res. **132**, 105336 (2021)
6. Kleinbekman, I.C., Mitici, M.A., Wei, P.: eVTOL arrival sequencing and scheduling for on-demand urban air mobility. In: 2018 IEEE/AIAA 37th Digital Avionics Systems Conference (DASC), pp. 1–7. IEEE (2018)
7. Mutapcic, A., Boyd, S.: Cutting-set methods for robust convex optimization with pessimizing oracles. Optim. Methods Softw. **24**(3), 381–406 (2009)
8. Pelegrín, M., D'Ambrosio, C.: Aircraft deconfliction via mathematical programming: review and insights. Transp. Sci. **56**(1), 118–140 (2022)
9. Pelegrín, M., D'Ambrosio, C., Delmas, R., Hamadi, Y.: Urban air mobility: from complex tactical conflict resolution to network design and fairness insights. Optim. Methods Softw. **38**(6), 1311–1343 (2023)
10. Straubinger, A., Rothfeld, R., Shamiyeh, M., Büchter, K.D., Kaiser, J., Plötner, K.O.: An overview of current research and developments in urban air mobility-setting the scene for UAM introduction. J. Air Transp. Manag. **87**, 101852 (2020)

11. Xue, M.: Urban air mobility conflict resolution: centralized or decentralized? In: AIAA Aviation 2020 Forum, p. 3192 (2020)
12. Yang, X., Wei, P.: Scalable multi-agent computational guidance with separation assurance for autonomous urban air mobility. J. Guid. Control. Dyn. **43**(8), 1473–1486 (2020)

Optimal Charging Station Location in a Linear Cycle Path with Deviations

Luca Pirolo[1], Pietro Belotti[1]([✉]), Federico Malucelli[1],
Rossella Moscarelli[2], and Paolo Pileri[2]

[1] Dip. di Elettronica, Informazione e Bioingegneria, Politecnico di Milano,
Via Ponzio 34, 20133 Milan, Italy
{luca.pirolo,pietro.belotti,federico.malucelli}@polimi.it
[2] Dip. di Architettura e Studi Urbani, Politecnico di Milano,
Via Bonardi 3, 20133 Milan, Italy
{rossella.moscarelli,paolo.pileri}@polimi.it

Abstract. Bicycle tourism is on the rise thanks to assisted-pedaling bikes, also known as e-bikes. While pedalling is still required on these bikes, they allow for longer rides through a battery-powered motor that has an autonomy of a few tens of kilometers. Batteries can then be recharged in one to two hours at recharging stations. Due to the waiting time, these stations should be installed at points of interest such as town centers or monuments for the cyclist to explore during recharge.

We consider the problem of installing charging stations (CSs) on a road or trail network in order to minimize the maximum distance between two CSs, subject to a budget constraint. Optimal placing of CSs for bike trail networks constitutes a known class of location problems; we focus on a special case where the graph representing the trail/road network is a *caterpillar* graph whose spine is a cycle path while the leaves are points of interest, connected to the trail via side roads. For this case, we show that the optimization problem can be solved to optimality by a binary search algorithm where a shortest path problem is solved at each iteration. We apply our approach to find the CS locations on a 210 km-long section of the VENTO bike trail in northern Italy.

Keywords: E-bike · Binary search · Shortest path

1 Introduction

Bicycles are the protagonist of a wave of change in mobility that is taking place in several cities across the globe, both big and small. The advent of assisted-pedaling bicycles, or e-bikes [6], which sport a battery-powered electric motor to provide extra riding power, has made this sustainable means of transportation even more enticing for both everyday commuters and tourists, the latter now capable of riding for long hours and enjoying the countryside or even mountain trails without a thorough training.

© The Author(s), under exclusive license to Springer Nature Switzerland AG 2024
A. Basu et al. (Eds.): ISCO 2024, LNCS 14594, pp. 361–375, 2024.
https://doi.org/10.1007/978-3-031-60924-4_27

Leisure riding is the focus of this paper. We consider a long linear cycle path such as VENTO[1] or the Danube cycle path[2]. Several cycle paths like these two usually run along the banks of a river, where stopping is of scarce touristic interest. However, from the main course of the cycle path, one has several possibilities to reach places of interest by making a small detour.

With the rapid growth of e-bike ridership, urban developers and city administrations face the problem of deploying a suitable charging infrastructure. The e-bike charging stations (CSs) should be placed in strategic positions so as to guarantee a coverage of the whole cycle path. However, since recharging a full e-bike requires a non-negligible time, the CSs should be positioned in places where alternative activities are possible, i.e., places with amenities such as restaurants, museums, swimming pools, and so forth. Moreover, the presence of a CS could induce e-cyclists to discover new places and generate positive externalities. One therefore faces the problem of finding the optimal locations of a set of e-bike CSs to optimize a given objective subject to several constraints arising from, e.g., budget and e-bike battery autonomy.

The problem of locating chargers is perhaps of much bigger impact for electric vehicles (EVs), because EVs need to recharge in order to move; several works on locating EV charging stations exist [5,12,19]. Note that the problem of locating e-bike chargers in a touristic setting cannot be compared to that for cars. First, the demand is not fixed as it is moving in the network. Moreover, the fact that riders tend to select their trajectories according to their specific interests generates a variety of different possible itineraries. Finally, while recharging a car requires a large amount of energy given its mass and required autonomy, the amount of energy for an e-bike is comparably negligible and thus does not require powerful grid outlets.

2   Optimally Locating Charging Stations on Cycle Paths

Consider a problem where a network of bike trails is given as a graph $G = (N, E)$, together with a path $p = (i_1, i_2, \ldots, i_m)$ with $i_k \in N \forall, k = 1, 2, \ldots, m$. A distance d_{ij} is associated with every edge $\{i, j\}$ of E. The cost of installing a CS at node i, denoted as c_i, and a budget b are also given.

Given a set S of nodes for installing a CS, we define a *stretch* as a pair $(i_{k'}, i_{k''})$ of nodes from p such that $1 \leq k' < k'' \leq m$, $i_{k'} \in S$, $i_{k''} \in S$, and $i_k \notin S, \forall k : k' < k < k''$. For simplicity, we assume that $i_1, i_m \in S$. The *stretch length* of $(i_{k'}, i_{k''})$ is defined as the total distance on p between $i_{k'}$ and $i_{k''}$, i.e., $\sum_{k=k'}^{k''-1} d_{i_k, i_{k+1}}$. Finally, the *longest stretch* on path p is the stretch whose length is maximum.

The problem we study in this paper consists of finding the location of CSs on the nodes of the network that minimizes the maximum stretch length while ensuring that the total CS installation cost is at most b. We call this problem

[1] https://www.cicloviavento.it.
[2] https://www.danube-cycle-path.com.

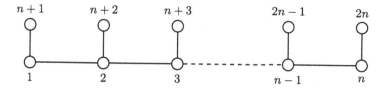

Fig. 1. Caterpillar graph structure: $1, 2, \ldots, n$ are the nodes forming the cycle path, whereas $n+1, n+2 \ldots, 2n$ are areas of interest, where CSs can be installed.

Capacitated MinMax Stretch Problem (CMMSP). The question posed by this problem is the following: in what nodes of the path p should a CS be installed so that (i) installation cost is within budget and (ii) the maximum distance that a cyclist may have to cover, without running out of battery, while entering and exiting the graph at any point (intermediate or not) of the path p, is minimum?

Our contribution is an algorithm for solving a special case of CMMSP: the trail network has the structure of a *caterpillar* graph (see Fig. 1), i.e., a graph that resembles the structure of a cycle path whose each node is connected to one or more nearby points of interest (POIs); second, we assume that (i) none of the POIs are on the cycle path and that (ii) to facilitate leisure activities while recharging the e-bike, a CS should only be installed at a POI, therefore only nodes that do not belong to the caterpillar spine can host a CS. Finally, in order to include the extreme case where a cyclist visits *all* POIs between her entrance and exit nodes, we consider a path p that begins at the first node of the caterpillar graph's spine and ends at the last one, and it visits all intermediate ones. The path p is therefore non-simple.

Because we aim at installing facilities in a subset of nodes and due to the *minmax* objective function, the problem can be classified as a facility location problem where a set of *centers* are sought on a path [3,9,17,18]. In the case where the CS installation cost is equal at every POI, the budget constraint reduces to a cardinality constraint and the problem falls into the class of k-center location [10,13]. The facility location problem literature is vast and k-center location problems on path graphs have been investigated [1,2,11]. Besides similar applications such as the installation of rechargers for EVs, rather than for e-bikes, some literature exists for path graphs with other applications [4]. To the best of our knowledge, our proposed algorithm is new and unlike any approach found in the literature for similar problems.

We formalize the problem in Sect. 3, where we provide a Mixed Integer Linear Optimization (MILO) model, then provide a reduction to a polynomially solvable problem in Sect. 4. In Sect. 5 we use real-world data from the VENTO cycle path and describe how we have created an instance of the CMMSP problem from a general Geographical Information System (GIS) database, which contains geographical data about the whole region rather than just the area surrounding the cycle path. Then in Sect. 6 we report on computational tests for finding an optimal solution on the same VENTO instance.

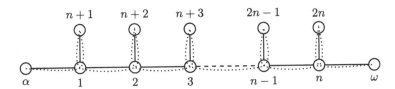

Fig. 2. Secondary caterpillar graph with fictitious source and destination nodes α and ω. The dotted curve is the path p through all POIs.

3 An Optimization Model

As mentioned above, specializing the CMMSP to a cycle path that is connected to a set of points of interest implies that $G = (N, E)$ is structured as a *caterpillar* graph, as shown in Fig. 1. Let $N = L \cup H$ where $L = \{1, 2, \ldots, n\}$ is a set of nodes that correspond to the locations along the cycle path from which we can deviate, and $H = \{n + 1, n + 2, \ldots, 2n\}$ is a set of nodes that correspond each to the tourist sites that may host a CS. The set of edges E is the union of two subsets: $E' = \{\{i, i + 1\}, i = 1, 2, \ldots, n - 1\}$, i.e., the edges forming the cycle path proper; and $E'' = \{\{i, n + i\}, i = 1, 2, \ldots, n\}$, which form the connection between each point i on the cycle path and a nearby POI.

The distances between consecutive nodes in L (i.e., $d_{i,i+1}, i = 1, \ldots, n - 1$) and the lengths of the deviations ($d_{i,n+i}, i = 1, \ldots, n$) are given. Due to the practical considerations outlined in the previous section, we also assume that the path p that is the basis for stretches is the path between the first and the last cycle path exits 1 and n, touching all POIs in between.

An optimization model can be developed by considering a *secondary* caterpillar graph with two extra nodes called α and ω and with edges $\{\alpha, 1\}$ and $\{n, \omega\}$, both having null distance, as shown in Fig. 2. Path $p = (\alpha, 1, n + 1, 1, 2, n + 2, 2, 3, \ldots, n - 1, n, 2n, n, \omega)$ is indicated with a dotted curve through all nodes from α to ω, traversing each node in $\{1, 2, \ldots, n\}$ twice.

Optimization Model: The Augmented Graph. Before we formulate an optimization model, we introduce an *augmented graph* $G^* = (N^*, A^*)$ as the directed acyclic graph defined with $N^* = \{\alpha, 1, 2, \ldots, n, \omega\}$ and $A^* = \{(i, j) : i, j \in N^*, i < j\}$, where, by a little abuse of notation, we assume $\alpha < i < \omega$ for all $i = 1, 2, \ldots, n$. Define now the *travelling length* l_{ij} on all $(i, j) \in A^*$ as the total distance of an $i - j$ path that traverses once every POI node in between:

$$
l_{ij} = \begin{cases} \sum_{h=1}^{j-1} d_{h,h+1} + 2\sum_{h=1}^{j-1} d_{h,n+h} \quad +d_{j,n+j} & \text{if } i = \alpha \\ d_{i,n+i} + \sum_{h=i}^{n-1} d_{h,h+1} + 2\sum_{h=i+1}^{n} d_{h,n+h} & \text{if } j = \omega \\ d_{i,n+i} + \sum_{h=i}^{j-1} d_{h,h+1} + 2\sum_{h=i+1}^{j-1} d_{h,n+h} \quad +d_{j,n+j} & \text{otherwise.} \end{cases}
$$

Every arc $(i, j) \in A^*$ represents a stretch on G with length l_{ij}. Arc (i, j) is associated a label pair (l_{ij}, f_{ij}), where l_{ij} is defined above while the *arc cost* f_{ij} is the cost c_{n+j} of installing a CS at the head node j of the arc (i, j), with

$f_{i,\omega} = 0 \forall, i \in N^*$. An example for a cycle path with $n = 4$ and for edge distances $d_{i,i+1} = 5 \forall, i = 1, 2, \ldots, n-1$ and $d_{i,n+i} = 2 \forall, i = 1, 2, \ldots, n$ is provided in Fig. 3, where each arc (i, j) has label (l_{ij}, f_{ij}). Note that in general, for a cycle path with n exit points and consequently n points of interest, the augmented graph contains $O(n^2)$ arcs.

We reduce the CMMSP to the problem of finding an $\alpha - \omega$ path p^* whose total cost, described with arc costs f_{ij}, is below the given budget b and such that $\max_{(i,j) \in p^*} l_{ij}$ is minimum. An optimal solution to CMMSP is then given by all nodes $n + j \in N$ such that there is $i \in N^* : (i, j) \in p^*$. Let us consider two sample solutions on the graph in Fig. 3, where for simplicity we use c_i in place of c_{n+i}. The first one is defined by the $\alpha - \omega$ path $(\alpha, 2, \omega)$, equivalent to installing a single CS at node 2; its two stretches $\alpha - 2$ and $2 - \omega$ have length $2 + 2 + 5 + 2 = 11$ and $2 + 5 + 2 + 2 + 5 + 2 + 2 = 20$, hence the maximum stretch length is 20 while its CS installation cost is the total travelling cost $c_2 + 0 = c_2$. The second example is path $(\alpha, 1, 2, 4, \omega)$, equivalent to installing a CS at nodes 1, 2, and 4. The stretch lengths are therefore $2, 2 + 5 + 2 = 9, 2 + 5 + 2 + 2 + 5 + 2 = 18$, and 2, their maximum being 18, while the total installation cost is $c_1 + c_2 + c_4$. Clearly the second solution has better objective function (maximum stretch length), but it has a larger installation cost which may be infeasible if $c_1 + c_2 + c_4 > b$.

An optimization model can be written with the following variables:

- $\delta \geq 0$: the maximum stretch length, to be minimized;
- $x_{ij} \in \{0, 1\}$: 1 if arc $(i, j) \in A^*$ is used, 0 otherwise.

An optimal solution $(\bar{\delta}, \bar{x})$ to this path problem is then translated to an optimal solution to the CMMSP as follows: install a CS at every node $n + j \in N$ of the original graph if there exists $(i, j) \in A^*$ such that $\bar{x}_{ij} = 1$. Below is a model for the CMMSP:

$$\min \quad \delta \tag{1}$$

$$\text{s.t.} \quad l_{ij} x_{ij} \leq \delta \qquad \forall (i, j) \in A^* \tag{2}$$

$$\sum_{(i,j) \in A^*} f_{ij} x_{ij} \leq b \tag{3}$$

$$\sum_{(i,j) \in FS(i)} x_{ij} - \sum_{(j,i) \in BS(i)} x_{ji} = \begin{cases} 1 & \text{if } i = \alpha \\ -1 & \text{if } i = \omega \\ 0 & \text{otherwise} \end{cases} \quad \forall i \in N^* \tag{4}$$

$$x_{ij} \in \{0, 1\} \qquad \forall (i, j) \in A^*. \tag{5}$$

The objective function indicates the *minmax* nature of our problem. Constraints (2) define δ as the maximum stretch length. Constraint (3) requires that the total cost of installing CSs be within budget. Finally, constraints (4) are flow conservation constraints that yield an $\alpha - \omega$ path; we use $FS(i)$ and $BS(i)$ for *forward star* and *backward star*, respectively, of a node $i \in N^*$.

The above MILO problem can be solved in polynomial time by another reduction procedure explained in the next section.

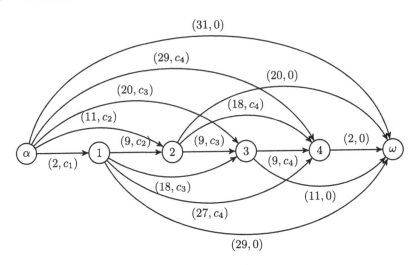

Fig. 3. Augmented graph for a sample caterpillar graph with $n = 4$ and edge distances $d_{i,n+i} = 2$ for $i = 1, 2, \ldots, 4$, $d_{1,2} = d_{2,3} = d_{3,4} = 5$, and $d_{\alpha,1} = d_{4,w} = 0$. Each arc has label (l_{ij}, f_{ij}) where l_{ij} is the stretch length and f_{ij} is the cost.

4   Reduction to Binary Search on Shortest Paths

Rather than solving the MILO problem (1)–(5) with an off-the-shelf solver, a simple observation allows for solving the problem in polynomial time: for a fixed value $\tilde{\delta}$ of δ, constraints (2) are equivalent to removing all arcs $(i, j) \in A^*$ such that $l_{ij} > \tilde{\delta}$, and the problem becomes equivalent to determining, via a shortest-path algorithm, whether or not there exists an $\alpha - w$ path whose total cost, in terms of the f_{ij}, is at most b. A binary search then suffices to find the optimal value of δ.

Denote as $F(G^*(\tilde{\delta}))$ the total cost of the shortest $\alpha - w$ path on the graph $G^*(\tilde{\delta}) := (N^*, A^*(\tilde{\delta}))$, where $A^*(\tilde{\delta}) := \{(i, j) \in A^* : l_{ij} \leq \tilde{\delta}\}$ and where the path length is based on arc costs f_{ij} as mentioned above. The case where α and w are not connected on $G^*(\tilde{\delta})$ is defined with $F(G^*(\tilde{\delta})) = +\infty$. Note that if $b = 0 < \min_{j \in N} c_{n+j}$, then the only feasible solution is $x_{\alpha,w} = 1, x_{ij} = 0 \forall, (i, j) \in A^* : i \neq \alpha \vee j \neq w$, with a stretch length of $\delta = l_{\alpha,w}$.

Algorithm 1 describes the binary search for solving CMMSP. Note that there are at most $|A^*|$ suitable values of $\tilde{\delta}$, since δ itself has value in the discrete set of l_{ij} values for $(i, j) \in A^*$. For this reason, the binary search loop requires at most $\log_2 |A^*| \in O(\log(n^2)) = O(2 \log n) = O(\log n)$ iterations, each with a complexity that is at most that of the shortest-path algorithm used.

Algorithm 1. Binary search for solving the CMMSP.

---

procedure BINSEARCHCMMSP(N^*, A^*, l, f, b)
    $\delta_{\min} \leftarrow \min_{(i,j) \in A^*} l_{ij}$
    $\delta_{\max} \leftarrow \max_{(i,j) \in A^*} l_{ij}$
    loop
        $\tilde{\delta} \leftarrow \max\{l_{ij} : (i,j) \in A^*, l_{ij} \leq \frac{\delta_{\min} + \delta_{\max}}{2}\}$
        $A^*(\tilde{\delta}) \leftarrow \{(i,j) \in A^* : l_{ij} \leq \tilde{\delta}\}$
        Compute shortest path p on $G^*(\tilde{\delta})$
        if $F(G^*(\tilde{\delta})) > b$ **then**
            $\delta_{\min} \leftarrow \min\{l_{ij} : (i,j) \in A^*, l_{ij} \geq \tilde{\delta}\}$ ▷ Solution over budget
        else
            $\delta_{\max} \leftarrow \max\{l_{ij} : (i,j) \in A^*, l_{ij} \leq \tilde{\delta}\}$ ▷ Seek shorter stretch
        end if
        if $\delta_{\min} = \delta_{\max}$ **then**   ▷ Current path defines optimal solution
            $S \leftarrow \{j \in N^* \setminus \{\omega\} : (i,j) \in p\}$
            return $(\delta_{\min}, S)$
        end if
    end loop
end procedure

---

5 Application to the VENTO Cycle Path

The VENTO (from VENezia–Venice and TOrino–Turin) is a cycle path about 700 km long that runs along the Po River, between Turin and Venice via Milan: it crosses four regions (Piedmont, Lombardy, Emilia Romagna, and Veneto), 13 districts/provinces and 118 municipalities. In 2016, it was included in the first Italian strategy to realize ten touristic cycle routes, the so-called "Ciclovie Turistiche Nazionali" [14]. Figure 4 shows the complete map of the cycle path. Note that the structure does not follow that of a single path but (i) it contains a long connecting trail from Pavia to Milan; and, most importantly, (ii) it splits into two parallel trails between Piacenza and Cremona.

For this article, we apply the algorithm described in the previous section to a portion of VENTO ranging from Piacenza to San Benedetto Po (province of Mantua), with a total length of about 210 km. Figure 5 shows in detail this portion, together with the selected POIs (which are selected based on well-defined criteria set forth within an EU project for this purpose). In order to apply our algorithm, only the southern of the parallel trails between Piacenza and Cremona has been considered, which makes the portion under analysis a path.

We have implemented and interfaced our algorithm to a Geographical Information System (GIS). The data used as input is in the form of three files in the GIS *shapefile layer* format. These files contain:

1. the cycle path: a *linestring* layer (i.e. a line formed as a sequence of segments) that contains the information on the path from the city of Piacenza to San Benedetto Po, but does not contain any node;

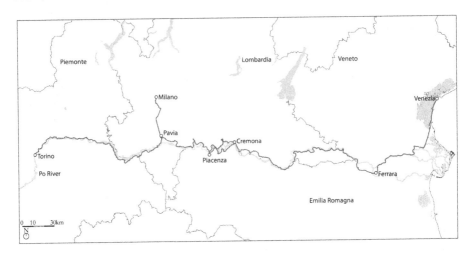

Fig. 4. The overall VENTO cycle route, from Turin to Venice.

2. the road network of the region containing the cycle path, defined as a directed graph with nodes and arcs;
3. the eligible destination points, selected with criteria such as artistic/cultural value, presence of bars/restaurants, etc.

All data and software used in this project are Open Source. In particular, data comes from OpenStreetMap [15] (OSM), which is a free and editable web map, built by volunteers and released under an open-content license. Therefore the process of creating the network is replicable and scalable, according to the size of the dataset and the location of the case study.

To deal with the creation of the cycle path graph, we have used QGIS v3.32.3. QGIS [16] is a free and open-source cross-platform desktop GIS application that supports viewing, editing, printing, and analyzing geospatial data using a user-friendly interface. QGIS is an official project of the Open Source Geospatial Foundation (OSGeo).

In order to create a suitable instance graph on which to apply our algorithm, the layers at hand must be manipulated with several algorithms, possibly computationally expensive, which we detail below.

Identify Cycleway/Network Intersections. Given that the cycle path is a linestring with no explicit connection to the road network, we must first transform the cycle path into a path graph that is embedded in the road graph. To this purpose, a *buffer* of 10 m around the cycleway layer is created. Then the terminal nodes within this buffer are extracted from the road network vector, and these nodes are adjoined to the cycle path, which then becomes a path connected to the road network.

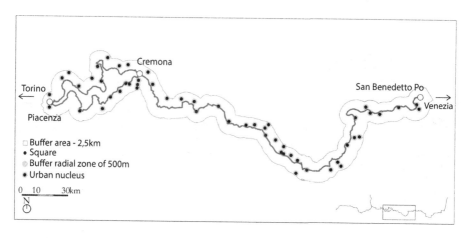

Fig. 5. Portion of cycle path under study with POIs within distance.

Creation of a 2500 m Buffer Area. The application at hand determines through touristic-based criteria that POIs are those that are at most 2500 m from the main cycle path. Hence we now restrict our attention to the subgraph induced by all nodes that can be reached via a 2500 m-long path from of the cycle path. This is done through the "service area" function in QGIS's Network Analysis library, which uses Dijkstra's shortest path algorithm to define the roads reachable with a total weight of 2500.

Identification of Nearby POIs and Connection to the Cycle Path. A circular buffer with a 500 m radius is created around the urban nuclei within the 2500 m buffer. Potential POIs, ranked on a set of criteria, are then selected and connected to the main cycle path, using functionality "v.net.distance" of QGIS's GRASS system library [7], which uses Dijkstra's shortest-path algorithm.

The resulting graph, depicted in Fig. 6, contains $n = 44$ nodes on the cycle path and about as many nearby POIs. Note that some of the latter is connected to the cycle path through the same exit point: the instance can be adapted by creating as many separate exit points, which have a distance of 0 from one another, and each exit point is connected independently to the corresponding POI.

6   Computational Tests

The experimental tests were conducted on a Lenovo ThinkPad with CPU Intel i5-1135G7 clocked at 2.40 GHz, with 16 GB of RAM and Windows 11 as operating system, equipped with a QGIS installation. All algorithms used for this work were either coded in Python 3.11.5 (binary search) or used Python modules such

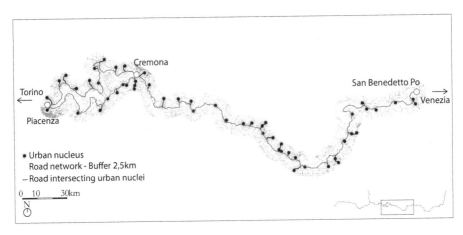

Fig. 6. VENTO cycle path portion with POIs and their connection (only the southern branch of the Piacenza-Cremona portion is considered).

as `networkx` [8]. In particular, the shortest path algorithm used in each binary search iteration is an implementation of Dijkstra's algorithm in `networkx`. While CSs can be of several types, for simplicity here we have decided to assign an equal price c_i for all POIs. This translates into a simplification of the budget constraint (3) into a cardinality constraint, and therefore the shortest-path step in our algorithm consists, for these tests, in finding an $\alpha - \omega$ path with the smallest number of arcs.

We now assess the performance of the algorithm on the VENTO instance for budget ranging in $B = \{2000, 4000, 6000, 8000, 16000, 24000, 32000, 64000\}$. For each $b \in B$ we have run the binary search algorithm presented in Sect. 4, and we report in Table 1 the run time for the binary search, the optimal value of δ and the corresponding optimal set of CSs.

Figures 7 and 8 report, for all values of $b \in B$, the shortest path (with thick red arrows) and the graph (with thin black arrows) containing all arcs (i, j) with $l_{ij} \leq \delta$ for the optimal value of δ found at the last iteration of the binary search, where $A^*(\delta)$ only contains arcs (i, j) with $l_{ij} \leq \delta$. For $b = 8000$, for instance, the optimal solution $\{7, 16, 27, 36\}$ consists of the POIs Castelvetro Piacentino, Motta Baluffi, Viadana, and Suzzara.

The decreasing run times observed in Table 1 for increasing b could be explained as follows: for large values of b, the binary search mainly updates (i.e., decreases) $\delta_{\max}$, at least at the initial iterations, due to the shortest path algorithm finding a solution with CS installation cost below b. This leads to sparser graphs, as all $(i, j) \in A^*$ with $l_{ij} > \tilde{\delta}$ are removed, and consequently faster runs of the chosen shortest-path algorithm. Small values of b instead yield, at least in the initial iterations of the binary search, an increase of $\delta_{\min}$ and hence comparably slower runs of the shortest-path algorithm.

Table 1. Run times and optimal value of δ for different values of the budget b.

Budget	Time [s]	δ_{opt} [km]	Optimal set of CSs
2000	0.31	133.160	$\{22\}$
4000	0.26	89.475	$\{13, 29\}$
6000	0.22	68.395	$\{10, 20, 33\}$
8000	0.20	54.784	$\{7, 16, 27, 36\}$
16000	0.14	33.280	$\{2, 10, 13, 18, 24, 30, 35, 38\}$
24000	0.14	24.102	$\{1, 3, 9, 12, 15, 18, 23, 27, 30, 34, 37, 38\}$
32000	0.12	19.025	$\{1, 3, 7, 11, 12, 14, 17, 20, 23, 27, 29, 31, 34, 37, 38, 40\}$
64000	0.12	18.581	$\{1, 3, 7, 11, 12, 13, 16, 18, 22, 25, 28, 30, 32, 35, 37, 38, 40\}$

The run times in the order of a fraction of a second are acceptable for the instance at hand, especially because they solve a long-term decision problem, i.e., that of choosing installation locations for CSs. Nevertheless, the algorithm presented here could be made more efficient especially for small values of b by observing that

- the augmented graph is acyclic, hence a more efficient shortest-path algorithm such as SPT_Acyclic could replace Dijkstra's; to the best of our knowledge, networkx implements only Dijkstra's and Bellman-Ford algorithms for finding the shortest-path tree;
- we need a single $\alpha - \omega$ path rather than the shortest-path tree, therefore one could use algorithms that stop as soon as ω's label becomes permanent; it might even be interesting to check whether the network simplex algorithm, when minimizing the installation cost $\sum_{(i,j) \in A^*} f_{ij} x_{ij}$ subject to the flow-conservation constraints (4), finds an optimal basis more efficiently.

Hence, for large CMMSP instances on caterpillar graphs an implementation that takes advantage of these observations could further improve on performance and perhaps even reverse the run time trend w.r.t. the values of b. However, the worst-case complexity is $O(\log_2 |A^*|) = O(\log n)$ iterations of the binary search, each solving a shortest-path problem on a directed acyclic graph with complexity $O(|N^*| + |A^*|) = O(n^2)$, for an overall polynomial-time complexity of $O(n^2 \log n)$, which makes our approach scalable to very large instances.

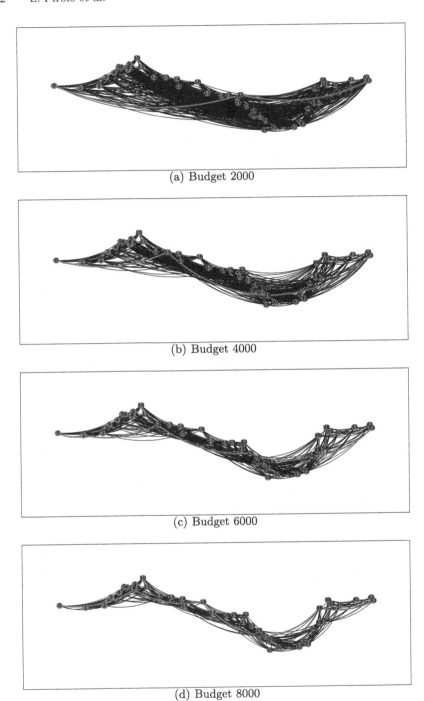

(a) Budget 2000

(b) Budget 4000

(c) Budget 6000

(d) Budget 8000

Fig. 7. Graph $G^*(\delta)$ (black arcs) and solution (red) for the optimal value of δ with the given budget for $b \in \{2000, 4000, 6000, 8000\}$. (Color figure online)

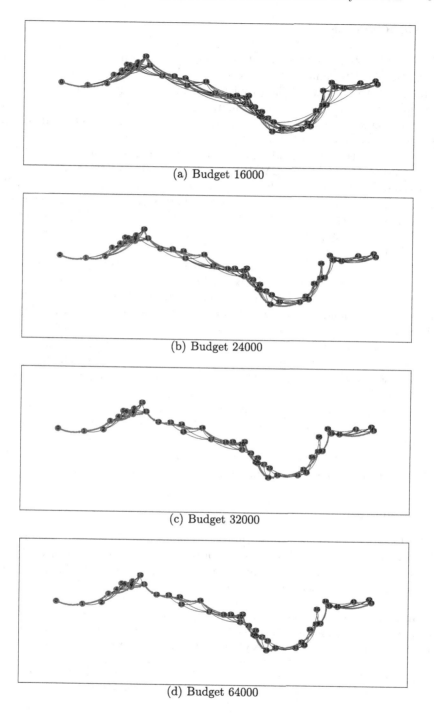

(a) Budget 16000

(b) Budget 24000

(c) Budget 32000

(d) Budget 64000

Fig. 8. Graph $G^*(\delta)$ (black arcs) and solution (red) for the optimal value of δ with the given budget for $b \in \{16000, 24000, 32000, 64000\}$. (Color figure online)

7   Conclusions

We have presented a minmax location problem on caterpillar graphs that finds application in determining the location of e-bike charging stations in the area surrounding a cycle path. The problem admits a Mixed Integer Linear Optimization model via a graph transformation, but we show that it can be solved in polynomial time thanks to a reduction step that yields a solution algorithm consisting in a binary search.

Aside from the theoretical result of finding a polynomial-time algorithm, in practice instances of the CMMSP can be large due to the length of cycle paths, some of which can span hundreds of kilometers and hence have potentially hundreds of points of interest in their surrounding areas. We apply our algorithm to a real-world case of the VENTO cycle path in Northern Italy and detail the procedure for solving it when starting from a GIS database for the area that contains the cycle path.

Possible research directions include the generalization of CMMSP to the case where only a subset of POIs are visited, and especially the case in which this subset is uncertain. Also, while the results in this paper hold for caterpillar graphs, the problem at hand has a parallel portion that has been only partially considered in our experiments, and it would be interesting to extend our results, if possible, to the case where the cycle path contains loops as is the case for VENTO. In addition, both stretch length and total installation costs are conflicting objective functions and it would be of interest to study a bi-objective optimization version of the problem.

Finally, the general CMMSP is an ongoing research direction. While the MILO model (1)–(5) can be easily adapted to a general bicycle trail network, i.e., on a general non-caterpillar graph G, and to more than one path p, in practice the problem becomes more difficult and we could not find a similar polynomial-time reduction scheme as the one presented here.

Contributions

FM, LP, and PB developed the optimization model and binary search algorithm; RM, PP, and LP identified the criteria for definition of POIs and collected data for the cycle path and surrounding areas; LP implemented the binary search algorithm and performed computational experiments.

Acknowledgment. This study was carried out within the MOST – Sustainable Mobility National Research Center and received funding from the European Union Next-GenerationEU (PIANO NAZIONALE DI RIPRESA E RESILIENZA (PNRR) – MISSIONE 4 COMPONENTE 2, INVESTIMENTO 1.4 – D.D. 1033 17/06/2022, CN00000023). This manuscript reflects only the authors' views and opinions, neither the European Union nor the European Commission can be considered responsible for them.

References

1. Ageev, A., Gimadi, E., Shtepa, A.: How fast can the uniform capacitated facility location problem be solved on path graphs. In: Burnaev, E., et al. (eds.) AIST 2021. LNCS, vol. 13217, pp. 303–314. Springer, Cham (2022). https://doi.org/10.1007/978-3-031-16500-9_25

2. Bhattacharya, B., De, M., Kameda, T., Roy, S., Sokol, V., Song, Z.: Back-up 2-center on a path/tree/cycle/unicycle. In: Cai, Z., Zelikovsky, A., Bourgeois, A. (eds.) COCOON 2014. LNCS, vol. 8591, pp. 417–428. Springer, Cham (2014). https://doi.org/10.1007/978-3-319-08783-2_36

3. Current, J., Daskin, M., Schilling, D.: Discrete network location models. Facility Locat. **1**, 81–118 (2002)

4. Demange, M., Gabrel, V., Haddad, M., Murat, C.: A robust p-center problem under pressure to locate shelters in wildfire context. EURO J. Comput. Optim. **8**(2), 103–139 (2020)

5. Dong, G., Ma, J., Wei, R., Haycox, J.: Electric vehicle charging point placement optimisation by exploiting spatial statistics and maximal coverage location models. Transp. Res. Part D: Transp. Environ. **67**, 77–88 (2019)

6. European Union: Regulation (EU) No. 168/2013 of the European Parliament and of the Council of 15 January 2013 on the approval and market surveillance of two- or three-wheel vehicles and quadricycles (2013). https://eur-lex.europa.eu/eli/reg/2013/168/2020-11-14. Accessed 28 Oct 2022

7. GRASS Development Team: Geographic Resources Analysis Support System (GRASS GIS) Software, Version 8.2. Open Source Geospatial Foundation (2022). https://doi.org/10.5281/zenodo.5176030

8. Hagberg, A.A., Schult, D.A., Swart, P.J.: Exploring network structure, dynamics, and function using NetworkX. In: Varoquaux, G., Vaught, T., Millman, J. (eds.) Proceedings of the 7th Python in Science Conference (SciPy2008), pp. 11–15 (2008)

9. Hakimi, S.L.: Optimum locations of switching centers and the absolute centers and medians of a graph. Oper. Res. **12**(3), 450–459 (1964). http://www.jstor.org/stable/168125

10. Hakimi, S.L., Schmeichel, E.F., Pierce, J.G.: On p-centers in networks. Transp. Sci. **12**(1), 1–15 (1978)

11. Huang, R.: A short note on locating facilities on a path to minimize load range equity measure. Ann. Oper. Res. **246**(1), 363–369 (2016)

12. Lam, A.Y., Leung, Y.W., Chu, X.: Electric vehicle charging station placement: formulation, complexity, and solutions. IEEE Trans. Smart Grid **5**(6), 2846–2856 (2014)

13. Minieka, E.: The m-center problem. SIAM Rev. **12**(1), 138–139 (1970)

14. Ministero delle Infrastrutture e dei Trasporti: Decreto ministeriale numero 517, 29 November 2018. https://www.gazzettaufficiale.it/eli/id/2019/01/22/19A00326/sg

15. OpenStreetMap contributors: Planet dump retrieved from 2023 (2023). https://www.openstreetmap.org

16. QGIS Development Team: QGIS Geographic Information System. QGIS Association (2024). https://www.qgis.org

17. Slater, P.J.: Centrality of paths and vertices in a graph: cores and pits. Technical report, Sandia National Labs, Albuquerque, NM, USA (1980)

18. Tansel, B.C., Francis, R.L., Lowe, T.J.: Location on networks: a survey. Part I: The p-center and p-median problems. Manag. Sci. **29**, 482–497 (1983)

19. Zhu, Z.H., Gao, Z.Y., Zheng, J.F., Du, H.M.: Charging station location problem of plug-in electric vehicles. J. Transp. Geogr. **52**, 11–22 (2016)

An Efficient Timing Algorithm for Drivers with Rest Periods

Giovanni Righini[✉][ID] and Marco Trubian[ID]

Department of Computer Science, University of Milan, Milan, Italy
{giovanni.righini,marco.trubian}@unimi.it

Abstract. We consider a timing problem arising from a vehicle routing context: it consists of optimally inserting rest periods of given duration into drivers' schedules, when the sequence of customers to visit is given, time windows are associated with customers and an upper limit is imposed on the driving time with no rest periods. We illustrate some properties that allow reformulating the problem in simpler terms, and provide the basis to design a very efficient exact optimization algorithm whose worst-case time complexity is $O(n \log n)$, where n is the number of customers to be visited.

Keywords: Timing problem · Time window constraints

1 Introduction

Timing problems arise when a sequence of activities requires optimal start times to optimize an objective function. Unlike scheduling problems, timing problems are simpler as they do not need defining the sequence optimally. However, they are crucial due to the need to solve them for each tentative sequence, often a large number, in applications like exact optimization algorithms (e.g., branch-and-bound) and approximation algorithms (e.g., local search). In these algorithms, a vast set of partial solutions is explored, each with a different partial sequence, requiring to solve a timing problem for upper and lower bound computations. Similarly, in approximation algorithms, exploring the neighborhood of a solution demands evaluating different sequences, each requiring a timing problem solution. Another typical need is real-time re-optimization of routes with rest periods, owing to unexpected delays that make planned solutions infeasible. Hence, while timing problems may have polynomial-time solutions, the search for highly efficient exact optimization algorithms, ideally with linear or near-linear time complexity, remains significant.

For a recent overview of timing problems, refer to Vidal et al. [2], covering various settings like production scheduling, network optimization, energy optimization, and statistical inference.

Our motivation originates from a real-world vehicle routing application, where driver routes are assessed within a local search framework. Each sequence necessitates finding corresponding timings rapidly. The routing problem is complicated by two factors: customer-associated time windows and restrictions on maximum allowed driving time without rest.

A. Basu et al. (Eds.): ISCO 2024, LNCS 14594, pp. 376–387, 2024.
https://doi.org/10.1007/978-3-031-60924-4_28

In this paper, we analyze the basic version of the timing problem and we demonstrate how finding the optimal solution for a sequence of visits to n customers has the same computational complexity as sorting n elements, i.e., $O(n \log n)$. In the conclusions, we illustrate how this result easily extends to more complex problem variations.

2 Problem Definition

The Timing Problem with Rest Periods (TPRP in the remainder) is defined as follows.

A sequence of customers to be visited is given. Let N be the indexed set of customers: $N = \{0, 1, 2, \ldots, n\}$, where 0 and n are the indices of the initial and final depots. Each customer $i \in N$ (including the initial and final depots) has an associated time window $[a_i, b_i]$. A solution is feasible only if the service at each customer starts within the corresponding time window. Traveling times are given for each pair of consecutive customers in the sequence. Let $d_i \geq 0$ be the time taken to travel from customer $i-1$ to customer i for each $i = 1, \ldots, n$. At each customer $i \in N$ it is required to work out some operations, whose duration $s_i \geq 0$ is known ($s_0 = s_n = 0$).

The maximum allowed driving time with no rest periods is indicated by T. The duration of a rest period is indicated by δ. All rest periods have the same duration, they cannot be interrupted and they cannot overlap with service to customers. The service time at customers is neither driving time nor rest time; while serving a customer, the driver's fatigue level remains constant. The same holds for waiting time, that occurs in case of early arrival at a customer (before the beginning of its time window).

The problem is to decide when to optimally insert the necessary rest periods. The objective is to minimize the total duration of the route.

3 Properties and Reformulations

Any instance of the TPRP can be equivalently reformulated (and simplified) as follows. We distinguish four mutually exclusive types of time periods: driving time, rest time, waiting time and service time.

Elimination of Service Time. Service times are fixed: they can be taken into account as if the driver's clock would stop while a customer is served. An equivalent instance is obtained by redefining the time windows as shown in Algorithm 1.

Elimination of Travel Time. Since traveling periods have a given duration and cannot overlap with rest periods and waiting periods, they can be eliminated. An equivalent instance is obtained by recomputing all time windows again, as shown in Algorithm 2.

Algorithm 1. Elimination of service time.

$\Delta \leftarrow 0$
for $i = 1, \ldots, n$ **do**
$\quad a_i \leftarrow a_i - \Delta$
$\quad b_i \leftarrow b_i - \Delta$
$\quad \Delta \leftarrow \Delta + s_i$

Algorithm 2. Elimination of travel time.

$\Delta \leftarrow 0$
for $i = 1, \ldots, n$ **do**
$\quad \Delta \leftarrow \Delta + d_i$
$\quad a_i \leftarrow a_i - \Delta$
$\quad b_i \leftarrow b_i - \Delta$

Symmetry. After reformulating the generic problem instance, the service at each customer is instantaneous and hence a variable t_i can represent both the arrival time and the departure time at/from each customer $i \in N$.

The constraint on rest periods imposes to insert a minimum number λ_i of rest periods before each customer $i \in N$:

$$\lambda_i = \left\lceil \frac{\sum_{j=1}^{i} d_j}{T} \right\rceil - 1.$$

For the same reason, symmetrically, the number of rest periods after each customer $i \in N$ is lower bounded by a limit μ_i:

$$\mu_i = \left\lceil \frac{\sum_{j=i+1}^{n} d_j}{T} \right\rceil - 1.$$

To minimize the total duration of the travel, the total number of rest periods must be minimum, that is equal to

$$\pi = \left\lceil \frac{\sum_{j=1}^{n} d_j}{T} \right\rceil - 1.$$

For each customer $i \in N$, either $\pi = \lambda_i + \mu_i$ or $\pi = \lambda_i + \mu_i + 1$.

Proof. Indicating for brevity $u = \dfrac{\sum_{j=1}^{i} d_j}{T}$, $v = \dfrac{\sum_{j=i+1}^{n} d_j}{T}$ and observing that $\dfrac{\sum_{j=1}^{n} d_j}{T} = \dfrac{\sum_{j=1}^{i} d_j}{T} + \dfrac{\sum_{j=i+1}^{n} d_j}{T}$, it holds $\lambda = \lceil u \rceil - 1$, $\mu = \lceil v \rceil - 1$ and $\pi = \lceil (u+v) \rceil - 1$. Hence, $u - 1 \le \lambda < u$, $v - 1 \le \mu < v$ and therefore, summing up the inequalities, $(u-1)+(v-1) \le \lambda + \mu < (u+v)$. Analogously, $(u+v) - 1 \le \pi < (u+v)$. Hence, by difference from the inequalities above, $-1 < \pi - (\lambda + \mu) < 2$. Since λ, μ and π are integers, strict inequalities imply $0 \le \pi - (\lambda + \mu) \le 1$ (Q.E.D.).

Example. In a problem instance A: $n = 2$, $d_1 = d_2 = 12$ and $T = 10$. Then $\pi = 2$ and for customer 1, $\lambda_1 = 1$ and $\mu_1 = 1$. In a problem instance B: $n = 2$, $d_1 = d_2 = 18$ and $T = 10$. Then $\pi = 3$ and for customer 1, $\lambda_1 = 1$ and $\mu_1 = 1$. The second rest period in instance B can be placed before or after customer 1.

We partition customers for which $\pi = \lambda_i + \mu_i$ from those for which $\pi = \lambda_i + \mu_i + 1$. We define U to be the set of the former ones and V the set of the latter ones. U and V are a partition of N. Now, we further partition U in subsets U_k for $k = 0, \ldots, \pi$, each one containing customers $i \in U$ for which $\lambda_i = k$. For sure, the depot 0 belongs to U_0 and depot n belongs to U_π. Analogously, for each value of $k = 1, \ldots, \pi$ let V_k be the subset of customers $i \in V$ such that $\lambda_i = k - 1$ and $\mu_i = \pi - k$. All subsets V_k are disjoint and they form a partition of V.

Elimination of Rest Periods. For all customers $i \in U$ the number of rest periods before and after the visit is forced to the pre-computed values λ_i and μ_i, respectively, and it can be eliminated, i.e. treated as a constant as for travel and service time.

For customers $i \in V_k$, for each value of $k = 1, \ldots, \pi$ only two cases are possible, that is with $\lambda_i = k - 1$ rest periods before the visit and $\mu_i + 1 = \pi - k + 1$ rest periods after it and with $\lambda_i + 1 = k$ rest periods before the visit and $\mu_i = \pi - k$ rest periods after it. This corresponds to generate pairs of time windows, indicated by $[a_i', b_i']$ and $[a_i'', b_i'']$ for each $i \in V_k$ for each $k = 1, \ldots, \pi$, as shown in Algorithm 3.

Algorithm 3. Elimination of rest periods.

$$
\begin{aligned}
&\textbf{for } i = 1, \ldots, n \textbf{ do} \\
&\quad \textbf{if } i \in U \textbf{ then} \\
&\qquad a_i \leftarrow a_i - \lambda_i \delta \\
&\qquad b_i \leftarrow b_i - \lambda_i \delta \\
&\quad \textbf{else} \\
&\qquad a_i' \leftarrow a_i - \lambda_i \delta \\
&\qquad b_i' \leftarrow b_i - \lambda_i \delta \\
&\qquad a_i'' \leftarrow a_i - (\lambda_i + 1)\delta \\
&\qquad b_i'' \leftarrow b_i - (\lambda_i + 1)\delta
\end{aligned}
$$

Tightening the Time Windows. Some parts of the time windows can be eliminated, because they cannot be used in any feasible solution. A too small value of a_i may correspond to a point in time that is too early, so that the vehicle cannot reach customer i in that moment. Symmetrically, a too large value of b_i may correspond to a point in time that is too late, because leaving customer i at time b_i would not allow the vehicle to reach the next customer in time. Time windows can be tightened by comparing consecutive time windows, as shown in Algorithm 4. All these preprocessing steps are illustrated by the example described in Table 1. It refers to an instance with 6 customers, where $T = 10$ and $\delta = 4$. The total distance to travel is 42; hence $\pi = 4$.

Table 1. An example with 6 customers besides the depot. Column 8 reports the time windows after the elimination of service times and travel times. Columns 9 − 11 report the time windows after the elimination of rest periods. Columns 12 − 14 report the tightened time windows.

i	d_i	$\sum_{j=1}^i d_j$	$\sum_{j=i+1}^n d_j$	λ_i	μ_i	Set	$[a,b]$	$[a'',b'']$	$[a',b']$	$[a,b]$	$[a'',b'']$	$[a',b']$	$[a,b]$
0	0		42	0	4	U_O	[0,50]			[0,50]			[0,18]
1	8	8	34	0	3	V_1	[12,25]	[4,17]	[8,21]		[4,17]	[8,18]	
2	10	18	24	1	2	V_2	[14,37]	[2,25]	[6,29]		[4,18]	[8,18]	
3	3	21	21	2	2	U_2	[16,30]		[8,22]			[8,18]	
4	10	31	11	3	1	U_3	[13,39]		[1,27]			[8,18]	
5	3	34	8	3	0	V_4	[26,38]	[6,18]	[10,22]		[8,18]	[10,22]	
6	2	36	6	3	0	V_4	[19,44]	[−1,24]	[3,28]		[8,24]	[10,28]	
7	6	42	0	4	0	U_4	[28,50]		[12,34]			[12,34]	

Algorithm 4. Time windows tightening.

1: $a \leftarrow a_0$	19: $b \leftarrow b_n$
2: **for** $i \in U_0$ **do**	20: **for** $k = \pi, \ldots, 1$ **do**
3: $\quad a_i \leftarrow \max\{a_i, a\}$	21: $\quad$ **for** $i \in U_k$ **do**
4: $\quad a \leftarrow \max\{a, a_i\}$	22: $\quad\quad b_i \leftarrow \min\{b_i, b\}$
5: $a'' \leftarrow a$	23: $\quad\quad b \leftarrow \min\{b, b_i\}$
6: $a' \leftarrow a$	24: $\quad b' \leftarrow b$
7: **for** $k = 1, \ldots, \pi$ **do**	25: $\quad b'' \leftarrow b$
8: $\quad$ **for** $i \in V_k$ **do**	26: $\quad$ **for** $i \in V_k$ **do**
9: $\quad\quad a_i' \leftarrow \max\{a_i', a'\}$	27: $\quad\quad b_i' \leftarrow \min\{b_i', b'\}$
10: $\quad\quad a' \leftarrow \max\{a', a_i'\}$	28: $\quad\quad b' \leftarrow \min\{b', b_i'\}$
11: $\quad\quad a_i'' \leftarrow \max\{a_i'', a''\}$	29: $\quad\quad b_i'' \leftarrow \min\{b_i'', b''\}$
12: $\quad\quad a'' \leftarrow \max\{a'', a_i''\}$	30: $\quad\quad b'' \leftarrow \min\{b'', b_i''\}$
13: $\quad a \leftarrow a''$	31: $\quad b \leftarrow b'$
14: $\quad$ **for** $i \in U_k$ **do**	32: **for** $i \in U_0$ **do**
15: $\quad\quad a_i \leftarrow \max\{a_i, a\}$	33: $\quad b_i \leftarrow \min\{b_i, b\}$
16: $\quad\quad a \leftarrow \max\{a, a_i\}$	34: $\quad b \leftarrow \min\{b, b_i\}$
17: $\quad a'' \leftarrow a$	
18: $\quad a' \leftarrow a$	

4 Minimization of the Total Waiting Time

After these pre-processing steps, only waiting times are left: travels, rest periods and customer services have got null duration. Therefore, the solution is to be found by defining a sequence of points in time t_i, representing the visit to each customer $i \in N$ in the modified instance, such that

1. $t_i \geq t_{i-1} \ \forall i = 1, \ldots, n$;
2. $a_i \leq t_i \leq b_i \ \forall i \in U$.

Hence, should $b_i < a_i$ occur for some $i \in U$, the instance would be infeasible.

The time window constraint for customers in V can be formulated as follows: for each $k = 1, \ldots, \pi$, the set V_k must be partitioned into two subsets V_k' and V_k'' of consecutive customers, where all customers in V_k' precede those in V_k'' so that

1. $a_i' \leq t_i \leq b_i' \ \forall i \in V_k'$;
2. $a_i'' \leq t_i \leq b_i'' \ \forall i \in V_k''$.

This choice corresponds to insert the rest period with index k after the customers in V_k' and before those in V_k''. As a special case, one of the two subsets can be left empty.

Special Case. If $a_n \leq b_0$, the modified instance may allow for an optimal solution of null duration, where $t_i = t \ \forall i \in N$ with $t \in [a_n, b_0]$, i.e. with no waiting time. This is guaranteed to happen if all time windows have an initial width not smaller than δ.

Proof. Owing to the non-decreasing order of the endpoints of the time windows ($a_{i+1} \geq a_i$ and $b_{i+1} \geq b_i$ for all $i \in N$), the inequality $t \leq b_0$ implies $t \leq b_i \ \forall i \in U$ and $t \leq b_i' \ \forall i \in V$; symmetrically, $t \geq a_n$ implies $t \geq a_i \ \forall i \in U$ and $t \geq a_i'' \ \forall i \in V$.

If all time windows have an initial width greater than or equal to δ, this guarantees that after the elimination of rest periods and before time windows tightening $b_i - a_i \geq \delta \ \forall i \in U$ and $b_i'' \geq a_i' \ \forall i \in V$. Therefore, any value of t between a_i'' and b_i' for $i \in V$ is feasible with respect to at least one of the two time windows of i. In other terms, no "holes" are generated between the two time windows of any customer $i \in V$. Furthermore, when going from i to $i+1$ it is never necessary to pass from a time window $[a_i'', b_i'']$ to a time window $[a_{i+1}', b_{i+1}']$, because it cannot happen that $a_i' > b_{i+1}''$.

Hence, solutions with no waiting time are feasible (Q.E.D.).

For instance, this happens in the example illustrated in Table 1, where $a_n = 12$ and $b_0 = 18$.

Finally, if $b_0 < a_n$, then there exists an optimal solution with waiting time equal to $b_0 - a_n$, whose feasibility is proven in the same way. It is always possible to set $t_0 = b_0$ and $t_n = a_n$ and to find non-decreasing values of t_i for all intermediate customers, complying with all time windows, since no "holes" exist between them.

General Case. In a more general case, where time windows may be initially smaller than δ, the above proof is no longer valid. In this case, empty intervals may occur between $[a_i'', b_i'']$ and $[a_i', b_i']$ for some $i \in V$. It is still true that $a_{i+1} \geq a_i$ and $b_{i+1} \geq b_i$ for all i, and possible "holes" may occur only between b_i'' and a_i' for some $i \in V$ (not before a_i'' nor after b_i').

To deal with this more general case, we exploit the constraint that all time windows $[a', b']$ must be traversed before all time windows $[a'', b'']$ for the customers in each set V_k. Hence, each set of customers V_k originates $|V_k| + 1$ possible sequences of time windows. Each of them corresponds to the choice of inserting the rest period in position $\ell = 0, \ldots, |V_k|$ in the ordered sequence of the customers in V_k. For each choice of $\ell = 0, \ldots, |V_k|$ the intersection of the time windows of the customers in V_k is $[a_\ell^k, b_\ell^k]$ with endpoints in

$$a_\ell^k = \max\{\max_{i \in V_k : i > \ell}\{a_i''\}, \max_{i \in V_k : i \leq \ell}\{a_i'\}\}$$

$$b_\ell^k = \min\{\min_{i \in V_k : i > \ell}\{b_i''\}, \min_{i \in V_k : i \leq \ell}\{b_i'\}\}.$$

If $a_\ell^k > b_\ell^k$ for some ℓ, then the corresponding time window does not exist. In this way, each subset V_k can be replaced by a single dummy customer with multiple time windows. If two or more of its time windows overlap, they can be merged in a single time window.

For instance, considering the two customers numbered 5 and 6 in the example above, both belonging to set V_4, there are three alternatives: if $\ell = 0$, then the rest period is inserted before customer 5 and $[a_\ell^k, b_\ell^k] = [8, 18]$; if $\ell = 1$, then the rest period is inserted between customers 5 and 6 and $[a_\ell^k, b_\ell^k] = [10, 22]$; if $\ell = 2$, then the rest period is inserted after customer 6 and $[a_\ell^k, b_\ell^k] = [10, 22]$. Since the three time windows overlap, the subset V_4 can be replaced by a dummy customer with a single time window $[8, 22]$.

On the contrary, in the example illustrated in Fig. 1 and Table 2 multiple time windows are generated. The example refers to an instance with 3 customers, where $T = 100$ and $\delta = 20$. The total distance to travel is 110; hence $\pi = 1$. All three customers belong to the same subset V_1, that can be traversed in four ways indexed by $\ell = 1, \ldots, 4$ corresponding to the four possible insertion positions of the rest period. The corresponding time windows are: $a_0^1 = [5, 12]$, $a_1^1 = [15, 19]$, a_2^1 empty, $a_3^1 = [25, 32]$. Hence the subset V_1 can be replaced by a dummy customer with three disjoint time windows.

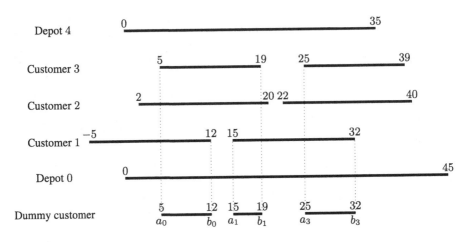

Fig. 1. Time windows $[a', b']$ and $[a'', b']$ of three customers replaced by multiple time windows of a single dummy customer.

Table 2. An example with 3 customers besides the depot. Column 8 reports the time windows after the elimination of service times and travel times. Columns 9–11 report the time windows after the elimination of rest periods. Columns 12–14 report the tightened time windows.

i	d_i	$\sum_{j=1}^{i} d_j$	$\sum_{j=i+1}^{n} d_j$	λ_i	μ_i	Set	$[a,b]$	$[a'',b'']$	$[a',b']$	$[a,b]$	$[a'',b'']$	$[a',b']$	$[a,b]$
0	0		110	0	1	U_O	[0, 45]			[0, 45]			[0, 32]
1	60	60	90	0	0	V_1	[15, 32]	[−5, 12]	[15, 32]		[0, 12]	[15, 32]	
2	10	70	70	0	0	V_1	[22, 40]	[2, 20]	[22, 40]		[2, 20]	[22, 35]	
3	20	90	60	0	0	V_1	[25, 39]	[5, 19]	[25, 39]		[5, 19]	[25, 35]	
4	20	110	0	1	0	U_1	[20, 55]			[0, 35]			[5, 35]

Subsets U_k are much easier to consider, because they can be feasibly traversed in one way, since no rest periods must be inserted in them. Hence, each subset U_k is simply replaced by a dummy customer with a single time window that is the intersection of the time windows of the real customers in U_k.

Now, the residual problem is to find an optimal path through the multiple time windows of the dummy customers originated by subsets U_k and V_k for $k = 1, \ldots, \pi$. In the remainder index j is used to indicate dummy customers; their number m is upper bounded by n and by $2\pi + 1$.

An algorithm to solve this problem is the following. Consider the "columns", made by sequences of consecutive customers with overlapping time windows. As a special case, a column can be made by a single customer. Once all columns have been listed, one searches for a shortest path on a weighted acyclic digraph with a node for each column; the weight of each arc can be determined easily, as shown hereafter.

The algorithm to list all columns is illustrated in Algorithm 5 and it uses the following data-structures:

- Data-structures to describe customers:
  - $ptop[j]$: vector with the index of the column terminating at customer j;
  - $pbot[j]$: vector with the index of the column starting at customer j.
- Data-structures to describe columns:
  - $top[c]$: vector with the last customer of each column c;
  - $bot[c]$: vector with the first customer of each column c;
  - $s[c]$: start time of column c;
  - $e[c]$: end time of column c.
- Data-structures to describe time windows:
  - $vertex[w]$: customer of time window w;
  - $dir[w]$: binary datum, where 1 means that w starts, 0 indicates that w ends;
  - $time[w]$: start time or end time of time window w.

Other indices that are used in Algorithm 5:

- $j = 0, \ldots, m$: customer index;
- nc: number of columns;
- $c = 1, \ldots, nc$: column index;
- nw: number of time windows;
- $w = 1, \ldots, nw$: time window index;
- t: current point in time.

The sub-routine $FindColumn(j)$ takes in input a customer index j and finds the column containing j among those already started and not yet terminated when the sub-routine is called.

Algorithm 5. The algorithm that lists all columns.

```
for j = 1, ..., m do                    for w = 1, ..., nw do
    ptop[i] ← 0                             j ← vertex[w]
    pbot[i] ← 0                             t ← time[w]
nc ← 0                                      if dir[w] = 1 then
/* Populate and sort the vectors vertex,        WOpen(j)
dir and time with start and end time of     else
all time windows */                             WClose(j)
```

Algorithm 6. Procedure $WOpen(j)$	**Algorithm 7.** Procedure $WClose(j)$

$nc \leftarrow nc + 1$
// Close the column above j, if any //
if $(j < m) \wedge (pbot[j+1] \neq 0)$ **then**
    $c \leftarrow pbot[j+1]$
    $e[c] \leftarrow t$
    $pbot[j+1] \leftarrow 0$
    $top[nc] \leftarrow top[c]$
    $Succ[c] \leftarrow nc$ (1)
else
    $top[nc] \leftarrow j$
    if $tail[j] \neq 0$ **then**
        $Succ[tail[j]] \leftarrow nc$ (2)
// Close the column under j, if any //
if $(j > 0) \wedge (ptop[j-1] \neq 0)$ **then**
    $c \leftarrow ptop[j-1]$
    $e[c] \leftarrow t$
    $ptop[j-1] \leftarrow 0$
    $bot[nc] \leftarrow bot[c]$
    $Succ[c] \leftarrow nc$ (3)
else
    $bot[nc] \leftarrow j$
    if $tail[j-1] \neq 0$ **then**
        $Succ[tail[j-1]]] \leftarrow nc$ (4)
// Open a new column including j //
$s[nc] \leftarrow t$
$ptop[top[nc]] \leftarrow nc$
$pbot[bot[nc]] \leftarrow nc$
// Update tails //
$tail[j] \leftarrow 0$
$tail[j-1] \leftarrow 0$
$tail[top[nc]] \leftarrow nc$

$c \leftarrow FindColumn(j)$
$h \leftarrow top[c]$
$k \leftarrow bot[c]$
// Close the column with j //
$e[c] \leftarrow t$
$ptop[h] \leftarrow 0$
$pbot[k] \leftarrow 0$
if $h > j$ **then**
    // Open a new column above j //
    $nc \leftarrow nc + 1$
    $top[nc] \leftarrow h$
    $bot[nc] \leftarrow j + 1$
    $s[nc] \leftarrow t$
    $ptop[h] \leftarrow nc$
    $pbot[j+1] \leftarrow nc$
    $Succ[c] \leftarrow nc$ (5)
    $tail[h] \leftarrow nc$
if $k < j$ **then**
    // Open a new column under j //
    $nc \leftarrow nc + 1$
    $top[nc] \leftarrow j - 1$
    $bot[nc] \leftarrow k$
    $s[nc] \leftarrow t$
    $ptop[j-1] \leftarrow nc$
    $pbot[k] \leftarrow nc$

A numerical example is shown in detail in Fig. 2 and Table 3.

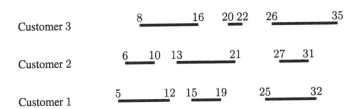

Fig. 2. Multiple time windows of three (dummy) customers.

Table 3. List of all columns obtained from the multiple time windows shown in Fig. 2.

w	1	2	3	4	5	6	7	8	9	10	11	12	13	14	15	16	17	18
$vertex$	1	2	3	2	1	2	1	3	1	3	2	3	1	3	2	2	1	3
$time$	5	6	8	10	12	13	15	16	19	20	21	22	25	26	27	31	32	35
dir	1	1	1	0	0	1	1	0	0	1	0	0	1	1	1	0	0	0

The datum $tail[j]$ for each customer j, when different from 0, is the index of the column that is the tail of an arc in the digraph. The corresponding head of the arc is identified when a column is found containing customer j or $j+1$. When this occurs the new column is identified as the head of the arc. All arcs found in this way are stored in the data-structure $Succ$ which has a component for each column, since in the acyclic digraph each node has at most one successor.

Figures 3 and 4 show the different cases that can occur when executing $WOpen(j)$ and $WClose(j)$, respectively.

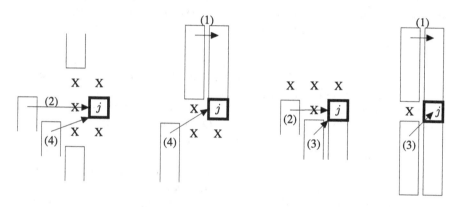

Fig. 3. Linking columns with $WOpen(j)$.

Once all columns have been listed, an acyclic digraph is defined with two nodes for each column (start and end of the corresponding time window). Columns that contain the initial depot 0 have a single node corresponding to the end of their time window, while those containing the final depot n have a single node corresponding to the beginning of their time window. The arc between the two nodes of a same column $c = 1, \ldots, n-1$ has weight $e[c] - s[c]$. The weight of any arc from the end node of a column c to the start node of $Succ[c]$ is set equal to $s[Succ[c]] - e[c]$. Each column c has at most one leaving arc, connecting it to the first successive column that contains customer $top[c]$ or $top[c] + 1$.

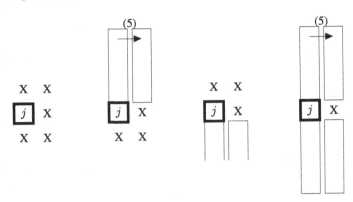

Fig. 4. Linking columns with $WClose(j)$.

4.1 Complexity

The number of dummy customers m is not larger than the number of real customers n. The number of time windows nw is also $O(n)$ by construction. Therefore the number of columns is also $O(n)$, because the number of columns increases by one or two units at every start time or end time of a time window. The number of arcs in the digraph is twice the number of columns and hence $O(n)$. The search for an optimal path in a weighted acyclic digraph has linear complexity in the number of arcs, i.e. $O(n)$. The procedure $FindColumns$ has complexity $O(\log n)$ if the list of the columns already started and not yet terminated is implemented as a binary tree, so that $O(\log n)$ is the complexity for inserting a new column (and this is done $O(n)$ times), as well as for deleting a column (this is also done $O(n)$ times) and also for finding the column containing a given customer (this is also done $O(n)$ times).

It remains to evaluate the complexity for sorting the time windows endpoints. If time windows are already sorted for each customer, it is necessary to merge $O(u)$ ordered lists, each one with $O(v)$ elements, where $uv = n$. Merging two sorted lists of $O(v)$ elements takes $O(v)$. Joining list of the same length we have $\log_2 u$ levels, each one with complexity $uv/2$, which yields $O(uv \log u)$ complexity. The worst case is with few windows per customer: when u tends to n and v tends to 1, the complexity tends to $O(n \log n)$.

If time windows are initially unsorted, then it is also needed to sort them, which takes $O(n \log n)$.

Therefore the worst-case time complexity of the algorithm is $O(n \log n)$ (which improves upon the complexity of the dynamic programming algorithm developed in [1]).

5   Extensions

The analysis of this version of the TPRP is a starting point to efficiently solve more complex variations.

Fixed Start Time. In this problem variation the driver is not allowed to select the optimal start time of the route, but he is forced to start at time 0. This problem is just a special case of the previous one, with $a_0 = b_0 = 0$.

Rest Periods with Different Duration. In this problem variation there are two different types of rest periods: short rest periods represent breaks during the day, while long rest periods represent overnight breaks. This problem variation can be solved in the same way as the basic case: it is only required to replace δ with δ_k in the rest period elimination at pre-processing. Also the computation of λ and μ values must be slightly modified: instead of multiplying the number of rest periods by δ, one must sum up the duration of the rest period, whose sequence is known.

Multiple Time Windows. An important variation is when customers may have two or more associated time windows. This can be solved as the general case of the TPRP with possibly disjoint multiple time windows for each customer.

Time-Dependent Travel Time. Another variation of the TPRP that is relevant in practice and remains to be addressed as a further development is that with time-dependent travel times. This allows to model different traffic conditions, intuitively making it profitable to place rest periods in correspondence of peak hours.

Acknowledgements. The authors acknowledge their fruitful collaboration with Giulia Novello, who developed a dynamic programming algorithm for the TPRP, and WorkWaveFleet that provided the problem description and several real or realistic instances.

References

1. Novello, G.: Mathematical optimization algorithms for the insertion of breaks in truck driver duties. Master thesis, University of Milan (2022)
2. Vidal, T., Crainic, T.G., Gendreau, M., Prins, C.: Timing problems and algorithms: time decisions for sequences of activities. Networks **65**(2), 102–128 (2015)

Fair Energy Allocation for Collective Self-consumption

Natalia Jorquera-Bravo[1,2](✉), Sourour Elloumi[1,2], Safia Kedad-Sidhoum[2], and Agnès Plateau[2]

[1] UMA, ENSTA Paris, Institut Polytechnique de Paris, 91120 Palaiseau, France
{natalia.jorquera,sourour.elloumi}@ensta-paris.fr
[2] CEDRIC, Conservatoire National des Arts et Métiers, 75003 Paris, France
{safia.kedad_sidhoum,aplateau}@cnam.fr

Abstract. This study explores a collective self-consumption community with several houses, a shared distributed energy resource (DER), and a common battery energy storage system. Each house has an energy demand over a discrete planning horizon, met either by using the DER, or the battery, or by purchasing electricity from the main power grid. Excess energy can be stored in the battery or sold back to the main grid. The objective is to determine a supply plan ensuring a fair allocation of renewable energy provided by the DER while minimizing the overall cost of the microgrid.

We investigate and discuss the formulation of such an optimization problem using mixed integer linear programming. We show some dominance properties that allow the model to be reformulated into a linear program. We study some fairness metrics like the proportional allocation rule and max-min fairness. Finally, we illustrate our proposal with a real case study in France with up to seven houses and a one-day time horizon with 15-min intervals.

Keywords: Energy Management · Fair Allocation · Production Planning

1 Introduction

In a time when renewable energy sources are gaining increasing prominence and communities seek to reduce their carbon footprint, collective self-consumption represents a fundamental transformation in the way individuals and communities access and use energy, offering a sustainable, economically viable, and environmentally friendly alternative to traditional energy distribution models [8]. Energy management under collective self-consumption empowers groups of individuals, neighborhoods, or even entire communities to unite and jointly harness, generate, and distribute locally produced renewable energy. It offers numerous advantages, including a reduced reliance on fossil fuels, and lower energy costs for participants. Furthermore, this approach fosters a sense of ownership and environmental responsibility among participants, thereby strengthening the sense of

A. Basu et al. (Eds.): ISCO 2024, LNCS 14594, pp. 388–401, 2024.
https://doi.org/10.1007/978-3-031-60924-4_29

community as they collectively embrace the energy transition [9]. In Europe, although collective self-consumption is regulated at the level of the European Union, it is governed by different rules depending on the country [8]. In France, for example, collective self-consumption must take place between producers and consumers living in a limited geographical area. They require a legal entity or moral person who is charged of organizing the self-consumption operation, and who maintains communication with the administrator of the public energy network. Thus, one of its most important tasks is to communicate the distribution of the produced energy to the distribution system operator, who is in charge of informing the energy suppliers. The energy suppliers will issue the electricity bills of each participant of the community [11]. For most of these communities, their operation can be divided into two major systems: a technical system and an economic system. The technical system must account for the actual energy distribution, while the economic system considers the monetary distribution that occurs after the energy has been processed. Consequently, the financial balance of each participant depends on the rules selected for the distribution of the generated energy, known as allocation rules [21].

The latest is a problem that has long been studied in the literature. However, its characteristics and formulation depend on the technical configuration, the goals and the social arrangements of the community [9]. The literature has focused mainly on microgrids where users have their own distributed energy resource (DER) and energy storage system, and where they are allowed to trade with community members [10,16,21]. These works focus on price selection for exchanges between users. On the other hand, microgrids where the DER is owned by all users are less studied. Ogando-Martínez et al. [18] present a linear programming (LP) model to evaluate different allocation strategies for an industrial community with a shared ownership of a photovoltaic system. However, they consider the allocation rule as a coefficient of the photovoltaic production, fixed before the optimization process, focusing mainly on the redistribution of the surplus energy of each user after an initial allocation. While in [23], fairness is measured by minimizing the discrepancy between the cost assigned to each house and the cost it would have incurred if it had been the only house in the microgrid. Another problem that could be considered related is the management of self-consumption of energy in microgrids that do not belong to a self-consumption community, i.e. a user is both a consumer and a producer of energy and must manage its production, consumption and storage by his own. In this problem we find two approaches, the first one focuses on demand response, aiming to plan the use of electrical appliances at home in such a way that fluctuations in electricity prices and electricity production can be exploited. For a complete review in this topic see [2]. Meanwhile, the second approach centers on the technical issue of renewable energy production and its distribution [12,15,17]. However, both perspectives are distant from our problem, since we consider non-flexible demands, and we focus on the economic system of the community, given that, once the DER has been installed all the energy produced is finally consumed by the nearest electrical device.

In this paper, we address the problem of energy supply planning for collective self-consumption communities, ensuring a fair allocation of renewable energy. We focus on communities with shared ownership and collective energy storage systems, studying different mathematical formulations, and definitions of a fair distribution of collectively produced green electricity, following the French regulations for self-consumption communities. This problem consists in finding an energy supply planning that not only ensures equitable distribution of renewable energy but also minimizes the overall expenses. Specifically, we consider a building in joint ownership, where the inhabitants of the property have decided to invest in photovoltaic panels and an electricity storage system. Each house has a demand that can be met either by using electricity from the photovoltaic panels, from the battery or from the main grid. The surplus energy of each user can be sold to the main grid or stored in the battery, and exchanges between users are not allowed. There is a legal entity, from now on the manager, in charge of managing the photovoltaic energy. This manager must report the distribution of photovoltaic electricity at each time period to the distribution system operator, who then informs the energy supplier. The energy supplier is then responsible for billing each user. It is assumed that, each member owns a smart meter that reports their actual consumption at each time period. The manager distributes all photovoltaic energy at each time period. A user cannot charge and discharge energy from the battery at the same time period, and a user cannot sell and purchase electricity from the main grid at the same time period.

The main contributions of the paper are the following. We first formulate the problem of energy supply planning, as a mixed integer linear programming (MILP) model. Then, we show some dominance properties for this problem that allow not only to relax the integrality constraints, but also to entirely remove binary variables from its formulation, leading to an LP model, and showing that the problem is solvable in polynomial time. Finally, we study the fairness aspect of this problem, by analyzing some fair allocation rules for green energy. We propose a mathematical formulation for each allocation rule studied, and show some dominance properties for these models. Some numerical experiments on a real case study in France are reported. The rest of this article is organized as follows, in Sect. 2 we analyze the energy supply planning problem in collective self-consumption without fairness, and in Sect. 3 we study some fair allocation rules and their mathematical formulation for the energy supply planning problem in collective self-consumption. In Sect. 4 we present some numerical experiments on a real case study in France.

2   Energy Supply Planning Problem

We consider a microgrid composed of a set J of smart houses, an array of photovoltaic panels, a shared battery, and a connection to the main power grid. The planning horizon is defined by a set T of discrete time periods of δ hours each, typically $\delta = 0.25$ h which corresponds to 15 min. Each house $j \in J$ has a known electricity demand $D_{j,t}$ that must be satisfied at each time period t.

The battery has a charge and discharge efficiency, e_c and e_d respectively, such that $0 < e_c e_d < 1$, i.e. energy is lost when charging and discharging the battery. The battery has an initial state of charge S_0, a maximum capacity $S^{\max}$, and a minimum storage limit $S^{\min}$. The charge and discharge rates have a maximum limit, $\bar{F}$ and $\underline{F}$, respectively. At each time period $t \in T$, photovoltaic panels produce at most C_t^{PV}. Moreover, electricity can be purchased at a price ν_t, discharged from the battery at a price μ_d or sold to the main grid at a price β whose value is set by the French Energy Regulatory Commission. The electricity sale price β is always strictly lower than the electricity purchase price ν_t for each t. An instance of the energy supply planning problem is therefore given by $(J, T, \delta, D, e_c, e_d, S_0, S^{\min}, S^{\max}, \bar{F}, \underline{F}, \nu, \mu_d, \beta, C^{PV})$.

In the following subsections, we first present a MILP model for this problem. And then, we present some dominance properties of this model that allow us to formulate this problem as an LP model.

2.1 Mathematical Formulation

We formulate the energy supply planning problem as (EPP), a MILP model.

$$
(EPP) \begin{cases}
\min \quad \delta \left(\displaystyle\sum_{j \in J, t \in T} \mu_d y_{j,t} + \nu_t i_{j,t} - \beta g_{j,t} \right) & (1a) \\[2ex]
\text{s.t.:} \\
\displaystyle\sum_{j \in J} p_{j,t} = C_t^{PV} & \forall t \in T & (1b) \\[2ex]
D_{j,t} = p_{j,t} + y_{j,t} + i_{j,t} - z_{j,t} - g_{j,t} & \forall j \in J, t \in T & (1c) \\[2ex]
s_t = s_{t-1} + \delta e_c \displaystyle\sum_{j \in J} z_{j,t} - \frac{\delta}{e_d} \sum_{j \in J} y_{j,t} & \forall t \in T & (1d) \\[2ex]
\displaystyle\sum_{j \in J} y_{j,t} \leqslant \underline{F} & \forall t \in T & (1e) \\[2ex]
\displaystyle\sum_{j \in J} z_{j,t} \leqslant \bar{F} & \forall t \in T & (1f) \\[2ex]
S^{\min} \leqslant s_t \leqslant S^{\max} & \forall t \in T & (1g) \\[1ex]
s_{|T|} = S_0 & & (1h) \\[1ex]
z_{j,t} \leqslant \bar{F} v_{j,t} & \forall j \in J, t \in T & (1i) \\[1ex]
y_{j,t} \leqslant \underline{F}(1 - v_{j,t}) & \forall j \in J, t \in T & (1j) \\[1ex]
i_{j,t} \leqslant M(1 - w_{j,t}) & \forall j \in J, t \in T & (1k) \\[1ex]
g_{j,t} \leqslant M w_{j,t} & \forall j \in J, t \in T & (1l) \\[1ex]
s_t, \, i_{j,t}, \, g_{j,t}, \, z_{j,t}, \, y_{j,t}, \, p_{j,t} \geqslant 0 & \forall j \in J, \, t \in T & (1m) \\[1ex]
v_{j,t}, \, w_{j,t} \in \{0,1\} & \forall j \in J, \, t \in T & (1n)
\end{cases}
$$

In (EPP) we have to decide for each house $j \in J$ and each time period $t \in T$ the electricity output from the photovoltaic panel $p_{j,t}$, from the battery $y_{j,t}$ (discharge), and from the main grid $i_{j,t}$. We also compute at each time period the amount of electricity that each house stores in the battery $z_{j,t}$ (charge), sells to the main grid $g_{j,t}$, and the state of charge of the battery s_t. Furthermore, we have variables $v_{j,t}$ and $w_{j,t}$, working variables that model the activation of the charging and discharging, and the activation of the purchasing and selling of electricity, respectively. Equation (1a) represents a well accepted function to compute the total cost of the microgrid [5] where all equipment capacities are considered as given, so that only operations and maintenance costs are included. The cost function consists of maintenance costs associated with battery discharge and the cost of purchasing electricity from the main grid. The income received from selling excess electricity is also taken into account. Note that no production or maintenance costs associated with the photovoltaic electricity are considered, this is because the production of electricity using photovoltaic panels for each time period is fixed to C_t^{PV}, independently of its allocation, so its operating costs become constant, and therefore are not considered in our problem.

Equations (1b) represent the electricity production of the photovoltaic panels. Constraints (1c) represent the electricity balance in the microgrid. Equations (1d) indicate the state of charge of the battery at the end of each time period, which depends on the state of charge of the previous period, the amount of electricity charged and the amount discharged in the same period. Constraints (1e) and (1f) represent the limits in the charge and discharge of the battery at each time period. Constraints (1g) set the capacity limits for the state of charge of the battery. Constraint (1h) imposes that the state of charge of the battery at the end of the time horizon must be equal to the initial state of charge. Constraints (1i) and (1j) indicate that for each house, at each time period, a charge and a discharge of the battery cannot take place at the same time. Similarly Constraints (1l) and (1k) indicate that a house cannot purchase and sell power to the grid in the same time period, where M is a parameter large enough.

2.2   Dominance Properties

We show below that the energy supply planning problem in collective self-consumption can be formulated as an LP model. To address this, we introduce a relaxed version of the model (EPP) where we remove Constraints (1i)–(1l) and (1n), obtaining an LP model that we call (EPP-L). Lemma 1 and Lemma 2 show some dominance properties of any optimal solution of (EPP-L).

Lemma 1. *Given an instance of the energy supply planning problem where $\nu_t > \beta$ for all $t \in T$, then in any optimal solution $x^* = (s^*, i^*, g^*, z^*, y^*, p^*)$ of (EPP-L), $i_{j,t}^* g_{j,t}^* = 0$ holds for all $j \in J, t \in T$.*

Proof. Let us assume, by contradiction, that there exists an optimal solution $x^* = (s^*, i^*, g^*, z^*, y^*, p^*)$ of (EPP-L) such that $\exists\, t' \in T, j' \in J : i_{j',t'}^* g_{j',t'}^* > 0$.

Let $c(x^*)$ be the cost of x^*. We show that we can build a feasible solution $\bar{x} = (\bar{s}, \bar{i}, \bar{g}, \bar{z}, \bar{y}, \bar{p})$ of (EPP-L) identical to x^* except for j', t' where $\bar{i}_{j',t'}\bar{g}_{j',t'} = 0$. We will prove that $c(\bar{x}) < c(x^*)$.

First we set $\bar{s} = s^*$, $\bar{p}_{j',t'} = p^*_{j',t'}$, $\bar{y}_{j',t'} = y^*_{j',t'}$ and $\bar{z}_{j',t'} = z^*_{j',t'}$. Then, let $\epsilon = \min\{i^*_{j',t'}, g^*_{j',t'}\}$, $\bar{i}_{j',t'} = i^*_{j',t'} - \epsilon$ and, $\bar{g}_{j',t'} = g^*_{j',t'} - \epsilon$. It follows that $\epsilon > 0$ and $\bar{i}_{j',t'}\bar{g}_{j',t'} = 0$.

Since we reduce the sold and purchased electricity by the same amount, all constraints, including the energy balance given by Eq. (1c) remain satisfied, and therefore $\bar{x}$ is a feasible solution of (EPP-L).

Furthermore, we have $c(x^*) - c(\bar{x}) = \delta\epsilon(\nu_{t'} - \beta)$. Since $\nu_t > \beta$ for all $t \in T$, we get $c(x^*) - c(\bar{x}) > 0$, and consequently, x^* is not optimal. Hence, we have reached a contradiction. □

Lemma 2. *Given an instance of the energy supply planning problem where $0 < e_c e_d < 1$, then in any optimal solution $x^* = (s^*, i^*, g^*, z^*, y^*, p^*)$ of (EPP-L), $y^*_{j,t} z^*_{j,t} = 0$ holds for all $j \in J, t \in T$*

Proof. In the same way as for Lemma 1, let us assume by contradiction that $x^* = (s^*, i^*, g^*, z^*, y^*, p^*)$ is an optimal solution to (EPP-L) such that $\exists t' \in T, j' \in J$: $y^*_{j',t'} z^*_{j',t'} > 0$. Let $c(x^*)$ be the cost of x^*. We build $\bar{x} = (\bar{s}, \bar{i}, \bar{g}, \bar{z}, \bar{y}, \bar{p})$, a feasible solution to (EPP-L), such that, $\bar{x}$ is identical to x^* except for variables related to house j' at time period t', where $\bar{y}_{j',t'} \bar{z}_{j',t'} = 0$. We will prove that $c(\bar{x}) < c(x^*)$.

First, we set $\bar{s}_{t'} = s^*_{t'}$, $\bar{i}_{j',t'} = i^*_{j',t'}$, and $\bar{p}_{j',t'} = p^*_{j',t'}$. Then, let $\epsilon = \min\{z^*_{j',t'}, \frac{y^*_{j',t'}}{e_c e_d}\}$, ϵ is positive since $z^*_{j',t'} y^*_{j',t'} > 0$. We set $\bar{z}_{j',t'} = z^*_{j',t'} - \epsilon$, $\bar{y}_{j',t'} = y^*_{j',t'} - e_c e_d \epsilon$, and $\bar{g}_{j',t'} = g^*_{j',t'} + (1 - e_c e_d)\epsilon$. From the definition of ϵ it follows that $\bar{z}_{j',t'}$ and $\bar{y}_{j',t'}$ are non-negative and that $\bar{z}_{j',t'} \bar{y}_{j',t'} = 0$. Also, $\bar{g}_{j',t'}$ is non-negative since $e_c e_d < 1$.

One can easily check that all the constraints but (1c) and (1d) are satisfied. Constraints (1c) are trivially satisfied for any pair (j,t) but (j', t'). Let us prove that they are also satisfied for (j', t').

$$
\bar{p}_{j',t'} + \bar{y}_{j',t'} + \bar{i}_{j',t'} - \bar{z}_{j',t'} - \bar{g}_{j',t'}
$$
$$
= p^*_{j',t'} + (y^*_{j',t'} - e_c e_d \epsilon) + i^*_{j',t'} - (z^*_{j',t'} - \epsilon) - (g^*_{j',t'} + (1 - e_c e_d)\epsilon)
$$
$$
= p^*_{j',t'} + y^*_{j',t'} + i^*_{j',t'} - z^*_{j',t'} - g^*_{j',t'} = D_{j',t'}
$$

Constraints (1d) are trivially satisfied for any $t \neq t'$. Let us prove that they are also satisfied for t'.

$$
\bar{s}_{t'-1} + \delta e_c \sum_{j \in J} \bar{z}_{j,t'} - \frac{\delta}{e_d} \sum_{j \in J} \bar{y}_{j,t'}
$$
$$
= s^*_{t'-1} + \delta e_c((\sum_{j \in J \setminus j'} z^*_{j,t'}) + z^*_{j',t'} - \epsilon) - \frac{\delta}{e_d}((\sum_{j \in J \setminus j'} y^*_{j,t'}) + y^*_{j',t'} - e_c e_d \epsilon)
$$
$$
= s^*_{t'-1} + \delta e_c \sum_{j \in J} z^*_{j,t'} - \frac{\delta}{e_d} \sum_{j \in J} y^*_{j,t'} = s^*_{t'} = \bar{s}_{t'}
$$

So, $\bar{x}$ is feasible. In this solution, we charge and discharge less electricity compared to x^*, and we also sell more electricity, so it is quite trivial that $\bar{x}$ is a better solution than x^*. We prove this analytically below.

$$(c(x^*)-c(\bar{x}))/\delta = \mu_d(y^*_{j',t'}-\bar{y}_{j',t'})-\beta(g^*_{j',t'}-\bar{g}_{j',t'}) = \mu_d(e_c e_d \epsilon)+\beta((1-e_c e_d)\epsilon) > 0,$$

which concludes the proof   □

Lemma 1 and Lemma 2 show that Constraints (1i)–(1l) are satisfied by optimality and therefore models (EPP) and (EPP-L) are equivalent. We can derive the following corollary.

Corollary 1. *Any instance of the energy supply planning problem such that $\nu_t > \beta$ for all $t \in T$ and $0 < e_c e_d < 1$ can be solved in polynomial time.*

3   Adding Fairness Considerations

So far, our analysis of energy management in microgrids has focused on overall system efficiency. However, in the context of shared resources, such as photovoltaic energy under joint ownership, we are faced with the challenge of allocating scarce resources. When considering only the overall efficiency, individual agents often find themselves in a position where they have to sacrifice personal benefits in order to enhance the common welfare. Consequently, the notions of fairness and equity have garnered significant attention as potential tools for addressing the disadvantages that some agents may face [1,14,19,22].

Several definitions of fairness have emerged in the context of resource allocation [4,7,20]. To implement these definitions in the resource allocation problem, allocation rules are used. Such rules can be defined as the vector $P \in \mathbb{R}^n$ composed of the resource quantities allocated to each user, where n is the number of users. The vector P must meet the following characteristics: non-negativity, demand limitation, and efficiency. In our problem, each house is a user, then $n = |J|$, the total amount of resources to be allocated is $\sum_{t\in T} C_t^{PV}$, and each component of vector P is defined as the aggregate amount of photovoltaic electricity allocated to each house $j \in J$ over the time horizon T, i.e. $P_j = \sum_{t\in T} p_{j,t}$. In the same way, we define D_j as the aggregate demand of each house $j \in J$ over the time horizon T, i.e. $D_j = \sum_{t\in T} D_{j,t}$.

The following subsections aim to define two resource allocation rules for the distribution of photovoltaic electricity in microgrids: proportional allocation rule, and max-min fairness. To conclude, we show that the dominance properties of (EPP-L) hold when adding the allocation rules, and therefore the fair energy allocation problem can be solved in polynomial time under mild instance assumptions.

3.1   Proportional Allocation Rule

The proportional allocation rule consists in allocating the resource to each user in proportion to its individual benefit for that resource, such that if a user's allocation is increased, then there exists at least one other user whose allocation decreases, and the loss of that user is proportionally greater than the gain of any other user. It has been proved in the literature that if, in this problem, the utility of each user is weighted according to its demand, then the allocation rule consists of allocating resources to each user in proportion to its demand [13]. This rule satisfies the following properties: each user receives the same portion of the aggregate demand, i.e. $\frac{P_j}{D_j} = \lambda$ for all $j \in J$; users with the same demand must receive the same amount of resource; scale invariance; and resource monotonicity, i.e. if we increase the amount of available resource, the users should receive at least the same allocation as before.

To implement this allocation rule in the energy supply planning problem, we add Eqs. (2) to (EPP), which state that the aggregate amount of photovoltaic electricity given to each house is proportional to its aggregate demand over the time horizon.

$$\frac{\sum_{t \in T} P_{j,t}}{\sum_{t \in T} D_{j,t}} = \frac{\sum_{t \in T} C_t^{PV}}{\sum_{l \in J, t \in T} D_{l,t}} \quad \forall j \in J \tag{2}$$

This allocation rule favors users with the highest demand, allocating large amounts of photovoltaic electricity to those with the highest aggregate demand.

3.2   Max-Min Fairness

The max-min fairness (MMF) allocation rule is based on the egalitarian notion, in the sense that each user receives the same amount of photovoltaic electricity. However, this can lead to a situation where a low-demand or weak user receives more photovoltaic electricity than it needs. In this situation, the rule is not to provide a user with more photovoltaic electricity than his aggregate demand and to distribute the remaining electricity among the other users, thus trying to protect the weaker users. This allocation rule can be calculated as follows, first we order the users according to their increasing demand, i.e. $D_1 \leqslant \ldots \leqslant D_n$, and then we allocate the resource following Eq. (3) provided in [3]. Consequently, the first user, who has the lowest demand, will receive $P_1 = \min\{D_1, \frac{\sum_{t \in T} C_t^{PV}}{|J|}\}$.

$$P_j = \min \left\{ D_j, \frac{\sum_{t \in T} C_t^{PV} - \sum_{k=1}^{j-1} P_k}{|J| - j + 1} \right\} \tag{3}$$

This rule satisfies the following property: to increase a user's allocation, we must decrease the allocation of another user whose MMF allocation was smaller than that of the beneficiary user. To implement this allocation rule to our problem, we add Constraints (4) and (5) to (EPP).

$$\sum_{t \in T} P_{j,t} \leqslant \sum_{t \in T} D_{j,t} \qquad \forall j \in J \tag{4}$$

$$\sum_{t \in T} P_{j,t} \leqslant \frac{\sum_{t \in T} C_t^{PV} - \sum_{k=1}^{j-1} \sum_{t \in T} p_{k,t}}{|J| - j + 1} \qquad \forall j \in J \tag{5}$$

3.3 Dominance Properties

We below show that the fair energy supply planning problem in collective self-consumption can be reformulated as an LP problem as proposed in the previous section. To this, we consider (EPP-L), and we incorporate an allocation rule, obtaining an LP model. Thus, we deduce Lemma 3.

Lemma 3. *Given an instance of the energy supply planning problem, where $\nu_t > \beta \; \forall t \in T$, then in any optimal solution $x^* = (s^*, i^*, g^*, z^*, y^*, p^*)$ of (EPP-L) with proportional allocation rule (or max-min fairness rule), $i_{j,t}^* g_{j,t}^* = 0 \; \forall j, t$, and $y_{j,t}^* z_{j,t}^* = 0 \forall j, t$.*

Proof. By incorporating Constraint (2) (or Constraints (4)–(5)) into (EPP-L), we are limiting the quantity of photovoltaic energy being supplied to each household. This allows us to apply the same proof approach as we did for Lemmas 1 and 2, where this variable does not play a significant role.

4   Numerical Experiments

All our models have been implemented in Julia 1.8.1, using the CPLEX 12.10 solver. We created 100 one-day instances from data collected by E4C's[1] DrahiX demonstrator between 15th July 2019 to 23rd October 2019. The building is separated into 7 zones, where each zone has a smart meter that allows to register its electricity consumption every 15 min. Thus, each zone represents a house j within our problem with a demand $D_{j,t}$ between 0 and 120 kW at each time period $t \in T$, and we index these zones in increasing order of their aggregate demand, i.e. $\sum_{t \in T} D_{1,t} \leqslant \ldots \leqslant \sum_{t \in T} D_{7,t}$. For the production of green electricity, this building has an array of photovoltaic panels with a maximum capacity of 37.5 kW at each time period, i.e. $C_t^{PV} \in [0, 37.5]$ for all $t \in T$. It also has a battery with a capacity $S^{\max} = 10.5$ kWh and $S^{\min} = 0.2$ kWh. A time horizon of 24 h with time periods of 15 min is used, then $|T| = 96$ and $\delta = 0.25$ h. To calculate the costs, we consider a random electricity purchase price ν_t between 0.1615 euros/kWh and 0.2228 euros/kWh, according to the price of electricity in the peak hours and low hours in France in 2023 (cf. [6]), as well as for the sale of electricity we use the price established for the sale of surplus electricity in French self-consumption communities, $\beta = 0.1$ euros/kWh. All the technical parameters of the battery, $e_c = e_d = 0.95$, $S_0 = S^{\min}$, $\bar{F} = \underline{F} = 4$ kW, and $\mu_d = 0.05$ euros/kWh, were obtained from Zhang et al. [23] since both batteries have similar features.

   In the subsections below, we illustrate the impact of the proposed allocation rules. First, we present the results obtained for one instance of the case study, considering the model (EPP-L) without fairness, then the results obtained using the proportional allocation rule, and finally, the max-min fairness allocation rule.

---

[1] Interdisciplinary center Energy4Climate of Institut Polytechnique de Paris and École des Ponts ParisTech, https://www.e4c.ip-paris.fr/#/fr/datahub/submission/introduction.

In this instance, we consider the following aggregate demands $D_1 = 14.6$ kW, $D_2 = 85.09$ kW, $D_3 = 142.41$ kW, $D_4 = 158.66$ kW, $D_5 = 249.05$ kW, $D_6 = 425.11$ kW, $D_7 = 540.34$ kW, and $\sum_{t\in T} C_t^{PV} = 632.607$ kW. For each model, we present two metrics, the autonomy, and the electricity sold to the main grid. We define the autonomy as the percentage of the aggregate demand that could be satisfied using photovoltaic electricity and battery power, i.e. the percentage of the aggregate demand that could be satisfied if the microgrid were disconnected from the main grid. While the electricity sold allows us to see how much of this photovoltaic electricity we are actually using. Finally, we present a comparison between the global costs obtained using the allocation rules, with respect to the cost obtained using the model (EPP-L) for each instance.

4.1 Energy Supply Planning Problem

Figure 1 shows the results obtained by solving (EPP-L) without fairness. Figure 1a shows the autonomy of each house. For each j, we plot the percentage of the aggregate demand that could be satisfied with the aggregate photovoltaic electricity $\frac{\sum_{t\in T} p_{j,t}}{\sum_{t\in T} D_{j,t}}$, and the percentage of the aggregate demand that could be satisfied with the aggregate discharge of the battery over the time horizon $\frac{\sum_{t\in T} y_{j,t}}{\sum_{t\in T} D_{j,t}}$. We can see that only one house is able to satisfy 100% of its aggregate demand with photovoltaic, while the others are around 40%. In Fig. 1b, for each house the yellow bar shows the amount of photovoltaic electricity allocated to the house, and the blue one, the amount of photovoltaic electricity sold to the main grid, both aggregated on the time horizon. As we can see, when we do not consider fairness in our model, we can have some houses that receive more than what they need, being forced to sell the surplus electricity to the main grid, even when some other houses could benefit from this electricity. An interesting observation here is that houses 3 and 4 have significantly different allocations despite having similar demands.

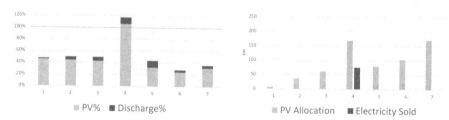

(a) Percentage of the aggregate demand satisfied with the allocated photovoltaic energy and the battery.

(b) Amount of electricity sold compared to amount of allocated photovoltaic.

Fig. 1. Results obtained by solving (EPP-L) aggregated by house over the time horizon. (Color figure online)

4.2 (EPP-L) with Proportional Rule

In Fig. 2a, we can see that all the houses are able to satisfy the same percentage of their aggregate demand with photovoltaic electricity, ensuring more than 40% of autonomy per house. In Fig. 2b, we observe that 5 out of the 7 houses are selling some electricity to the main grid. This is mainly due to the mismatch between photovoltaic electricity production and consumption, i.e. there are periods where there is no photovoltaic electricity production, but there is still demand, while in periods of maximum production, it exceeds demand and storage capacity, so photovoltaic electricity must be sold to the main grid. Furthermore, in order to ensure the same autonomy to all houses, we are forced to allocate more photovoltaic electricity to the houses with a higher demand. This promotes the consumption of electricity, which is contrary to the ecological objectives of the self-consumption community.

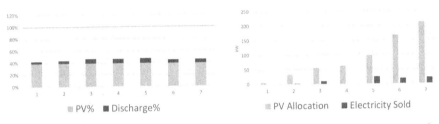

(a) Percentage of the aggregate demand satisfied with the allocated photovoltaic energy and the battery.

(b) Amount of electricity sold compared to amount of allocated photovoltaic energy.

Fig. 2. Results obtained using the proportional rule aggregated by house over the time horizon.

4.3 (EPP-L) with Max-Min Fairness

In Fig. 3a, we can see that houses with a small aggregate demand are able to satisfy their total demand with photovoltaic electricity. In Fig. 3b, as we have seen before, some houses that are not able to satisfy their total demand with photovoltaic electricity, are selling electricity to the main grid. Furthermore, houses with less than 100% autonomy, i.e. houses 3 to 7, receive the same amount of photovoltaic electricity, which is higher than the amount received by those who achieve 100% autonomy.

4.4 Price of Fairness

In Fig. 4, we can see the percentage increase in total cost when using the fair allocation rules, compared to the cost obtained when solving (EPP-L) without fairness, for each daily instance created between the 15th July and 23rd October of 2019. Here we can see that the proportional rule does not present an increase in costs, while MMF increases the global cost by about 3% only.

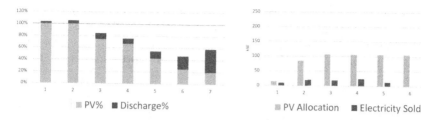

(a) Percentage of the aggregate demand satisfied with the allocated photovoltaic energy and the battery.   (b) Amount of electricity sold compared to amount of allocated photovoltaic energy.

Fig. 3. Results obtained using the max-min fairness rule aggregated by house over the time horizon.

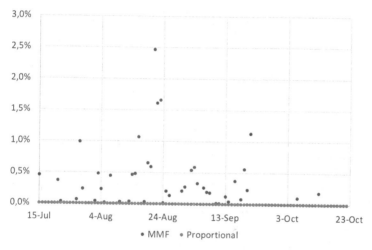

Fig. 4. Percentage increase in total cost when using the proportional and max-min fairness rules.

5   Conclusion

This work addressed the energy supply planning problem in collective self-consumption. First we presented a mixed integer linear formulation of the problem, and we reformulated it as a linear program based on some dominance properties. Then we presented two fairness allocation rules from the literature, and we proposed their implementation in the energy supply planning problem. We also showed how the dominance properties from the energy planning problem hold for the problem considering fairness.

We illustrated our proposal in a real case study in France, using consumption profiles from E4C's DrahiX demonstrator building. We were able to show the impact of the different allocation rules on the distribution of photovoltaic energy and on the economical efficiency of the community.

To continue this work, we want to study the impact of time aggregation in allocation, and the impact of uncertainty in predictions of photovoltaic production or demand. Furthermore, we want to study the α-Fairness allocation rule, which is a collection of different allocations that captures different utility functions.

Acknowledgments. The authors would like to thank Energy4Climate (E4C) Interdisciplinary Center for having partially funded this study.

References

1. Aziz, H.: Developments in multi-agent fair allocation. In: Proceedings of the AAAI Conference on Artificial Intelligence, vol. 34, pp. 13563–13568 (2020)
2. Beaudin, M., Zareipour, H.: Home energy management systems: a review of modelling and complexity. Renew. Sustain. Energy Rev. **45**, 318–335 (2015)
3. Bertsekas, D., Gallager, R.: Data Networks, Series 4. Prentice-Hall, Upper Saddle River (1987)
4. Bertsimas, D., Farias, V.F., Trichakis, N.: The price of fairness. Oper. Res. **59**(1), 17–31 (2011)
5. Calderon-Obaldia, F., Badosa, J., Migan-Dubois, A., Bourdin, V.: A two-step energy management method guided by day-ahead quantile solar forecasts: cross-impacts on four services for smart-buildings. Energies **13**(22), 5882 (2020)
6. EDF: Tarif base EDF: Grille de tarif réglementé en 2023. Technical report (2023). https://prix-elec.com/tarifs/evolution
7. Foley, D.K.: Resource allocation and the public sector. Yale University (1966)
8. Frieden, D., Tuerk, A., Neumann, C., d'Herbemont, S., Roberts, J.: Collective self-consumption and energy communities: trends and challenges in the transposition of the EU framework. COMPILE, Graz, Austria (2020)
9. Gjorgievski, V.Z., Cundeva, S., Georghiou, G.E.: Social arrangements, technical designs and impacts of energy communities: a review. Renew. Energy **169**, 1138–1156 (2021)
10. Gjorgievski, V.Z., Cundeva, S., Markovska, N., Georghiou, G.E.: Virtual net-billing: a fair energy sharing method for collective self-consumption. Energy **254**, 124246 (2022)
11. For Government: Code de l'énergie: Chapitre V: L'autoconsommation (Articles L315-1 à L315-8)
12. Hanna, R., Kleissl, J., Nottrott, A., Ferry, M.: Energy dispatch schedule optimization for demand charge reduction using a photovoltaic-battery storage system with solar forecasting. Sol. Energy **103**, 269–287 (2014)
13. Kelly, F.P., Maulloo, A.K., Tan, D.K.H.: Rate control for communication networks: shadow prices, proportional fairness and stability. J. Oper. Res. Soc. **49**(3), 237–252 (1998)
14. Kumar, A., Kleinberg, J.: Fairness measures for resource allocation. In: Proceedings 41st Annual Symposium on Foundations of Computer Science, pp. 75–85. IEEE (2000)
15. Langenmayr, U., Wang, W., Jochem, P.: Unit commitment of photovoltaic-battery systems: an advanced approach considering uncertainties from load, electric vehicles, and photovoltaic. Appl. Energy **280**, 115972 (2020)

16. Mustika, A.D., Rigo-Mariani, R., Debusschere, V., Pachurka, A.: A two-stage management strategy for the optimal operation and billing in an energy community with collective self-consumption. Appl. Energy 310, 118484 (2022)
17. Nottrott, A., Kleissl, J., Washom, B.: Energy dispatch schedule optimization and cost benefit analysis for grid-connected, photovoltaic-battery storage systems. Renew. Energy 55, 230–240 (2013)
18. Ogando-Martínez, A., García-Santiago, X., Díaz García, S., Echevarría Camarero, F., Blázquez Gil, G., Carrasco Ortega, P.: Optimization of energy allocation strategies in Spanish collective self-consumption photovoltaic systems. Sustainability 15(12), 9244 (2023)
19. Ogryczak, W., Luss, H., Pióro, M., Nace, D., Tomaszewski, A., et al.: Fair optimization and networks: a survey. J. Appl. Math. 2014, 1–25 (2014)
20. Rawls, J.: A Theory of Justice. Cambridge (Mass.) (1971)
21. Roy, A., Olivier, J.C., Auger, F., Auvity, B., Bourguet, S., Schaeffer, E.: A comparison of energy allocation rules for a collective self-consumption operation in an industrial multi-energy microgrid. J. Clean. Prod. 389, 136001 (2023)
22. Thomson, W.: Chapter twenty-one - fair allocation rules. In: Arrow, K.J., Sen, A., Suzumura, K. (eds.) Handbook of Social Choice and Welfare, Handbook of Social Choice and Welfare, vol. 2, pp. 393–506. Elsevier (2011)
23. Zhang, D., Liu, S., Papageorgiou, L.G.: Fair cost distribution among smart homes with microgrid. Energy Convers. Manage. 80, 498–508 (2014)

Day-Ahead Lot-Sizing Under Uncertainty: An Application to Green Hydrogen Production

Victor Spitzer[1,3], Céline Gicquel[1], Evgeny Gurevsky[2(✉)], and François Sanson[3]

[1] LISN, Université Paris-Saclay, Gif-sur-Yvette, France
[2] LS2N, Université de Nantes, Nantes, France
evgeny.gurevsky@univ-nantes.fr
[3] Lhyfe, Nantes, France

Abstract. This work investigates the short-term production planning of green hydrogen obtained through water electrolysis using electricity from a wind power source and a connection to the national electricity grid. Electricity consumption on the grid has to be declared a day ahead of production and cannot be adjusted afterwards, while future availability of the wind power source is uncertain. This production problem can be reduced to a two-stage stochastic lot-sizing model, and a cohesive framework is introduced to solve it efficiently. First, the innovative use of a variational auto-encoder to estimate the conditional wind power uncertainty and generate scenarios is investigated. Then, a time-efficient Benders decomposition approach is proposed, in which special features of our problem are exploited to speed up its resolution. Finally, a novel application of an adaptive partition-based approach and a stabilization method further improve the solving time of the decomposition scheme. A realistic simulation demonstrates the benefits of the presented framework.

Keywords: Lot-sizing · Data-driven stochastic programming · Benders decomposition · Adaptive partition · Stabilization method · Green hydrogen · Wind power uncertainty

1 Introduction

Hydrogen plays a crucial role in industrial processes such as glass and ammonia production, and is mainly generated from fossil fuels. Thus, green hydrogen production via water electrolysis powered by renewable energy sources is essential to decarbonize those industries, consuming only electricity and water in the process (see, *e.g.*, [1]). However, managing this electrolysis process poses challenges, such as ensuring that hydrogen hourly demand is met despite the uncertain availability of renewable energy sources. This work focuses on short-term electrolytic hydrogen production planning, considering a real-life case with a production site connected to a wind farm and the electricity grid. The wind farm supplies fluctuating renewable electricity at negligible cost, while grid electricity incurs a higher cost. Wind power forecasts, although imprecise, provide information on the future availability of this energy source. The purchase of electricity from the grid has to be planned and declared a day ahead of production, *i.e.*, before the exact availability of wind power is known. Finally, on-site hydrogen storage offers some degree of flexibility for production planning, as it allows to produce hydrogen in advance.

The problem is similar to the widely studied single-item lot-sizing problem under uncertainty (see, *e.g.*, Brahimi et al. [2]), and this work extends prior research to include uncertainties in power supply from a local wind farm. Existing works in unit-commitment address such uncertainties (see, *e.g.*, van Ackooij et al. [3]), but for complex problems where decomposition schemes for a large number of uncertainty scenarios are impractical. The use of Benders decomposition scheme is scarce in the lot-sizing literature. Adulyasak et al. [4] and Witthayapraphakorn et al. [5] apply it to lot-sizing problems under demand uncertainty, identifying a network flow structure in the dual second-stage mathematical model. In this work, a similar structure obtained from a novel problem approximation is leveraged to achieve a computational gain. Moreover, an adaptive partition-based approach as well as a stabilization method are subsequently introduced, which to the best of our knowledge have not yet been documented for lot-sizing problems. The estimation of the error distribution in wind power forecasting is also a well studied problem (see, *e.g.*, Pinson [6]). Cramer et al. [7] present a comparison of documented approaches for wind power scenario generation, and conclude that deep learning generative methods may offer more benefits for day-ahead stochastic production problems. In the present work, we use such a method called «Variational Auto-Encoder», known for the simplicity of its implementation (see, *e.g.*, Hernandez Capel and Dumas [8]), and illustrate its ability to generate reliable scenarios in realistic settings. Its use is innovative in its application to solving stochastic programming problems.

To the best of our knowledge, this work is the first one to propose a cohesive and time-efficient framework integrating stochastic programming and machine learning to solve a two-stage lot-sizing problem under uncertainty regarding the available energy source.

After a short problem description, a stochastic programming model with two decision stages is introduced in Sect. 2. Section 3 presents a probabilistic neural network used to generate wind power scenarios for uncertainty representation in the problem modeling. In Sect. 4, the problem is addressed by a Benders decomposition scheme, in which the mixed-integer sub-problems are approximately solved by a linear time algorithm, and the number of optimality cuts and iterations is reduced to improve computational performance. Realistic numerical experiments simulating the production site management over a year demonstrate the satisfactory results of the proposed approach in Sect. 5.

2   Problem Description

2.1   Context and Assumptions

This work is motivated by industrial use-cases of the Lhyfe company such as the Bouin production plant in France, or the renewable hydrogen demonstrator from Siemens Gamesa in Zealand, Denmark. Hybrid production sites, connected to both renewable assets and the distribution grid, benefit from the cleanest possible energy from these renewable sources, while ensuring production despite intermittency thanks to the grid connection. Such a configuration enables the electrolytic hydrogen industry to take advantage of the ability of electrolyzers to adjust in near real-time to variations in energy supply and costs.

The cost of wind power is considered to be negligible compared to electricity market prices: we aim at using as much as possible the renewable energy source to produce hydrogen with the least greenhouse gas emissions. In practice, electricity prices from renewable assets are negotiated ahead of production and are of constant value. Note that a realistic representation can be retrieved from this model by subtracting the constant renewable cost from electricity market prices.

In practice, the limited hydrogen storage capacity is mainly used to reduce production costs by anticipating the lack of wind power or the increase in electricity market prices in the forthcoming hours. Consequently, the stored hydrogen is mostly produced and consumed during the same day, and the hydrogen available in storage at the beginning of a given day is almost always consumed during that day. Thus, the hydrogen stock level at the beginning of a day has an impact on the production plan for that day, but not on the production plan for the days beyond. In other words, there is little benefit to plan production more than two days in advance.

In the following, the uncertainty is only considered for the first day of the production planning, and the production of the second day is incorporated in the model in order to anticipate the impact of this uncertainty on future production costs. The two-day production planning problem is thus modeled as a two-stage stochastic program. Note that one may consider a longer planning horizon while using this approach, either by only considering uncertainty for the first day or by dynamically estimating future production costs, as described in Sect. 4.1.

The presented problem remains relatively general, as the proposed method can be adapted to fit a broad range of applications, some of which are discussed in Sect. 6. Therefore, this work aims at proposing a framework to address a wide variety of use-cases, rather than focusing on a single one.

2.2   Mathematical Modeling

This work aims at minimizing the costs relative to the daily declaration of the grid electricity exploitation, for a production site equipped with a single electrolyzer. As discussed above, in our case, the production decisions relative to a given day do not affect the production plan beyond the following day. Therefore, the planning horizon extends over two days, each one divided into 24 time steps (hours), denoted $T = \{1, 2, \ldots, 24\}$. In practice, production planning is carried out on the basis of a rolling horizon, $i.e.$, only first-day decisions are actually implemented, and second-day decisions are incorporated into the model to ensure that the future consequences of the first-day decisions are correctly evaluated.

At each period «day-hour» (d,t), for $d \in \{1,2\}$ and $t \in T$, the site has to meet a hourly demand for hydrogen, noted $q_{d,t}$. In order to produce hydrogen, the electrolyzer should be activated, which consumes a fixed amount of power p^{on}. It has a constant power-to-H_2 conversion efficiency, noted h, and can produce up to q^{max} kg of H_2 per time-step (corresponding to a maximum power consumption $p^{max} = q^{max}/h$ per hour). The storage has a maximal capacity s^{max} and its initial stock level is noted s^0. It is assumed that the demand $q_{d,t}$ at any hour t and day d is lower than the production capacity q^{max}.

The site is connected simultaneously to a wind farm and the national power grid. The purchase cost of wind power is negligible, but its availability at (d,t), noted $\widetilde{w}_{d,t}$, is limited, time-varying and uncertain. However, there is a deterministically known guaranteed minimum wind power at (d,t), denoted by $\overline{w}_{d,t}$. The purchase cost of electricity from the grid at (d,t), $c_{d,t}^g$, is also assumed to be known with certainty. Any purchase from the grid has to be declared a day ahead of production, i.e., before period $(d,1)$ for day d and, once declared, upward or downward adjustments of the electricity consumed during day d are forbidden.

A two-stage scenario-based stochastic programming approach is introduced to handle this problem. The first stage represents decisions to be made before the actual wind power availability is known. The second stage corresponds to decisions that can be postponed until after the realization of this uncertain parameter.

Recall that, once declared at the beginning of day d, the amount of grid electricity to be purchased cannot be adjusted during that day. This means that the demand might not be met on time if we rely on uncertain wind power, i.e., on $\widetilde{w}_d - \overline{w}_d$, to produce it. Thus, to avoid any shortage, the production of day d, intended to satisfy the demand of that same day, is not allowed to rely on wind power production beyond the minimum guaranteed output $\overline{w}_d$. Wind power production exceeding this output, i.e., $\widetilde{w}_d - \overline{w}_d$, is only exploited to produce hydrogen for storage, which is then used to meet future demand on day $d+1$.

The first decision stage corresponds to setting up the production planning, and determining the day-ahead declaration of grid power purchase for day 1. The following variables are introduced for each period $(1,t)$ of day 1. First, the binary variable $z_{1,t}$ represents the active/inactive state of the electrolyzer. Then, the total power consumed for hydrogen production is denoted $x_{1,t}$. It decomposes into $x_{1,t}^w$, the amount of guaranteed wind power used, and $x_{1,t}^g$, the amount of power purchased from the grid. The stock of hydrogen at the end of the period $(1,t)$ and available to meet hydrogen demand on day 1 is represented by $s_{1,t}$.

In this work, the wind power uncertainty during day 1 is incorporated into the modeling of the problem. This uncertainty is represented by a set $\mathcal{S}$ of discrete scenarios of known probability distribution. A fixed scenario $\xi \in \mathcal{S}$ corresponds to a potential realization $w_{\xi,1,t}$ of the available wind power for all time-steps $(1,t)$, $t \in T$. Note that the uncertainty related to day 2 is ignored: it has no significant impact on this two-day planning since it only affects the third day of production.

The second-stage decision variables concern both day 1 and day 2. Regarding day 1, a variable $\tilde{s}_{\xi,1,t}$ is introduced for each time period $(1,t)$ and scenario $\xi \in \mathcal{S}$, to represent an additional quantity of hydrogen produced using the real wind power generated. This hydrogen is kept in stock, and it is only available to meet customer demand at the beginning of the second day, once its exact value is known. Regarding day 2, we use the same variables as the ones introduced for day 1, *i.e.*, variables $z_{\xi,2,t}$, $x_{\xi,2,t}$, $x^w_{\xi,2,t}$, $x^g_{\xi,2,t}$ and $s_{\xi,2,t}$. Note that these variables are now indexed by the scenario ξ and the index of day 2.

Before presenting the two-stage stochastic programming model, the set of constraints that the day-ahead production planning and declaration variables should respect is formally described for each day d. Given an initial stock s^{in}_d and a guaranteed minimum wind power $\overline{w}_d$, we have:

$$x_{d,t} = x^w_{d,t} + x^g_{d,t}, \qquad \forall t \in T \qquad (1)$$

$$x_{d,t} \le z_{d,t} \cdot p^{max}, \qquad \forall t \in T \qquad (2)$$

$$x^w_{d,t} \le \overline{w}_{d,t}, \qquad \forall t \in T \qquad (3)$$

$$s_{d,1} = s^{in}_d + (x_{d,1} - z_{d,1} \cdot p^{on}) \cdot h - q_{d,1}, \qquad (4)$$

$$s_{d,t} = s_{d,t-1} + (x_{d,t} - z_{d,t} \cdot p^{on}) \cdot h - q_{d,t}, \qquad \forall t \in T \setminus \{1\} \qquad (5)$$

$$s_{d,t} \le s^{max}, \qquad \forall t \in T \qquad (6)$$

$$x_{d,t}, x^w_{d,t}, x^g_{d,t}, s_{d,t} \ge 0, \qquad \forall t \in T \qquad (7)$$

$$z_{d,t} \in \{0,1\}, \qquad \forall t \in T \qquad (8)$$

Constraints (1)–(2) compute the total power exploited by the electrolyzer and ensure that it complies with its activation/deactivation state. Constraints (3) limit the use of wind power to its minimum guaranteed quantity. Constraints (4)–(5) are the hydrogen stock balance equations which, together with the non-negativity requirements on the variables $s_{d,t}$, guarantee on-time demand satisfaction. Note that the quantity of hydrogen produced in (d,t) is calculated as $(x_{d,t} - z_{d,t} \cdot p^{on}) \cdot h$ to account for the fixed activation power p^{on}. Inequalities (6) limit the amount of hydrogen stored to the maximum storage capacity. Let $X_d(s^{in}_d, \overline{w}_d)$ be the set of all production plans $(x_d, x^w_d, x^g_d, s_d, z_d)$ for day d complying with constraints (1)–(8), where each plan component is a $|T|$-dimensional vector.

Let the first-stage decision variables be denoted $\chi = (x_1, x^w_1, x^g_1, s_1, z_1)$, then the two-stage stochastic programming model can be formulated as follows:

$$\min \sum_{t \in T} c^g_{1,t} \cdot x^g_{1,t} + \mathop{\mathbb{E}}_{\xi \in S} [\mathcal{Q}(\chi, \xi)] \qquad (9)$$

$$\text{s.t. } \chi \in X_1(s^0, \overline{w}_1) \qquad (10)$$

Objective function (9) aims at minimizing the grid power purchasing costs for day 1 and the expected value, over all scenarios, of the grid power purchasing costs for day 2. Here, $\mathcal{Q}(\chi, \xi)$ returns the optimal value of the second-stage sub-problem, which computes the second-stage cost in the case where the first-stage decisions are equal to χ and the scenario ξ realizes. This sub-problem involves the excess stock variables relative to day 1, and the day-ahead production planning and power purchase decisions relative to day 2 for this scenario. It is thus formulated as:

$$\mathcal{Q}(\chi,\xi) := \min \sum_{t \in T} c_{2,t}^g \cdot x_{\xi,2,t}^g \tag{11}$$

$$\text{s.t. } \tilde{s}_{\xi,1,1} \leq (\tilde{w}_{\xi,1,1} \cdot z_{1,1} - x_{1,1}^w) \cdot h, \tag{12}$$

$$\tilde{s}_{\xi,1,t} - \tilde{s}_{\xi,1,t-1} \leq (\tilde{w}_{\xi,1,t} \cdot z_{1,t} - x_{1,t}^w) \cdot h, \qquad \forall t \in T \backslash \{1\} \tag{13}$$

$$\tilde{s}_{\xi,1,1} \leq q^{\max} - (x_{1,1} - z_{1,1} \cdot p^{\mathrm{on}}) \cdot h, \tag{14}$$

$$\tilde{s}_{\xi,1,t} - \tilde{s}_{\xi,1,t-1} \leq q^{\max} - (x_{1,t} - z_{1,t} \cdot p^{\mathrm{on}}) \cdot h, \qquad \forall t \in T \backslash \{1\} \tag{15}$$

$$\tilde{s}_{\xi,1,t} \leq s^{\max} - s_{1,t}, \qquad \forall t \in T \tag{16}$$

$$\tilde{s}_{\xi,1,t} \geq 0, \qquad \forall t \in T \tag{17}$$

$$(x_{\xi,2}, x_{\xi,2}^w, x_{\xi,2}^g, s_{\xi,2}, z_{\xi,2}) \in X_2(s_{1,24} + \tilde{s}_{\xi,1,24}, \overline{w}_2) \tag{18}$$

Inequalities (12)–(17) deal with the management of the hydrogen produced during day 1, using the generated wind power, and available to meet demand only at the beginning of day 2. Note that constraints (12)–(15) directly represent the ability to generate additional stock according to the additional available wind power and production capacity. They can thus be considered as inventory balance equations. Finally, inequalities (16) ensure that the maximum hydrogen storage capacity is respected, and constraint (18) reflects the need to build a feasible production planning for day 2. Note also that the initial incoming stock for day 2, $s_{1,24} + \tilde{s}_{\xi,1,24}$, depends on the scenario ξ, *i.e.*, on wind power realization during day 1.

Note that the only coupling between the first and second stages comes from the final stock of day 1, $s_{1,24} + \tilde{s}_{\xi,1,24}$, being the incoming stock of day 2. Therefore, the recourse $\mathcal{Q}(\chi,\xi)$ is relatively complete. The second-stage problem is feasible for any first-stage decision χ and any scenario ξ, since a production plan may be found for any incoming stock value.

3 Uncertainty Modeling

Establishing a wind forecast error distribution is a complex problem (see, *e.g.*, Pinson [6]). First, the uncertainty depends on the weather conditions: here the wind power forecast is already known, and we consider that the error distribution depends on the forecast itself. Furthermore, these errors are strongly correlated over time and depend on the overall shape taken by the forecast across the whole horizon. Fortunately, error models can be identified and learnt from recurrent patterns from past forecasts that resemble those being studied. For instance, in the case of sudden increases in wind power, forecasts may often correctly identify the time periods of such events, but not their exact amplitude.

State-of-the-art approaches for scenario-based probabilistic energy forecasts include Gaussian copula, auto-regressive models and more recently, deep learning generative methods. Cramer et al. [7] compare these approaches to generate wind power scenarios for day-ahead stochastic production problems. The authors conclude that deep learning generative methods may present more benefits, using a «Normalizing Flow» technique. In the present work, a similar approach is chosen, in the form of a probabilistic neural network called «Variational Auto-Encoder» (see, *e.g.*, Kingma and Welling [9]). As shown by Hernandez Capel and Dumas [8], this method might provide poorer results for complex tasks but it is far simpler to implement. It is, however, sufficient for the relatively simple task of deriving scenarios from a forecast input, rather than from weather data alone.

This machine learning method first transforms a time series representing the forecast, used as input parameter, into the mean and variance of a multivariate normal distribution by the use of a traditional, non-linear neural network. Samples obtained from this distribution are then transformed back into output parameters in a similar fashion, thus obtaining the desired wind power forecast scenarios. The overall process corresponds to a single, probabilistic neural network trained in a unified manner. The neural networks before and after sampling are trained jointly to recognize forecast error patterns across the training dataset.

This method is trained on a dataset comparing predicted and actual wind power. Through scenario sampling, it constructs a discrete approximation of the error distribution for any given forecast, based on the observed errors of similar forecasts in the historical data. In this case study, the forecast error distribution is estimated for the first day of the planning horizon. A comparison of actual wind power with the forecasted and scenario-sampled ones is illustrated below. The neural network may fail to correctly estimate the distribution, if the situation at hand has not yet been observed in the historical data upon which it has been trained (Figs. 1 and 2).

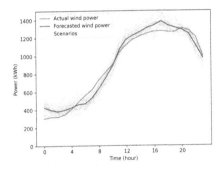

Fig. 1. First example of test data

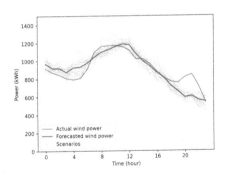

Fig. 2. Second example of test data

4   Time-Efficient Benders Decomposition Algorithm

A large number $|\mathcal{S}|$ of scenarios are needed to accurately represent the uncertainty distribution. As the size of the mixed-integer linear problem is broadly proportional to $|\mathcal{S}|$, solving it directly as a whole leads to significant numerical difficulties. However, it has a block-decomposable structure that makes it suitable for a Benders decomposition approach (see, *e.g.*, Rahmaniani et al. [10]).

In our case, the Benders decomposition approach relies on the formulation (9)–(10) involving only the first-stage decision variables $\chi = (x_1, x_1^w, x_1^g, s_1, z_1)$ and an approximation of a closed-form expression of the function $\mathcal{Q}(\cdot, \xi)$ for each scenario $\xi \in \mathcal{S}$, denoted hereafter Q_ξ for simplicity. The objective of a Benders decomposition algorithm thus consists in iteratively constructing such approximations taking the form of a convex piece-wise linear function.

At iteration j of the Benders decomposition algorithm, we therefore first solve the following relaxation of problem (9)–(10), which is called the master problem:

$$\min \sum_{t \in T} c_{1,t}^g \cdot x_{1,t}^g + \mathop{\mathbb{E}}_{\xi \in \mathcal{S}} [Q_\xi] \tag{19}$$

$$\text{s.t. } \chi \in X_1(s^0, \overline{w}_1), \tag{20}$$

$$Q_\xi \geq \pi_m^\xi \chi + \rho_m^\xi, \qquad \forall m \in \{1, \ldots, j-1\}, \ \forall \xi \in \mathcal{S} \tag{21}$$

Once the first-stage decisions χ^j are made for iteration j, the $|\mathcal{S}|$ single-scenario sub-problems $Q(\chi^j, \xi)$ are solved, and their solution is used to generate a new optimality cut of the form (21) which is added to the formulation of the master problem (19)–(21) in order to improve the current under-approximation of function $\mathcal{Q}(\cdot, \xi)$. This decomposition algorithm thus iteratively solves the master problem and a sequence of single-scenario sub-problems until convergence.

However, such an approach requires to repeatedly solve a large number of single-scenario sub-problems, and a master problem whose size gradually increases due to the iterative addition of optimality cuts. Thus, to be numerically tractable, the single-scenario sub-problems and the master problem have to be solved in a reasonable time. Both issues are discussed in the following.

4.1 Approximate Resolution of the Second-Stage Sub-problems

Each sub-problem $\mathcal{Q}(\chi, \xi)$ is a mixed-integer linear program involving binary activation variables. A Benders decomposition approach, based on the exact resolution of a set of mixed-integer linear sub-problems at each iteration, is likely to require prohibitive computation times. Instead, a particular feature of our problem is exploited to speed up the resolution of each sub-problem $\mathcal{Q}(\chi, \xi)$ at each iteration. Namely, the set of constraints $X_2(s_{1,24} + \tilde{s}_{\xi,1,24}, \overline{w}_2)$, relative to the production planning of day 2, depends only on the final stock of day 1 and is therefore scenario-independent. Consequently, $\mathcal{Q}(\chi, \xi)$ can be reformulated as follows:

$$\mathcal{Q}(\chi, \xi) := \min \Gamma(s_{1,24} + \tilde{s}_{\xi,1,24}) \tag{22}$$

$$\text{s.t. } (12)–(17) \tag{23}$$

Here, the function $\Gamma(\cdot)$ gives the value of the second-day production cost as a function of the final stock level at the end of day 1. Note that the function $\Gamma(\cdot)$ depends only on the stock level but neither on the scenario nor on other variables. Moreover, $\Gamma(\cdot)$ is a non-increasing function that would be convex without the binary activation variables. Indeed, the greater the input stock at the beginning of day 2, the less hydrogen production is necessary during that day, and the lower the corresponding production cost. In addition, this input stock is used to avoid production from the most expensive production hours to the cheapest ones. Hence the incremental decrease in production cost on day 2 tends to become smaller as the level of incoming stock increases, with occasional variations due to the electrolyzer activation state.

The resolution time of each sub-problem $\mathcal{Q}(\chi, \xi)$ is therefore reduced by solving an approximate continuous problem $\widehat{\mathcal{Q}}(\chi, \xi)$, whose model relies on a piecewise linear and convex approximation $\widehat{\Gamma}(\cdot)$ of the function $\Gamma(\cdot)$. To construct this approximation as a pre-optimization step, we start by decomposing the interval $[0, s^{\max}]$ into a set $N = \{1, \ldots, n\}$ of sub-intervals. Let δ_i be the width of interval $i \in N$. We then calculate the value of $\Gamma(\cdot)$ at each corresponding break-point in the approximation by solving the second-day production problem for each of these values. Finally, we construct the convex hull $\widehat{\Gamma}(\cdot)$ of the resulting image points to obtain $\widehat{\Gamma}(0)$, $\widehat{\Gamma}(\delta_1)$, $\widehat{\Gamma}(\delta_1 + \delta_2), \ldots, \widehat{\Gamma}(\sum_{i=1}^{n} \delta_i)$.

The slope of the piece-wise linear approximation of $\widehat{\Gamma}(\cdot)$ over interval $i \in N$ is given by $\kappa_i = \left(\widehat{\Gamma}(\sum_{j=1}^{i-1} \delta_j) - \widehat{\Gamma}(\sum_{j=1}^{i} \delta_j) \right)/\delta_i$. Since this convex piece-wise linear function appears in the objective of a minimization problem, it can be expressed by introducing a set of continuous variables w_i, $i \in N$:

$$\widehat{\mathcal{Q}}(\chi, \xi) := \min \widehat{\Gamma}(0) - \sum_{i \in N} w_i \cdot \kappa_i \tag{24}$$

$$\text{s.t. } (12)\text{--}(17), \tag{25}$$

$$\sum_{i \in N} w_i \leq \tilde{s}_{\xi,1,24} + s_{1,24}, \tag{26}$$

$$0 \leq w_i \leq \delta_i, \qquad \forall i \in N \tag{27}$$

Let a, b and c be the non-negative, constant right-hand side vectors of constraints (12)–(13), (14)–(15) and (16). The dual of the approximated sub-problem (24)–(27) is formulated as follows:

$$\max \widehat{\Gamma}(0) - \left(\sum_{t \in T} [a_t \cdot \alpha_t + b_t \cdot \beta_t + c_t \cdot \eta_t] + s_{1,24} \cdot \lambda + \sum_{i \in N} \delta_i \cdot \mu_i \right) \tag{28}$$

$$\text{s.t. } \alpha_t + \beta_t + \eta_t \geq \alpha_{t+1} + \beta_{t+1}, \qquad \forall t \in T \backslash \{24\} \tag{29}$$

$$\alpha_{24} + \beta_{24} + \eta_{24} \geq \lambda, \tag{30}$$

$$\lambda + \mu_i \geq \kappa_i, \qquad \forall i \in N \tag{31}$$

$$\alpha_t, \beta_t, \eta_t, \lambda, \mu_i \geq 0, \qquad \forall t \in T, \forall i \in N \tag{32}$$

Here, α, β, η and μ are respectively the dual vectors of variables for constraints (12)–(13), (14)–(15), (16) and (27), while λ is the dual variable for constraint (26). This problem may be solved in linear time. Suppose that for a given scenario $\xi \in \mathcal{S}$, the optimal solution λ^ξ to this problem is known. Indeed, the optimal values $(\alpha^\xi, \beta^\xi, \eta^\xi)$ at time-step t depend on those at time-step $t+1$, and the values at time-step 24 depend on λ^ξ. One may use dynamic programming to deduce these values, by solving a shortest path problem on the directed acyclic graph represented below. There, each variable whose corresponding node is part of the shortest path has its optimal value equal to λ^ξ, and any other variable have its value equal to zero.

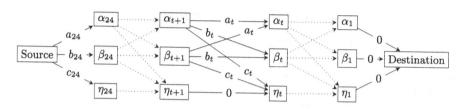

Fig. 3. Directed acyclic graph for the second-stage approximate dual problem

As for the optimal values μ^ξ, they are directly deduced so that:

$$\mu_i^\xi = \max\{0, \kappa_i - \lambda^\xi\}, \quad \forall i \in N \tag{33}$$

Hence, since the optimal values $(\alpha^\xi, \beta^\xi, \eta^\xi)$ depend linearly on the optimal value of λ, the only possible extreme points of this latter variable belong to the set $\{0, \kappa_1, \ldots, \kappa_n\}$, and the variable λ is of optimal value in this set. Thus, to solve this approximate sub-problem, it is sufficient to solve the aforementioned shortest path problem once, and then enumerate the possible values of the variable λ. For a solution value $\mathcal{P}$ of the shortest path problem, the approximate dual sub-problem is equivalent to:

$$\min \left\{ (\mathcal{P} + s_{1,24}) \cdot \lambda + \sum_{i \in N} \delta_i \cdot \max\{0, \kappa_i - \lambda\} : \lambda \in \{0, \kappa_1, \ldots, \kappa_n\} \right\} \tag{34}$$

Handling the sub-problem requires both this enumeration and solving the shortest path problem, and is therefore of linear complexity $\mathcal{O}(|N| + |T|)$. Note that the set of optimal solutions is finite, since the number of paths in Fig. 3, and the possible optimal values of λ, are finite. An optimal solution $(\alpha^\xi, \beta^\xi, \eta^\xi, \lambda^\xi, \mu^\xi)$ of the approximate dual sub-problem is finally used to infer parameters (π^ξ, ρ^ξ) like so:

$$\rho^\xi = \widehat{\Gamma}(0) - \left(\sum_{t \in T} \left[q^{max} \cdot \beta_t^\xi + s^{max} \cdot \eta_t^\xi \right] + \sum_{i \in N} \delta_i \cdot \mu_i^\xi \right) \tag{35}$$

$$\pi^\xi \cdot \chi = \sum_{t \in T} \left[s_{1,t} \cdot \eta_t^\xi - (\widetilde{w}_{\xi,1,t} \cdot z_{1,t} - x_{1,t}^w) \cdot h \cdot \alpha_t^\xi \right]$$

$$+ \sum_{t \in T} \left[(x_{1,t} - z_{1,t} \cdot p^{\mathrm{on}}) \cdot h \cdot \beta_t^{\xi} \right] - s_{1,24} \cdot \lambda^{\xi} \qquad (36)$$

The resulting optimality cut is then added to the set of constraints (21) in the master problem.

Therefore the second-stage problem can be solved with good precision in an efficient manner, as guaranteed by its linear complexity. Although the network flow structure of the second-stage problem was already known for two-stage stochastic lot-sizing problems, this particular application accounting for the presented approximation can be considered to be an original contribution.

4.2   Accelerating the Benders Decomposition Scheme

This sub-section tackles the difficulties met when solving the master problem once a large number of optimality cuts has been generated. It aims at reducing the number of generated optimality cuts at each iteration, and the number of iterations to achieve convergence.

First, it is possible to exploit scenarios with identical second-stage dual solutions to reduce the number of generated optimality cuts. An adaptive partition-based Benders decomposition is implemented (see, e.g., Song and Luedtke [11]). At each iteration j we solve the master problem, as well as the second-stage dual problem for each scenario $\xi \in S$, and retrieve the corresponding second-stage dual solution. We use this information to define a partition of the scenario set S into $K \leq |S|$ non-empty subsets S_k^j, $k \in \{1, \ldots, K\}$, gathering scenarios of identical dual solution. Then, we replace the optimality cuts presented in (21) and (35)–(36) by a single one per non-empty subset $S_k^j \neq \emptyset$ of this partition:

$$\mathbb{E}_{\xi \in S_k^j} [Q_\xi] \geq \mathbb{E}_{\xi \in S_k^j} \left[\pi_j^\xi \chi + \rho_j^\xi \right] \qquad (37)$$

This approximate optimality cut remains precise: all scenarios in that set share very similar parameters (π_j^ξ, ρ_j^ξ) since they all have an identical dual solution $(\alpha^k, \beta^k, \eta^k, \lambda^k, \mu^k)$. Thus, the right-hand term of the approximate optimality cut (37) is expressed as:

$$\mathbb{E}_{\xi \in S_k^j} \left[\rho_j^k \right] = \widehat{\Gamma}(0) - \left(\sum_{t \in T} \left[q^{\mathrm{max}} \cdot \beta_t^k + s^{\mathrm{max}} \cdot \eta_t^k \right] + \sum_{i \in N} \delta_i \cdot \mu_i^k \right) \qquad (38)$$

$$\mathbb{E}_{\xi \in S_k^j} \left[\pi_j^\xi \cdot \chi \right] = \sum_{t \in T} \left[s_{1,t} \cdot \eta_t^k - \left(\mathbb{E}_{\xi \in S_k^j} \left[\widetilde{w}_{\xi,1,t} \right] \cdot z_{1,t} - x_{1,t}^w \right) \cdot h \cdot \alpha_t^k \right]$$
$$+ \sum_{t \in T} \left[(x_{1,t} - z_{1,t} \cdot p^{\mathrm{on}}) \cdot h \cdot \beta_t^k \right] - s_{1,24} \cdot \lambda^k \qquad (39)$$

In other words, all scenarios in that set share an identical optimality cut except for the available wind power $\widetilde{w}$ that depends on the considered scenario. Rather than generating multiple and extremely similar optimality cuts, a single one is added to the model with an available wind power approximated by its average value across that set. This approach enables a significant reduction in the number of optimality cuts added to the master problem. In fact, scenarios of different values having the same impact on certain first-stage decisions are bundled together into a single constraint.

In practice, the partition-based acceleration improves convergence in the early iterations of the decomposition scheme, with a fast decrease of the precision gap. Yet because of the mentioned approximation on the available uncertain wind power, once a small gap is achieved, the scheme could take more time than expected to converge. Thus one may consider dialing down the partitioning according to the observed gap at each iteration, for example using partial refinement as presented by Song and Luedtke [11].

Finally, a trust-region stabilization method is introduced to reduce the number of iterations of the decomposition scheme (see, *e.g.*, van Ackooij et al. [12]). It forces the master problem solution of the next iteration to lie in the neighborhood of the previous one. Such method improves the performance of the decomposition scheme by decreasing its instability, *i.e.*, the fact that two successive first-stage decisions can be "very far apart". It is here more specifically applied to the binary activation variable z: at iteration $j + 1$, there can be no more than C changes in the variable values by adding the following constraint to the master problem:

$$\sum_{\{t:z_{1,t}^j=1\}} (1 - z_{1,t}) + \sum_{\{t:z_{1,t}^j=0\}} z_{1,t} \leq C \tag{40}$$

In practice, the optimal production plan is mainly dependent on the choice to activate the production unit at each period. Thus, one may limit the change in activation variable values from a planning to another to effectively force them to share a similar shape, reducing the instability of the decomposition scheme in the process.

These acceleration methods result in a master problem of reduced size that has to be solved for a smaller number of iterations, without loosing the finite-optimal convergence of the Benders decomposition algorithm. To the best of our knowledge, none were used until now to improve a Benders decomposition scheme applied to a stochastic lot-sizing problem.

5   Numerical Experiments

This section assesses the performance of the stochastic programming approach to reduce the production overcosts induced by wind power forecast uncertainty.

To that end, a production site is considered with the following parameter values: $p^{\text{on}} = 200$ kWh, $p^{\text{max}} = 1000$ kWh, $q^{\text{max}} = 15$ kg, $h = 0.015$ kg/kWh, $s^{\text{max}} = 70$ kg and $q = 9$ kg. We simulate the day-to-day operation of this site over a rolling one-year period. Electricity prices of the national grid network correspond to those of France in the year 2016. Actual and forecasted wind power values are obtained from realistic, open-access data by Pan et al. [13]. The comparison between actual and predicted wind power is carried out over two years: the scenario generation method is trained on the first year and the simulation is performed on the second year.

For each day of the simulated year, we calculate the production plan provided by the stochastic model (9)–(10), the one provided by a deterministic oracle model having full knowledge of the future wind power availability, and the one provided by a deterministic naive model using the forecast but not considering any uncertainty. The stochastic model is evaluated for $|S| = 1000$ scenarios, a stabilization parameter $C = 3$ and $|N| = 10$ pieces approximating the function $\Gamma(\cdot)$. For each model, the decisions relative to the first day are implemented, and the actual production cost and resulting storage value are recorded.

Table 1 displays the simulation results, comparing the overcosts related to the wind power forecast error over the year. This overcost is defined as a percentage of the theoretical minimum costs indicated by the oracle model. The stochastic model performs better than the naive one with 0.8% less overcosts in total for the year, which represents 21% of the loss caused by the uncertainty.

Table 1. Overcosts related to uncertainty as part (%) of the theoretical optimum

	Jan.	Feb.	Mar.	Apr.	May	Jun.	Jul.	Aug.	Sep.	Oct.	Nov.	Dec.	Year average
Naive	3.1	4.9	3.5	3.4	4.3	**3.7**	4	4	3.8	3.6	3.8	3.7	3.8
Stochastic	**1.9**	**2.2**	**2.9**	**2.8**	**3.1**	3.9	**2.8**	4.	**2.6**	**3.5**	**3.1**	**3.6**	**3.**

We also demonstrate the benefits of the partition-based and stabilized decomposition in reducing the runtime of the stochastic model. The day-by-day runtime across the simulation over a year for $|S| = 1000$ is observed to compare the decomposition scheme enhanced with one of these methods, both or none. The computational experiments were conducted on a CentOS Linux machine with 16 GB of RAM and an Intel Core CPU i7-8565U processor at 1.80 GHz. All the mixed-integer linear programs were solved with the commercial solver IBM CPLEX12.10, for which a single thread was used. The algorithms were realized in Python, and CPLEX was called via the framework of Pyomo. The cumulative runtime over the simulation for each method is shown in Fig. 4.

Using the stabilization and partition-based methods jointly in the decomposition scheme appears to be most beneficial in reducing the runtime. Overall when using both acceleration methods, the solving time is reduced by two third compared to the standard decomposition scheme. Take note that although the partition-based decomposition scheme alone doesn't exhibit significantly better performance compared to the standard decomposition scheme, it becomes highly beneficial when integrated with the stabilization scheme, reducing by a third the average solving time compared to a scheme with stabilization only.

The presented numerical experiments evaluate the proposed stochastic approach in a real-life setting and for a wide variety of electricity grid prices and wind

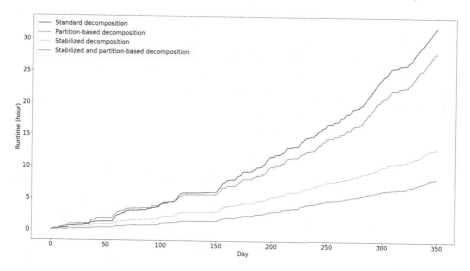

Fig. 4. Cumulative runtime comparison of the stochastic models

power availability. They demonstrate the method's ability to improve production performance within a reasonable timeframe, for a broad range of practical cases. The data used to carry out this simulation (grid prices, wind power, scenarios) is available for reproducibility purposes[1].

6   Conclusion

This work addresses day-ahead production planning with uncertain availability of a wind power source and a historical dataset of forecast error. It is modelled as a two-stage stochastic lot-sizing problem, solved by a Benders decomposition scheme with an innovative approximation of the second-stage model and the novel use of acceleration methods. An accurate method for generating wind power scenarios and a time-efficient resolution provide a significant reduction in production costs, as demonstrated by a realistic simulation of the production site.

Possible extensions to more realistic models could be achieved from this work. The presented problem is solved in a similar fashion when accounting for other uncertain energy sources (*e.g.* photovoltaic), or for the deterministic participation to other energy markets (*e.g.* intra-day market). The electrolysis production efficiency could be more precisely represented by a piecewise-linear concave approximation with continuous representation (see, *e.g.*, Baumhof et al. [14]), with minor changes to the network flow model of the second-stage approximate dual problem. We may also extend the work to consider a set of parallel hydrogen production units. Finally, demand uncertainty might bear strong resemblance to

---

[1] https://github.com/VSpitzer/Stochastic-day-ahead-lot-sizing.git.

the studied uncertain power availability, as both seem equivalent to a variation in production capacity.

Further research work would include the consideration of uncertainty relative to day-ahead electricity grid prices, as it mainly impacts the first-stage decision. In addition, other stabilization methods could further improve the time efficiency of the decomposition scheme.

References

1. Global Hydrogen Review 2023. In: International Energy Agency (IEA), Paris (2023)
2. Brahimi, N., Absi, N., Gendreau, M., Dauzère-Pérès, S., Nordli, A.: Single-item dynamic lot-sizing problems: an updated survey. Eur. J. Oper. Res. **263**(3), 838–863 (2017)
3. van Ackooij, W., Danti Lopez, I., Frangioni, A., Lacalandra, F., Tahanan, M.: Large-scale unit commitment under uncertainty: an updated literature survey. Ann. Oper. Res. **271**(1), 11–85 (2018)
4. Adulyasak, Y., Cordeau, J.-F., Jans, R.: Benders decomposition for production routing under demand uncertainty. Oper. Res. **63**(4), 851–867 (2015)
5. Witthayapraphakorn, A., Charnsethikul, P.: Benders decomposition with special purpose method for the sub problem in lot sizing problem under uncertain demand. Oper. Res. Perspect. **6**, 100096 (2019)
6. Pinson, P.: Estimation of the uncertainty in wind power forecasting. Ph.D. Thesis, École des Mines de Paris (2006)
7. Cramer, E., Paeleke, L., Mitsos, A., Dahmen, M.: Normalizing flow-based day-ahead wind power scenario generation for profitable and reliable delivery commitments by wind farm operators. Comput. Chem. Eng. **166**, 107923 (2022)
8. Hernandez Capel, E., Dumas, J.: Denoising diffusion probabilistic models for probabilistic energy forecasting. In: IEEE PowerTech 2023 Conference, Belgrade, Serbia, 25–29 June 2023 (2023)
9. Kingma, D.P., Welling, M.: Auto-encoding variational Bayes. In: 2nd International Conference on Learning Representations (ICLR 2014), Banff, Canada, 14–16 April 2014 (2014)
10. Rahmaniani, R., Crainic, T.G., Gendreau, M., Rei, W.: The Benders decomposition algorithm: a literature review. Eur. J. Oper. Res. **259**(3), 801–817 (2017)
11. Song, Y., Luedtke, J.: An adaptive partition-based approach for solving two-stage stochastic programs with fixed recourse. SIAM J. Optim. **25**(3), 1344–1367 (2015)
12. van Ackooij, W., Frangioni, A., de Oliveira, W.: Inexact stabilized Benders' decomposition approaches with application to chance-constrained problems with finite support. Comput. Optim. Appl. **65**(3), 637–669 (2016)
13. Pan, K., Matusz, K., Lalanne, C.: Predicting excess wind electricity in Ireland: machine learning against climate change. In: Towards Data Science, 3 June 2021 (2021)
14. Baumhof, M.T., Raheli, E., Johnsen, A.G., Kazempour, J.: Optimization of hybrid power plants: when is a detailed electrolyzer model necessary? In: IEEE PowerTech 2023 Conference, Belgrade, Serbia, 25–29 June 2023 (2023)

Author Index

Printed in the United States
by Baker & Taylor Publisher Services